Improving poultry meat quality

It is widely recognised that agriculture is a significant contributor to global warming and climate change. Agriculture needs to reduce its environmental impact and adapt to current climate change whilst still feeding a growing population, i.e. become more 'climate-smart'. Burleigh Dodds Science Publishing is playing its part in achieving this by bringing together key research on making the production of the world's most important crops and livestock products more sustainable.

Based on extensive research, our publications specifically target the challenge of climate-smart agriculture. In this way we are using 'smart publishing' to help achieve climate-smart agriculture.

Burleigh Dodds Science Publishing is an independent and innovative publisher delivering high quality customer-focused agricultural science content in both print and online formats for the academic and research communities. Our aim is to build a foundation of knowledge on which researchers can build to meet the challenge of climate-smart agriculture.

For more information about Burleigh Dodds Science Publishing simply call us on +44 (0) 1223 839365, email info@bdspublishing.com or alternatively please visit our website at www.bdspublishing.com.

Related titles:

Advances in poultry genetics and genomics
Print (ISBN 978-1-78676-324-2); Online (ISBN 978-1-78676-326-6, 978-1-78676-327-3)

Achieving sustainable production of poultry meat Volume 1: Safety, quality and sustainability
Print (ISBN 978-1-78676-064-7); Online (ISBN 978-1-78676-066-1, 978-1-78676-067-8)

Achieving sustainable production of poultry meat Volume 2: Breeding and nutrition
Print (ISBN 978-1-78676-068-5); Online (ISBN 978-1-78676-070-8, 978-1-78676-071-5)

Achieving sustainable production of poultry meat Volume 3: Health and welfare
Print (ISBN 978-1-78676-072-2); Online (ISBN 978-1-78676-075-3, 978-1-78676-074-6)

Chapters are available individually from our online bookshop:
https://shop.bdspublishing.com

climate•SMART•publishing
IN AGRICULTURAL SCIENCE

BURLEIGH DODDS SERIES IN AGRICULTURAL SCIENCE
NUMBER 127

Improving poultry meat quality

Edited by Professor Massimiliano Petracci, Alma Mater Studiorum - Università di Bologna, Italy and Dr Mario Estévez, Universidad de Extremadura, Spain

burleigh dodds
SCIENCE PUBLISHING

Published by Burleigh Dodds Science Publishing Limited
82 High Street, Sawston, Cambridge CB22 3HJ, UK
www.bdspublishing.com

Burleigh Dodds Science Publishing, 1518 Walnut Street, Suite 900, Philadelphia, PA 19102-3406, USA

First published 2023 by Burleigh Dodds Science Publishing Limited

Library of Congress Control Number: 2022949225

British Library Cataloguing in Publication Data
A catalogue record for this book is available from the British Library

ISBN 978-1-80146-103-0 (print)
ISBN 978-1-80146-106-1 (PDF)
ISBN 978-1-80146-105-4 (ePub)
ISSN 2059-6936 (print)
ISSN 2059-6944 (online)

DOI 10.19103/AS.2022.0112

Typeset by Deanta Global Publishing Services, Dublin, Ireland

Contents

Series list

Title | Series number

Achieving sustainable cultivation of maize - Vol 1 001
From improved varieties to local applications
Edited by: Dr Dave Watson, CGIAR Maize Research Program Manager, CIMMYT, Mexico

Achieving sustainable cultivation of maize - Vol 2 002
Cultivation techniques, pest and disease control
Edited by: Dr Dave Watson, CGIAR Maize Research Program Manager, CIMMYT, Mexico

Achieving sustainable cultivation of rice - Vol 1 003
Breeding for higher yield and quality
Edited by: Prof. Takuji Sasaki, Tokyo University of Agriculture, Japan

Achieving sustainable cultivation of rice - Vol 2 004
Cultivation, pest and disease management
Edited by: Prof. Takuji Sasaki, Tokyo University of Agriculture, Japan

Achieving sustainable cultivation of wheat - Vol 1 005
Breeding, quality traits, pests and diseases
Edited by: Prof. Peter Langridge, The University of Adelaide, Australia

Achieving sustainable cultivation of wheat - Vol 2 006
Cultivation techniques
Edited by: Prof. Peter Langridge, The University of Adelaide, Australia

Achieving sustainable cultivation of tomatoes 007
Edited by: Dr Autar Mattoo, USDA-ARS, USA and Prof. Avtar Handa, Purdue University, USA

Achieving sustainable production of milk - Vol 1 008
Milk composition, genetics and breeding
Edited by: Dr Nico van Belzen, International Dairy Federation (IDF), Belgium

Achieving sustainable production of milk - Vol 2 009
Safety, quality and sustainability
Edited by: Dr Nico van Belzen, International Dairy Federation (IDF), Belgium

Achieving sustainable production of milk - Vol 3 010
Dairy herd management and welfare
Edited by: Prof. John Webster, University of Bristol, UK

Ensuring safety and quality in the production of beef - Vol 1 011
Safety
Edited by: Prof. Gary Acuff, Texas A&M University, USA and Prof. James Dickson, Iowa State University, USA

Ensuring safety and quality in the production of beef - Vol 2 012
Quality
Edited by: Prof. Michael Dikeman, Kansas State University, USA

Achieving sustainable production of poultry meat - Vol 1 013
Safety, quality and sustainability
Edited by: Prof. Steven C. Ricke, University of Arkansas, USA

Achieving sustainable production of poultry meat - Vol 2 014
Breeding and nutrition
Edited by: Prof. Todd Applegate, University of Georgia, USA

Introduction

Recent years have seen a meteoric rise in the number of consumers wishing to know more about the journey their food takes from farm to fork. Consumer expectations on the sensory and nutritional qualities of food products have risen as a result of this. Poultry producers and processors must balance these quality parameters when making decisions aimed to improve production efficiency or food yield.

This volume addresses the wealth of recent research on the genetic and environmental factors affecting the development of quality traits in poultry meat and their potential implications for breeding, husbandry and postharvest processing. The chapters are split into three parts: Part 1 chapters cover poultry muscle development and meat quality, focusing on aspects such as poultry muscle morphology and its impact on meat quality, understanding the genetics of muscle development and the impact of dietary backgrounds. Chapters in Part 2 focus on the individual quality attributes of poultry meat such as colour, texture and flavour development. Part 3 chapters discuss different poultry myopathies and shelf life, focusing specifically on quality defects such as deep pectoral myopathy, dorsal cranial myopathy, pale soft exudative meat, white striping, wooden breast, spaghetti meat and intramuscular connective tissue defect. A chapter on factors affecting the shelf-life of poultry meat is also included.

Part 1 Poultry muscle development and meat quality

Part 1 of the book opens with a chapter that focuses on the advances in understanding the development and morphology of the poultry breast muscle, focusing specifically on its impact on meat quality. Chapter 1 begins by identifying the different types of muscle fibre metabolism, then moves on to examine post-hatch skeletal muscle growth. The chapter also discusses the heterogeneity of satellite cells and their role in the development/growth of pectoralis major muscle. It reviews the role of satellite cells and the extracellular matrix in breast muscle structure associated with various myopathies as well as the conversion of satellite cells to an adipogenic lineage. The chapter concludes by analysing the effect of selection practices on breast muscle yield, comparing both fast and slow muscle growth.

The next chapter focuses on understanding the genetics of poultry muscle development. Chapter 2 first examines the origin and genetic regulation of muscle precursors, then moves on to discuss the genetic control of muscle precursor migration, specifically from the somite phase to the final phase the limb. The chapter also reviews the genetic regulation of satellite cells during post-hatch muscle growth. Sections on the various signalling pathways are also provided, focusing on the control of skeletal muscle development and muscle mass, myostatin/activin signalling, bone morphogenetic protein signalling as well as GH-GHR-IGFs axis, PI3K-Akt pathway and FOXO signalling.

The final chapter of Part 1 examines the nutritional strategies and management practices currently in place to improve poultry meat quality. Chapter 3 begins by highlighting the nutritional and feeding strategies to modulate meat composition, which is then followed by an overview of the strategies to improve meat quality and mitigate the incidence of breast meat abnormalities. A section on the consequences of heat stress on poultry meat quality and the nutritional strategies that can limit product alterations is also provided. The chapter also provides a discussion on the effect of management practices on meat quality, focusing specifically on the case of caponisation.

Part 2 Individual quality attributes

Part 2 begins with a chapter that reviews understanding colour development in poultry meat. Chapter 4 first reviews how different concentrations of myoglobin can influence the colour of poultry meat, which is then followed by a more in-depth analysis of myoglobin primary structure and its oxidation properties. The role of different biochemical factors and their effect on poultry meat colour is also discussed. The chapter moves on to review the application of metabolomics and proteomics in poultry colour research as well as the importance of understanding how pinkness can occur in cooked meat. Postharvest practices to improve meat colour are also provided. The chapter includes two case studies, the first reviews the impact of packaging on ground chicken colour and the second looks at molecular mechanisms that regulate poultry colour characteristics. A section on tools to assess poultry quality is also included.

The next chapter looks at texture development in poultry meat, focusing on advances and developmental strategies in meat tenderness. Meat tenderness is a complex quality trait to handle as each consumer has their own level of acceptance and preferences. It should also be noted that texture development is a very intricate process that is multifaceted. Chapter 5 begins by highlighting how texture development in the end product is important to the consumer. It then goes on to review the various factors that can affect poultry meat tenderness, such as rigor mortis, contractile state of the muscle, water holding capacity, collagen composition and age and strain effects. Tenderness assessment methods are also discussed, focusing specifically on visual inspection as a means to assess tenderness and instrumental methods.

The final chapter of Part 2 focuses on advances in understanding flavour development in poultry meat. Chapter 6 first examines the flavour chemistry of poultry meat, focusing on the changes that occur during the process of cooking the meat. The chapter then draws attention to the main flavour precursors of poultry meat: water-soluble components and lipids. Factors influencing the flavour formation in poultry meat are also discussed, specifically factors during the pre-slaughter and post-mortem stages of the poultry product. The chapter also highlights various strategies for preserving or enhancing the flavour of poultry meat, such as selecting the suitable breed/strain of chicken, improving lipid class and fatty acid composition and analysing the intramuscular contents of amino acids and nucleotides. The diet of the bird, the use of alternative production systems, aging, cooking, irradiation, high-pressure and plasma treatment and the use of additives in poultry products are also discussed.

Part 3 Poultry myopathies and shelf life

The first chapter of Part 3 focuses on breast meat abnormalities associated with ischaemic necrosis, focusing specifically on dorsal cranial myopathy and deep pectoral myopathy. Chapter 7 summarises the most relevant findings for both myopathies, focusing on the morphological and microscopic characteristics as well as incidence levels, their aetiology and associated factors that can influence this, such as seasonal influence and nutritional factors. To provide further background information, the chapter includes a section on muscle abnormalities in Latin American countries such as Brazil, Argentina, Peru, Chile, Colombia, Bolivia and Mexico as the growth of broiler meat production is highest.

The subject of Chapter 8 is quality defects associated with poultry muscle development, specifically pale, soft exudative meat. The chapter first reviews the causes of the occurrence of pale, soft and exudative meat, focusing specifically on genetic factors, ante-mortem stress as well as slaughter and post-mortem factors. It then goes on to examine the characteristics of the condition, drawing specific attention to physico-chemical and microbial properties as well as the technological properties of meat that is affected by the condition. The chapter describes genetic and epigenetic strategies to reduce and mitigate the impact of pale, soft exudative meat as well as the pre-slaughter strategies that can be used. Processing strategies such as the use of functional ingredients and technologies to reduce and mitigate the condition's impact on poultry meat are described. A section on the methods to detect pale, soft and exudative meat is also provided.

Chapter 9 focuses on another harmful quality defect that affects poultry muscle development, white striping. The chapter begins by describing the impacts of white striping on quality traits of poultry meat, such as its physicochemical properties, nutritional quality, technological properties, sensory quality and consumer acceptance and shelf life. It moves on to review the biological changes associated with the development of white striping. Factors contributing to the development of white striping are also included, discussing both genetic and non-genetic factors. The chapter also examines potential strategies for reducing white striping such as feed modification, farm management, processing and breeding.

The subject of Chapter 10 is wooden breast, another quality defect associated with poultry muscle development. The chapter first describes the possible factors underlying the occurrence of wooden breast, then moves on to review the main features characterising wooden breast. It looks at the criteria and methods for wooden breast evaluation and the histological characteristics of abnormal wooden breast muscles. The effects of wooden breast on poultry meat quality are also examined, such as its implications on proximal composition and nutritional value as well as the impact on technological properties of poultry meat. A section on the potential strategies to reduce the incidence of wooden breast and/or mitigate the effects of such condition on meat quality is also included, focusing specifically on two main approaches - nutritional strategies and the postharvest solutions to counteract reduced sensory and technological properties of poultry meat affected by wooden breast.

Chapter 11 looks at the quality defect known as spaghetti meat. This condition causes significant economic losses to the industry due to the affected breast meat being downgraded. The chapter begins by analysing the characteristics of spaghetti meat, specifically focusing on the macroscopic and microscopic traits, as well as connective tissue characteristics and chemical composition. It moves on to review the impact of spaghetti meat myopathy on meat quality and functionality, which is then followed by an analysis of pathological mechanisms and predisposing factors for spaghetti meat.

The final chapter of the book examines the factors affecting the shelf-life of poultry meat. Chapter 12 first highlights the importance of understanding both microbial and oxidative spoilage of poultry meat, focusing specifically on pre- and post-slaughter oxidative stress. This is then followed by a review of the impact of oxidative stress on poultry meat shelf life, specifically the effect oxidation has on the appearance of poultry meat. Strategies for extending poultry meat shelf life are also discussed, such as dietary strategies, the use of antioxidants and antimicrobials in meat products and the packaging of these final products as well.

Part 1

Poultry muscle development and meat quality

Chapter 1

Advances in understanding the development and morphology of the poultry breast muscle: impact on meat quality

Sandra G. Velleman, The Ohio State University, USA

1 Introduction

Meat quality is the result of the morphological structure of muscle and cellular mechanisms regulating skeletal muscle growth and development. Meat-type poultry have been selected for increased growth, carcass weight, breast muscle yield, reduced time to reach processing weight, and improved feed conversion (Havenstein et al., 1994a,b, 2003; Collins et al., 2014). An emphasis has been placed by the poultry breeding industry especially on increased growth rate, breast must yield and conformation. Despite positive changes in growth rate,

http://dx.doi.org/10.19103/AS.2022.0112.01

feed conversion, and muscling, meat quality problems in the breast muscle have arisen in both meat-type broilers and turkeys.

In broilers, three recently identified myopathies, Wooden Breast, White Striping, and Spaghetti Meat, are receiving considerable research and industry attention due to their negative effects on breast meat quality. Wooden Breast was first described by Sihvo et al. (2014) as a breast muscle myopathy resulting in muscle discoloration, turbid viscous coating of the outer surface of the muscle, and a palpably hard consistency. It has been estimated that 48–71% of commercial broilers have mild phenotypic Wooden Breast (Sihvo et al., 2017). In addition, to the visual changes associated with Wooden Breast, the breast muscle has reduced myofibrillar protein levels, decreased muscle yield, and reduced water-holding capacity. The Wooden Breast affected meat product often is downgraded or condemned from human consumption. White Striping is characterized by white striations of connective tissue and fat running parallel with the breast muscle myofibers. Almost all commercial broilers in the United States have some White Striping whereas in 2012 only 5% of the broiler breast muscle was affected (Kuttapan et al., 2016). In contrast to Wooden Breast, White Striping does not affect product quality, but consumers do not view excessive White Striping positively in making a purchase. Spaghetti Meat is characterized by the separation of muscle fiber bundles in the pectoralis major (p. major) muscle. The name Spaghetti Meat was chosen as the muscle fiber cylindrical rods are visible and resemble pasta noodles. In contrast to Wooden Breast and White Striping, following processing the p. major muscle can be pulled apart by hand. As described by Baldi et al. (2018), the endomysial and perimysial connective tissue layer in Spaghetti Meat degrades in a progressive manner. Meat quality is negatively impacted by this myopathy.

In fast growing heavy weight meat-type turkeys, more degenerating muscle fibers have been observed in lines selected for growth and heavy breast muscle weight (Wilson et al., 1990; Velleman et al., 2003). In broilers, almost all heavy weight meat-type birds have structural lesions in the breast muscle changing the biochemical composition of the meat, and negatively impacting water holding and binding capacity of the meat (Mazzoni et al., 2015). To date Wooden Breast, White Striping, and Spaghetti Meat have not been reported in turkeys. However, other breast muscle myopathies have been reported in turkeys including deep pectoral myopathy (Wilson et al., 1990; Sosnicki and Wilson, 1991), focal myopathy (Sosnicki, 1993), and pale, soft, and exudative (Pietrazak et al., 1997).

Many of the avian myopathies of economic concern are associated with the breast muscle (p. major muscle), and not the leg or thigh muscles. The primary difference between the p. major muscle and leg/thigh muscles is the type of metabolism to provide the energy for contractile needs of the tissue. Understanding why the p. major muscle is prone to the development of myopathies requires knowledge of muscle fiber type metabolism, posthatch hypertrophic growth mechanisms mediated by the adult myoblasts (satellite cells), development of the muscle morphological cell structure, and sensitivity of the satellite cells to extrinsic stimuli.

2 Types of muscle fiber metabolism

To provide the energy requirements needed, three basic types of skeletal muscle fibers exist which are categorized by contractile speed: (1) red slow twitch oxidative Type I muscle fibers, (2) white fast twitch glycolytic Type IIB muscle fibers, and (3) intermediate fast twitch oxidative/glycolytic Type IIA muscle fibers. Muscle contraction is fueled by ATP hydrolysis with the mechanisms of contraction and relaxation being similar across all skeletal muscle types. The ATP can be derived from aerobic, anaerobic, or a mix between aerobic and anaerobic respiration. In both commercial turkeys and chickens, the breast muscle is a Type IIB glycolytic muscle and utilizes anaerobic respiration.

Despite the contractile process being similar between aerobic and anaerobic muscle fibers, there are distinct differences in the contractile protein isoforms, metabolism, and morphological structure of the muscle. Aerobic, oxidative, muscle fibers contract more frequently and for longer sustained periods of time compared to anaerobic fibers. Anaerobic muscles like the breast muscle have muscle fibers that are larger in diameter and myosin is hydrolyzed at a faster rate to sustain quick bursts of contractions. In addition, the circulatory supply is significantly less in an anaerobic muscle compared to an aerobic muscle. In fast-growing heavy weight meat-type birds, circulatory supply in the breast muscle is further limited due to reduced available connective tissue spacing (Velleman et al., 2003).

3 Posthatch skeletal muscle growth

After hatch, skeletal muscle growth is due to the enlargement of the existing myofibers, hypertrophy. The mechanism of this hypertrophy was largely determined from studies with chickens. Smith (1963) demonstrated increases in myofiber number in chickens prior to hatch, and subsequent growth after hatch due to hypertrophy. Thus, myofiber number is determined prior to hatch. Posthatch muscle growth is dependent upon another cell population called satellite cells. Satellite cells fuse and donate their nuclei to existing myofibers (Moss and LeBlond, 1971) to increase myonuclear DNA content which is required for protein synthesis and accretion, and hypertrophic skeletal myofiber growth. The hypertrophy of muscle fibers from increased protein synthesis results in myofiber growth through the enlargement of existing fibers (Moss and LeBlond, 1971; Campion, 1984).

Historically, early experiments with isolated chick embryonic myoblasts revealed that the myonuclei in the myotubes were incapable of mitotic division and that another cell was responsible for DNA accretion (Stockdale and Holtzer, 1961). In fact, during embryonic development, myoblasts proliferate, align, and fuse to form multinucleated myotubes which subsequently form myofibers. After fusion, the myoblasts withdraw from the cell cycle (Moss and LeBlond, 1971). The discovery of satellite cells by Mauro (1961), based on electron microscopic examination, paved the way to clear the enigma as to what cell type was responsible for posthatch muscle growth. Those cells, unique to skeletal muscle in all vertebrates, were termed 'satellite cells' due to their peripheral location on myofibers, residing in a niche between the sarcolemma and basal lamina of the myofibers. The satellite cell mass consisted mostly of nuclei, and there was little cytoplasm and only a few organelles, such as mitochondria or Golgi apparatus.

Early evidence of a role for satellite cells in postnatal growth was reported by Moss and LeBlond (1971), where tritiated thymidine label injected into young growing rats was followed in the muscle tissue. The label was localized to the satellite cells, indicating their division in the myofibers; later, the label began to appear within myonuclei, indicating that the satellite cells had fused with the adjacent fibers. These observations established a key role for satellite cells in myofiber DNA accretion, especially in young growing animals. Moss (1968) established the idea that DNA accretion occurs only via satellite cell incorporation into the growing myofiber, and that it is based on a constant ratio of myofiber diameter to DNA content. At hatch, the satellite cell population makes up about 30% of the total myonuclei; however, their proliferative capacity and numbers diminish rapidly to 1–5% toward the end of the growing period (Hawke and Garry, 2001), when they become quiescent (Schultz et al., 1978) and reside in their niche. In heavy-strain chickens, satellite cell myogenic activity persists for the first 8 days posthatch and peaks on day 3 posthatch (Halevy et al., 2000).

In birds, during early growth, satellite cell myogenic activity is largely influenced by nutritional and/or environmental conditions such as light or temperature, all of which affect the pool of these cells for future hypertrophy. In chicken and turkey poults, changes in nutrition (Halevy et al., 2000; Mozdziak et al., 2002; Powell et al., 2013; Velleman et al., 2019) or in environmental conditions (Halevy et al., 2006a,b; Piestun et al., 2009; Hadad et al., 2014; Patael et al., 2019) during the early posthatch period impaired or enhanced the proliferation and differentiation activity of satellite cells, resulting in changes in the cell reservoir and long-term effects on muscle hypertrophy and growth. Recent studies have shown long-term promotive effects on satellite cells in late-term chick embryos that experienced nutritional (Kornasio et al., 2011) or environmental manipulations during the third wave of myogenesis (Halevy et al., 2006b; Piestun et al., 2009; Zhang et al., 2012).

Despite the inactive state of satellite cells and their diminished numbers in mature muscle, upon injury, excessive exercise, overload, or various types of stress, quiescent satellite cells residing in their niche do have the capacity to become active and undergo myogenesis and fusion into existing or newly formed myofibers to regenerate damaged skeletal muscle (Bischoff, 1975; Hawke and Garry, 2001). As such, satellite cells serve as a reservoir for muscle precursor cells, and are therefore classified as 'adult skeletal muscle stem cells' (Collins et al., 2005). They fulfill the definition of a stem cell in that they undergo asymmetric division in which one daughter cell will continue the myogenic program and the other will replenish the muscle precursor cell reservoir. In addition, under specific external signals, satellite cells can transform from the myogenic lineage to adipogenic, chondrogenic, or fibrogenic lineages (Asakura et al., 2001; Shefer et al., 2004).

4 Heterogeneity of satellite cells

Satellite cells are a heterogenous population of stem cells. The heterogeneity of satellite cells can take many different forms with satellite cells from different muscle fiber types expressing the genes of the fiber type it was isolated from (Feldman and Stockdale, 1991; Lagord et al., 1998; Huang et al., 2006; Manzano et al., 2011), and variable rates of proliferation and differentiation in satellite cells isolated from the same muscle (McFarland et al., 1995). Satellite cell heterogeneity is not well understood to date, but the heterogeneity affects the function of satellite cells with regard to muscle growth and regeneration, and recent evidence strongly suggests the ability to transdifferentiate to an adipogenic lineage (Xu et al., 2021a).

The heterogeneity of satellite cells exists even in a pure single fiber type muscle like the turkey and chicken p. major muscle which contains homogenous Type IIb fibers. Using a robotic manipulator, McFarland et al. (1995) isolated 73 different satellite cells from the p. major muscle of 6-week-old tom turkeys. Each individual cell was expanded and then studied for their proliferation, differentiation, and growth factor responsiveness. The characterization showed a distribution of cellular behaviors from fast growing to slow growing satellite cells. The fastest growing satellite cell reached 65% confluency during proliferation in 17 days whereas the slowest growing satellite cell clone took 30 days. Differentiation was affected in a similar manner to proliferation. Growth factors are potent stimulators or inhibitors of satellite cell proliferation and differentiation. In the faster growing satellite cells, responsiveness to fibroblast growth factor-2, insulin-like growth factor-I, platelet-derived growth factor-BB, and transforming growth factor-beta were augmented. These changes in growth factor responsiveness have stimulatory or inhibitory effects on muscle growth and regeneration of muscle fibers due to damage.

To further complicate satellite cell heterogeneity, there is evidence that the terminal fate of a satellite cell is not fixed. Daughter cells derived from Pax7+/MyoD+ cells were both Pax7+/MyoD+ and Pax7+/MyoD-. These findings suggest that in addition to unique subpopulations of satellite cells that individual satellite cell groups may have multiple fates. Thus, it is likely that satellite cell heterogeneity is also dynamic within a subpopulation of cells.

5 Morphological structure of the pectoralis major muscle: role of satellite cells

Selection for increased muscle mass accretion can be based on growth during the embryonic phase of fiber formation, hyperplasia, and/or on posthatch fiber enlargement, hypertrophy, through satellite cell-mediated mechanisms. Since hyperplasia occurs during the embryonic phase of development, muscle growth is usually determined posthatch. Hypertrophy-based selection can result in muscle fibers that are three to five times larger in cross-sectional area and are frequently referred to as giant fibers (Dransfield and Sosnicki, 1999). Giant fibers can remain in a permanently contracted state, hypercontracted.

The morphological structure of muscle is characterized by the formation of muscle fiber bundles. Bundle size can be variable and range from 50 fibers to 300 fibers and the size of the fibers composing the bundle can also vary. Important to the structure of muscle are the connective tissue layers: epimysium, perimysium, and endomysium. The epimysium surrounds the entire muscle, the perimysium muscle fiber bundles, and the endomysium individual muscle fibers. These connective tissue layers are joined together at the myotendinous junction providing structural support for muscle. Figure 1 contains a representative image of the morphological structure of the avian p. major muscle.

Changes in the structure or cell biology of muscle can impact meat quality in either a positive or negative manner. In faster growing heavy weight meat-type birds, more degenerating fibers are present (Wilson et al., 1990; Velleman et al., 2003). Typical with muscle fiber degeneration in the breast muscle is a significant reduction or absence of perimysial and/or endomysial connective tissue spacing. When the muscle fibers begin to juxtapose each other, degeneration is associated with a loss of connective tissue space. This loss of connective tissue space is frequently accompanied by the presence of excessive hypertrophic growth and giant fibers. Berri et al. (2007) reported in

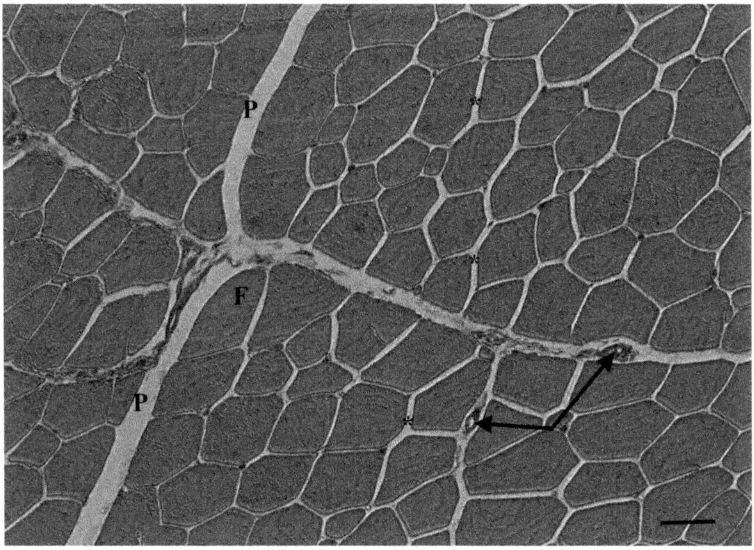

Figure 1 Representative image showing turkey pectoralis major structure with well-defined myofibers (F), perimysial (P), and endomysial (*) connective tissue spacing. Capillaries are highlighted by arrows. Scale bar = 100 µm. Source: Reproduced from Velleman and Clark (2015).

broilers that increased breast muscle mass was associated with larger muscle fiber cross-sectional area.

The reduction in available connective tissue spacing also limits the space for capillaries. The breast muscle being an anaerobic muscle already has less blood supply than its aerobic counterparts. The reduced capillary supply impacts oxygenation, satellite cell-mediated repair and regeneration processes, and removal of lactic acid from glycolytic respiration. Oxidative stress resulting from reduced blood supply can have serious detrimental effects on the breast muscle and breast meat product. For example, Wooden Breast (WB), a necrotic/fibrotic myopathy, in commercial broilers is characterized by severe hardening of the breast muscle leading to product downgrades and in severe cases condemnation (Sihvo et al., 2014) is associated with severe oxidative stress (Abasht et al., 2016).

When the muscle fibers and bundles begin to touch one another, myofiber degeneration commences and evokes satellite cell-mediated repair mechanisms (Velleman et al., 2003). Ideally when a muscle fiber is regenerated, it should be identical to the originating fiber. Similar to hypertrophy, satellite cells need to be reactivated to enter the cell cycle, proliferate, and differentiate. As part of this activation process, satellite cells require close proximity to circulatory supply in the niche surrounding the satellite cells. For satellite activation as shown by Christov et al. (2007) and Rhoads et al. (2009), signaling is required between the satellite cells and vascular system. Satellite cells will proliferate close to capillaries (Luque et al., 1995; Christov et al., 2007; Rhoads et al., 2009). Thus, the regeneration of muscle requires satellite cells to be in close proximity to capillaries. In fact, in humans, satellite cells need to be within 21 μm of capillaries for regeneration to occur (Christov et al., 2007). Furthermore, angiogenesis is also stimulated by satellite cells which likely involves the presence of growth factors like fibroblast growth factor, hepatocyte growth factor, and vascular endothelial growth factor which is secreted by the satellite cells into their niche (Rhoads et al., 2009). With limited or even without satellite cell-mediated regeneration of muscle, muscle fiber structure will be compromised and lead to the replacement of muscle with connective tissue and fat, fibrosis. Breast muscle myopathies especially in broilers resulting from necrosis and subsequent fibrosis have significant economic impact and affect meat quality.

6 Role of satellite cells and extracellular matrix in breast muscle structure associated with myopathies

Wooden Breast-affected p. major muscle is characterized by muscle fiber necrosis, inflammatory cell infiltration, and fibrosis (Sihvo et al., 2014; Velleman and Clark, 2015). To regenerate the necrotic and damaged muscle fibers, satellite cells will be activated to proliferate and differentiate into a new muscle fiber identical to the originating one. The regeneration process is precisely regulated and requires the construction of parallel arrays of microfibrils which are composed of repeating structures of the contractile unit of muscle, the sarcomere. If the process is altered, muscle fiber regeneration will be interfered with as well as muscle function, and meat quality will be reduced. The primary proteins of meat are myosin and actin. If actin and myosin concentrations are reduced than muscle fiber protein levels will be negatively impacted in the final meat product.

The myofiber is composed of parallel arrays of microfibrils. The microfibrils contain repeating units of sarcomeres. Sarcomeres contain three primary components: Z-lines, and thick and thin filaments. The Z-lines are composed of several proteins and define the boundaries of the sarcomere. Actin composes the thin filaments whereas the thick filaments contain myosin. The interaction between myosin and actin results in the contraction of the sarcomere. As the sarcomere contracts, the thin filament area (I band) is reduced due to the increased overlap with the thick filaments. Hypercontraction of the muscle is frequently observed with myopathies such as WB (Velleman and Clark, 2015; Clark and Velleman, 2017) and results when extreme contraction occurs and is not released. Velleman et al. (2018) showed that Wooden Breast-affected muscle had smaller diameter myofibrils and a lack of normal sarcomere structure ranging in severity. Figure 2 contains representative images of sarcomere structure in the avian p. major muscle. In the most severe cases, no sarcomeric structure was observable by transmission electron microscopy.

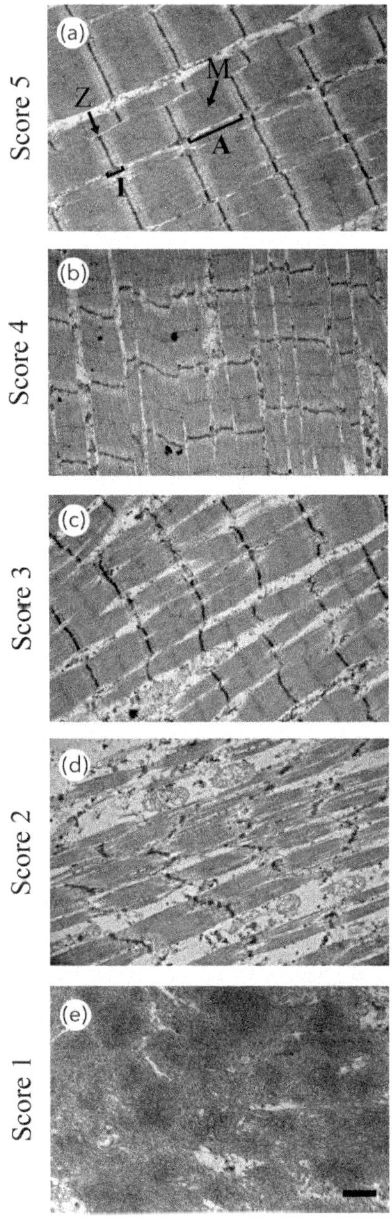

Figure 2 Representative transmission electron microscopy images for sarcomere organization scoring from 1 to 5. (a) Score of 5 shows sarcomeres with normal structure; (b) score of 4 means mildly altered; (c) score of 3 means further disorganized from a score of 4 but still moderate; (d) score of 2 means severely altered sarcomere structure; and (e) score of 1 means complete absence of sarcomere structure. A = A zone; I = I zone; M = M line; Z = Z line. Scale bar = 1 μm. Source: Reproduced from Velleman et al. (2018).

Furthermore, the smaller diameter myofibrils had the severest disruption of sarcomere structure (Velleman et al., 2018). In normal p. major muscle, during the posthatch period, myofiber diameter should increase through satellite cell-mediated hypertrophic growth (Moss and LeBlond, 1971) and normal sarcomere structure should be evident within the myofibrils.

Since the satellite cell-mediated regeneration process should restore damaged muscle fibers to their original state, the presence of a high percentage of smaller diameter fibers in WB muscle is representative of myofibers undergoing regeneration (Clark and Velleman, 2017; Velleman et al., 2018). Despite the smaller diameter myofibers with a high degree of myofibril sarcomeric disorganization, satellite cell-mediated regeneration is activated as evidenced by the increased expression of myogenic transcriptional regulatory factors modulating satellite cell proliferation and differentiation (Velleman and Clark, 2015). Since selection for breast muscle yield in most meat-type commercial broiler lines has led to an increase in myofiber diameter through hypertrophic growth, it is possible that the reduction in circulatory supply which is necessary for satellite cell-mediated regeneration is associated with the suppression in myofiber regeneration.

The degeneration, necrosis, and reduced regeneration of myofibers lead to an increased deposition of extracellular matrix proteins such as collagen and proteoglycans. The extracellular matrix network outside the cells in the connective tissue spaces of muscle determines the structural stability of the muscle, stretch, and water-holding capacity. The fibrosis process is a direct response to the necrosis and lack of regeneration of myofibers. Fibrosis results primarily in the overproduction of fibrillar collagens Types I and III. The degree of stiffness of the meat in birds with WB is not due to the total concentration of the fibrillar collagens but their organization and degree of crosslinking as shown by Velleman et al. (2017) and Tonniges et al. (2019). Thus, the morphometric organization of the fibrillar collagens plays a major role in regulating the tenderness of the meat product.

By definition, all collagens are composed of three polypeptide chains with the amino acid repeat Glycine-X-Y, where X and Y are any amino acids but are frequently lysine and proline. The three peptide chains form a right-handed helix. When the collagen helix is exocytosed from the cell, collagen fibril formation will take place followed by maturation into collagen fibers which is the functional form of the collagen. The collagen will now align into a quarter stagger array with the fibrils being stabilized by the formation of reversible divalent crosslinks. After fibril formation, the fibrils will come together and form collagen fibers. The divalent crosslinks are replaced with trivalent nonreversible hydroxylysylpyridinolone (HP) crosslinks. The HP crosslink increases with age and results in an increase in collagen fibril diameter through the linking of multiple collagen helices together. Collagen fibrils linked together with the HP crosslink are a major factor in the phenotypic detection of WB by palpation (Velleman et al., 2017; Tonniges et al., 2019). Furthermore, the HP crosslink increases with age so as the bird ages, the p. major muscle will become stiffer.

To further understand the impact of fibrillar collagen organization on breast meat quality, a series of studies were undertaken by the Velleman research group (Velleman and Clark, 2015; Velleman et al., 2017; Tonniges et al., 2019). In these studies, collagen fibril structure in three fast-growing meat-type broiler lines with distinctly different degrees of phenotypically detectable WB were investigated. The lines for the purposes of this discussion will be termed A, B, and C. Line C does not have any Wooden Breast-affected birds. In contrast, Line A has a high degree of phenotypically detectable WB and Line B is intermediate between Lines A and C. In light microscopy studies, Line A collagen fibers were observed to be packed in parallel whereas in Line B, the perimysial collagen was not packed but diffuse in structure (Fig. 3, Velleman and Clark, 2015; Velleman et al., 2017). Histological examination of the p. major muscle of the birds comprising each line showed that 70% of Line B had muscle fiber necrosis and fibrosis with only a small percentage of the birds having phenotypically detectable WB by palpation. These findings indicate that standard palpation practices used by the broiler industry are insufficient to detect necrotic/fibrotic myopathies that do not stiffen the breast muscle.

Further transmission electron microscopy analysis by Velleman et al. (2017) measured D periodicity and collagen fiber diameter. When collagen aligns into the quarter stagger array, the alignment is not a random process. There are areas of gaps and overlaps between the collagen triple helices. The length of one overlap zone is 67 nm and this is termed the D period. In Line A, Wooden Breast-affected perimysial collagen had a smaller average collagen fiber diameter and longer D periods. No such difference was measured in Line B. In fact, the perimysial collagen in Line B Wooden Breast-affected muscle was similar to unaffected perimysial collagen in Line B (Tables 1 and 2). The changes observed in Line A are suggestive of collagen functional changes with the potential for altering protein binding sites and collagen flexibility.

More importantly, the studies conducted by the Velleman research group on the three different fast-growing broiler lines demonstrated multiple organizations of fibrillar collagen with the occurrence of fibrotic myopathies. Fibrotic myopathies with highly crosslinked fibrillar collagen will be phenotypically detected by palpation with a rigid breast muscle. In contrast, collagen fibrils that have low levels of collagen crosslinking and are diffuse in structure, the breast muscle will remain phenotypically soft. Despite not being phenotypically detectable, the necrosis and fibrosis of the p. major muscle will alter breast meat nutritional quality by increasing fat content and reducing the amount of protein.

The primary fibrillar collagens in the p. major muscle are Types I and III collagen. The expression of these fibrillar collagens in skeletal muscle is age

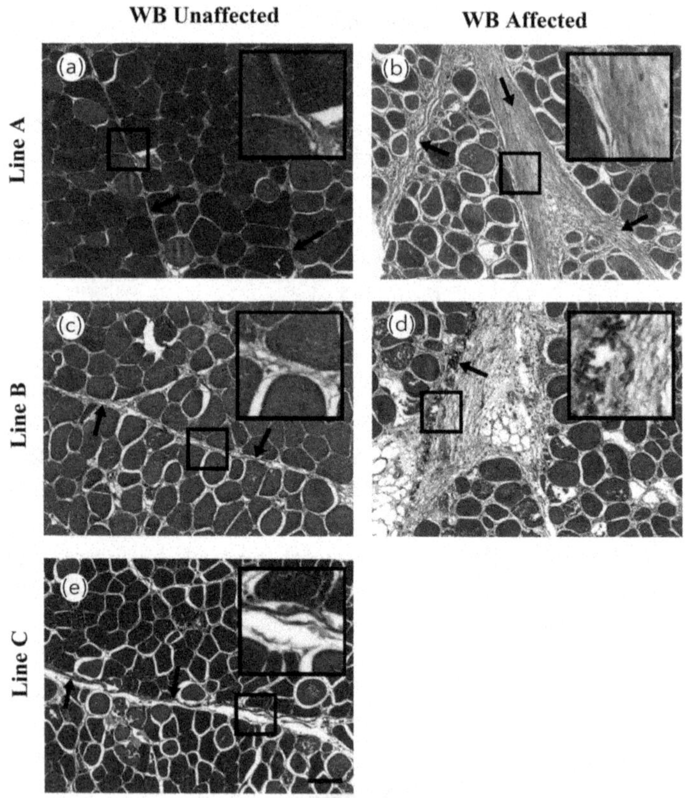

Figure 3 Masson trichrome staining of collagen organization in Wooden Breast (WB)-unaffected and -affected pectoralis major muscle. (a, c, e) Representative images of WB-unaffected pectoralis major muscle from Lines A, B, and C, respectively. (b, d) Representative images of WB-affected p. major muscle from Lines A and B, respectively. The arrows highlight fibrillar collagen and the boxes contain enlargements of the fibrillar collagen. Scale bar = 100 μm. Source: Reproduced from Velleman et al. (2017).

dependent with Type III collagen being the predominant collagen during early development and with age there is a shift to Type I collagen (Bornstein and Sage, 1980; Light and Champion, 1984; Kovanen and Suominen, 1989). Collagen fibrils can be homotypic containing one type of collagen or heterotypic containing multiple collagen types. McCormick (1994) hypothesized that the composition of collagen fibrils can influence meat tenderness through affecting stiffness. With regard to poultry breast meat quality, the impact of collagen fiber composition has not been studied to date. However, Type III collagen has been reported in highly aligned and tightly packed collagen fibrils similar to the organization of collagen fibrils in the WB myopathy (Brisson et al, 2015; McConnell et al., 2016).

Table 1 Effect of Wooden Breast (WB) myopathy and commercial broiler line on average D-period length (nm) of collagen fibrils in the endomysium, perimysium, and fibrotic areas of the breast muscle from three different lines

Location	Line A	Line B	Line C
Unaffected			
Endomysium	54.7 ± 0.1[c,y]	62.6 ± 0.1[a,x]	61.0 ± 0.1[b,x]
Perimysium	59.3 ± 1.1[a,x]	57.8 ± 2.0[a,xy]	58.7 ± 1.9[a,x]
WB affected			
Endomysium	61.2 ± 1.1[a,x]	62.8 ± 2.1[a,x]	*
Fibrotic	55.1 ± 0.7[b,y]	56.9 ± 0.7[a,y]	*

[a–c] Means ± SEM within row (Location) without a common letter are significantly different ($P \leq 0.10$).
[x–y] Means ± SEM within column (Line) without a common letter are significantly different ($P \leq 0.10$).

* Line C did not exhibit WB myopathy. Source: Reproduced from Velleman et al. (2017).

Table 2 Effect of Wooden Breast (WB) myopathy and commercial broiler line on average diameter (nm) of collagen fibrils in endomysium, perimysium, and fibrotic areas of the breast muscle from three different lines

Location	Line A	Line B	Line C
Unaffected			
Endomysium	42.4 ± 1.2[a,y]	37.5 ± 1.2[b,y,z]	40.9 ± 1.2[a,b,x]
Perimysium	50.6 ± 1.7[a,x]	48.4 ± 3.6[a,x,y]	44.8 ± 1.3[a,x]
WB affected			
Endomysium	38.2 ± 1.1[a,z]	37.4 ± 1.1[a,z]	*
Fibrotic	44.9 ± 1.2[a,y]	49.9 ± 3.1[a,x]	*

[a–b] Means ± SEM within row (Location) without a common letter are significantly different ($P \leq 0.10$).
[x–y] Means ± SEM within column (Line) without a common letter are significantly different ($P \leq 0.10$).

* Line C did not exhibit WB myopathy. Source: Reproduced from Velleman et al. (2017).

Spaghetti Meat is another necrotic/fibrotic myopathy in fast-growing heavy weight broilers that is less well described compared to WB, but also results in meat product downgrades (Baldi et al., 2018). Breast meat affected with Spaghetti Meat results in fillets that are soft with separated muscle fibers giving a spaghetti-like phenotype. Similar to WB, Spaghetti Meat breast muscle is characterized by greater collagen, fat and water content, and decreased protein (Baldi et al., 2018; Petracci et al., 2019). Sanden et al. (2021) using Fourier transform infrared microspectroscopy and imaging compared WB and Spaghetti Meat collagen structure. WB-affected breast muscle was characterized by a mixture of thick and thin collagen fibers whereas Spaghetti Meat had thinner, shorter, and fewer collagen fibers. Furthermore, the Spaghetti Meat fillets had a reduced amount of mature collagen fibers. Future meat quality-based studies will need to proceed beyond just measuring total collagen concentration as that is an insufficient measure of fiber composition and structure, and frequently total concentration does not vary between normal and breast meat with a degenerative myopathy (Mazzoni et al., 2015).

7 Satellite cell conversion to an adipogenic lineage

Satellite cells are a multipotential mesenchymal stem cell population with plasticity to commit to myogenesis or alternative cellular fates such as adipogenesis or osteogenesis (Asakura et al., 2001; Shefer et al., 2004). The first week posthatch is the period of maximal satellite cell mitotic activity (Halevy et al., 2000; Mozdziak et al., 2002). During this period of maximal mitotic activity, satellite cells are sensitive to extrinsic stimuli-like temperature (Xu et al., 2021b) and nutrition (Powell et al., 2013, 2014; Harthan et al., 2014) that can alter their cellular fate as well as proliferation and differentiation. Using a 2-day feed deprivation, Halevy et al. (2000) showed the timing of the feed deprivation was critical in its effect on satellite cell behavior. The closer the feed deprivation was to hatch up to 4 days, satellite cell mitotic activity and number was reduced, and the weight of the breast muscle never reached the weight of the full fed birds at processing age. Removing feed at 4–6 days of age did not alter satellite cell mitotic activity or breast muscle weight at market age.

Feed restrictions are frequently used by the industry and newly hatched chicks and poults can be exposed to hot and cold temperatures during handling and transportation. In addition, with a potential to reduce final breast muscle weight, these extrinsic stimuli immediately after hatch can also change the expression of muscle-specific genes and cause the conversion of the satellite cells to an adipogenic cell fate increasing fat within the breast muscle (Velleman et al., 2010, 2014a,b). Xu et al. (2021a) reported that proliferating satellite cells are more sensitive to temperature both hot and cold than differentiating satellite cells in the expression of adipogenic genes, and in the turkey selection for growth has increased the thermal sensitivity of satellite cells. Thus, when the cells have their highest mitotic activity immediately after hatch with greater proliferation taking place, this is the time that the satellite cells can express adipogenic genes, which may significantly affect the composition of the breast muscle and its overall structure. In support of the adipogenic potential of the satellite cells being the greatest immediately after hatch, Velleman et al. (2014b) showed that feed restrictions administered the first week posthatch had increased intramuscular lipid accumulation whereas moving the feed restriction to the second week posthatch after the period of maximal satellite cell activity eliminated the increase in intramuscular fat depots.

Breast muscle fillets with severe myofiber degeneration have significantly higher lipid levels (Mazzoni et al., 2015). Lake and Abasht (2020) have hypothesized that the dysregulation of lipid and glucose metabolism are key underlying causes of WB and perhaps other myopathies. In WB-affected birds, glycolysis is inhibited which results in an upregulation of lipid metabolism in birds at 2–3 weeks of age (Papah et al., 2018; Lake et al., 2019). The increase in lipid metabolism is supported by an upregulation in the expression of fatty acid metabolism genes (Lake et al., 2019). The accumulation of excessive lipid in the breast muscle is thought to result in a lipid toxicity leading to an immune response as well as contributing to the oxidative stress in the breast muscle. According to Schrauwen and Hesselink (2004) and Anderson and Neufer (2006), when fatty acids exceed the storage and oxidative capacity of the cell, they will remain within the cell where it can have damaging effects. The p. major muscle is a Type II B anaerobic muscle with limited mitochondria and circulatory supply. The dysregulation of lipid metabolism may also be associated with the fat accumulation characteristic of White Striping.

8 Effect of selection for breast muscle yield: fast compared to slower growth

A move is occurring to more heritage type or slower-growing meat-type birds which, in part, is consumer driven as meat quality is thought to be better and/or to reduce the incidence of meat quality affecting myopathies like WB. Despite the low heritability of body weight to the WB myopathy, the view still persists that heavy body weight is a primary cause of WB (Bailey et al., 2015). Regardless of the growth of the bird, mechanisms of muscle growth must be included in any selection scheme. As discussed previously in this chapter, the embryonic formation of myofibers and posthatch muscle fiber hypertrophy are distinct processes. Myofiber formation is regulated by myoblasts, which withdraw from the cell cycle after the individual myoblasts fuse. Posthatch growth through hypertrophy is mediated by the satellite cells. Muscle survivability requires sufficient endomysial spacing between individual myofibers and perimysial spacing between fiber bundles. If the myofibers or muscle fiber bundles begin to touch, degeneration commences (Velleman et al., 2003). When the available connective tissue space is reduced, this limits circulatory supply to the muscle which will affect satellite cell activity. Fast-growing broilers have been prone to these changes as selection has been largely based on the conformation of the breast muscle and posthatch hypertrophic muscle mass accumulation. Wilson et al. (1990) postulated in turkeys that if current selections strategies continued that breast muscle size had potentially exceeded its capacity for survivability.

If attention is paid to the morphological structure and cellular growth mechanisms, a breast muscle structure can be maintained that does not increase the possibility of myopathies or negatively impact meat quality. Furthermore, providing a sustainable environment for muscle growth while maintaining the appropriate structure will reduce the conversion of satellite cells to non-myogenic lineages like expressing adipogenic genes. To achieve this goal, focus needs to also be placed on the embryonic period of muscle fiber formation and limit fiber hypertrophy from exceeding the available space. Ultimately, selection practices need to include the morphological structure of the breast muscle and prevent muscle fibers and fiber bundles from touching which leads to degeneration and the need for satellite cell repair and regeneration processes.

Satellite cells can be suppressed from regenerating myofibers in damage-mediated events like in WB. In WB which is characterized by continual myofiber damage and inflammation, the damaged fibers are replaced with fat and connective tissue especially crosslinked fibrillar collagen. The suppression is supported by WB-affected muscle having significantly greater amounts of smaller myofibers. Transmission electron microscopy showed the smaller myofibers to have lack or disorganization of the sarcomeres (Velleman et al., 2018). Thus, the WB myopathy likely represses sarcomere assembly and satellite cell-mediated regeneration.

9 Conclusion

The avian breast muscle is characterized by distinctive phases of muscle growth that will alter muscle structure, function, and ultimately meat quality. Selection for breast muscle yield must take into consideration embryonic fiber formation, posthatch hypertrophic growth, and the overall morphological structure to reduce the incidence of degenerative myopathies and product downgrades. To have a high-quality poultry breast meat product, selection practices must include the cell biology and overall structure of the breast muscle.

10 Future trends in research

Technological advances in transcriptome analysis through RNA-Seq methodologies and analytical chemistry approaches to quantify metabolites with single cells, tissues, or biofluids will enhance traditional genetic selection approaches. For example, RNA-Seq transcriptome analysis has already been used to determine signal transduction pathway differences in the breast muscle between growth selected and nongrowth selected turkey lines (Reed et al., 2017a,b; Barnes et al., 2019). Changes in the expression of signal transduction pathways will alter both the cell biology and growth of the breast muscle. Furthermore, metabolomics research can identify metabolites that are involved in cellular mechanisms and relate these differences to genome changes that have occurred from selection (Clark et al., 2020). Taken together, these emerging technologies can provide new cellular-based selection tools in the breeding of commercial birds for increased growth and other performance characteristics.

11 Where to look for further information

- Velleman, S. G. and Petracci, M. (eds.) (2021). Avian Muscle Development and Growth Mechanisms: Association with Muscle Myopathies and Meat Quality. Lausanne: Frontiers Media SA. doi: 10.3389/978-2-88966-314-9.

12 References

Abasht, B., Mutryn, M. F., Michalke, R. D. and Lee, W. R. (2016). Oxidative stress and metabolic perturbations in wooden breast disorder in chickens. *PLoS ONE* 11(4), e0153750.

Anderson, E. J. and Neufer, P. D. (2006). Type II skeletal myofibers possess unique properties that potentiate mitochondrial H_2O_2 generation. *Am. J. Physiol. Cell Physiol.* 290(3), C844–C851.

Asakura, A., Komaki, M. and Rudnicki, M. (2001). Muscle satellite cells are multipotential stem cells that exhibit myogenic, osteogenic, and adipogenic differentiation. *Differentiation* 68(4–5), 245–253.

Bailey, R. A., Watson, K. A., Bilgili, S. F. and Avendano, S. (2015). The genetic basis of pectoralis major myopathies in modern broiler chicken lines. *Poult. Sci.* 94(12), 2870–2879.

Baldi, G., Soglia, F., Mazzoni, M., Sirri, F., Canonico, L., Babini, E., Laghi, L., Cavani, C. and Petracci, M. (2018). Implications of white striping and spaghetti meat abnormalities on meat quality and histological features in broilers. *Animal* 12(1), 164–173.

Barnes, N. E., Mendoza, K. M., Strasburg, G. M., Velleman, S. G. and Reed, K. M. (2019). Thermal challenge alters the transcriptional profile of the breast muscle in Turkey poults. *Poult. Sci.* 98(1), 74–91.

Berri, C., Le Bihan-Duval, E., Debut, M., Santé-Lhoutellier, V., Baéza, E., Gigaud, V., Jégo, Y. and Duclos, M. J. (2007). Consequence of muscle hypertrophy on characteristics of pectoralis major muscle and breast meat quality of broiler chickens. *J. Anim. Sci.* 85(8), 2005–2011.

Bischoff, R. (1975). Regeneration of single muscle fibers in vitro. *Anat. Rec.* 182(2), 215–235.

Bornstein, P. and Sage, H. (1980). Structurally distinct collagen types. *Annu. Rev. Biochem.* 49, 957–1003.

Brisson, B. K., Mauldin, E. A., Lei, W., Vogel, L. K., Power, A. M., Lo, A., Dopkin, D., Khanna, C., Wells, R. G., Puré, E. and Volk, S. W. (2015). Type III collagen directs stromal organization and limits metastasis in a murine model of breast cancer. *Am. J. Pathol.* 185(5), 1471–1486.

Campion, D. R. (1984). The muscle satellite cell: A review. *Int. Rev. Cytol.* 87, 225–251.

Christov, C., Chrétien, F., Abou-Khalil, R., Bassez, G., Vallet, G., Authier, F. J., Bassaglia, Y., Shinin, V., Tajbakhsh, S., Chazaud, B. and Gherardi, R. K. (2007). Muscle satellite cells and endothelia cells: Close neighbors and privileged partner. *Mol. Biol. Cell* 18(4), 1397–1409.

Clark, D. L. and Velleman, S. G. (2017). Spatial influence on breast muscle morphological structure, myofiber size, and gene expression associated with the wooden breast myopathy in broilers. *Poult. Sci.* 95(12), 2930–2945.

Clark, D. L., Velleman, S. G., Bernier, M., McCormick, J. and Blakeslee, J. J. (2020). The effect of selection for 16 week body weight on turkey serum metabolome. *Poult. Sci.* 99(1), 517–525.

Collins, C. A., Olsen, I., Zammit, P. S., Heslop, L., Petrie, A., Partridge, T. A. and Morgan, J. E. (2005). Stem cell function, self-renewal, and behavioral heterogeneity of cells from the adult muscle satellite cell niche. *Cell* 122(2), 289–301.

Collins, K. E., Kiepper, B. H., Ritz, C. W., McLendon, B. L. and Wilson, J. L. (2014). Growth, livability, feed consumption, and carcass composition of the Athens Canadian random bred 1955 meat-type chicken versus the 2012 high-yielding Cobb 500 broiler. *Poult. Sci.* 93(12), 2953–2962.

Dransfield, E. and Sosnicki, A. A. (1999). Relationship between muscle growth and poultry meat quality. *Poult. Sci.* 78(5), 743–746.

Feldman, J. L. and Stockdale, F. E. (1991). Skeletal muscle satellite cell diversity: Satellite cells form fibers of different types in cell culture. *Dev. Biol.* 143(2), 320–334.

Hadad, Y., Halevy, O. and Cahner, A. (2014). Featherless and leathered broilers under control versus hot conditions. 1. Breast meat yield and quality. *Poult. Sci.* 93(5), 1067–1075.

Halevy, O., Geyra, A., Barak, M., Uni, Z. and Sklan, D. (2000). Early posthatch starvation decreases satellite cell proliferation and skeletal muscle growth in chicks. *J. Nutr.* 130(4), 858–864.

Halevy, O., Yahav, S. and Rozenboim, I. (2006a). Enhancement of meat production by environmental manipulation in embryo and broilers. *Worlds Poult. Sci. J.* 62, 485–497.

Halevy, O., Pietsun, Y., Rozenboim, I. and Yablonka-Reuveni, Z. (2006b). In-ovo exposure to monochromatic green light promotes skeletal muscle cell proliferation and synchronizes myofiber development in posthatch chicks. *Am. J. Physiol.* 290, R1062–R1070.

Harthan, L. B., McFarland, D. C. and Velleman, S. G. (2014). The effect of nutritional status and myogenic satellite cell age on proliferation, differentiation, and expression of myogenic transcriptional regulatory factors and heparan sulfate proteoglycans syndecan-4 and glypican-1. *Poult. Sci.* 93(1), 174–186.

Havenstein, G. B., Ferket, P. R. and Qureshi, M. A. (2003). Carcass composition and yield of 1957 versus 2001 broilers when fed representative 1957 and 2001 broiler diets. *Poult. Sci.* 82(10), 1509–1518.

Havenstein, G. B., Ferket, P. R., Scheideler, S. E. and Larson, B. T. (1994a). Growth, livability, and feed conversion of 1957 vs. 1991 broilers when fed "typical" 1957 and 1991 broiler diets. *Poult. Sci.* 73(12), 1785–1794.

Havenstein, G. B., Ferket, P. R., Scheideler, S. E. and Rives, D. V. (1994b). Carcass composition and yield of 1991 vs. 1957 broilers when fed "typical" 1957 and 1991 broiler diets'. *Poult. Sci.* 73(12), 1795–1804.

Hawke, T. J. and Garry, D. J. (2001). Myogenic satellite cells: Physiology to molecular biology. *J. Appl. Physiol. (1985)* 91(2), 534–551.

Huang, Y. C., Dennis, R. G. and Barr, K. (2006). Cultured slow vs. fast skeletal muscle cells differ in physiology and responsiveness to stimulation. *Am. J. Physiol. Cell Physiol.* 291(1), C11–C17.

Kovanen, V. and Suominen, H. (1989). Age- and training related changes in the collagen metabolism of rat skeletal muscle. *Eur. J. Appl. Physiol. Occup. Physiol.* 58(7), 765–771.

Kornasio, R., Halevy, O., Kedar, O. and Uni, Z. (2011). Effect of in ovo feeding and its interaction with timing of first feed on glycogen reserves, muscle growth, and body weight. *Poult. Sci.* 90(7), 1467–1477.

Kuttapan, V. A., Hargis, B. M. and Owens, C. M. (2016). White striping and woody breast myopathies in the modern poultry industry: A review. *Poult. Sci.* 95(11), 2724–2733.

Lagord, C., Soulet, L., Bonavaud, S., Bassaglia, Y., Rey, C., Barlovatz-Meimon, G., Gautron, J. and Martelly, I. (1998). Differential myogenicity of satellite cells isolated from extensor digitorium longus (EDL) and soleus rat muscles revealed in vitro. *Cell Tissue Res.* 291(3), 455–468.

Lake, J. A., Papah, M. B. and Abasht, B. (2019). Increased expression of lipid metabolism genes in early stages of wooden breast links myopathy of broilers to metabolic syndrome in humans. *Genes* 10(10), 746.

Lake, J. A. and Abasht, B. (2020). Glucolipotoxicity: A proposed etiology for wooden breast and related myopathies in commercial broiler chickens. *Front. Physiol.* 11, 169.

Light, N. and Champion, A. E. (1984). Characterization of muscle epimysium, perimysium and endomysium collagens. *Biochem. J.* 219(3), 1017–1026.

Luque, E., Pena, J., Martin, P., Jimena, I. and Vaamonde, R. (1995). Capillary supply during development of individual regenerating muscle fibers. *Anat. Histol. Embryol.* 24(2), 87–89.

Manzano, R., Toivonen, J. M., Calvo, A. C., Miana-Mena, F. J., Zaragoza, P., Muñoz, M. J., Montarras, D. and Osta, R. (2011). Sex, fiber-type, and age dependent in vitro proliferation of mouse muscle satellite cells. *J. Cell. Biochem.* 112(10), 2825–2836.

Mauro, A. (1961). Satellite cell of skeletal muscle fibers, *J. Biophys. Biochem. Cytol.* 9, 493-495.

Mazzoni, M., Petracci, M., Meluzzi, A., Cavani, C., Clavenzani, P. and Sirri, F. (2015). Relationship between pectoralis major muscle histology and quality traits of chicken meat. *Poult. Sci.* 94(1), 123–130.

McConnell, J. C., O'Connell, O. V., Brennan, K., Weiping, L., Howe, M., Joseph, L., Knight, D., O'Cualain, R., Lim, Y., Leek, A., Waddington, R., Rogan, J., Astley, S. M., Gandhi, A., Kirwan, C. C., Sherratt, M. J. and Streuli, C. H. (2016). Increased peri-ductal collagen micro-organization may contribute to raised mammographic density. *Breast Cancer Res.* 18(1), 5.

McCormick, R. J. (1994). Structure and properties of tissues. In: Kinsman, D. M., Kotula, A. W. and Breidenstein, B. C. (Eds.), *Journal of Muscle Foods*. Boston, MA: Springer, pp. 25–62.

McFarland, D. C., Gilkerson, K. K., Pesall, J. E., Walker, J. S. and Yun, Y. (1995). Heterogeneity in growth characteristics of satellite cell populations. *Cytobios* 82, 21-27.

Moss, F. P. and LeBlond, C. P. (1971). Satellite cells are the source of nuclei in muscles of growing rats. *Anat. Rec.* 170(4), 421–435.

Moss, F. P. (1968). The relationship between the dimensions of the fibres and the number of nuclei during normal growth of skeletal muscle in domestic fowl. *Am. J. Anat.* 170, 421-436.

Mozdziak, P. E., Walsh, T. J. and McCoy, D. W. (2002). The effect of early posthatch nutrition on satellite cell mitotic activity. *Poult. Sci.* 81(11), 1703–1708.

Papah, M. B., Brannick, E. M., Schmidt, C. J. and Abasht, B. (2018). Gene expression profiling of the early pathogenesis of wooden breast disease in commercial broiler chickens using RNA-sequencing. *PLoS ONE* 13(12), e0207346.

Patael, T., Piestun, Y., Soffer, A., Mordechay, S., Yahav, S., Velleman, S. G. and Halevy, O. (2019). Early posthatch thermal stress causes long-term adverse effects on pectoralis muscle development in broilers. *Poult. Sci.* 98(8), 3268–3277.

Petracci, M., Soglia, F., Madruga, M., Carvalho, L., Ida, E. and Estévez, M. (2019). Wooden-breast, white striping, and spaghetti meat: Causes, consequences and consumer perception of emerging broiler meat abnormalities. *Compr. Rev. Food Sci. Food Saf.* 18(2), 565–583.

Piestun, Y., Harel, M., Barak, M., Yahav, S. and Halevy, O. (2009). Thermal manipulations in late-term chick embryos affect skeletal muscle development and promote myoblast proliferation and muscle hypertrophy. *J. Appl. Physiol. (1985)* 106(1), 233–240.

Pietrazak, M., Greaser, M. L. and Sosnicki, A. A. (1997). Effect of rapid rigor mortis processes on protein functionality in pectoralis major muscle of domestic Turkeys. *J. Anim. Sci.* 75(8), 2106–2116.

Powell, D. J., McFarland, D. C., Cowieson, A. J., Muir, W. I. and Velleman, S. G. (2013). The effect of nutritional status on myogenic satellite cell proliferation and differentiation. *Poult. Sci.* 92(8), 2163–2173.

Powell, D. J., McFarland, D. C., Cowieson, A. J., Muir, W. I. and Velleman, S. G. (2014). The effect of nutritional status on myogenic gene expression of satellite cells derived from different muscle types. *Poult. Sci.* 93(9), 2278–2288.

Reed, K. M., Mendoza, K. M., Abrahante, J. E., Barnes, N. E., Velleman, S. G. and Strasburg, G. M. (2017a). Response of turkey muscle satellite cells to thermal challenge. I. Transcriptome effects in proliferating cells. *BMC Genomics* 18(1), 352.

Reed, K. M., Mendoza, K. M., Strasburg, G. M. and Velleman, S. G. (2017b). Response of turkey muscle satellite cells to thermal challenge. II. Transcriptome effects in differentiating cells. *Front. Physiol.* 8, 948. DOI: 10.3389/fphys.2017.00948.

Rhoads, R. P., Johnson, R. M., Rathbone, C. R., Liu, X., Temm-Grove, C., Sheehan, S. M., Hoying, J. B. and Allen, R. E. (2009). Satellite cell-mediated angiogenesis in vitro coincides with a functional hypoxia-inducible factor pathway. *Am. J. Physiol. Cell Physiol.* 296(6), C1321–C1328.

Sanden, K. W., Bőcker, U., Ofstad, R., Pedersen, M. E., Høst, V., Afseth, N. K., Rønning, S. B. and Pleshko, N. (2021). Characterization of collagen structure in normal, wooden breast spaghetti meat chicken fillets by FITR microspectroscopy and histology. *Foods* 10(3), 548.

Schrauwen, P. and Hesselink, M. K. C. (2004). Oxidative capacity, lipotoxicity, and mitochondrial damage in type 2 diabetes. *Diabetes* 53(6), 1412–1417.

Schultz, E., Gibson, M. C. and Champion, T. (1978). Satellite cells are mitotically quiescent in mature mouse muscle: An EM and radioautographic study. *J. Exp. Zool.* 206(3), 451–456.

Shefer, G., Wleklinski-Lee, M. and Yablonka-Reuveni, Z. (2004). Skeletal muscle satellite cells can spontaneously enter an alternative mesenchymal pathway. *J. Cell Sci.* 117(22), 5393–5404.

Sihvo, H. K., Immonen, K. and Puolanne, E. (2014). Myodegeneration with fibrosis and regeneration in the pectoralis major muscle of broilers'. *Vet. Pathol.* 51(3), 619–623.

Sihvo, H. K., Lindén, J., Airas, N., Immonen, K., Valaja, J. and Puolanne, E. (2017). Wooden breast myodegeneration of pectoralis major muscle over the growth period in broilers. *Vet. Pathol.* 54(1), 119–128.

Smith, J. H. (1963). Relation to body size to muscle cell size and number in the chicken. *Poult. Sci.* 42, 619–625.

Sosnicki, A. A. (1993). Focal myonecrosis effects in turkey skeletal muscle tissue. In: *Reciprocal meat conference proceedings* (Vol. 46), pp. 97-102.

Sosnicki, A. A. and Wilson, B. W. (1991). Pathology of turkey skeletal muscle: Implications for the poultry industry. *Food Struct.* 10, 317-326.

Stockdale, F. E. and Holtzer, H. (1961). DNA synthesis and myogenesis. *Exp. Cell Res.* 24, 508-520.

Tonniges, J. R., Clark, D. L. and Velleman, S. G. (2019). The effect of the wooden breast fibrotic myopathy in broilers on fibrillar collagen organization and decorin-collagen binding. *Avian Dis.* 63(1), 48-60.

Velleman, S. G. and Clark, D. L. (2015). Histopathologic and myogenic gene expression changes associated with wooden breast in broiler breast muscles. *Avian Dis.* 59(3), 410-418.

Velleman, S. G., Anderson, J. W., Coy, C. S. and Nestor, K. E. (2003). Effect of selection for growth rate on muscle damage during turkey breast muscle development. *Poult. Sci.* 82(7), 1069-1074.

Velleman, S. G., Clark, D. L. and Tonniges, J. R. (2017). Fibrillar collagen organization associated with the broiler wooden breast fibrotic myopathy. *Avian Dis.* 61(4), 481-490.

Velleman, S. G., Clark, D. L. and Tonniges, J. R. (2018). The effect of the wooden breast myopathy on sarcomere structure and organization. *Avian Dis.* 62(1), 28-35.

Velleman, S. G., Clark, D. L. and Tonniges, J. R. (2019). The effect of nutrient restriction on the proliferation and differentiation of turkey pectoralis major satellite cells differing in age and growth rate. *Poult. Sci.* 98(4), 1893-1902.

Velleman, S. G., Coy, C. S. and Emmerson, D. A. (2014a). Effect of the timing of posthatch feed restrictions on broiler breast muscle development and muscle transcriptional regulatory factor gene expression. *Poult. Sci.* 93(6), 1484-1494.

Velleman, S. G., Coy, C. S. and Emmerson, D. A. (2014b). Effect of the timing of posthatch feed restrictions on the deposition of fat during broiler breast muscle development. *Poult. Sci.* 93(10), 2622-2627.

Velleman, S. G., Nestor, K. E., Coy, C. S., Harford, I. and Anthony, N. B. (2010). Effect of posthatch feed restriction on broiler breast muscle development and muscle transcriptional regulatory factor gene and heparan sulfate proteoglycan expression. *Int. J. Poult. Sci.* 9(5), 417-425.

Wilson, B. W., Nieberg, P. S., Buhr, R. J., Kelly, B. J. and Shultz, F. T. (1990). Turkey muscle growth and focal myopathy. *Poult. Sci.* 69(9), 1553-1562.

Xu, J., Strasburg, G. M., Reed, K. M. and Velleman, S. G. (2021a). Effect of temperature and growth selection on intracellular lipid accumulation and adipogenic gene expression in turkey pectoralis major muscle satellite cells. *Front. Physiol.* 12, 667814.

Xu, J., Strasburg, G. M., Reed, K. M. and Velleman, S. G. (2021b). Response of turkey pectoralis major muscle satellite cells to hot and cold thermal stress: Effect of growth selection on satellite cell proliferation and differentiation. *Comp. Biochem. Physiol. A Mol. Integr. Physiol.* 252, 110823.

Zhang, Z. Y., Jia, G. Q., Zuo, J. J., Zhang, Y., Lei, J., Ren, L. and Feng, D. Y. (2012). Effects of constant and cyclic heat stress on muscle metabolism and meat quality of broiler breast fillet and thigh meat. *Poult. Sci.* 91(11), 2931-2937.

Chapter 2

Understanding the genetics of poultry muscle development

Wen Luo, Qinghua Nie and Xiquan Zhang, South China Agricultural University, China

1 Introduction

Muscle mass is mainly determined by muscle fiber number, muscle fiber diameter and muscle fiber type. Different muscle fibers come in different sizes, shapes and contractile protein content. These differences determine both the quantity and quality of meat (Sobolewska et al., 2011). Enhancing meat production involves promoting muscle cell development and muscle fiber structure conversion. Skeletal muscle development and its regulation have been widely studied in recent decades because these processes help in understanding meat production, cell differentiation, molecular regulation, muscle dystrophy and regeneration.

From somite to muscle fibers, the formation of skeletal muscle should progress in a programmed sequence of cell proliferation, differentiation and migration. These programs are highly conserved among mammalian and poultry embryos. Many major discoveries in muscle development

http://dx.doi.org/10.19103/AS.2022.0112.02

depend on poultry model systems, especially the chick. However, most of these findings are related to embryogenesis. Although there are several reviews summarizing the latest research progress in avian myogenesis (Hirst and Marcelle, 2015; Luo et al., 2017; Scaal and Marcelle, 2018; Sobolewska et al., 2011), these reviews mainly focus on the developmental processes of skeletal muscle without exploring how genetic regulation of muscle development affects meat production. Here we will review current knowledge of the development of skeletal muscle in the poultry embryo model system. Special focus will be placed on how the genetic factors regulate skeletal muscle development and affect poultry meat production and quality.

2 The origin and genetic regulation of muscle precursors

As in other mammals, poultry skeletal muscle is derived from an embryonic structure known as the somite. Somites then separate into the sclerotome and dermomyotome. Two distinct signals regulate the formation of the sclerotome and dermomyotome. Shh triggers the epithelial mesenchyme transition of the somite ventral portion to form the sclerotome (Fan and Tessier-Lavigne, 1994; Munsterberg et al., 1995), while Wnt triggers and maintains the epithelial organization of the dermomyotome (Geetha-Loganathan et al., 2006). As indicated by its name, the primitive skeletal muscle is derived from the dermomyotome.

Skeletal muscle precursor cells within the dermomyotome are characterized by the expression of Pax3 and Pax7, paired-homeobox transcription factors that are critical for myogenic lineage specification during development. Pax3 and Pax7 are negative regulators of one another (Galli et al., 2008). Pax3 acts in primary myogenesis and plays an essential role in both the formation and specification of myogenic cells during embryogenesis (Mansouri et al., 1996). Pax7 plays a major role in later phases of muscle formation, as well as during muscle regeneration. All muscle progenitors are absent, and muscle growth is consequently arrested when Pax3 and Pax7 are both deficient (Hirst and Marcelle, 2015).

Cells delaminate and migrate from edges of the dermomyotome, generating differentiated and postmitotic myocytes that organize into the primary myotome (Hirst and Marcelle, 2015). At least two distinct signals can induce the formation of the myotome. Wnt1 and Wnt3a, in combination with low levels of Shh, induce the myogenic differentiation of cells located in the epaxial domain of the dermomyotome (Musumeci et al., 2015). Wnt7a, in combination with BMP4 and FGF5, defines the hypaxial myotome in the ventral-lateral cells of the dermomyotome (Cheng et al., 2004; Pourquie et al., 1996). Wnt1 and Wnt3a induce *MyoD* expression and myogenesis via activation of *Myf5* gene expression, whereas Wnt7a directly induces *MyoD* expression in the dorsal ectoderm (Musumeci et al., 2015).

In addition to myotomal cells, three other types of premyoblastic cells can arise from dermomyotome: embryonic myoblasts, fetal myoblasts and adult myoblasts (also known as satellite cells or muscle stem cells). These cells are distinct myogenic populations. Both embryonic myoblasts and myotomal cells migrate to a programmed location, then proliferate and differentiate into myoblasts that fuse to form primary myotubes. The fetal myoblasts give rise to secondary myotubes which require nervous input for their formation and differentiation. The tremendous increase in the number and size of muscle fibers during the late-stage embryo development is primarily from secondary myotubes.

The classification of embryonic, fetal and adult myoblasts is based on their *in vitro* characteristics. These myogenic cells differ in their gene expression, appearance, response to extrinsic signaling molecules and morphology of myofibers they generate (Biressi et al., 2007a; Hutcheson et al., 2009). Embryonic, fetal and adult myoblasts are all originated from $Pax3^{+}Pax7^{-}$ myogenic progenitors (Messina and Cossu, 2009). Embryonic myoblasts are $Pax3^{+}$-derived cells, while fetal and adult myoblasts are Pax3-derived $Pax3^{-}Pax7^{+}$ cells. Embryonic and fetal myoblasts also differ in the expression of several transcription factors and surface molecules, such as Pax3, Muscle creatine kinase and Integrin α7 (Biressi et al., 2007a). These different myoblast populations give rise to different muscle fibers at different times when they migrate to a specific region (Fig. 1).

3 From somite to limb: genetic control of muscle precursor migration

The pectoralis major muscle, leg muscle and wing muscle are the most important parts of the carcass for poultry meat sales. Where do these muscles come from during the embryo development? Using the chick chimera technique and the systematic transplantation of one somite at a time, it has been demonstrated that both the wing muscles and the pectoralis major muscle arise from somites 16-21 (Beresford et al., 1978; Beresford, 1979; Zhi et al., 1996). Leg muscles originate from progenitors present in somites 26-33 (Lance-Jones, 1988). All muscles receive progenitors from at least three different somite segments (Scaal and Marcelle, 2018).

Limb muscle precursor cells migrate from the dermomyotome as single cells into the limb bud, form distinct muscle groups and attach to the developing skeletal elements via tendons (He et al., 2003). A detailed description of the migration, differentiation and regulation of skeletal muscle precursors for limbs can be found in Christ and Brand-Saberi (2002). This process is summarized in Fig. 2 and briefly described in the following paragraphs.

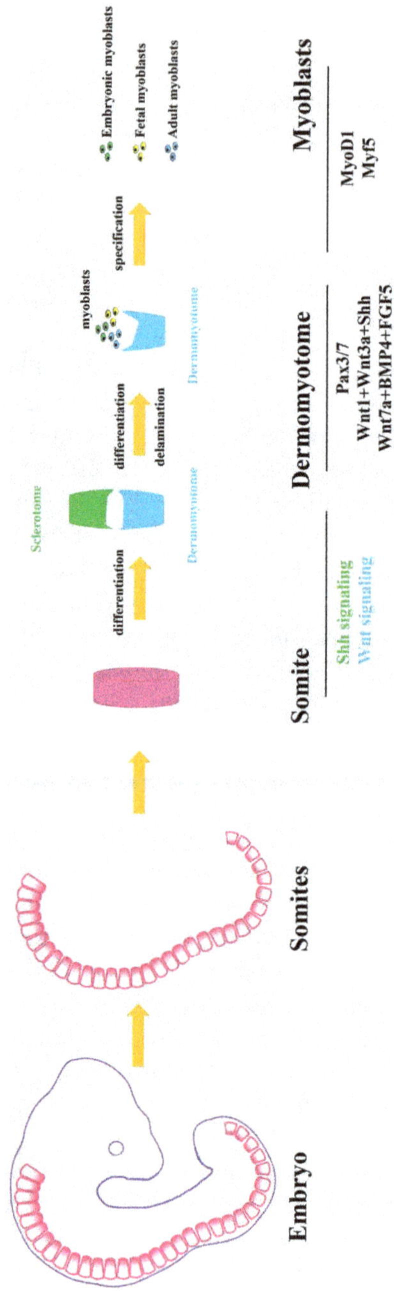

Figure 1 The origin and major genetic regulators of muscle precursors.

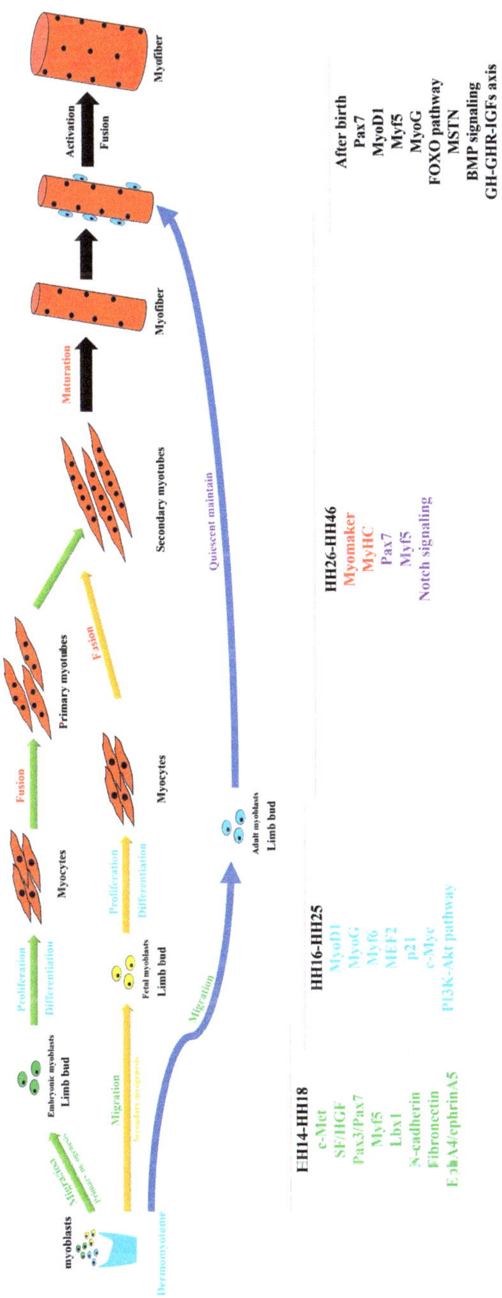

Figure 2 Developmental processes and genetic regulation from skeletal muscle precursors to muscle fiber.

The migration of limb muscle precursors is initiated by interaction between the tyrosine kinase receptor c-Met and its ligand scatter factor/hepatocyte growth factor (SF/HGF) (Bladt et al., 1995). Via expression of the transcription factors Pax3, Pax7 and myf5, precursor cells adopt a mesenchymal morphology without losing their myogenic specification. These transcription factors also enable precursor cell motility and proliferation during migration. Several genes have been found to be involved in the regulation of limb muscle precursor cell migration, such as:

- the homeobox gene *Lbx1* (Dietrich et al., 1998; Jagla et al., 1995);
- cell adhesion molecule N-cadherin (Brand-Saberi et al., 1996; George-Weinstein et al., 1997);
- fibronectin (Brand-Saberi et al., 1993); and
- the interaction of EphA4 and its ligand ephrinA5 (Swartz et al., 2001).

After having reached their target sites, myogenesis is initiated by the activation of the muscle determination factors MyoD, myogenin and MRF4 (Christ and Brand-Saberi, 2002).

The migration of muscle precursor cells into the limb bud region begins at Hamburger and Hamilton stage 14/15 and is completed by stage 18 (Schramm and Solursh, 1990). Myoblast overt differentiation does not begin until stage HH25 (Sweeney et al., 1989) while fusion begins at stage HH26 (Hilfer et al., 1973). Stages 14-25 are thus critical for myoblast proliferation. The maximum postnatal number of limb muscle fibers can be predetermined by the number of muscle precursors migrated to the limb bud and by the number of times that their offspring divide.

Pectoral muscles are derived from the wing premuscle mass. The superficial shoulder girdles muscles, which include pectoral and latissimus dorsi muscles, undergo a two-step 'in-out' developmental process (Masyuk et al., 2014; Masyuk et al., 2014; Valasek et al., 2011). The muscle precursors first migrate from the lateral lip of the dermomyotome into the wing bud and then migrate in a retrograde manner from the wing premuscle mass to the trunk (Pu et al., 2016). Wing premuscle masses are discontinuous with the trunk before HH23 (Valasek et al., 2011). However, progenitors from the wing premuscle mass are detectable in the pectoral girdle region at stage HH26 (Masyuk et al., 2014). This suggests that the 'in-out' mechanism is deployed in the wind region of the chicken. The 'in-out' mechanism of pectoral muscle development depends particularly on CXCR4/SDF-1 signaling (Masyuk et al., 2014).

After muscle precursor cells migrate to programmed regions, they proliferate and differentiate into myoblasts. *Pax3* and *Myf5* genes are expressed in proliferating muscle precursor cells, while *MyoD1* and muscle proteins are expressed in myoblasts. An increased proliferation rate of these cells results in stimulation of muscle growth and enlargement of muscle size. Muscle fiber formation is achieved by two waves of myoblast differentiation, primary and secondary myogenesis (Bren-Mattison et al., 2011). Embryonic myoblasts are dominant in primary myogenesis, while fetal myoblasts are dominant in secondary myogenesis (Biressi et al., 2007b).

From chicken embryonic day 3 to day 7, embryonic myoblasts proliferate and differentiate into small fibers, which are called primary myofibers (also known as primary myotubes *in vitro*) with few nuclei (Zappelli et al., 1996). The primary myofibers act as a scaffold for the formation of secondary myofibers. Later, during chick embryonic day 6 to day 13, fetal myoblasts form large multinucleated secondary myofibers (also known as secondary myotubes *in vitro*) adjacent to the established primary myofibers (Bren-Mattison et al., 2011). Fetal myoblasts can also fuse into primary myofibers to increase fiber size (Messina and Cossu, 2009). In the case of limb muscle, most monocyte myoblasts have fused into myofibers after chick embryonic day 14, while it might take at least 2 days further in pectoral muscle (Figs. 3 and 4).

Primary and secondary myofibers differ in the expression of myosin heavy chain (MyHC) isoforms. In embryonic chickens, all primary myofibers express slow (type I) MyHC isoforms (Marcucio and Noden, 1999), while fast Ca2(+)-ATPase antibody is present in nearly all secondary myofibers (Fredette and Landmesser, 1991). Slow- and fast-contracting primary myofibers are

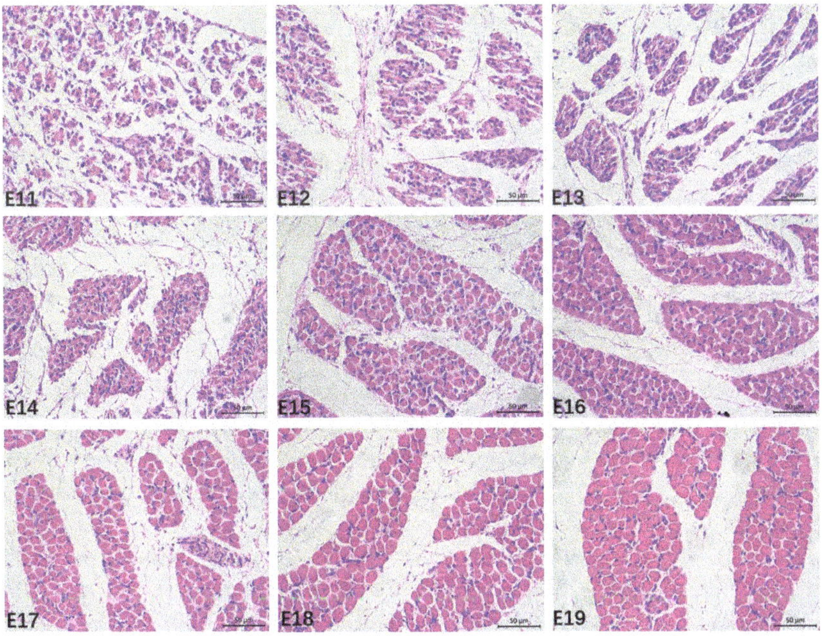

Figure 3 Cross-section images of chicken embryonic leg muscle.

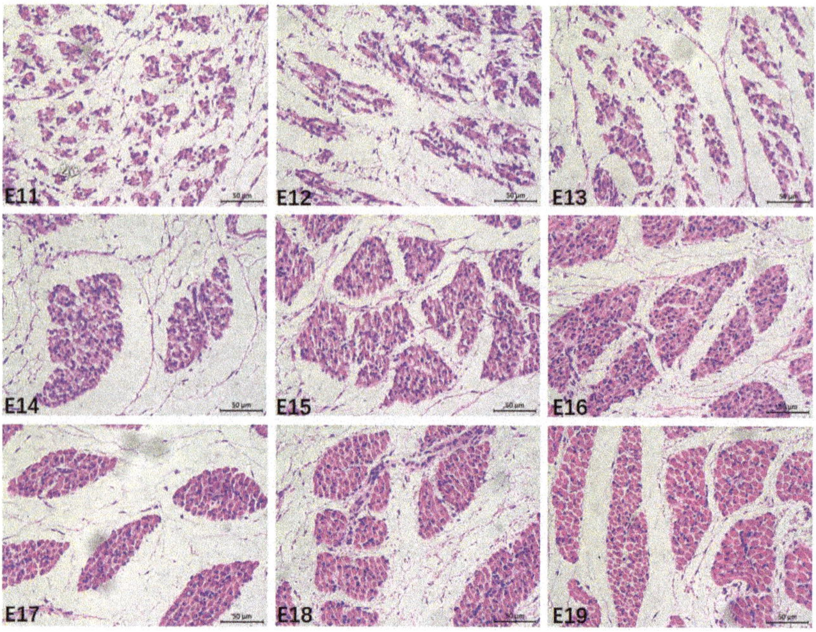

Figure 4 Cross-section images of chicken embryonic pectoral muscle.

established within individual muscles of limbs and the trunk during embryonic development. Distinct fiber types can be distinguished as early as stage HH29 (McLennan, 1983). Patterning of slow and fast fibers within embryonic muscles is established by myogenic precursors. Studies have shown that either a slow or a fast fiber-type lineage occurs in myogenic precursors while still within the somite (Nikovits et al., 2001). After birth, the contractile speed and isoform content of muscle fibers can be modified by innervation, activity and hormonal influences (Pette and Staron, 2000).

4 Genetic regulation of satellite cells during post-hatch muscle growth

New myofiber formation is completed by birth in most muscles (Smith, 1963). However, several fetal myoblasts maintain proliferation and fusion to form new myofibers in the late embryonic stages and post-hatch (Li et al., 2015). An increase in total muscle fiber number has been confirmed during the early post-hatch stage in chicken (Halevy et al., 2006; Halevy et al., 2004; Sobolewska et al., 2011). However, the increase in the number of muscle fibers is limited.

Post-hatch muscle fiber growth mainly occurs through muscle fiber enlargement via hypertrophy, which results from an increase in protein synthesis and the recruitment of satellite cell nuclei (Velleman, 2007; Wilson

Table 1 Pathways, genes and noncoding RNAs involved in poultry skeletal muscle development

Skeletal muscle developmental processes	Signaling pathways	Genes	Noncoding RNAs	References
Muscle precursors differentiation	Shh-gli1 pathway; Wnt1/3a/4 pathway; Notch-HES1 signaling; beta-catenin-Wnt signaling; Pax3/Paraxis-Myf5 signaling; Wnt6-MyoD1 signaling; BMP signaling	*MSTN; Activin A; hox; PAX3/PAX7; EMX2; BMP4; WNT4; WNT6; Hoxa-11/Hoxa-13; Follistatin; N-cadherin; FGF18; ATOH8; Myf5; noggin; MXRA5*	miR-125a-5p; miR-1; miR-206	Brauner et al. (2010); Bren-Mattison and Olwin (2002); Geetha-Loganathan et al. (2005); Stockdale et al. (2000); Wagner et al. (2000); Wang et al. (2005); Alvares et al. (2003); Amthor et al., (2002); Amthor et al., (1999); Amthor et al., (2006); Balakrishnan-Renuka et al., (2014); Bell et al. (2000); Berthoud et al. (2004); Dietrich (1999); El-Magd et al. (2019); Esteve and Bovolenta (1999); Galli et al. (2008); Geetha-Loganathan et al. (2006); Geetha-Loganathan et al. (2005); George-Weinstein et al. (1997); Hashimoto et al. (1999); He et al. (2005); Holowacz et al. (2006); Kiefer and Hauschka (2001); Mennerich et al. (1998); Mok et al. (2014); Rallis et al. (2001); Reijntjes et al. (2007); Robins and Capehart (2018); Takata et al. (2007); Tamura et al. (2003); Tonegawa and Takahashi (1998); Weimer et al. (2013); Wilm et al. (2004); Yang and Ordahl (2006); Fafeur et al. (1997); Riley et al. (2013)
Muscle precursors migration	CXCR4-SDF-1 signaling; FGF-FGFR signaling	*MSX1; EphA4; SF/HGF; Lmx1b; Wnt11; lbx1; ETS1*		Itoh et al. (1996); Masyuk et al. (2014); Bendall et al. (1999); Chen et al. (2004); Fafeur et al. (1997); Heymann et al. (1996); Masselink et al. (2017); Morosan-Puopolo et al. (2014); Schweizer et al. (2004); Swartz et al. (2001)
Myoblast proliferatior	MAPK pathway; FOXO pathway; p53 pathway; Insulin pathway; TGF-β/SMAD signaling pathway; PI3K/Akt pathway; Cell cycle pathway; Wnt pathway; Mtor pathway	*ANKRD9; MUSTN1; LMO7; Dact1; PDLIM5; MSTN; TRMT61A; KLF4; PDLIM3; TGF-β2; IGF2BP3; LRP1; MDFIC; IGFBP2; c-Myc; FOXO3; PRRX1; CDH11; SESN1; IGF2; Bcl2; E2F1; HDAC2; IGF-1; c-JUN; CARP; Smad3; FGF2; DAP1; FGFR1*	circSVIL; circHIPK3; miR-146b-5p; let-7b; Lnc-IRS1; miR-214; miR-205a; miR-16; miR-146b-3p; miR-2954; miR-20a-5p; miR-20b-5p; miR-16-5p; miR-7; miR-223; miR-27b-3p; miR-133a-3p; lncRNA-Six1; miR-29b-1-5p; miR-34b-5p; miR-203; circITSN2; circ-FGFR2; lnc-SMUL	Cai et al. (2018); He et al. (2021); Huang et al. (2019); Jia et al. (2017); Jingting et al. (2017); Ju et al. (2021); Li et al. (2020b); Luo et al. (2019); Vignale et al. (2015); Xu and Wu (2000); Yin et al. (2020d); Yu et al. (2015); Contriciani et al. (2021); Duan et al. (2020); Guo et al. (2018); Hu et al. (2021); Jia et al. (2017); Lee et al. (2019); Li et al. (2017); Li et al. (2021); Lin et al. (2017); Lu et al. (2013); Luo et al. (2016); Luo et al. (2014); Lv et al. (2019); Ma et al. (2014); Parakati and DiMario (2013); Shahjahan et al. (2016); Shen et al. (2021a); Shin et al. (2013); Shin et al. (2015); Tripathi et al. (2012); Wang et al. (2018); Wang et al. (2019); Zhang et al. (2020); Zhang et al. (2021); Cai et al. (2021); Cai et al. (2018); Cao et al. (2020); Chen et al. (2019); Chen et al. (2018); Dong et al. (2021); Duan et al. (2020); Guo et al. (2018); Huang et al. (2019); Jia et al. (2017); Lee et al. (2020); Li et al. (2017); Li et al. (2021); Z. Li et al. (2019); Lin et al. (2017); Luo et al. (2016); Luo et al. (2014); Ma et al. (2018); Ouyang et al. (2018); Shen et al. (2021a); Wang et al. (2018); Wang et al. (2019); Zhang et al. (2020); Zhang et al. (2021)

(*Continued*)

Table 1 *(Continued)*

Skeletal muscle developmental processes	Signaling pathways	Genes	Noncoding RNAs	References
Myoblast differentiation and fusion	Calcium pathway; PPAR pathway; p38-MAPK pathway; Focal adhesion pathway; TGF-β pathway	*KLHL30; MYH1B; PDLIM5; KLF4; FTO; Islr; Fmod; PDLIM3; TGF-β2; IGF2BP3; MEF2D; CSRP3; MDFIC; TP63; Myoz3; MYOD; Ebp1; MEF2C; DAP1; MSTN; Myf5; MyoD*	circSVIL; circHIPK3; miR-146b-5p; let-7b; Lnc-IRS1; miR-214; miR-205a; miR-146b-3p; miR-2954; miR-20a-5p; miR-20b-5p; miR-7; miR-223; miR-27b-3p; miR-133a-3p; lncRNA-Six1; miR-29b-1-5p; miR-34b-5p; miR-203; miR-130b-3p; miR-301b-3p; circITSN2; circ-FGFR2; lnc-SMUL	Han et al. (2019); He et al. (2021); Huang et al. (2020); Huang et al. (2019); Jingting et al. (2017); Ju et al. (2021); Ye et al. (2017); Yin et al. (2020a); Yin et al. (2020b); Yin et al. (2020c); Yin et al. (2020d); Yu et al. (2017); Chen et al. (2021); Cui et al. (2020); Li et al. (2017); Luo et al. (2018); Luo et al. (2014); Ouyang et al. (2020); Shin et al. (2013); Tripathi et al. (2012); Xu and Wu (2000); Yu et al. (2021); Zhang et al. (2020); Cai et al. (2021); Cai et al. (2018); Cao et al. (2020); Chen et al. (2019); Chen et al. (2018); Dong et al. (2021); Duan et al. (2020); Guo et al. (2018); Huang et al. (2019); Jia et al. (2017); Lee et al. (2020); Li et al. (2017); Li et al. (2021); Z. Li et al. (2019); Lin et al. (2017); Luo et al. (2016); Luo et al. (2014); Ma et al. (2018); Ouyang et al. (2018); Shen et al. (2021a); Wang et al. (2018); Wang et al. (2019); Zhang et al. (2020); Zhang et al. (2021)
Satellite cell development	p38-MAPK pathway	*PDLIM5; ITGB1; MUSTN1; GPC-1; PDGF; GDF-8; CFL2; CSRP3; TGF-β1; HDAC4; Myf5; GH; GHR; IGF-1; CTRP3; Sdc4; Sdc1; MTMR3*	miR-99a-5p; miR-128-3p; miR-9-5p; miR-128; miR-24; circTMTC1; circPPP1R13B	He et al. (2021); Cao et al. (2020); Day et al. (2009); Halevy et al. (1996); Han et al. (2019); He et al. (2021); Hu et al. (2021); Li et al. (2009); Liu et al. (2012b); McFarland et al. (2006); Ran et al. (2021); Shin et al. (2012); Song et al. (2012); Velleman et al. (2018); Velleman et al. (2007); Velleman et al. (2006); Velleman and McFarland (2004); Wang et al. (2021); Ye et al. (1996); Zhao et al. (2020); Cao et al. (2020); Shen et al. (2019); Shen et al. (2021b); Velleman and Harding (2017)
Muscle fiber hypertrophy	MAPK pathway; GH-GHR-IGFs signaling; Focal adhesion kinase; IGF1-calcineurin-NFATc3-dependent pathway; ER stress-related pathway; PI3K-Akt pathway	*c-Myc; PAX7; SRE1; TEF-1; SDH; CSRP3; LMOD2; MUSTN1; MYBPC1; MRF4; MyoG; MSTN; IGF-1; MYOD1; Atrogin-1; CnAα; NFATc3; BMP4; rc*	Linc-2949; Linc-1369; miR-128; miR-24	Chen et al. (2012); Fluck et al. (1999); Luo et al. (2019); Rehfeldt et al. (2011); Shu et al. (2015); Sun et al. (2013); Yin et al. (2020a); Zanou and Gailly (2013); Allouh and Aldirawi (2012); Al-Musawi et al. (2012); Carson et al. (1996); Cerruti et al. (1997); Chen et al. (2012); Gu et al. (2013); Kikuchi et al. (1980); Li et al. (2020a); Liu et al. (2012a); Liu et al. (2012b); Luo et al. (2019); Shu et al. (2015); Luo et al. (2019); Velleman and Harding (2017)
Muscle fiber atrophy	FAK-ERK signaling; FAK-Akt signaling; FOXO pathway; TGF-β pathway	*TMEM182; atrogin-1/MAFbx; SMURF2; Atrogin-1*	LncIRS1; SMUL	Chen et al. (2012); Luo et al. (2021); Yin et al. (2020a); Cai et al. (2021); Chen et al. (2012); Z. Li et al. (2019); Luo et al. (2021); Nakashima et al. (2016); Cai et al. (2021); Z. Li et al. (2019)
Muscle fiber-type transition		*MyoD1; ANKRD2; MSTN; PGC-1α; Six1*	miR-1611; lncRNA-Six1	Ijiri et al. (2009a); Ijiri et al. (2009b); Li et al. (2016); Shu et al. (2014); Stamenkovic et al. (2020); Ma et al. (2018)

et al., 1992). Satellite cells appear (embryonic day 18) and become dominant in late chicken embryogenesis (Hartley et al., 1992). Satellite cell activation and fusion will either enlarge muscle fiber size or form a new myofiber after birth. It has been found that satellite cells are most numerous on days 2 and 3 post-hatch, with a subsequent decrease on day 8 (Halevy et al., 2006). Fast-growing broilers have a greater number of activated satellite cells than slow-growing broilers in the early post-hatch period (Geiger et al., 2018), while broilers with greater muscle hypertrophy have less satellite cells with diminished ability to function in adulthood (Daughtry et al., 2017). Aggressive selection for muscle growth in broilers may exhaust satellite cell pools (Daughtry et al., 2017).

Satellite cells in poultry adults comprise a population of slow-cycling $Pax7^{+}Myf5^{-}$ stem cells and a fast-dividing population of $Pax7^{+}Myf5^{+}$ stem cells (Kuang et al., 2007; Picard and Marcelle, 2013). Slow-cycling cells display self-renewing capabilities that are critical for maintaining the long-term regenerating capacity of muscle (Kuang et al., 2007; Ono et al., 2012). Fast-dividing cells activate satellite cells, giving rise to a higher number of myogenic differentiated cells (Motohashi and Asakura, 2014).

Several signaling pathways are likely to be important players in the balance between proliferation and myogenic differentiation of satellite cells, such as Notch, TGF-beta, EGF and FGF signaling. Notch and EGF signaling promote the self-renewal of the satellite cell pool (Van Ho et al., 2011; Vasyutina et al., 2007). Myostatin, a member of the TGF-beta family, inhibits the activation, proliferation and differentiation of myogenic satellite cells (McFarland et al., 2007). Blocking FGF signaling leads to a significant decrease in satellite cell differentiation (Lagha et al., 2008). MicroRNAs and growth factor responsiveness to IGF-1 and platelet-derived growth factor-BB are also found to be involved in the regulation of satellite cell proliferation and differentiation (Velleman, 2019).

Hypertrophic muscle growth is mediated by the addition of nuclei to the myofibers via fusion of mono-nucleated satellite cells. Satellite cell numbers decrease with age and size in broilers (Daughtry et al., 2017; Gibson and Schultz, 1983). Mitotic activity of satellite cells is greatest during the first week post-hatch (Halevy et al., 2000). Manipulating the activity of satellite cells in the first week post-hatch affects muscle quality. It has been shown that various light treatments affect skeletal muscle growth by regulating satellite cell proliferation in broilers (Halevy et al., 1998). Stimulation with monochromatic green light during incubation enhanced post hatch muscle growth by affecting satellite cell proliferation (Bai et al., 2016). Early post-hatch thermal stress can also affect skeletal muscle growth by influencing satellite cell proliferation (Piestun et al., 2015, 2017). Satellite cell activity and function have also been correlated with muscle diseases, especially white striping muscle myopathy. The detailed function of satellite cells in poultry skeletal muscle growth and muscle myopathy has been discussed in a number of excellent recent reviews (Ayansola et al., 2021; Velleman, 2019; Xie et al., 2017).

5 Signaling pathways controlling skeletal muscle development and muscle mass in poultry

From somite to skeletal muscle fiber, many genes, noncoding RNAs and signaling pathways are involved in the regulation of muscle development (Table 1), including cell delamination, migration, proliferation, differentiation, fusion and maturation. However, only some of these pathways significantly affect skeletal muscle production. We summarize here some critical signaling pathways controlling skeletal muscle mass in poultry. A brief introduction to their functional roles in the regulation of muscle developmental process is also included.

6 Signaling pathways: myostatin/activin signaling pathway and bone morphogenetic protein signaling

6.1 *Myostatin and activin signaling pathway*

Myostatin (also known as *GDF-8*) is the most significant gene with dramatic effects on skeletal muscle mass in a wide range of mammalian and avian species. Many polymorphisms in the *myostatin* gene are associated with growth traits in chicken (Bhattacharya and Chatterjee, 2013; Dushyanth et al., 2016; Zhang et al., 2011). *Myostatin* gene knockdown by RNA interference has been found to increase body weight in chicken (Bhattacharya et al., 2017). The knockout of *myostatin* gene results in significantly larger skeletal muscles in the breast and leg of chickens (Kim et al., 2020). *Myostatin* gene knockout has been found to increase muscle production via muscle fiber hypertrophy, hyperplasia or their combination (Kim et al., 2020).

Myostatin inhibits skeletal muscle production by repressing muscle precursor proliferation, satellite cell activation, myoblast proliferation and differentiation, and protein synthesis (Aiello et al., 2018; McFarland et al., 2007). Ectopic expression of *myostatin* down-regulates *Pax3*, a key marker of proliferating muscle precursors (Amthor et al., 2002). p21, an inhibitor of *MyoD*-expressing myoblasts, is upregulated by *myostatin* (Thomas et al., 2000). By activating its downstream signaling pathway, myostatin significantly affects muscle hypertrophy and muscle differentiation. Mature myostatin protein binds to the receptor ActRIIB, a high-affinity type-2 activin receptor on muscle cell membranes. This results in the activation of type-1 activin receptor serine kinases, ALK4 or ALK5, which causes phosphorylate Smads 2/3 to inhibit muscle growth gene transcription and muscle growth (Han and Mitch, 2011). Myostatin is a member of the TGF-β family. As well as mature myostatin proteins, other TGF-β family members, such as activin A and GDF11, can also bind to ActRIIB and activate the same intracellular signaling pathway (Han and Mitch, 2011).

Myostatin gene knockout can also reduce abdominal fat deposition in chicken. Based on its function and expression in muscle and adipose tissues, *myostatin* gene has been a valuable candidate to improve meat production and feed conversion rate in poultry. The signaling pathway mediated by myostatin and activin also plays a potentially significant role in muscle-related disease (Sartori et al., 2014).

6.2 Bone morphogenetic protein signaling

Unlike the negative role of myostatin in skeletal muscle development and growth, bone morphogenetic proteins (BMPs) are positive regulators of muscle growth and are dominant over myostatin signaling (Sartori et al., 2014). Blocking BMP signaling reverted the hypertrophic phenotype of myostatin-knockout mice, suggesting a genetic epistasis between myostatin and BMP pathways in muscle (Sartori et al., 2013).

BMP signaling has significant roles in determining muscle fiber numbers. Blocking BMP signaling in chicken limbs reduces the number of fetal muscle progenitors and muscle fibers, while ectopic activation of BMP signaling increases their number and results in larger muscles (Wang et al., 2010). This function is very valuable in poultry breeding. Commercial selection for muscle yield traditionally occurs at ages targeting hypertrophic rather than hyperplastic growth (Orlowski et al., 2021). This has promoted giant muscle fibers and may be responsible for the development of certain muscle myopathies (Mason et al., 2020). Developing a novel method of selection focusing on increase in muscle fiber number will be important for future broiler breeding. BMP signaling may be a critical candidate in this breeding target.

BMP signaling also controls skeletal muscle mass by balancing protein synthesis and degradation. Inhibition of BMP signaling causes muscle atrophy. The BMP-Smad1/5/8-Smad4 pathway negatively regulates MUSA1, a protein that is required for muscle loss (Sartori et al., 2013). BMP signaling also increases activated satellite cell numbers affecting poultry muscle growth after birth via phosphorylated Smad protein (Friedrichs et al., 2011). However, research on the BMP signaling in poultry is limited. Considering its critical roles in the regulation of muscle fiber number and size, further research on BMP signaling in poultry is needed.

7 Signaling pathways: GH-GHR-IGFs axis, PI3K-Akt pathway and FOXO signaling

7.1 GH-GHR-IGFs axis

The growth hormone/growth hormone receptor/insulin-like growth factor (GH-GHR-IGFs) pathway of the somatotropic axis is the major controlling factor for growth rate and body size in growing animals. GH, which regulates body growth mainly at the postnatal stage, is a peptide hormone released from anterior pituitary somatotroph cells. When released into the bloodstream, GH directly binds to GHR which is ubiquitously expressed. GH binding results in the dimerization of two GHRs. Intracellular signaling involves the Janus kinase 2, signal transducer and activator of transcription 5 (JAK2/STAT5), and Src family kinases (Velloso, 2008).

GH also stimulates the synthesis of IGF-1 in most tissues, especially in the liver which is the major tissue responsible for the production and secretion of circulating IGF-1. The growth-promoting effects of GH are mainly mediated by IGF-1 (Walenkamp and Wit, 2007). Unlike GH, IGF-1 is critical for intrauterine growth. At least two cell surface receptors, the IGF-1 receptor (IGF1R) and the insulin receptor, can be bound by IGF-1. Its primary action is mediated by binding to IGF1R. Binding to IGF1R initiates intracellular signaling. Many signaling pathways involved in regulating muscle mass can be initiated by IGF-1, such as the PI3K-AKT, MAPK, mTOR and FOXO signaling pathways (Velloso, 2008). The function and stimulation roles of IGF-1 in skeletal muscle growth depend on its downstream pathways.

The chicken *GH* gene is diverse among breeds (Nie et al., 2005). Many single nucleotide polymorphisms (SNP) in the chicken *GH* gene are associated with growth and carcass traits (Nie et al., 2005). GH regulates muscle growth mainly at the postnatal stage. GH deficiency does not significantly affect the birth size and muscle formation but can lead to reduced body growth and retarded muscle development during childhood and adolescence (Velloso, 2008). GH can also promote skeletal muscle cell fusion and increase muscle fiber size via an IGF-1-independent effect (Sotiropoulos et al., 2006).

The regulation of chicken skeletal muscle growth by the chicken *GHR* gene is mainly dependent on its downstream signaling pathways. Mutations in the chicken *GHR* gene are related to dwarfism in chicken (Lin et al., 2018), with 30–40% reduction in body weight and smaller muscle fiber size than those of normal chickens (Hull et al., 1993; Luo et al., 2016). *GHR* gene mutation also reduced muscle fiber number in chicken (Knizetova, 1993), suggesting that the *GHR* gene and its downstream signaling pathways may regulate muscle fiber formation during embryogenesis. The absence of functional GHR was reported to cause an increase in systemic GH levels and a decrease in circulating the IGF-1 level, which resulted in retarded chicken growth (Agarwal et al., 1994).

Using integrative analyzed transcriptome expression profiles between normal and GHR-mutation chickens, we found that the GHR-mutation inhibits chicken skeletal muscle development mainly by regulating PI3K-Akt, MAPK and Wnt pathways (Luo et al., 2016). Both PI3K-Akt and MAPK pathways are involved in the IGF-1 signaling pathway, indicating that the function of *GHR* in the regulation of skeletal muscle depends on IGF-1. Chickens selected for increased breast muscle yield have a higher IGF-1 level (Guernec et al., 2003). IGF-1 expression mainly depends on the stimulation of GH. The latent transcription factor signal transducer and activator of transcription (STAT) 5B is a key agent in GH-stimulated growth. GH-activated STAT5B promotes transcription of IGF-1 in the liver and other organs (Varco-Merth and Rotwein, 2014). In skeletal muscle growth, the circulating, autocrine or paracrine IGF-1 binds to the cell membrane and stimulates downstream signaling pathways to promote protein synthesis, cell proliferation and differentiation, resulting in the increase of skeletal muscle mass (Velloso, 2008).

7.2 PI3K-Akt pathway

It has been shown that IGF-1 promotes myoblast proliferation and skeletal muscle development of embryonic chickens via the PI3K-Akt signaling pathway (Yu et al., 2015). The pro-hypertrophy activity of IGF-1 also comes predominantly through its ability to activate the PI3K-Akt signaling pathway (Glass, 2010). During chicken skeletal muscle fiber formation and hypertrophy, several noncoding RNAs, such as long-noncoding RNA LncIRS1, microRNA miR-148a-3p and miR-9-5p, and circular RNA PPP1R13B, are involved in the regulation of satellite cell proliferation and differentiation, as well as protein synthesis and degradation, by targeting the PI3K-Akt signaling pathway (Li et al., 2019; Shen et al., 2021b; Yin et al., 2020c).

IGF-1-PI3K-Akt signaling can significantly inhibit the effects of myostatin on muscle growth. Myostatin-inducing muscle atrophy is mediated by Smad2/3 proteins. Smad2/3 promotes protein degradation and muscle atrophy not only by directly inhibiting the transcription of muscle growth genes but also via inhibition of Akt phosphorylation (Trendelenburg et al., 2009). The activation of IGF-1-PI3K-Akt signaling would eliminate the inhibition role of Smad2/3 on Akt, leading to the repression of FOXO proteins and E3 ubiquitin ligases, MuRF1 and MAFbx, affecting skeletal muscle atrophy (Bodine et al., 2001). During short-term fasting in chicken, MuRF1 and MAFbx expression were increased by fasting, and these increases were downregulated by refeeding (Li et al., 2011). pAkt and pS6 contents were also decreased in chicken skeletal muscle (Saneyasu et al., 2017), suggesting that the PI3K-Akt signaling is disrupted during fasting with protein degradation increased.

PI3K-Akt and its downstream proteins, such as the mammalian target of rapamycin (mTOR) and p70 S6 Kinase, are critical for the regulation of cell size (Kozma and Thomas, 2002; Montagne et al., 1999). Commercial broilers have larger-sized muscle fibers than slow-growing native chickens. This might be due to increased activity of the PI3K-Akt signaling pathway. Many genes in the PI3K-Akt pathway are differentially expressed between fast-growing and slow-growing chickens (Luo et al., 2016; Zhou et al., 2015).

7.3 Forkhead box O signaling

The forkhead box O (FOXO) transcription factors are negatively regulated by the PI3K-Akt pathway and are considered to affect skeletal muscle atrophy. Foxo1 is a negative regulator of skeletal muscle differentiation and is crucial for muscle fiber-type specification (Xu et al., 2017). Foxo3 can act on the promoter of atrogin-1, a ubiquitin ligase necessary for muscle atrophy, to cause *atrogin-1* transcription and skeletal muscle atrophy (Sandri et al., 2004). Foxo4 regulates myoblast cycle progression by G2 arrest (Furukawa-Hibi et al., 2002). However, unlike other FOXO family members, Foxo6 maintained myotubes and protected against atrophy (Zhang et al., 2021).

During the growth of chicken breast muscle, FOXO signaling is the core pathway for muscle development (Li et al., 2019). Many genes and miRNAs target the FOXO signaling pathway during skeletal muscle growth in the late postnatal stage for chickens (Li et al., 2019). Using the genome-wide association study technique, we found that many SNPs surrounding the *FOXO1a* gene region had strong effects on chicken growth and muscle mass (Xie et al., 2012). These SNPs may be the main factors leading to the difference in meat production between commercial broilers and native chickens.

8 Conclusion and future trends

Skeletal muscle development is a complex process involving numerous regulation pathways for gene expression, molecular interactions and cell remodeling. The chicken embryo has been an important experimental system for developmental biology, especially for skeletal muscle development. Knowledge of poultry skeletal muscle development derived from developmental biology studies provides an important resource for poultry industry researchers. Understanding this process is critical to the yield and quality of poultry meat. We have summarized the developmental process of skeletal muscle from somites to muscle fiber in poultry and identified several functional genes and signaling pathways involved in this process.

However, there are still many unsolved problems in skeletal muscle development at present, especially the genetic regulation of muscle fiber number. Oversized skeletal muscle fiber can affect muscle quality and cause myopathy (MacRae et al., 2006). It is necessary to select the number of muscle fibers rather than the size to improve the meat production capacity of broilers. This requires a more comprehensive understanding of the genetic regulation of muscle fiber number.

Several new techniques have emerged that may help us to further understand muscle growth and development, such as single-cell omics and spatial transcriptome studies. Studies at the single cell level are helpful to understand heterogeneity among cells, the molecular map of developing muscle cell and the connections and interactions between cells (Miao et al., 2021; Rai et al., 2021). A new generation of gene editing methods, such as single base editing and prime editing, will help identify functional genes involved in poultry skeletal muscle development. Many quantitative trait variations (QTVs) have been found to be related to chicken muscle mass, but most of these QTVs are located in noncoding region. A precise base editing method may be helpful in understanding the function of these QTVs. Finally, application of artificial intelligence methods, such as machine learning and deep learning, in multi-omics data mining will also play an important role in predicting gene function and determining core regulatory factors during muscle development (Picard et al., 2021; Reel et al., 2021).

9 Where to look for further information

9.1 Further reading

- Bottinelli, R. and Reggiani, C. (2006). Skeletal muscle plasticity in health and disease: From genes to whole muscle. *Advances in Muscle Research*, Springer Dordecht. DOI: 10.1007/1-4020-5177-8.
- Avian Muscle Development and Growth Mechanisms: Association with Muscle Myopathies and Meat Quality. https://www.frontiersin.org/research-topics/10177/avian-muscle-development-and-growth-mechanisms-association-with-muscle-myopathies-and-meat-quality.
- Velleman, S. G. and McFarland, D. C. (2015). Chapter 16 - Skeletal muscle. *Sturkie's Avian Physiology (Sixth Edition)*, Academic Press. pp. 379–402.
- Du, M., Tong, J., Zhao, J., Underwood, K. R., Zhu, M., Ford, S. P. and Nathanielsz, P. W. (2009). Fetal programming of skeletal muscle development in ruminant animals. *J Anim Sci*. 2010 Apr;88(13 Suppl): E51–60. DOI: 10.2527/jas.2009-2311.
- Kalinski, M. and Drozdovska, S. (2019). Genetic and epigenetic determinants of muscle mass. *Sports, Exercise, and Nutritional Genomics: Current Status and Future Directions*, Academic Press. pp. 251–272.
- Robinson, D. C. L. and Dilworth, F. J. (2017). Epigenetic regulation of adult myogenesis. *Current Top Dev Biol*. 2018; 126: 235–284. DOI: 10.1016/bs.ctdb.2017.08.002.

9.2 Key journals/conferences

- Poultry Science (https://www.sciencedirect.com/journal/poultry-science).
- Skeletal muscle (https://skeletalmusclejournal.biomedcentral.com/).
- ICMAG (International Conference on Muscle Adaptation and Genetics).
- ICoMST (International Congress of Meat Science and Technology).
- WMSC (World Muscle Science Congress).

9.3 Major international research projects

- FAANG (Functional Annotation of Animal Genomes) project is to produce comprehensive maps of functional elements in the genomes of domesticated animal species: https://www.faang.org/.

10 References

Agarwal, S. K., Cogburn, L. A. and Burnside, J. (1994). Dysfunctional growth hormone receptor in a strain of sex-linked dwarf chicken: evidence for a mutation in the intracellular domain. *J. Endocrinol.* 142(3), 427–434.

Aiello, D., Patel, K. and Lasagna, E. (2018). The myostatin gene: an overview of mechanisms of action and its relevance to livestock animals. *Anim. Genet.* 49(6), 505–519.

Allouh, M. Z. and Aldirawi, M. H. (2012). Influence of mesterolone on satellite cell distribution and fiber morphology within maturing chicken pectoralis muscle. *Anat. Rec. (Hoboken)* 295(5), 792–799.

Al-Musawi, S. L., Stickland, N. C. and Bayol, S. A. (2012). In ovo temperature manipulation differentially influences limb musculoskeletal development in two lines of chick embryos selected for divergent growth rates. *J. Exp. Biol.* 215(9), 1594–1604.

Alvares, L. E., Schubert, F. R., Thorpe, C., Mootoosamy, R. C., Cheng, L., Parkyn, G., Lumsden, A. and Dietrich, S. (2003). Intrinsic, Hox-dependent cues determine the fate of skeletal muscle precursors. *Dev. Cell* 5(3), 379–390.

Amthor, H., Christ, B., Rashid-Doubell, F., Kemp, C. F., Lang, E. and Patel, K. (2002). Follistatin regulates bone morphogenetic protein-7 (BMP-7) activity to stimulate embryonic muscle growth. *Dev. Biol.* 243(1), 115–127.

Amthor, H., Christ, B. and Patel, K. (1999). A molecular mechanism enabling continuous embryonic muscle growth - a balance between proliferation and differentiation. *Development* 126(5), 1041–1053.

Amthor, H., Otto, A., Macharia, R., McKinnell, I. and Patel, K. (2006). Myostatin imposes reversible quiescence on embryonic muscle precursors. *Dev. Dyn.* 235(3), 672–680.

Ayansola, H., Liao, C., Dong, Y., Yu, X., Zhang, B. and Wang, B. (2021). Prospect of early vascular tone and satellite cell modulations on white striping muscle myopathy. *Poult. Sci.* 100(3), 100945.

Bai, X., Wang, Y., Wang, Z., Cao, J., Dong, Y. and Chen, Y. (2016). In ovo exposure to monochromatic lights affect posthatch muscle growth and satellite cell proliferation of chicks: role of IGF-1. *Growth Factors* 34(3–4), 107–118.

Balakrishnan-Renuka, A., Morosan-Puopolo, G., Yusuf, F., Abduelmula, A., Chen, J., Zoidl, G., Philippi, S., Dai, F. and Brand-Saberi, B. (2014). ATOH8, a regulator of skeletal myogenesis in the hypaxial myotome of the trunk. *Histochem. Cell Biol.* 141(3), 289–300.

Bell, K. M., Western, P. S. and Sinclair, A. H. (2000). SOX8 expression during chick embryogenesis. *Mech. Dev.* 94(1–2), 257–260.

Bendall, A. J., Ding, J., Hu, G., Shen, M. M. and Abate-Shen, C. (1999). Msx1 antagonizes the myogenic activity of Pax3 in migrating limb muscle precursors. *Development* 126(22), 4965–4976.

Beresford, B., Le Lievre, C. and Rathbone, M. P. (1978). Chimaera studies of the origin and formation of the pectoral musculature of the avian embryo. *J. Exp. Zool.* 205(2), 321–326.

Beresford, B. J. (1979). *Experimental Analysis of the Embryonic Origin and Development of the Pectoralis Major Muscle of the Chicken*. McMaster University, Hamilton, CA.

Berthoud, V. M., Singh, R., Minogue, P. J., Ragsdale, C. W. and Beyer, E. C. (2004). Highly restricted pattern of connexin36 expression in chick somite development. *Anat. Embryol. (Berl)* 209(1), 11–18.

Bhattacharya, T. K. and Chatterjee, R. N. (2013). Polymorphism of the myostatin gene and its association with growth traits in chicken. *Poult. Sci.* 92(4), 910–915.

Bhattacharya, T. K., Shukla, R., Chatterjee, R. N. and Dushyanth, K. (2017). Knock down of the myostatin gene by RNA interference increased body weight in chicken. *J. Biotechnol.* 241, 61–68.

Biressi, S., Tagliafico, E., Lamorte, G., Monteverde, S., Tenedini, E., Roncaglia, E., Ferrari, S., Ferrari, S., Cusella De Angelis, M. G., Tajbakhsh, S. and Cossu, G. (2007). Intrinsic phenotypic diversity of embryonic and fetal myoblasts is revealed by genome-wide gene expression analysis on purified cells. *Dev. Biol.* 304(2), 633–651.

Bladt, F., Riethmacher, D., Isenmann, S., Aguzzi, A. and Birchmeier, C. (1995). Essential role for the c-met receptor in the migration of myogenic precursor cells into the limb bud. *Nature* 376(6543), 768–771.

Bodine, S. C., Latres, E., Baumhueter, S., Lai, V. K., Nunez, L., Clarke, B. A., Poueymirou, W. T., Panaro, F. J., Na, E., Dharmarajan, K., Pan, Z. Q., Valenzuela, D. M., DeChiara, T. M., Stitt, T. N., Yancopoulos, G. D. and Glass, D. J. (2001). Identification of ubiquitin ligases required for skeletal muscle atrophy. *Science* 294(5547), 1704–1708.

Brand-Saberi, B., Ebensperger, C., Wilting, J., Balling, R. and Christ, B. (1993). The ventralizing effect of the notochord on somite differentiation in chick embryos. *Anat. Embryol. (Berl)* 188(3), 239–245.

Brand-Saberi, B., Gamel, A. J., Krenn, V., Muller, T. S., Wilting, J. and Christ, B. (1996). N-cadherin is involved in myoblast migration and muscle differentiation in the avian limb bud. *Dev. Biol.* 178(1), 160–173.

Brauner, I., Spicer, D. B., Krull, C. E. and Venuti, J. M. (2010). Identification of responsive cells in the developing somite supports a role for beta-catenin-dependent Wnt signaling in maintaining the DML myogenic progenitor pool. *Dev. Dyn.* 239(1), 222–236.

Bren-Mattison, Y., Hausburg, M. and Olwin, B. B. (2011). Growth of limb muscle is dependent on skeletal-derived Indian hedgehog. *Dev. Biol.* 356(2), 486–495.

Bren-Mattison, Y. and Olwin, B. B. (2002). Sonic hedgehog inhibits the terminal differentiation of limb myoblasts committed to the slow muscle lineage. *Dev. Biol.* 242(2), 130–148.

Cai, B., Li, Z., Ma, M., Zhang, J., Kong, S., Abdalla, B. A., Xu, H., Jebessa, E., Zhang, X., Lawal, R. A. and Nie, Q. (2021). Long noncoding RNA SMUL suppresses SMURF2 production-mediated muscle atrophy via nonsense-mediated mRNA decay. *Mol. Ther. Nucleic Acids* 23, 512–526.

Cai, B., Ma, M., Chen, B., Li, Z., Abdalla, B. A., Nie, Q. and Zhang, X. (2018). MiR-16-5p targets SESN1 to regulate the p53 signaling pathway, affecting myoblast proliferation and apoptosis, and is involved in myoblast differentiation. *Cell Death Dis.* 9(3), 367.

Cao, X., Tang, S., Du, F., Li, H., Shen, X., Li, D., Wang, Y., Zhang, Z., Xia, L., Zhu, Q. and Yin, H. (2020). miR-99a-5p regulates the proliferation and differentiation of skeletal muscle satellite cells by targeting MTMR3 in chicken. *Genes (Basel)* 11(4), 369.

Carson, J. A., Schwartz, R. J. and Booth, F. W. (1996). SRF and TEF-1 control of chicken skeletal alpha-actin gene during slow-muscle hypertrophy. *Am. J. Physiol.* 270(6 Pt 1), C1624–C1633.

Cerruti, S. S., Castagnaro, M. and Cheng, K. M. (1997). Histological changes caused by the rc mutation in chickens. *J. Comp. Pathol.* 116(4), 329–338.

Chen, B., Yu, J., Guo, L., Byers, M. S., Wang, Z., Chen, X., Xu, H. and Nie, Q. (2019). Circular RNA circHIPK3 promotes the proliferation and differentiation of chicken myoblast cells by sponging miR-30a-3p. *Cells* 8(2), 177.

Chen, G., Yin, Y., Lin, Z., Wen, H., Chen, J. and Luo, W. (2021). Transcriptome profile analysis reveals KLHL30 as an essential regulator for myoblast differentiation. *Biochem. Biophys. Res. Commun.* 559, 84–91.

Chen, W., Tangara, M., Xu, J. and Peng, J. (2012). Developmental transition of pectoralis muscle from atrophy in late-term duck embryos to hypertrophy in neonates. *Exp. Physiol.* 97(7), 861–872.

Chen, X., Ouyang, H., Wang, Z., Chen, B. and Nie, Q. (2018). A novel circular RNA generated by FGFR2 gene promotes myoblast proliferation and differentiation by sponging miR-133a-5p and miR-29b-1-5p. *Cells* 7(11), 199.

Chen, Y., Gutmann, D. H., Haipek, C. A., Martinsen, B. J., Bronner-Fraser, M. and Krull, C. E. (2004). Characterization of chicken Nf2/merlin indicates regulatory roles in cell proliferation and migration. *Dev. Dyn.* 229(3), 541–554.

Cheng, L., Alvares, L. E., Ahmed, M. U., El-Hanfy, A. S. and Dietrich, S. (2004). The epaxial-hypaxial subdivision of the avian somite. *Dev. Biol.* 274(2), 348–369.

Christ, B. and Brand-Saberi, B. (2002). Limb muscle development. *Int. J. Dev. Biol.* 46(7), 905–914.

Contriciani, R. E., da Veiga, F. C., do Amaral, M. J., Castelucci, B. G., de Sousa, L. M., de Jesus, M. B., Consonni, S. R., Collares-Buzato, C. B., Mermelstein, C., Dietrich, S. and Alvares, L. E. (2021). Dact1 is expressed during chicken and mouse skeletal myogenesis and modulated in human muscle diseases. *Comp. Biochem. Physiol. B Biochem. Mol. Biol.* 256, 110645.

Cui, C., Han, S., Shen, X., He, H., Chen, Y., Zhao, J., Wei, Y., Wang, Y., Zhu, Q., Li, D. and Yin, H. (2020). ISLR regulates skeletal muscle atrophy via IGF1-PI3K/Akt-Foxo signaling pathway. *Cell Tissue Res.* 381(3), 479–492.

Daughtry, M. R., Berio, E., Shen, Z., Suess, E. J. R., Shah, N., Geiger, A. E., Berguson, E. R., Dalloul, R. A., Persia, M. E., Shi, H. and Gerrard, D. E. (2017). Satellite cell-mediated breast muscle regeneration decreases with broiler size. *Poult. Sci.* 96(9), 3457–3464.

Day, K., Paterson, B. and Yablonka-Reuveni, Z. (2009). A distinct profile of myogenic regulatory factor detection within Pax7+ cells at S phase supports a unique role of Myf5 during posthatch chicken myogenesis. *Dev. Dyn.* 238(4), 1001–1009.

Dietrich, S. (1999). Regulation of hypaxial muscle development. *Cell Tissue Res.* 296(1), 175–182.

Dietrich, S., Schubert, F. R., Healy, C., Sharpe, P. T. and Lumsden, A. (1998). Specification of the hypaxial musculature. *Development* 125(12), 2235–2249.

Dong, X., Cheng, Y., Qiao, L., Wang, X., Zeng, C. and Feng, Y. (2021). Male-biased gga-miR-2954 regulates myoblast proliferation and differentiation of chicken embryos by targeting YY1. *Genes (Basel)* 12(9), 1325.

Duan, Y., Wu, Y., Yin, X., Li, T., Chen, F., Wu, P., Zhang, S., Wang, J. and Zhang, G. (2020). MicroRNA-214 inhibits chicken myoblasts proliferation, promotes their differentiation, and targets the TRMT61A gene. *Genes* 11(12), 1400.

Dushyanth, K., Bhattacharya, T. K., Shukla, R., Chatterjee, R. N., Sitaramamma, T., Paswan, C. and Guru Vishnu, P. (2016). Gene expression and polymorphism of myostatin gene and its association with growth traits in chicken. *Anim. Biotechnol.* 27(4), 269–277.

El-Magd, M. A., Elsayed, S. A., El-Shetry, E. S., Abdelfattah-Hassan, A., Saleh, A. A., Allen, S., McGonnell, I. and Patel, K. (2019). The role of chick Ebf genes in the mediolateral patterning of the somites. *Genesis* 57(11–12), e23339.

Esteve, P. and Bovolenta, P. (1999). cSix4, a member of the six gene family of transcription factors, is expressed during placode and somite development. *Mech. Dev.* 85(1–2), 161–165.

Fafeur, V., Tulasne, D., Quéva, C., Vercamer, C., Dimster, V., Mattot, V., Stéhelin, D., Desbiens, X. and Vandenbunder, B. (1997). The ETS1 transcription factor is expressed during epithelial-mesenchymal transitions in the chick embryo and is activated in scatter factor-stimulated MDCK epithelial cells. *Cell Growth Differ.* 8(6), 655–665.

Fan, C. M. and Tessier-Lavigne, M. (1994). Patterning of mammalian somites by surface ectoderm and notochord: evidence for sclerotome induction by a hedgehog homolog. *Cell* 79(7), 1175–1186.

Fluck, M., Carson, J. A., Gordon, S. E., Ziemiecki, A. and Booth, F. W. (1999). Focal adhesion proteins FAK and paxillin increase in hypertrophied skeletal muscle. *Am. J. Physiol.* 277(1), C152–C162.

Fredette, B. J. and Landmesser, L. T. (1991). Relationship of primary and secondary myogenesis to fiber type development in embryonic chick muscle. *Dev. Biol.* 143(1), 1–18.

Friedrichs, M., Wirsdoerfer, F., Flohe, S. B., Schneider, S., Wuelling, M. and Vortkamp, A. (2011). BMP signaling balances proliferation and differentiation of muscle satellite cell descendants. *BMC Cell Biol.* 12, 26.

Furukawa-Hibi, Y., Yoshida-Araki, K., Ohta, T., Ikeda, K. and Motoyama, N. (2002). FOXO forkhead transcription factors induce G(2)-M checkpoint in response to oxidative stress. *J. Biol. Chem.* 277(30), 26729–26732.

Galli, L. M., Knight, S. R., Barnes, T. L., Doak, A. K., Kadzik, R. S. and Burrus, L. W. (2008). Identification and characterization of subpopulations of Pax3 and Pax7 expressing cells in developing chick somites and limb buds. *Dev. Dyn.* 237(7), 1862–1874.

Geetha-Loganathan, P., Nimmagadda, S., Huang, R., Christ, B. and Scaal, M. (2006). Regulation of ectodermal Wnt6 expression by the neural tube is transduced by dermomyotomal Wnt11: a mechanism of dermomyotomal lip sustainment. *Development* 133(15), 2897–2904.

Geetha-Loganathan, P., Nimmagadda, S., Prols, F., Patel, K., Scaal, M., Huang, R. and Christ, B. (2005). Ectodermal Wnt-6 promotes Myf5-dependent avian limb myogenesis. *Dev. Biol.* 288(1), 221–233.

Geiger, A. E., Daughtry, M. R., Gow, C. M., Siegel, P. B., Shi, H. and Gerrard, D. E. (2018). Long-term selection of chickens for body weight alters muscle satellite cell behaviors. *Poult. Sci.* 97(7), 2557–2567.

George-Weinstein, M., Gerhart, J., Blitz, J., Simak, E. and Knudsen, K. A. (1997). N-cadherin promotes the commitment and differentiation of skeletal muscle precursor cells. *Dev. Biol.* 185(1), 14–24.

Gibson, M. C. and Schultz, E. (1983). Age-related differences in absolute numbers of skeletal muscle satellite cells. *Muscle Nerve* 6(8), 574–580.

Glass, D. J. (2010). PI3 kinase regulation of skeletal muscle hypertrophy and atrophy. *Curr. Top. Microbiol. Immunol.* 346, 267–278.

Gu, L. H., Xu, T. S., Huang, W., Xie, M., Shi, W. B., Sun, S. D. and Hou, S. S. (2013). Developmental characteristics of pectoralis muscle in Pekin duck embryos. *Genet. Mol. Res.* 12(4), 6733–6742.

Guernec, A., Berri, C., Chevalier, B., Wacrenier-Cere, N., Le Bihan-Duval, E. and Duclos, M. J. (2003). Muscle development, insulin-like growth factor-I and myostatin mRNA levels in chickens selected for increased breast muscle yield. *Growth Horm. IGF Res.* 13(1), 8–18.

Guo, L., Huang, W., Chen, B., Jebessa Bekele, E., Chen, X., Cai, B. and Nie, Q. (2018). gga-mir-133a-3p regulates myoblasts proliferation and differentiation by targeting PRRX1. *Front. Genet.* 9, 577.

Halevy, O., Biran, I. and Rozenboim, I. (1998). Various light source treatments affect body and skeletal muscle growth by affecting skeletal muscle satellite cell proliferation in broilers. *Comp. Biochem. Physiol. A Mol. Integr. Physiol.* 120(2), 317–323.

Halevy, O., Geyra, A., Barak, M., Uni, Z. and Sklan, D. (2000). Early posthatch starvation decreases satellite cell proliferation and skeletal muscle growth in chicks. *J. Nutr.* 130(4), 858–864.

Halevy, O., Hodik, V. and Mett, A. (1996). The effects of growth hormone on avian skeletal muscle satellite cell proliferation and differentiation. *Gen. Comp. Endocrinol.* 101(1), 43–52.

Halevy, O., Piestun, Y., Allouh, M. Z., Rosser, B. W., Rinkevich, Y., Reshef, R., Rozenboim, I., Wleklinski-Lee, M. and Yablonka-Reuveni, Z. (2004). Pattern of Pax7 expression during myogenesis in the posthatch chicken establishes a model for satellite cell differentiation and renewal. *Dev. Dyn.* 231(3), 489–502.

Halevy, O., Yahav, S. and Rozenboim, I. (2006). Enhancement of meat production by environmental manipulations in embryo and young broilers. *Worlds Poult. Sci. J.* 62(3), 485–497.

Han, H. Q. and Mitch, W. E. (2011). Targeting the myostatin signaling pathway to treat muscle wasting diseases. *Curr. Opin. Support. Palliat. Care* 5(4), 334–341.

Han, S., Cui, C., Wang, Y., He, H., Liu, Z., Shen, X., Chen, Y., Li, D., Zhu, Q. and Yin, H. (2019). Knockdown of CSRP3 inhibits differentiation of chicken satellite cells by promoting TGF-beta/Smad3 signaling. *Gene* 707, 36–43.

Hartley, R. S., Bandman, E. and Yablonka-Reuveni, Z. (1992). Skeletal muscle satellite cells appear during late chicken embryogenesis. *Dev. Biol.* 153(2), 206–216.

Hashimoto, K., Yokouchi, Y., Yamamoto, M. and Kuroiwa, A. (1999). Distinct signaling molecules control Hoxa-11 and Hoxa-13 expression in the muscle precursor and mesenchyme of the chick limb bud. *Development* 126(12), 2771–2783.

He, L., Papoutsi, M., Huang, R., Tomarev, S. I., Christ, B., Kurz, H. and Wilting, J. (2003). Three different fates of cells migrating from somites into the limb bud. *Anat. Embryol. (Berl)* 207(1), 29–34.

He, H., Yin, H., Yu, X., Zhang, Y., Ma, M., Li, D. and Zhu, Q. (2021). PDLIM5 affects chicken skeletal muscle satellite cell proliferation and differentiation via the p38-MAPK pathway. *Animals (Basel)* 11(4), 1016.

He, L., Vichev, K., Macharia, R., Huang, R., Christ, B., Patel, K. and Amthor, H. (2005). Activin A inhibits formation of skeletal muscle during chick development. *Anat. Embryol. (Berl)* 209(5), 401–407.

Heymann, S., Koudrova, M., Arnold, H., Köster, M. and Braun, T. (1996). Regulation and function of SF/HGF during migration of limb muscle precursor cells in chicken. *Dev. Biol.* 180(2), 566–578.

Hilfer, S. R., Searls, R. L. and Fonte, V. G. (1973). An ultrastructural study of early myogenesis in the chick wing bud. *Dev. Biol.* 30(2), 374–391.

Hirst, C. E. and Marcelle, C. (2015). The avian embryo as a model system for skeletal myogenesis. *Results Probl. Cell Differ.* 56, 99–122.

Holowacz, T., Zeng, L. and Lassar, A. B. (2006). Asymmetric localization of numb in the chick somite and the influence of myogenic signals. *Dev. Dyn.* 235(3), 633–645.

Hu, Z., Xu, H., Lu, Y., He, Q., Yan, C., Zhao, X., Tian, Y., Yang, C., Zhang, Z., Qiu, M. and Wang, Y. (2021). MUSTN1 is an indispensable factor in the proliferation, differentiation and apoptosis of skeletal muscle satellite cells in chicken. *Exp. Cell Res.* 407(2), 112833.

Huang, H., Liu, L., Li, C., Liang, Z., Huang, Z., Wang, Q., Li, S. and Zhao, Z. (2020). Fat mass- and obesity-associated (FTO) gene promoted myoblast differentiation through the focal adhesion pathway in chicken. *3 Biotech* 10(9), 403.

Huang, W., Guo, L., Zhao, M., Zhang, D., Xu, H. and Nie, Q. (2019). The inhibition on MDFIC and PI3K/AKT pathway caused by miR-146b-3p triggers suppression of myoblast proliferation and differentiation and promotion of apoptosis. *Cells* 8(7), 656.

Hull, K. L., Fraser, R. A., Marsh, J. A. and Harvey, S. (1993). Growth hormone receptor gene expression in sex-linked dwarf Leghorn chickens: evidence against a gene deletion. *J. Endocrinol.* 137(1), 91–98.

Hutcheson, D. A., Zhao, J., Merrell, A., Haldar, M. and Kardon, G. (2009). Embryonic and fetal limb myogenic cells are derived from developmentally distinct progenitors and have different requirements for beta-catenin. *Genes Dev.* 23(8), 997–1013.

Ijiri, D., Kanai, Y. and Hirabayashi, M. (2009a). Possible roles of myostatin and PGC-1alpha in the increase of skeletal muscle and transformation of fiber type in cold-exposed chicks: expression of myostatin and PGC-1alpha in chicks exposed to cold. *Domest. Anim. Endocrinol.* 37(1), 12–22.

Ijiri, D., Miura, M., Kanai, Y. and Hirabayashi, M. (2009b). Increased mass of slow-type skeletal muscles and depressed myostatin gene expression in cold-tolerant chicks. *Zoolog. Sci.* 26(4), 277–283.

Itoh, N., Mima, T. and Mikawa, T. (1996). Loss of fibroblast growth factor receptors is necessary for terminal differentiation of embryonic limb muscle. *Development* 122(1), 291–300.

Jagla, K., Dolle, P., Mattei, M. G., Jagla, T., Schuhbaur, B., Dretzen, G., Bellard, F. and Bellard, M. (1995). Mouse Lbx1 and human LBX1 define a novel mammalian homeobox gene family related to the Drosophila lady bird genes. *Mech. Dev.* 53(3), 345–356.

Jia, X., Ouyang, H., Abdalla, B. A., Xu, H., Nie, Q. and Zhang, X. (2017). miR-16 controls myoblast proliferation and apoptosis through directly suppressing Bcl2 and FOXO1 activities. *Biochim. Biophys. Acta Gene Regul. Mech.* 1860(6), 674-684.

Jingting, S., Qin, X., Yanju, S., Ming, Z., Yunjie, T., Gaige, J., Zhongwei, S. and Jianmin, Z. (2017). Oxidative and glycolytic skeletal muscles show marked differences in gene expression profile in Chinese Qingyuan partridge chickens. *PLoS ONE* 12(8), e0183118.

Ju, X., Liu, Y., Shan, Y., Ji, G., Zhang, M., Tu, Y., Zou, J., Chen, X., Geng, Z. and Shu, J. (2021). Analysis of potential regulatory LncRNAs and CircRNAs in the oxidative myofiber and glycolytic myofiber of chickens. *Sci. Rep.* 11(1), 20861.

Kiefer, J. C. and Hauschka, S. D. (2001). Myf-5 is transiently expressed in nonmuscle mesoderm and exhibits dynamic regional changes within the presegmented mesoderm and somites I-IV. *Dev. Biol.* 232(1), 77-90.

Kikuchi, T., Doerr, L. and Ashmore, C. R. (1980). A possible mechanism of phenotypic expression of normal and dystrophic genomes on succinic dehydrogenase activity and fiber size within a single myofiber of muscle transplants. *J. Neurol. Sci.* 45(2-3), 273-286.

Kim, G. D., Lee, J. H., Song, S., Kim, S. W., Han, J. S., Shin, S. P., Park, B. C. and Park, T. S. (2020). Generation of myostatin-knockout chickens mediated by D10A-Cas9 nickase. *FASEB J.* 34(4), 5688-5696.

Knizetova, H. (1993). Effects of the sex-linked dwarf gene (dw) on skeletal muscle cellularity in broiler chickens. *Br. Poult. Sci.* 34(3), 479-485.

Kozma, S. C. and Thomas, G. (2002). Regulation of cell size in growth, development and human disease: PI3K, PKB and S6K. *BioEssays* 24(1), 65-71.

Kuang, S., Kuroda, K., Le Grand, F. and Rudnicki, M. A. (2007). Asymmetric self-renewal and commitment of satellite stem cells in muscle. *Cell* 129(5), 999-1010.

Lagha, M., Kormish, J. D., Rocancourt, D., Manceau, M., Epstein, J. A., Zaret, K. S., Relaix, F. and Buckingham, M. E. (2008). Pax3 regulation of FGF signaling affects the progression of embryonic progenitor cells into the myogenic program. *Genes Dev.* 22(13), 1828-1837.

Lance-Jones, C. (1988). The somitic level of origin of embryonic chick hindlimb muscles. *Dev. Biol.* 126(2), 394-407.

Lee, J. H., Kim, S. W., Han, J. S., Shin, S. P., Lee, S. I. and Park, T. S. (2020). Functional analyses of miRNA-146b-5p during myogenic proliferation and differentiation in chicken myoblasts. *BMC Mol. Cell Biol.* 21(1), 40.

Lee, J. H., Park, J. W., Kang, K. S. and Park, T. S. (2019). Forkhead box O3 promotes cell proliferation and inhibits myotube differentiation in chicken myoblast cells. *Br. Poult. Sci.* 60(1), 23-30.

Li, D., Pan, Z., Zhang, K., Yu, M., Yu, D., Lu, Y., Wang, J., Zhang, J., Zhang, K. and Du, W. (2020a). Identification of the differentially expressed genes of muscle growth and intramuscular fat metabolism in the development stage of yellow broilers. *Genes (Basel)* 11(3), 244.

Li, D., Wang, Q., Shi, K., Lu, Y., Yu, D., Shi, X., Du, W. and Yu, M. (2020b). Testosterone promotes the proliferation of chicken embryonic myoblasts via androgen receptor mediated PI3K/Akt signaling pathway. *Int. J. Mol. Sci.* 21(3), 1152.

Li, G., Luo, W., Abdalla, B. A., Ouyang, H., Yu, J., Hu, F., Nie, Q. and Zhang, X. (2017). miRNA-223 upregulated by MYOD inhibits myoblast proliferation by repressing

IGF2 and facilitates myoblast differentiation by inhibiting ZEB1. *Cell Death Dis.* 8(10), e3094.

Li, H.-F., Shu, J.t., Shan, Y.-J., Chen, W., Song, C. and Xu, W. (2016). Myofiber development during embryonic to neonatal development in duck breeds differing in muscle growth rates. *J. Integr. Agric.* 15(2), 403–413.

Li, X., McFarland, D. C. and Velleman, S. G. (2009). Transforming growth factor-beta1-induced satellite cell apoptosis in chickens is associated with beta1 integrin-mediated focal adhesion kinase activation. *Poult. Sci.* 88(8), 1725–1734.

Li, Y., Zhai, B., Yuan, P., Fan, S., Jin, W., Li, W., Sun, G., Tian, Y., Liu, X., Kang, X. and Li, G. (2021). MiR-29b-1-5p regulates the proliferation and differentiation of chicken primary myoblasts and analysis of its effective targets. *Poult. Sci.* 101(2), 101557.

Li, Z., Cai, B., Abdalla, B. A., Zhu, X., Zheng, M., Han, P., Nie, Q. and Zhang, X. (2019). LncIRS1 controls muscle atrophy via sponging miR-15 family to activate IGF1-PI3K/AKT pathway. *J. Cachexia Sarcopenia Muscle* 10(2), 391–410.

Lin, S., Li, C., Li, C. and Zhang, X. (2018). Growth hormone receptor mutations related to individual dwarfism. *Int. J. Mol. Sci.* 19(5), 1433.

Lin, S., Luo, W., Ye, Y., Bekele, E. J., Nie, Q., Li, Y. and Zhang, X. (2017). Let-7b regulates myoblast proliferation by inhibiting IGF2BP3 expression in dwarf and normal chicken. *Front. Physiol.* 8, 477.

Liu, H. H., Wang, J. W., Yu, H. Y., Zhang, R. P., Chen, X., Jin, H. B., Dai, F., Li, L. and Xu, F. (2012a). Injection of duck recombinant follistatin fusion protein into duck muscle tissues stimulates satellite cell proliferation and muscle fiber hypertrophy. *Appl. Microbiol. Biotechnol.* 94(5), 1255–1263.

Liu, H. H., Wang, J. W., Zhang, R. P., Chen, X., Yu, H. Y., Jin, H. B., Li, L., Han, C. C., Xu, F., Kang, B., He, H. and Xu, H. Y. (2012b). In ovo feeding of IGF-1 to ducks influences neonatal skeletal muscle hypertrophy and muscle mass growth upon satellite cell activation. *J. Cell. Physiol.* 227(4), 1465–1475.

Lu, Y., Chen, S. and Yang, N. (2013). Expression and methylation of FGF2, TGF-beta and their downstream mediators during different developmental stages of leg muscles in chicken. *PLoS ONE* 8(11), e79495.

Luo, W., Abdalla, B. A., Nie, Q. and Zhang, X. (2017). The genetic regulation of skeletal muscle development: insights from chicken studies. *Front. Agri. Sci. Eng.* 4(3), 295–304.

Luo, W., Chen, J., Li, L., Ren, X., Cheng, T., Lu, S., Lawal, R. A., Nie, Q., Zhang, X. and Hanotte, O. (2019). c-Myc inhibits myoblast differentiation and promotes myoblast proliferation and muscle fibre hypertrophy by regulating the expression of its target genes, miRNAs and lincRNAs. *Cell Death Differ.* 26(3), 426–442.

Luo, W., Li, G., Yi, Z., Nie, Q. and Zhang, X. (2016). E2F1-miR-20a-5p/20b-5p auto-regulatory feedback loop involved in myoblast proliferation and differentiation. *Sci. Rep.-UK* 6(1), 27904.

Luo, W., Lin, Z., Chen, J., Chen, G., Zhang, S., Liu, M., Li, H., He, D., Liang, S., Luo, Q., Zhang, D., Nie, Q. and Zhang, X. (2021). TMEM182 interacts with integrin beta 1 and regulates myoblast differentiation and muscle regeneration. *J. Cachexia Sarcopenia Muscle* 12(6), 1704–1723.

Luo, W., Ren, X., Chen, J., Li, L., Lu, S., Chen, T., Nie, Q. and Zhang, X. (2018). TP63 transcripts play opposite roles in chicken skeletal muscle differentiation. *Front. Physiol.* 9, 1298.

Luo, W., Wu, H., Ye, Y., Li, Z., Hao, S., Kong, L., Zheng, X., Lin, S., Nie, Q. and Zhang, X. (2014). The transient expression of miR-203 and its inhibiting effects on skeletal muscle cell proliferation and differentiation. *Cell Death Dis.* 5(7), e1347.

Lv, C., Niu, S., Yan, S., Bai, C., Yu, X., Hou, J., Gao, W., Zhang, J., Zhao, Z., Yang, C. and Zhang, Y. (2019). Low-density lipoprotein receptor-related protein 1 regulates muscle fiber development in cooperation with related genes to affect meat quality. *Poult. Sci.* 98(9), 3418–3425.

Ma, G., Wang, H., Gu, X., Li, W., Zhang, X., Cui, L., Li, Y., Zhang, Y., Zhao, B. and Li, K. (2014). CARP, a myostatin-downregulated gene in CFM Cells, is a novel essential positive regulator of myogenesis. *Int. J. Biol. Sci.* 10(3), 309–320.

Ma, M., Cai, B., Jiang, L., Abdalla, B. A., Li, Z., Nie, Q. and Zhang, X. (2018). lncRNA-Six1 is a target of miR-1611 that functions as a ceRNA to regulate Six1 protein expression and fiber type switching in chicken myogenesis. *Cells* 7(12), 243.

MacRae, V. E., Mahon, M., Gilpin, S., Sandercock, D. A. and Mitchell, M. A. (2006). Skeletal muscle fibre growth and growth associated myopathy in the domestic chicken (Gallus domesticus). *Br. Poult. Sci.* 47(3), 264–272.

Mansouri, A., Hallonet, M. and Gruss, P. (1996). Pax genes and their roles in cell differentiation and development. *Curr. Opin. Cell Biol.* 8(6), 851-857.

Marcucio, R. S. and Noden, D. M. (1999). Myotube heterogeneity in developing chick craniofacial skeletal muscles. *Dev. Dyn.* 214(3), 178–194.

Mason, J. G., Gilley, A. D., Orlowski, S. K. and Anthony, N. B. (2020). Divergent selection for relative breast yield at 4 D posthatch and the effect on embryonic and early posthatch development. *Poult. Sci.* 99(6), 2888–2894.

Masselink, W., Masaki, M., Sieiro, D., Marcelle, C. and Currie, P. D. (2017). Phosphorylation of Lbx1 controls lateral myoblast migration into the limb. *Dev. Biol.* 430(2), 302–309.

Masyuk, M., Abduelmula, A., Morosan-Puopolo, G., Odemis, V., Rehimi, R., Khalida, N., Yusuf, F., Engele, J., Tamamura, H., Theiss, C. and Brand-Saberi, B. (2014). Retrograde migration of pectoral girdle muscle precursors depends on CXCR4/SDF-1 signaling. *Histochem. Cell Biol.* 142(5), 473–488.

McFarland, D. C., Velleman, S. G., Pesall, J. E. and Liu, C. (2006). Effect of myostatin on turkey myogenic satellite cells and embryonic myoblasts. *Comp. Biochem. Physiol. A Mol. Integr. Physiol.* 144(4), 501-508.

McFarland, D. C., Velleman, S. G., Pesall, J. E. and Liu, C. (2007). The role of myostatin in chicken (Gallus domesticus) myogenic satellite cell proliferation and differentiation. *Gen. Comp. Endocrinol.* 151(3), 351-357.

McLennan, I. S. (1983). Differentiation of muscle fiber types in the chicken hindlimb. *Dev. Biol.* 97(1), 222-228.

Mennerich, D., Schäfer, K. and Braun, T. (1998). Pax-3 is necessary but not sufficient for lbx1 expression in myogenic precursor cells of the limb. *Mech. Dev.* 73(2), 147–158.

Messina, G. and Cossu, G. (2009). The origin of embryonic and fetal myoblasts: a role of Pax3 and Pax7. *Genes Dev.* 23(8), 902-905.

Miao, Z., Humphreys, B. D., McMahon, A. P. and Kim, J. (2021). Multi-omics integration in the age of million single-cell data. *Nat. Rev. Nephrol.* 17(11), 710–724.

Mok, G. F., Cardenas, R., Anderton, H., Campbell, K. H. and Sweetman, D. (2014). Interactions between FGF18 and retinoic acid regulate differentiation of chick embryo limb myoblasts. *Dev. Biol.* 396(2), 214–223.

Montagne, J., Stewart, M. J., Stocker, H., Hafen, E., Kozma, S. C. and Thomas, G. (1999). Drosophila S6 kinase: a regulator of cell size. *Science* 285(5436), 2126–2129.

Morosan-Puopolo, G., Balakrishnan-Renuka, A., Yusuf, F., Chen, J., Dai, F., Zoidl, G., Lüdtke, T. H., Kispert, A., Theiss, C., Abdelsabour-Khalaf, M. and Brand-Saberi, B. (2014). Wnt11 is required for oriented migration of dermogenic progenitor cells from the dorsomedial lip of the avian dermomyotome. *PLoS ONE* 9(3), e92679.

Motohashi, N. and Asakura, A. (2014). Muscle satellite cell heterogeneity and self-renewal. *Front. Cell Dev. Biol.* 2(1), 1.

Munsterberg, A. E., Kitajewski, J., Bumcrot, D. A., McMahon, A. P. and Lassar, A. B. (1995). Combinatorial signaling by Sonic hedgehog and Wnt family members induces myogenic bHLH gene expression in the somite. *Genes Dev.* 9(23), 2911–2922.

Musumeci, G., Castrogiovanni, P., Coleman, R., Szychlinska, M. A., Salvatorelli, L., Parenti, R., Magro, G. and Imbesi, R. (2015). Somitogenesis: From somite to skeletal muscle. *Acta Histochem.* 117(4–5), 313–328.

Nakashima, K., Ishida, A., Ijiri, D. and Ohtsuka, A. (2016). Effect of dexamethasone on the expression of atrogin-1/MAFbx in chick skeletal muscle. *Anim. Sci. J.* 87(3), 405–410.

Nie, Q., Sun, B., Zhang, D., Luo, C., Ishag, N. A., Lei, M., Yang, G. and Zhang, X. (2005). High diversity of the chicken growth hormone gene and effects on growth and carcass traits. *J. Hered.* 96(6), 698–703.

Nikovits, W. J., Cann, G. M., Huang, R., Christ, B. and Stockdale, F. E. (2001). Patterning of fast and slow fibers within embryonic muscles is established independently of signals from the surrounding mesenchyme. *Development* 128(13), 2537–2544.

Ono, Y., Masuda, S., Nam, H. S., Benezra, R., Miyagoe-Suzuki, Y. and Takeda, S. (2012). Slow-dividing satellite cells retain long-term self-renewal ability in adult muscle. *J. Cell Sci.* 125(5), 1309–1317.

Orlowski, S. K., Dridi, S., Greene, E. S., Coy, C. S., Velleman, S. G. and Anthony, N. B. (2021). Histological analysis and gene expression of satellite cell markers in the pectoralis major muscle in broiler lines divergently selected for percent 4-day breast yield. *Front. Physiol.* 12, 712095.

Ouyang, H., Chen, X., Li, W., Li, Z., Nie, Q. and Zhang, X. (2018). Circular RNA circSVIL promotes myoblast proliferation and differentiation by sponging miR-203 in chicken. *Front. Genet.* 9, 172.

Ouyang, H., Yu, J., Chen, X., Wang, Z. and Nie, Q. (2020). A novel transcript of MEF2D promotes myoblast differentiation and its variations associated with growth traits in chicken. *PeerJ* 8, e8351.

Parakati, R. and DiMario, J. X. (2013). Repression of myoblast proliferation and fibroblast growth factor receptor 1 promoter activity by KLF10 protein. *J. Biol. Chem.* 288(19), 13876–13884.

Pette, D. and Staron, R. S. (2000). Myosin isoforms, muscle fiber types, and transitions. *Microsc. Res. Tech.* 50(6), 500–509.

Picard, C. A. and Marcelle, C. (2013). Two distinct muscle progenitor populations coexist throughout amniote development. *Dev. Biol.* 373(1), 141–148.

Picard, M., Scott-Boyer, M. P., Bodein, A., Perin, O. and Droit, A. (2021). Integration strategies of multi-omics data for machine learning analysis. *Comput. Struct. Biotechnol. J.* 19, 3735–3746.

Piestun, Y., Patael, T., Yahav, S., Velleman, S. G. and Halevy, O. (2017). Early posthatch thermal stress affects breast muscle development and satellite cell growth and characteristics in broilers. *Poult. Sci.* 96(8), 2877–2888.

Piestun, Y., Yahav, S. and Halevy, O. (2015). Thermal manipulation during embryogenesis affects myoblast proliferation and skeletal muscle growth in meat-type chickens. *Poult. Sci.* 94(10), 2528–2536.

Pourquie, O., Fan, C. M., Coltey, M., Hirsinger, E., Watanabe, Y., Breant, C., Francis-West, P., Brickell, P., Tessier-Lavigne, M. and Le Douarin, N. M. (1996). Lateral and axial signals involved in avian somite patterning: a role for BMP4. *Cell* 84(3), 461–471.

Pu, Q., Huang, R. and Brand-Saberi, B. (2016). Development of the shoulder girdle musculature. *Dev. Dyn.* 245(3), 342–350.

Rai, M. F., Wu, C. L., Capellini, T. D., Guilak, F., Dicks, A. R., Muthuirulan, P., Grandi, F., Bhutani, N. and Westendorf, J. J. (2021). Single cell omics for musculoskeletal research. *Curr. Osteoporos. Rep.* 19(2), 131–140.

Rallis, C., Stamataki, D., Pontikakis, S., Mankoo, B. S. and Karagogeos, D. (2001). Isolation of the avian homologue of the homeobox gene Mox2 and analysis of its expression pattern in developing somites and limbs. *Mech. Dev.* 104(1–2), 121–124.

Ran, J., Li, J., Yin, L., Zhang, D., Yu, C., Du, H., Jiang, X., Yang, C. and Liu, Y. (2021). Comparative analysis of skeletal muscle DNA methylation and transcriptome of the chicken embryo at different developmental stages. *Front. Physiol.* 12, 697121.

Reel, P. S., Reel, S., Pearson, E., Trucco, E. and Jefferson, E. (2021). Using machine learning approaches for multi-omics data analysis: a review. *Biotechnol. Adv.* 49, 107739.

Rehfeldt, C., Te Pas, M. F., Wimmers, K., Brameld, J. M., Nissen, P. M., Berri, C., Valente, L. M., Power, D. M., Picard, B., Stickland, N. C. and Oksbjerg, N. (2011). Advances in research on the prenatal development of skeletal muscle in animals in relation to the quality of muscle-based food. I. Regulation of myogenesis and environmental impact. *Animal* 5(5), 703–717.

Reijntjes, S., Stricker, S. and Mankoo, B. S. (2007). A comparative analysis of Meox1 and Meox2 in the developing somites and limbs of the chick embryo. *Int. J. Dev. Biol.* 51(8), 753–759.

Riley, M. F., Bochter, M. S., Wahi, K., Nuovo, G. J. and Cole, S. E. (2013). Mir-125a-5p-mediated regulation of Lfng is essential for the avian segmentation clock. *Dev. Cell* 24(5), 554–561.

Robins, J. E. and Capehart, A. A. (2018). Matrix remodeling associated 5 expression in trunk and limb during avian development. *Int. J. Dev. Biol.* 62(4–5), 335–340.

Sandri, M., Sandri, C., Gilbert, A., Skurk, C., Calabria, E., Picard, A., Walsh, K., Schiaffino, S., Lecker, S. H. and Goldberg, A. L. (2004). Foxo transcription factors induce the atrophy-related ubiquitin ligase atrogin-1 and cause skeletal muscle atrophy. *Cell* 117(3), 399–412.

Saneyasu, T., Tsuchii, N., Nakano, Y., Kitashiro, A., Tsuchihashi, T., Shindo, H., Honda, K. and Kamisoyama, H. (2017). Effects of short-term fasting on the Akt-mediated pathway involved in protein metabolism in chicken skeletal muscle. *Domest. Anim. Endocrinol.* 61, 54–61.

Sartori, R., Gregorevic, P. and Sandri, M. (2014). TGFbeta and BMP signaling in skeletal muscle: potential significance for muscle-related disease. *Trends Endocrinol. Metab.* 25(9), 464–471.

Sartori, R., Schirwis, E., Blaauw, B., Bortolanza, S., Zhao, J., Enzo, E., Stantzou, A., Mouisel, E., Toniolo, L., Ferry, A., Stricker, S., Goldberg, A. L., Dupont, S., Piccolo, S., Amthor, H. and Sandri, M. (2013). BMP signaling controls muscle mass. *Nat. Genet.* 45(11), 1309–1318.

Scaal, M. and Marcelle, C. (2018). Chick muscle development. *Int. J. Dev. Biol.* 62(1-2), 127-136.

Schramm, C. and Solursh, M. (1990). The formation of premuscle masses during chick wing bud development. *Anat. Embryol. (Berl)* 182(3), 235-247.

Schweizer, H., Johnson, R. L. and Brand-Saberi, B. (2004). Characterization of migration behavior of myogenic precursor cells in the limb bud with respect to Lmx1b expression. *Anat. Embryol. (Berl)* 208(1), 7-18.

Shahjahan, M., Liu, R., Zhao, G., Wang, F., Zheng, M., Zhang, J., Song, J. and Wen, J. (2016). Identification of histone deacetylase 2 as a functional gene for skeletal muscle development in chickens. *Asian-Australas. J. Anim. Sci.* 29(4), 479-486.

Shen, X., Liu, Z., Cao, X., He, H., Han, S., Chen, Y., Cui, C., Zhao, J., Li, D., Wang, Y., Zhu, Q. and Yin, H. (2019). Circular RNA profiling identified an abundant circular RNA circTMTC1 that inhibits chicken skeletal muscle satellite cell differentiation by sponging miR-128-3p. *Int. J. Biol. Sci.* 15(10), 2265-2281.

Shen, X., Wei, Y., Liu, W., You, G., Tang, S., Su, Z., Du, M., He, J., Zhao, J., Tian, Y., Zhang, Y., Ma, M., Zhu, Q. and Yin, H. (2021a). A novel circular RNA circITSN2 targets the miR-218-5p/LMO7 axis to promote chicken embryonic myoblast proliferation and differentiation. *Front. Cell Dev. Biol.* 9, 748844.

Shen, X., Wei, Y., You, G., Liu, W., Amevor, F. K., Zhang, Y., He, H., Ma, M., Zhang, Y., Li, D., Zhu, Q. and Yin, H. (2021b). Circular PPP1R13B RNA promotes chicken skeletal muscle satellite cell proliferation and differentiation via targeting miR-9-5p. *Animals (Basel)* 11(8), 2396.

Shin, J., McFarland, D. C., Strasburg, G. M. and Velleman, S. G. (2013). Function of death-associated protein 1 in proliferation, differentiation, and apoptosis of chicken satellite cells. *Muscle Nerve* 48(5), 777-790.

Shin, J., McFarland, D. C. and Velleman, S. G. (2012). Heparan sulfate proteoglycans, syndecan-4 and glypican-1, differentially regulate myogenic regulatory transcription factors and paired box 7 expression during turkey satellite cell myogenesis: implications for muscle growth. *Poult. Sci.* 91(1), 201-207.

Shin, S., Song, Y., Ahn, J., Kim, E., Chen, P., Yang, S., Suh, Y. and Lee, K. (2015). A novel mechanism of myostatin regulation by its alternative splicing variant during myogenesis in avian species. *Am. J. Physiol. Cell Physiol.* 309(10), C650-C659.

Shu, J., Li, H., Shan, Y., Xu, W., Chen, W., Song, C. and Song, W. (2015). Expression profile of IGF-I-calcineurin-NFATc3-dependent pathway genes in skeletal muscle during early development between duck breeds differing in growth rates. *Dev. Genes Evol.* 225(3), 139-148.

Shu, J. T., Xu, W. J., Zhang, M., Song, W. T., Shan, Y. J., Song, C., Zhu, W. Q., Zhang, X. Y. and Li, H. F. (2014). Transcriptional co-activator PGC-1alpha gene is associated with chicken skeletal muscle fiber types. *Genet. Mol. Res.* 13(1), 895-905.

Smith, J. H. (1963). Relation of body size to muscle cell size and number in the chicken. *Poult. Sci.* 42(2), 283-290.

Sobolewska, A., Elminowska Wenda, G., Bogucka, J., Szpinda, M., Walasik, K., Bednarczyk, M. and Paruszewska-Achtel, M. (2011). Myogenesis--possibilities of its stimulation in chickens. *Folia Biol. (Krakow)* 59(3-4), 85-90.

Song, Y., McFarland, D. C. and Velleman, S. G. (2012). Syndecan-4 cytoplasmic domain regulation of turkey satellite cell focal adhesions and apoptosis. *Mol. Biol. Rep.* 39(8), 8251-8264.

Sotiropoulos, A., Ohanna, M., Kedzia, C., Menon, R. K., Kopchick, J. J., Kelly, P. A. and Pende, M. (2006). Growth hormone promotes skeletal muscle cell fusion independent of insulin-like growth factor 1 up-regulation. *Proc. Natl. Acad. Sci. U. S. A.* 103(19), 7315–7320.

Stamenkovic, N., Jasnic, J., Novkovic, M., Milosevic, E., Boskovic, S., Kojic, A., Popic, K., Stankovic, M., Wang, Y., Milenkovic, S., Radojkovic, D., Ma, G. and Kojic, S. (2020). Cloning and expression profiling of muscle regulator ANKRD2 in domestic chicken Gallus gallus. *Histochem. Cell Biol.* 154(4), 383–396.

Stockdale, F. E., Nikovits, W. J. and Christ, B. (2000). Molecular and cellular biology of avian somite development. *Dev. Dyn.* 219(3), 304–321.

Sun, L., Lu, K., Liu, H., Wang, H., Li, X., Yang, C., Li, L. and Wang, J. (2013). The effects of endoplasmic reticulum stress response on duck decorin stimulate myotube hypertrophy in myoblasts. *Mol. Cell. Biochem.* 377(1–2), 151–161.

Swartz, M. E., Eberhart, J., Pasquale, E. B. and Krull, C. E. (2001). EphA4/ephrin-A5 interactions in muscle precursor cell migration in the avian forelimb. *Development* 128(23), 4669–4680.

Sweeney, L. J., Kennedy, J. M., Zak, R., Kokjohn, K. and Kelley, S. W. (1989). Evidence for expression of a common myosin heavy chain phenotype in future fast and slow skeletal muscle during initial stages of avian embryogenesis. *Dev. Biol.* 133(2), 361–374.

Takata, H., Terada, K., Oka, H., Sunada, Y., Moriguchi, T. and Nohno, T. (2007). Involvement of Wnt4 signaling during myogenic proliferation and differentiation of skeletal muscle. *Dev. Dyn.* 236(10), 2800–2807.

Tamura, K., Amano, T., Satoh, T., Saito, D., Yonei-Tamura, S. and Yajima, H. (2003). Expression of rigf, a member of avian VEGF family, correlates with vascular patterning in the developing chick limb bud. *Mech. Dev.* 120(2), 199–209.

Thomas, M., Langley, B., Berry, C., Sharma, M., Kirk, S., Bass, J. and Kambadur, R. (2000). Myostatin, a negative regulator of muscle growth, functions by inhibiting myoblast proliferation. *J. Biol. Chem.* 275(51), 40235–40243.

Tonegawa, A. and Takahashi, Y. (1998). Somitogenesis controlled by Noggin. *Dev. Biol.* 202(2), 172–182.

Trendelenburg, A. U., Meyer, A., Rohner, D., Boyle, J., Hatakeyama, S. and Glass, D. J. (2009). Myostatin reduces Akt/TORC1/p70S6K signaling, inhibiting myoblast differentiation and myotube size. *Am. J. Physiol. Cell Physiol.* 296(6), C1258–C1270.

Tripathi, A. K., Aparnathi, M. K., Vyavahare, S. S., Ramani, U. V., Rank, D. N. and Joshi, C. G. (2012). Myostatin gene silencing by RNA interference in chicken embryo fibroblast cells. *J. Biotechnol.* 160(3–4), 140–145.

Valasek, P., Theis, S., DeLaurier, A., Hinits, Y., Luke, G. N., Otto, A. M., Minchin, J., He, L., Christ, B., Brooks, G., Sang, H., Evans, D. J., Logan, M., Huang, R. and Patel, K. (2011). Cellular and molecular investigations into the development of the pectoral girdle. *Dev. Biol.* 357(1), 108–116.

Van Ho, A. T., Hayashi, S., Brohl, D., Aurade, F., Rattenbach, R. and Relaix, F. (2011). Neural crest cell lineage restricts skeletal muscle progenitor cell differentiation through Neuregulin1-ErbB3 signaling. *Dev. Cell* 21(2), 273–287.

Varco-Merth, B. and Rotwein, P. (2014). Differential effects of STAT proteins on growth hormone-mediated IGF-I gene expression. *Am. J. Physiol. Endocrinol. Metab.* 307(9), E847–E855.

Vasyutina, E., Lenhard, D. C., Wende, H., Erdmann, B., Epstein, J. A. and Birchmeier, C. (2007). RBP-J (Rbpsuh) is essential to maintain muscle progenitor cells and to generate satellite cells. *Proc. Natl. Acad. Sci. U. S. A.* 104(11), 4443–4448.

Velleman, S. G. (2007). Muscle development in the embryo and hatchling. *Poult. Sci.* 86(5), 1050–1054.

Velleman, S. G. (2019). Recent developments in breast muscle myopathies associated with growth in poultry. *Annu. Rev. Anim. Biosci.* 7, 289–308.

Velleman, S. G., Clark, D. L. and Tonniges, J. R. (2018). The effect of syndecan-4 and glypican-1 knockdown on the proliferation and differentiation of turkey satellite cells differing in age and growth rates. *Comp. Biochem. Physiol. A Mol. Integr. Physiol.* 223, 33–41.

Velleman, S. G., Coy, C. S. and McFarland, D. C. (2007). Effect of syndecan-1, syndecan-4, and glypican-1 on turkey muscle satellite cell proliferation, differentiation, and responsiveness to fibroblast growth factor 2. *Poult. Sci.* 86(7), 1406–1413.

Velleman, S. G., Liu, C., Coy, C. S. and McFarland, D. C. (2006). Effects of glypican-1 on turkey skeletal muscle cell proliferation, differentiation and fibroblast growth factor 2 responsiveness. *Dev. Growth Differ.* 48(4), 271–276.

Velleman, S. G. and Harding, R. L. (2017). Regulation of turkey myogenic satellite cell migration by microRNAs miR-128 and miR-24. *Poult. Sci.* 96(6), 1910–1917.

Velleman, S. G. and McFarland, D. C. (2004). Beta1 integrin mediation of myogenic differentiation: implications for satellite cell differentiation. *Poult. Sci.* 83(2), 245–252.

Velloso, C. P. (2008). Regulation of muscle mass by growth hormone and IGF-I. *Br. J. Pharmacol.* 154(3), 557–568.

Vignale, K., Greene, E. S., Caldas, J. V., England, J. A., Boonsinchai, N., Sodsee, P., Pollock, E. D., Dridi, S. and Coon, C. N. (2015). 25-Hydroxycholecalciferol enhances male broiler breast meat yield through the mTOR pathway. *J. Nutr.* 145(5), 855–863.

Wagner, J., Schmidt, C., Nikowits, W. J. and Christ, B. (2000). Compartmentalization of the somite and myogenesis in chick embryos are influenced by wnt expression. *Dev. Biol.* 228(1), 86–94.

Walenkamp, M. J. and Wit, J. M. (2007). Genetic disorders in the GH IGF-I axis in mouse and man. *Eur. J. Endocrinol.* 157 (Suppl. 1), S15–S26.

Wang, B., He, L., Ehehalt, F., Geetha-Loganathan, P., Nimmagadda, S., Christ, B., Scaal, M. and Huang, R. (2005). The formation of the avian scapula blade takes place in the hypaxial domain of the somites and requires somatopleure-derived BMP signals. *Dev. Biol.* 287(1), 11–18.

Wang, H., He, K., Zeng, X., Zhou, X., Yan, F., Yang, S. and Zhao, A. (2021). Isolation and identification of goose skeletal muscle satellite cells and preliminary study on the function of C1q and tumor necrosis factor-related protein 3 gene. *Anim. Biosci.* 34(6), 1078–1087.

Wang, H., Noulet, F., Edom-Vovard, F., Tozer, S., Le Grand, F. and Duprez, D. (2010). Bmp signaling at the tips of skeletal muscles regulates the number of fetal muscle progenitors and satellite cells during development. *Dev. Cell* 18(4), 643–654.

Wang, Z., Ouyang, H., Chen, X., Yu, J., Abdalla, B. A., Chen, B. and Nie, Q. (2018). Gga-miR-205a affecting myoblast proliferation and differentiation by targeting CDH11. *Front. Genet.* 9, 414.

Wang, Z., Zhang, X., Li, Z., Abdalla, B. A., Chen, Y. and Nie, Q. (2019). MiR-34b-5p mediates the proliferation and differentiation of myoblasts by targeting IGFBP2. *Cells* 8(4), 360.

Weimer, K., Theobald, J., Campbell, K. S., Esser, K. A. and DiMario, J. X. (2013). Genome-wide expression analysis and EMX2 gene expression in embryonic myoblasts committed to diverse skeletal muscle fiber type fates. *Dev. Dyn.* 242(8), 1001–1020.

Wilm, B., James, R. G., Schultheiss, T. M. and Hogan, B. L. (2004). The forkhead genes, Foxc1 and Foxc2, regulate paraxial versus intermediate mesoderm cell fate. *Dev. Biol.* 271(1), 176–189.

Wilson, S. J., McEwan, J. C., Sheard, P. W. and Harris, A. J. (1992). Early stages of myogenesis in a large mammal: formation of successive generations of myotubes in sheep tibialis cranialis muscle. *J. Muscle Res. Cell Motil.* 13(5), 534–550.

Xie, L., Luo, C., Zhang, C., Zhang, R., Tang, J., Nie, Q., Ma, L., Hu, X., Li, N., Da, Y. and Zhang, X. (2012). Genome-wide association study identified a narrow chromosome 1 region associated with chicken growth traits. *PLoS ONE* 7(2), e30910.

Xie, X., Wu, S. P., Tsai, M. J. and Tsai, S. (2017). The role of COUP-TFII in striated muscle development and disease. *Curr. Top. Dev. Biol.* 125, 375–403.

Xu, M., Chen, X., Chen, D., Yu, B. and Huang, Z. (2017). FoxO1: a novel insight into its molecular mechanisms in the regulation of skeletal muscle differentiation and fiber type specification. *Oncotarget* 8(6), 10662–10674.

Xu, Q. and Wu, Z. (2000). The insulin-like growth factor-phosphatidylinositol 3-kinase-Akt signaling pathway regulates myogenin expression in normal myogenic cells but not in rhabdomyosarcoma-derived RD cells. *J. Biol. Chem.* 275(47), 36750–36757.

Yang, Y. and Ordahl, C. P. (2006). The pattern of MyoD and contractile protein localization in primary epaxial myotome reflects the dynamic progression of nascent myoblast differentiation. *Dev. Dyn.* 235(2), 382–394.

Ye, M., Ye, F., He, L., Luo, B., Yang, F., Cui, C., Zhao, X., Yin, H., Li, D., Xu, H., Wang, Y. and Zhu, Q. (2017). Transcriptomic analysis of chicken Myozenin 3 regulation reveals its potential role in cell proliferation. *PLoS ONE* 12(12), e0189476.

Ye, W. V., McFarland, D. C., Gilkerson, K. K. and Pesall, J. E. (1996). The role of platelet-derived growth factor in turkey skeletal muscle development. *Cytobios* 88(352), 53–62.

Yin, H., Cui, C., Han, S., Chen, Y., Zhao, J., He, H., Li, D. and Zhu, Q. (2020a). Fibromodulin modulates chicken skeletal muscle development via the transforming growth factor-beta signaling pathway. *Animals (Basel)* 10(9), 1477.

Yin, H., He, H., Shen, X., Tang, S., Zhao, J., Cao, X., Han, S., Cui, C., Chen, Y., Wei, Y., Wang, Y., Li, D. and Zhu, Q. (2020b). MicroRNA profiling reveals an abundant miR-200a-3p promotes skeletal muscle satellite cell development by targeting TGF-β2 and regulating the TGF-β2/SMAD signaling pathway. *Int. J. Mol. Sci.* 21(9), 3274.

Yin, H., He, H., Shen, X., Zhao, J., Cao, X., Han, S., Cui, C., Chen, Y., Wei, Y., Xia, L., Wang, Y., Li, D. and Zhu, Q. (2020c). miR-9-5p inhibits skeletal muscle satellite cell proliferation and differentiation by targeting IGF2BP3 through the IGF2-PI3K/Akt signaling pathway. *Int. J. Mol. Sci.* 21(5), 1655.

Yin, H., Zhao, J., He, H., Chen, Y., Wang, Y., Li, D. and Zhu, Q. (2020d). Gga-miR-3525 targets PDLIM3 through the MAPK signaling pathway to regulate the proliferation and differentiation of skeletal muscle satellite cells. *Int. J. Mol. Sci.* 21(15), 5573.

Yu, J. A., Wang, Z., Yang, X., Ma, M., Li, Z. and Nie, Q. (2021). LncRNA-FKBP1C regulates muscle fiber type switching by affecting the stability of MYH1B. *Cell Death Discov.* 7(1), 73.

Yu, M., Wang, H., Liu, Z., Lu, Y., Yu, D., Li, D. and Du, W. (2017). Ebp1 regulates myogenic differentiation of myoblast cells via SMAD2/3 signaling pathway. *Dev. Growth Differ.* 59(6), 540–551.

Yu, M., Wang, H., Xu, Y., Yu, D., Li, D., Liu, X. and Du, W. (2015). Insulin-like growth factor-1 (IGF-1) promotes myoblast proliferation and skeletal muscle growth of embryonic chickens via the PI3K/Akt signalling pathway. *Cell Biol. Int.* 39(8), 910–922.

Zanou, N. and Gailly, P. (2013). Skeletal muscle hypertrophy and regeneration: interplay between the myogenic regulatory factors (MRFs) and insulin-like growth factors (IGFs) pathways. *Cell. Mol. Life Sci.* 70(21), 4117–4130.

Zappelli, F., Willems, D., Osada, S., Ohno, S., Wetsel, W. C., Molinaro, M., Cossu, G. and Bouche, M. (1996). The inhibition of differentiation caused by TGFbeta in fetal myoblasts is dependent upon selective expression of PKCtheta: a possible molecular basis for myoblast diversification during limb histogenesis. *Dev. Biol.* 180(1), 156–164.

Zhang, G., Chen, F., Wu, P., Li, T., He, M., Yin, X., Shi, H., Duan, Y., Zhang, T., Wang, J., Xie, K. and Dai, G. (2020). MicroRNA-7 targets the KLF4 gene to regulate the proliferation and differentiation of chicken primary myoblasts. *Front. Genet.* 11, 842.

Zhang, G., Ding, F., Wang, J., Dai, G., Xie, K., Zhang, L., Wang, W. and Zhou, S. (2011). Polymorphism in exons of the myostatin gene and its relationship with body weight traits in the Bian chicken. *Biochem. Genet.* 49(1–2), 9–19.

Zhang, G., He, M., Wu, P., Zhang, X., Zhou, K., Li, T., Zhang, T., Xie, K., Dai, G. and Wang, J. (2021). MicroRNA-27b-3p targets the myostatin gene to regulate myoblast proliferation and is involved in myoblast differentiation. *Cells* 10(2), 423.

Zhao, J., Shen, X., Cao, X., He, H., Han, S., Chen, Y., Cui, C., Wei, Y., Wang, Y., Li, D., Zhu, Q. and Yin, H. (2020). HDAC4 regulates the proliferation, differentiation and apoptosis of chicken skeletal muscle satellite cells. *Animals (Basel)* 10(1), 84.

Zhi, Q., Huang, R., Christ, B. and Brand-Saberi, B. (1996). Participation of individual brachial somites in skeletal muscles of the avian distal wing. *Anat. Embryol. (Berl)* 194(4), 327–339.

Zhou, N., Lee, W. R. and Abasht, B. (2015). Messenger RNA sequencing and pathway analysis provide novel insights into the biological basis of chickens' feed efficiency. *BMC Genomics* 16(1), 195.

Chapter 3

Nutritional strategies and management practices to improve poultry meat quality

Marco Zampiga and Federico Sirri, Alma Mater Studiorum – Università di Bologna, Italy

1 Introduction

The estimated growth in world population is increasing global food demand, particularly of protein. From a historical perspective, meat has represented the main protein source for humans, contributing to biosocial well-being and sustenance (Leroy and Praet, 2015; Leroy and De Smet, 2019). Among the different types of meat, poultry meat has experienced a significant production increase in the past 50 years, and future forecasts seem to confirm that the same trend will be observed in the next decade (Windhorst, 2017). Many factors have contributed to the universal appreciation of poultry meat, including its affordability, nutritional and sensory properties, versatility and ease of preparation as well as the absence of religious limitations related to its consumption (Baldi et al., 2020). In addition, poultry meat production is considered relatively efficient and sustainable in terms of resources utilization and environmental impact, which are crucial aspects for fulfilling the increasing protein demand in an efficient and sustainable way (Zampiga et al., 2021a).

http://dx.doi.org/10.19103/AS.2022.0112.03

Modern consumers are increasingly aware of the nutritional value and quality traits of foods, with special regard to the importance of high-quality protein for healthy life and well-being. Poultry meat is generally recognized as 'healthy' mostly because of its nutritional profile which is characterized by high protein, low fat and cholesterol content and balanced ω-6/ω-3 polyunsaturated fatty acid (PUFA) ratio (Petracci et al., 2015). Poultry meat quality could be influenced by both genetic (e.g. bird genetic background and metabolic features) and environmental related factors (e.g. nutrition, management and microclimatic conditions) as well as their interaction. Among the environmental factors, nutrition has been extensively regarded as one of the most important determinants of poultry meat quality, being able to induce changes in the proximate composition and technological properties of poultry meat (Cavani et al., 2009). Furthermore, other factors such as bird exposure to high environmental temperatures or castration (technically called 'caponization' in chickens) were associated with significant modifications of meat quality traits (Calik, 2014; Zaboli et al., 2019). The aim of this chapter is, therefore, to summarize the most relevant findings concerning the effects of nutrition and feeding strategies on the different facets of the meat quality concept as well as to discuss the consequences of some environmental and management factors, such as heat stress and caponization, on this crucial topic for the modern poultry industry.

2 Nutritional and feeding strategies to modulate meat composition

Poultry meat has a relatively stable crude protein content (23–25% for breast and 18% for thigh meat) and amino acid profile, with glutamine, asparagine, lysine (Lys), leucine, arginine (Arg) and alanine being the most relevant amino acids (Baéza et al., 2021). Overall, the diet could only slightly modulate the amino acid profile of poultry meat, while the crude protein content is mostly affected by other production factors such as the slaughter age. The lipid fraction represents the most variable one being influenced by different factors including nutrition and feeding (Baéza et al., 2021). For instance, the intramuscular lipid content can be modified by the energy-to-protein ratio of the diet or by a suboptimal intake of essential amino acids (e.g. Lys and methionine (Met)). As far as lipids are concerned, the fatty acid profile is known to be the most responsive fraction to nutritional treatments since the relationship between the fatty acid profile of the diet and that of the forthcoming meat has been established more than 50 years ago (Marion and Woodroof, 1963). Nutrition can affect the mineral content of poultry meat only to a limited extent (approximately 1%, mainly calcium, phosphorus and potassium) provided that the diet covers the animal requirements of such nutrients. On the contrary, the quantity of some micronutrients such as vitamins and trace elements largely depends on the dietary intake (Baéza et al., 2021). Considering these aspects, the next sections are devoted to the presentation and discussion of the main nutritional strategies able to 'enrich' the fatty acid profile as well as the vitamin and trace mineral contents of poultry meat, with the aim of improving the nutritional quality and shelf-life of this important food product.

2.1 Lipids

One of the most interesting and famous approaches related to nutrition is modifying the fatty acid composition of poultry meat with the aim of increasing the concentrations of certain fatty acid families that can provide potential health benefits to the consumers. For instance, a critical aspect of current Western diets is the elevated concentrations of ω-6 PUFAs, which results in an imbalance of the ω-6/ω-3 ratio that is approximately five to six times higher than the optimal recommendations for humans (i.e. <4; Simopoulos, 2002). The consumption of diets presenting excessive amounts of ω-6 PUFAs and high ω-6/ω-3 ratios has been associated with a greater occurrence of cardiovascular failures, cancer, obesity as well as inflammatory and autoimmune diseases (Simopoulos, 2004, 2016). The ω-3 PUFA intake in Western countries is relatively low due to the scarce consumption of fish, which is the richest dietary source of ω-3 long-chain PUFAs (Simopoulos, 2002). It should be considered that humans are rather inefficient in synthesizing long-chain PUFAs, and thus these compounds should be necessarily assumed with the diet (Givens and Gibbs, 2008). An alternative strategy to increase the ω-3 PUFA intake could be the enrichment of staple foods that are frequently and widely consumed, such as the chicken meat.

As previously reported, the fatty acid profile of chicken meat could be easily modified by changing the fatty acid composition of the diet (Rymer and Givens, 2005). Overall, two different approaches can be followed to increase the ω-3 PUFA content of broiler meat. In the first approach, the precursor α-linolenic acid (ALA, C18:3) is provided with the diet and then converted in the liver into ω-3 long-chain PUFA, such as eicosapentaenoic acid (EPA, C20:5) and docosahexaenoic acid (DHA, C22:6), by means of the action of Δ5- and Δ6-desaturase and elongase enzymes (Lopez-Ferrer et al., 2001a). In the second one, broiler diets are supplemented with the ω-3 PUFA derivatives, which are subsequently absorbed during the digestive processes and stored in the tissues (Lopez-Ferrer et al., 2001b). Birds, and in general all terrestrial animals, have scarce capacity of synthesizing long-chain PUFAs, which suggests that the optimal solution would be enriching the diet directly with ω-3 PUFA sources (Rymer and Givens, 2005). For instance, fish oil presents remarkable quantities of ω-3 PUFAs such as EPA and DHA, and its dietary utilization to enrich chicken meat has been extensively evaluated in the past. Overall, the results available in the literature indicate that providing fish oil in broiler diets can be considered an effective strategy to increase the concentrations of ω-3 PUFAs, in particular EPA and DHA, in chicken meat (Lopez-Ferrer et al., 2001a; Farhoomand and Checaniazer, 2009; Mirghelenj et al., 2009; Morales-Barrera et al., 2013). Despite the high efficacy for ω-3 enrichment, the use of fish oil in broiler nutrition has some important drawbacks that include its high cost, the reduced feed and meat oxidative stability, the presence of undesirable flavors (i.e. fishy flavor) and the sustainability concerns associated with natural fish stock depletion due to overfishing (Naylor et al., 2009). Alternative solutions to alleviate such pressure on natural marine resources include the utilization of terrestrial sources of ω-3 PUFAs, providing precursors that could be metabolized by the bird into ω-3 long-chain PUFAs, or other marine organisms whose production is characterized by low environmental impact (e.g. microalgae).

Among the terrestrial ω-3 PUFA sources, linseed (*Linum usitatissimum*) has received great attention because of its considerably high ALA concentration (Herchi et al., 2014). More than 20 years ago, Lopez-Ferrer et al. (2001b) demonstrated that the substitution of tallow with increasing concentrations of linseed oil (20–40 g/kg of feed) was able to elevate the total amount of ω-3 PUFAs and reduce the ω-6/ω-3 ratio in chicken thigh meat. Similarly, Jia et al. (2010) showed that the administration of linseed-containing diets (120 g/kg of feed) significantly elevated the EPA and DHA content of chicken breast meat. More recently, positive results in terms of meat enrichment were obtained by feeding broilers diets supplemented with 1 mL nanoencapsulated linseed oil per kg of body weight (Abbasi et al., 2019). The inclusion of defatted linseed meal (100 g/kg of feed) in broiler diets from 21 days to 35 days of age significantly increased the ω-3 PUFA content of thigh meat, in particular that of EPA and DHA (Kumar et al., 2020). Feeding turkeys with diets containing linseed oil (25 g/kg of feed) in replacement of soybean oil in the last 3 weeks before slaughter allowed to obtain meat with the recommended ω-6/ω-3 PUFA ratio for humans (Jankowski et al., 2017).

Another valuable nutritional approach for enriching the ω-3 PUFA content of chicken meat is the use of microalgae, which are the primary natural producers of ω-3 PUFAs (Rymer et al., 2010). Algal biomass has demonstrated comparable efficacy to fish oil in terms of ω-3 PUFA enrichment in broiler meat, with no significant consequence on meat oxidative stability (Rymer et al., 2010). Analogously, the dietary administration of *Schizochytrium limacinum* microalgae linearly increased the concentration of EPA and DHA and reduced the ω-6/ω-3 ratio in chicken breast meat (Long et. al., 2018). Positive effects of *Schizochytrium* ssp. seaweeds on the ω-3 PUFA content of thigh meat were also reported by Petrolli et al. (2019). Gatrell et al. (2015), testing dietary levels of defatted green microalgal biomass (*Nannochloropsis oceanica*) up to 16%, observed a linear increase in total ω-3 PUFA, EPA and DHA content in both breast and thigh meat at 42 days of age. Similarly, the dietary supplementation of *Spirulina platensis* (5 g/kg of feed for 6 weeks) determined the positive effects on the fatty acid profile of broiler thigh meat by increasing the proportion of EPA and DHA (Bonos et al., 2016). Finally, it is noteworthy to mention that the ω-3 meat enrichment is also remarkably influenced by the genetic background of the bird. Indeed, studies have shown that slow-growing genotypes generally present higher activity of hepatic Δ5- and Δ6-desaturase and elongase enzymes (Dal Bosco et al., 2012), which are involved in the metabolism of ALA to obtain ω-3 long-chain PUFAs. Such observations were confirmed by Boschetti et al. (2016), who demonstrated that the healthier profile of meat from slow-growing strains could be associated with a differential expression of the fatty acid desaturase 1 and 2 genes as well as with greater activity of Δ5- and Δ6-desaturase. In addition, provided that birds have access to outdoor pasture, slow-growing chickens exhibit a different feeding behavior (i.e. feed-to-pasture consumption ratio) than fast-growing ones, which can result in a higher intake of ALA and antioxidant compounds that can play a protective role on meat lipid oxidation (Sirri et al., 2011; Dal Bosco et al., 2021). On the other hand, Baéza et al. (2013) reported lower effectiveness of ω-3 PUFA deposition in Label Rouge chickens than in Ross 708 broilers when both were fed on diets enriched with ω–3 PUFAs.

Conjugated linoleic acid (CLA) is the name given to a mixture of geometrical and positional conjugated isomers of the linoleic acid (C18:2 *cis*-9, *cis*-12), with double bonds in position 9 and 11 or 10 and 12 in either *cis* or *trans* configuration (e.g. *cis*-9, *trans*-11 and *trans*-10, *cis*-12; Schmid et al., 2006). Beneficial effects of CLA intake on human health, including cancer growth, atherosclerosis, diabetes and obesity, have been documented in the past decades (Salas-Salvado et al., 2006). CLA naturally results from bacterial isomerization and/or biohydrogenation of PUFAs in the rumen, and thus CLA concentration is generally higher in meat from ruminant than monogastric species. Indeed, the highest CLA amount can be found in lamb and beef meat (4.3–19.0 mg/g and 1.2–10.0 mg/g of lipid, respectively), while meat from non-ruminant species usually presents no more than 1 mg CLA/g of lipid (summarized by Schmid et al., 2006). Nonetheless, the amount of CLA in chicken meat can be increased through appropriate nutritional strategies, such as the administration of diets containing CLA or trans-fatty acids (for instance trans-vaccenic acid) that can serve as substrate for the endogenous synthesis of CLA. Du et al. (2002) observed greater CLA content in breast fillets by feeding broilers with diets presenting 2% and 3% of CLA from 3 weeks to 8 weeks of age. Similarly, Sirri et al. (2003), testing the administration of either 2% or 4% CLA in grower diet, observed that CLA deposition in both breast and thigh meat increased as the dietary CLA concentration increased. Other studies confirm the linear relationship between the CLA content in the diet and its deposition in broiler muscle (Szymczyk et al., 2001; Bölükbaşi, 2006). It has been recently demonstrated that 1% CLA supplementation in broiler diets based on dried distiller grains was able to increase the concentration of *cis*-9, *trans*-11 and *trans*-10, *cis*-12 in thigh meat (Jiang et al., 2014). The dietary administration of CLA has also been associated with variations of the fatty acid profile of the meat, in particular with the reduction of MUFA and the increase of SFA content (Szymczyk et al., 2001; Du and Ahn, 2002; Sirri et al., 2003; Bölükbaşi, 2006; Jiang et al., 2014). This is due to the CLA-induced inhibition of the Δ-9 desaturase enzyme which is responsible for the desaturation of SFA. Among the several MUFA that were lowered by such dietary treatment, the reduction of arachidonic acid (C20:4) content could be particularly relevant from a consumer health standpoint because of the role of this fatty acid in the synthesis of pro-inflammatory eicosanoids.

Overall, nutritional strategies represent a valuable approach to modify the lipid fraction of poultry meat, particularly its fatty acid profile, yielding products whose consumption can provide beneficial effects on consumers' health. At the same time, it is widely known that changes in the fatty acid profile can also alter the oxidative stability and shelf-life of the final product, as occurs in the case of ω-3 enrichment (Pompeu et al., 2018). In this context, nutrition can be an effective approach to increase also the amount of micronutrients with antioxidant properties in poultry meat, such as vitamins and trace minerals.

2.2 Vitamins and trace minerals

Vitamin E (i.e. α-tocopherol) is generally recognized as the most important antioxidant in biological systems due to its ability in reducing lipid peroxidation and neutralizing free radicals (Marković et al., 2017). Therefore, the dietary supplementation with α-tocopheryl acetate, the synthetic analogue of vitamin E, has become a widespread practice to reduce oxidative stress and, in turn, to improve meat quality traits (Cheng et al., 2016). The meta-analysis study carried out by Pompeu et al. (2018) evidenced that the dietary supplementation of α-tocopheryl acetate in broiler diets is an effective strategy to increase the vitamin E content in the muscle as well as to protect meat lipid from oxidation. Leskovec et al. (2018) reported that 200 IU vitamin E/kg of feed elevated the α-tocopherol content in breast meat and inhibited lipid peroxidation in fresh, frozen, and cooked meat of broilers fed on diets presenting high ω-3 PUFA content (i.e. 50 g/kg of linseed oil from 21 days to 40 days; Leskovec et al., 2019). Furthermore, positive effects on breast meat enrichment and lipid oxidation were observed by supplementing broilers diets with either water-miscible or lipid-soluble vitamin E (Pitargue et al., 2019). Finally, Niu et al. (2018) showed that the oxidative stability of breast lipids during storage was positively influenced by vitamin E administration in feed (100 mg/kg or 200 mg/kg vitamin E up to 42 days). The antioxidant effects exerted by dietary vitamin E seem to be mediated by the enhanced activity of some antioxidant enzymes, such as superoxide dismutase and glutathione peroxidase. In particular, the activity of glutathione peroxidase in both breast and thigh muscle was reported to be higher in response to the dietary supplementation of α-tocopheryl acetate (Kakhki et al., 2016). Niu et al. (2018) found that the mRNA expression of superoxide dismutase and glutathione peroxidase genes in the liver was linearly augmented by increasing dosages of dietary vitamin E, whereas serum superoxide dismutase activity was significantly elevated by 200 mg of vitamin E per kg of diet (Yesilbag et al., 2011). In turkeys, the vitamin E content of the *Pectoralis* and *Sartorius* muscles can be improved by providing high levels of vitamin E through the diet (Mercier et al., 1998). Interestingly, the dietary treatment with 400 ppm α-tocopheryl-acetate significantly delayed lipid oxidation and decreased the carbonyl content in the *Sartorius* muscle after 9 days of refrigerated storage (Mercier et al., 1998). Similarly, Sheldon et al. (1997) reported that increasing dietary vitamin E levels beyond National Research Council (NRC 1994) recommendation (10× and 25×) significantly influenced the oxidative stability and functionality of turkey breast meat.

Ascorbic acid, also known as vitamin C, is another antioxidant compound in biological systems. The antioxidant properties of vitamin C can be related to its reducing and electron carrier functions as well as to the capacity of converting back oxidized forms of α-tocopherol (Whitehead and Keller, 2003). Vitamin C is considered not essential in poultry as, in physiological conditions, the endogenous synthesis is thought to be sufficient to cover the requirements. Thus, the dietary supplementation of vitamin C is not a common practice in commercial poultry nutrition. It is generally held that vitamin C has limited deposition capacity in broiler tissues, although Skřivan et al. (2012) reported that the dietary administration of vitamin C (280 mg/kg and 560 mg/kg of diet for 5 weeks) increased the content of this vitamin in thigh meat following a dose-dependent manner. On the other hand, the role played by vitamin C on the oxidative stability of poultry meat is still controversial. Skřivan et al. (2012) pointed out that dietary vitamin C supplementation (280 mg/kg diet for 5 weeks) improved the stability of lipids in thigh meat during refrigerated storage (4°C for 5 days). Conversely, Yu et al. (2021) reported that neither thiobarbituric acid-reactive substance (TBARS) concentration in breast meat nor the liver antioxidant status was significantly modified by the administration of diets enriched with vitamin C (200 mg/kg diet from 21 to 35 days). Another study (Leskovec et al., 2018) indicated that the dietary supplementation of ascorbic acid (250 mg vitamin C/kg from 21 to 40 days) and linseed oil (5%) determined no significant effects on plasma content of vitamin C, malondialdehyde, α- and γ-tocopherol, antioxidant capacity of water-soluble compounds as well as blood glutathione peroxidase and superoxide dismutase. However, the dietary treatment increased the antioxidant capacity of lipid-soluble compounds, which has been related to the interaction between ascorbic acid and vitamin E. It has been postulated that the beneficial effects of dietary vitamin C on oxidation might not be so evident when birds are raised in normal or low stressful conditions (Yu et al., 2021). Nevertheless, vitamin C supplementation showed much interesting results when birds were exposed to challenging conditions that can exacerbate oxidative stress, such as copper toxicity (Cinar et al., 2014), dexamethasone treatment (El-Senousey et al., 2018) or heat-stress exposure (Abidin and Khatoon, 2013; Shakeri et al., 2020b).

The enrichment of chicken meat with trace minerals such as selenium (Se) and zinc has also been tested. Selenium (Se) has important antioxidant functions being an essential constituent of the glutathione peroxidase and thioredoxin reductase enzymes (Marković et al., 2017) and detoxifying the hydroperoxide and H_2O_2 produced by the superoxide dismutase (Surai, 2002). Moreover, improving the total Se content of meat through appropriate nutritional strategies could enhance the human intake of Se, which is particularly important for populations in Se-deficient areas. Se accumulation in chicken tissues showed a dose-dependent response (Marković et al., 2018), which corroborates the findings of Silva et al. (2019) who observed a linear increase of Se deposition in tissues of broiler chickens by increasing the dietary dosages of Se from 0.3 mg/kg to 0.6 mg/kg feed. Nowadays, different sources of Se, such as inorganic, organic, and nano-forms, are available as feed supplements for poultry. The results available in the literature seem to indicate that the organic forms of Se can be more effective than inorganic ones for muscle Se enrichment (Briens et al., 2013, 2014; Oliveira et al., 2014; Couloigner et al., 2015). Nevertheless, it has been demonstrated that the dietary supplementation of Se increases the chicken antioxidant systems regardless of the source (Khan et al., 2018; Marković et al., 2018; Hou et al., 2020; Zepeda-Velazquez et al., 2020). According to the literature review of Marković et al. (2018), nano-Se and organic Se have similar or higher efficacy than inorganic Se in upregulating selenoenzymes as well as greater bioavailability. Recent studies have shown that Se-enriched probiotics have greater capacity than sodium selenite in improving broiler meat quality and oxidative stability either in thermoneutral or in heat-stress conditions (Khan et al., 2018; Hou et al., 2020). Furthermore, the dietary administration of Se-enriched yeast, selenomethionine, or nano-red element Se (0.3 mg/kg feed for 40 day) enhanced the glutathione peroxidase activity and reduced the carbonyl content in the breast muscle if compared with sodium Se (Li et al., 2018). Also Se–chitosan, a synthetic Se polysaccharide, has shown great potential in improving meat quality traits, including oxidative stability and functional properties (Bami et al., 2021). Juniper et al. (2011) found greater Se content and glutathione peroxidase activity in the breast muscle of turkeys receiving diets supplemented with Se-enriched yeast, but this improvement yielded no significant outcome in terms of meat quality. Both sodium selenite and Se-enriched yeast determined a significant decrease of TBARS in raw and stored turkey meat (Mikulski et al., 2009). Although it is noteworthy to consider that the dietary Se concentration in broiler feed is regulated in the European Union (maximum 0.5 mg Se/kg of feed, with no more than 0.2 mg Se/kg for selenized yeasts and DL-selenomethionine; EFSA, 2012, 2014), some works showed that increasing dietary Se beyond the NRC (1994) recommendation (i.e. 0.15 mg Se/kg feed) could have beneficial effects in terms of meat quality (Liao et al., 2012; Bami et al., 2021). Taken together, Estévez (2015) concluded that supplementation with α-tocopherol (around 200 mg tocopherol/kg feed) and ascorbate (up to 1000 mg ascorbate/kg feed), either alone or combined with other antioxidant compounds such as Se, magnesium, and zinc, can be useful to improve oxidative stability of chicken meat.

As for zinc (Zn), Bou et al. (2004) observed a similar amount of Zn in meat from broilers fed on diets containing up to 200 mg Zn/kg of feed. In a following study, the administration of diets presenting either 300 mg Zn/kg or 600 mg Zn/kg of feed did not exert significant effects on the Zn content of breast and thigh meat, although it increased the Se content (Bou et al., 2005). Recent findings of Dukare et al. (2021) suggest that the dietary Zn levels do not influence the Zn content of breast muscle, although 80 ppm of Zn increased the concentration of this mineral in the liver. Conversely, Saleh et al. (2018) stated that Zn concentration in the breast muscle can be increased in a dose-dependent manner through the dietary supplementation of Zn–Met (25 mg/kg, 50 mg/kg and 100 mg/kg). In line with the latter study, Kakhki et al. (2016) demonstrated that breast and thigh muscle could be enriched with Zn by providing diets with 50 mg Zn–Met/kg feed. With regard to the source, nano zinc sources seem to be more effective than inorganic ones considering Zn deposition in muscle tissue (Dukare et al., 2021).

3 Nutritional and feeding strategies to improve meat quality and mitigate the incidence of breast meat abnormalities

3.1 Technological properties and sensory traits

Along with the chemical composition, technological properties represent a key aspect related to the concept of meat quality. As reported by Mir et al. (2017), the importance of meat technological properties such as pH, water-holding capacity (WHC), color, collagen content and protein solubility is unquestionable for processors, considering either fresh retail or value-added meat products. On the other hand, consumers' quality perception is influenced by some sensory traits (e.g. color, flavor, texture, juiciness, wateriness, firmness and tenderness) which are, either directly or indirectly, related to meat technological properties (Mir et al., 2017). Within this context, nutritional and feeding strategies could represent a valuable approach for improving the technological and sensory properties of poultry meat.

Considering the dietary macronutrients, the modulation of crude protein and amino acid content of broiler diets has been associated with important variations of meat technological and sensory properties. For instance, Belloir et al. (2017) observed a significant increase in ultimate pH and decrease of meat lightness and drip loss when dietary crude protein was reduced from 19% to 16%. On the other hand, a 2% reduction of dietary crude protein content (23% to 21% and 20% to 18%, respectively, during 1–21 days and 22–42 days) did not significantly affect pH, meat color, WHC, cooking loss or drip loss of breast meat (Park and Kim, 2019). Finally, other papers showed that variation of the dietary crude protein content can modulate the glycolytic potential of the breast muscle, which in turn affected meat pH and related quality traits (Jlali et al., 2012; Zhao et al., 2012). As for amino acids, the effects of dietary Lys on meat technological properties have stimulated great interest within the scientific community. Increased levels of digestible lysine (dig. Lys) in broiler diets (from 0.83% up to 1.03%) were associated with higher final breast pH, which in turn influenced WHC and drip loss (Berri et al., 2008). Recently, Belloir et al. (2019) evaluated the meat quality traits of broilers receiving diets with different dosages of dig. Lys (from 7.0 g/kg feed to 11.5 g/kg feed) associated with other essential amino acids (90%, 100% and 110% of the ideal amino acid profile proposed by Mack et al., 1999). Results showed that both dietary Lys and amino acid concentration affected breast meat quality traits, including ultimate pH, color and drip loss. Moreover, a significant interaction between the experimental factors was observed for most of the meat quality traits. Diets with low amino acid concentrations combined with reduced Lys levels yielded meat with higher ultimate pH (>6.0), whereas a more acid meat (<5.8) was obtained providing higher concentrations of amino acids associated with low Lys. The authors suggested that these outcomes could be related to potential changes of the energy metabolism, including amino acid catabolism, glucose production and glycogen storage within the muscle. Finally, dietary Lys and amino acid concentration also affected meat lightness (L*) and drip loss in the same fashion observed for pH, which could be explained by the high correlation between these quality parameters (Belloir et al., 2019). Furthermore, Guardia et al. (2014) showed that changing the dietary amino acid content in last 3 days before slaughter can affect ultimate pH and color attributes of breast meat. The highest pH value was detected in the breast muscle of broilers fed diets with high Lys content associated with low or adequate concentrations of other essential amino acids. In contrast, feeding Lys-deficient diets with high amounts of other essential amino acids yielded meat with the lowest pH values, which corroborates the findings of Belloir et al. (2019). Lys supplementation up to 120% of NRC (1994) recommendation increased sarcoplasmic protein solubility and meat pH, while it decreased cooking loss, shear force and lightness (Zhai et al., 2016). Interestingly, a significant interaction between dietary Lys and Met was reported for cooking loss, with breast meat from birds fed the highest Lys and Met concentrations (120% and 140% of NRC (1994) recommendation) showing the lowest losses (Zhai et al., 2016). Soglia et al. (2021) found that dietary Lys restriction (-13%) from 10 to 34 days of bird age increased protein solubility as well as the anserine content of breast meat, which could have played a role in protecting proteins from denaturation by buffering the acidic end products generated during early postmortem acidification. The reduction of Lys concentration tested in the study of Watanabe et al. (2017) (90% of NRC recommendation) determined higher levels of free amino acid and glutamate as well as an improvement of some organoleptic parameters (e.g. umami taste and overall intensity).

Met is a sulfur-containing amino acid that is commonly supplemented in broiler diets, being the first limiting essential amino acid in corn-soybean meal diets (Baker et al., 2006). Met is involved in crucial biological pathways related to meat quality, including protein synthesis, lipid metabolism and antioxidant response (Jankowski et al., 2014; Estévez et al., 2020). Conde-Aguilera et al. (2016) showed that reducing Met content in broiler diets from 7 days to 14 days and 21 days to 25 days (-45% and -30%, respectively, for Met:Lys and total sulfur amino acid:Lys ratio compared to the recommended dosages) induced significant variations on some colorimetric attributes of the *Pectoralis major* muscle (i.e. increased redness and hue angle). Moreover, it has been reported that drip loss, pH and sensory shelf-life of chicken fillets are positively correlated with dietary Met levels (0.10%, 0.25%, and 0.40% in a Met + cysteine-deficient diet; Albrecht et al., 2017). Similarly, Met supplementation was associated with higher pH values and lower drip loss of breast meat (Albrecht et al., 2019). Furthermore, the colorimetric variables (lightness, redness and yellowness) were reported to be negatively influenced by Met supplementation (Albrecht et al., 2019). A significant relationship between dietary Met and meat pH was observed also in other studies (Aksu et al., 2007; Liu et al., 2010). Wen et al. (2017) also revealed that low-Met diets (0.35% and 0.31% from 1 day to 21 days and 22 days to 42 days, respectively) negatively affected meat quality and oxidative status in a strain-dependent manner, particularly for fast-growing broilers. On the other hand, increasing dietary Met beyond the concentrations that are generally considered adequate (0.65% and 0.57% vs. 0.50% and 0.44% from 1 day to 21 days and 22 days to 42 days, respectively) yielded only slight differences on breast meat technological properties, antioxidant activity and chemical composition (Wen et al., 2017). Similarly, Zhai et al. (2016) found only minimal variations in proximate composition as well as pH, texture and color profile of breast meat in response to dietary Met supplementation. Finally, different studies reported that the two mostly used Met sources, DL-Met and

DL-2-hydroxy-4-methylthiobutanoic acid, have similar effects on the main quality traits of broiler meat (Albrecht et al., 2017, 2019).

Betaine is considered a valid substitute for Met, providing the methyl groups that are necessary for the conversion of homocysteine to Met (Alirezaei et al., 2012). Fu et al. (2016) evaluated the effects of partial or complete replacement of DL-Met in broiler diets with betaine (total Met: 5.0 g/kg and 4.2 g/kg in 1–21 days and 22–42 days, respectively). The most relevant outcome observed by the authors is that betaine utilization impaired some quality parameters of breast meat, reducing pH value, WHC and the concentration of some amino acids in a dose-dependent manner, with the greatest effects observed at 100% replacement level. Another study showed that the on-top supplementation of betaine (1 g/kg, 2 g/kg or 3g/kg of feed) can affect some physicochemical properties of breast meat according to the dosage, for instance by increasing the drip loss while reducing the shear force value and the amino acid content (Nutautaitė et al., 2020). Overall, the main effects of dietary betaine seem to be related to its antioxidant properties and to the ability to preserve the cellular antioxidant stores (Nutautaitė et al., 2020). A similar conclusion can be drawn from the work of Park and Kim (2019), who observed an enhancement of glutathione peroxidase activity in response to betaine supplementation in diets with different crude protein levels, although minimal benefits in terms of meat quality were obtained. Conversely, Fu et al. (2016) reported a reduction of glutathione peroxidase activity in the breast muscle of broilers fed diets in which betaine replaced supplemental Met. Finally, some studies focused on the potential role of dietary Arg in modulating the technological properties of chicken meat. Corzo et al. (2003) reported that fillet lightness (L*) and yellowness (b*) increased when the dig. Arg:Lys ratio was elevated from 0.91 to 1.45 between 42 days and 56 days of bird age. Jiao et al. (2010), testing dietary Arg dosages ranging from 80% to 140% of NRC recommendations, observed a significant effect on L* value, cooking loss and ultimate pH of breast meat. Technological properties of breast meat were not significantly affected by increasing the dig. Arg:Lys ratios beyond current recommendations (+10%, 20% or 30%; Zampiga et al., 2018, 2019). However, positive effects of high Arg:Lys ratio were detected on breast myopathies occurrence, as discussed in the next section of this chapter.

With regard to vitamins, the results of the meta-analysis study carried out by Pompeu et al. (2018) indicate that the dietary supplementation of vitamin E determines no significant change in meat pH. Similarly, neither pH value nor colorimetric attributes of breast meat were significantly modified by the administration of 250 mg vitamin E *per* kg of diet (Fernandes et al., 2013). In line with these observations, Leskovec et al. (2019) pointed out that no variations on meat WHC, pH and color attributes were observed in response to the dietary supplementation of vitamin E. Yesilbag et al. (2011) reported that increasing the dietary dosages of α-tocopherol acetate from 50 mg/kg to 200 mg/kg of feed induced slight but still significant variations in the pH value of breast and thigh meat without changing their color attributes. In addition, the treatments did not affect taste, odor, color, tenderness and overall acceptability of the meat (Yesilbag et al., 2011). It has been previously reported that 250 mg vitamin E/kg of feed can enhance consumer-perceived texture quality and juiciness (Kennedy et al., 2005). Similarly, Zdanowska-Sąsiadek et al. (2016) showed that meat sensory attributes such as juiciness and tenderness as well as WHC can be improved of vitamin E (200 mg DL-α-tocopheryl acetate/kg of feed). Furthermore, feed supplementation with 225 mg of vitamin E/kg of diet decreased the rancid aroma and flavor scores in frozen thigh meat (Bou et al., 2001), whereas α-tocopheryl acetate supplementation (70 mg/kg and 140 mg/kg) did not modify the acceptability of cooked thigh meat at 15 days after slaughtering and after 5 months of storage at −20°C (Bou et al., 2004).

In a recent study, Yu et al. (2021) demonstrated that the administration of diets supplemented with vitamin C (200 mg/kg) determined no significant improvements in breast meat quality traits such as pH, WHC and color profile. The results of the dose−response study carried out by Fernandes et al. (2013) showed that pH, color and WHC were not significantly modified by dietary vitamin C dosages ranging from 150 mg/kg to 450 mg/kg. No substantial effects of dietary ascorbic acid on meat pH and WHC were also reported by Leskovec et al. (2019). Increased dietary dosages of vitamin C (475 mg/kg feed) did not improve color, odor, flavor and overall acceptability of breast and thigh meat (Skřivan et al., 2012). Analogously, Bou et al. (2001) revealed that the dietary administration of ascorbic acid (110 mg/kg) only marginally affects the rancidity flavor and aroma as well as the overall acceptability of chicken thigh meat stored at −20°C for 13 months (Bou et al., 2001).

As for trace minerals, Oliveira et al. (2014) found that cooking loss of chicken fillets could be reduced by increasing the amount of Se yeast from 0.15 ppm to 0.60 ppm, although pH and shear force were not affected by the treatment. According to Mohammadi et al. (2019), dietary Se has only minimal effects on pH value and WHC capacity of both thigh and breast meat. The results reported by Leskovec et al. (2019) seem to confirm that Se supplementation in broiler diets (0.2 mg Se/kg of feed) does not have relevant influence on the breast meat quality traits. In addition, the dietary administration of Se exhibited no substantial effects on consumer acceptability scores of cooked thigh meat after 74 days and 18 months of frozen storage (Bou et al., 2005). The outcomes of the study carried out by Haug et al. (2007) confirm the slight impact of dietary Se on the sensory quality of cooked breast meat stored at −20°C. Salim et al. (2011) showed that feed supplementation with 40 ppm or 80 ppm of organic Zn can increase the WHC of chicken fillets. However, other meat quality parameters and sensory traits of both breast and thigh meat were not modified by the dietary Zn concentrations tested in that study. Finally, the overall consumer acceptability of thigh meat samples stored at −20°C for 5 months was not significantly influenced by the dietary administration of 200 mg Zn per kg of feed (Bou et al., 2004).

Herbs, spices and their extracts also known as phytocompounds have been used for centuries to improve the sensory characteristics and shelf-life of foods because of their antioxidant and antimicrobial properties. Currently, the consumers' attention toward the use of natural products in livestock feeding has further stimulated the interest on the efficacy of these products as substitutes for synthetic antioxidants. In this context, particular emphasis has been given to herbs belonging to the Lamiaceae family which includes species with recognized antioxidant properties, such as rosemary (*Rosmarinus officinalis*), oregano (*Origanum vulgare*), green tea (*Camellia sinensis*) and lavender (*Lavandula stoechas*). Yesilbag et al. (2011) showed that supplementing broiler chickens diets with rosemary leaves and volatile oil could improve meat quality by decreasing lipid oxidation. However, the sensory analysis carried out in that study indicated that the highest inclusion level of both rosemary leaves (11.5 g/kg) and volatile oil (200 mg/kg) worsened meat taste and overall acceptability. The dietary administration of rosemary essential oil (300 mg/kg) had some beneficial effects in preventing lipid oxidation in thigh but not breast meat (Mohammadi et al., 2019). However, the WHC was not significantly affected by the dietary treatment considering either breast or thigh meat (Mohammadi et al., 2019). Moreover, also Loetscher et al. (2013) observed that rosemary leaf supplementation in broiler diets (25 g/kg) is effective in reducing lipid oxidation and in elevating total tocopherol content in breast meat. On the other hand, feed supplementation with oregano improved the oxidative stability of breast and thigh meat during refrigerated storage (Florou-Paneri et al., 2006). Interestingly, the authors claimed that the oxidative stability offered by oregano was superior to that exhibited by vitamin C. Botsoglou et al. (2003) reported that the lipid oxidation in long-term frozen stored chicken meat can be reduced through the dietary administration of oregano essential oil, even if the treatment with α-tocopheryl acetate was much more effective. Recent works seem to confirm the benefits exerted by dietary oregano supplementation on the antioxidant status in broilers, although the consequences on the main meat quality traits are negligible (Ri et al., 2017). Seidavi et al. (2020) concluded that the use of green tea in poultry diets can have some beneficial outcomes on broiler meat quality and oxidative stability, whereas Kiyma et al. (2017) stated that essential oils from lavender can enhance superoxide dismutase activity in the breast muscle, but this was not coupled with relevant improvements in meat technological traits. Dietary oregano and garlic (*Allium sativum)* essential oils had only minimal effects on proximate composition, technological properties and oxidative stability of broiler meat even if some sensory attributes were influenced positively (Kirkpinar et al., 2014). Although the antioxidant properties of garlic in chickens were demonstrated (Jakubcova et al., 2014), the dietary supplementation of garlic powder yielded minimal benefits in terms of broiler meat quality (Abdullah et al., 2010). Similarly, Lukanov et al. (2015) showed that garlic powder supplementation had negligible effects on chicken meat quality traits. However, according to the authors, some benefits on meat tenderness can be achieved with 0.4% and 0.8% of garlic powder in feed. Also, Choi et al. (2010) suggested that the combination of garlic powder and α-tocopherol was effective for enhancing lipid and meat color stability, although it provided limited effects on meat WHC and shear force. Finally, the positive effects of the dietary supplementation of turmeric (*Curcuma longa*) rhizome powder on the malondialdehyde content in thigh meat during refrigerated storage were documented by Daneshyar et al. (2011). Some studies reported that turmeric rhizome powder can improve the antioxidant capacity and WHC of chicken meat (Wang et al., 2015), even when supplemented in diets including flaxseed meal (Kumar et al., 2020).

3.2 Breast meat abnormalities

In the last decade, a greater occurrence of growth-related quality defects affecting the breast muscle of fast-growing broilers has been observed in commercial poultry abattoirs worldwide. These abnormalities, the detailed description of which can be found elsewhere in this book, were defined by meat scientists as white striping (WS), wooden breast (WB) and spaghetti meat (SM). Briefly, the main distinctive trait of the WS defect is the presence of white striations following the same direction of the muscle fibers (Kuttappan et al., 2012a). Breast muscles presenting the WB abnormality are characterized by hardened consistency, pale appearance and polyphasic myodegenerations with fibrosis at the histological level (Sihvo et al., 2014). Finally, the increased tendency toward muscle fiber bundle separation resulting in an impaired integrity of the *P. major* muscle is typical of the SM defect (Baldi e tal., 2021). It is important to mention that all these abnormalities do not represent a threat for human health but can affect the organoleptic, technological and nutritional quality of breast meat, reducing consumer willingness to buy and the reutilization of affected meat for further processing (Petracci et al., 2019).

Although it is generally held that the artificial selection for growth rate and breast muscle yield is a major determinant of myopathies occurrence in modern broiler lines (Alnahhas et al., 2016; Griffin et al., 2018), a great number of nutritional approaches have been tested to mitigate the incidence and severity of these abnormalities (Table 1). In detail, some studies investigated the impact of feed restriction programs as a strategy to limit the onset of

Table 1 Main nutritional and feeding strategies tested to mitigate the occurrence of growth-related breast meat abnormalities (white striping, wooden breast and spaghetti meat)

Strategy	Reference	Features of the study	Slaughter age (days)	Effect on myopathy occurrence			Overall effect on growth and processing performance
				WS	WB	SM	
Feed restriction	Meloche et al. (2018a)	Reduction of feed allocation at 85% and 95% of *ad libitum* consumption	32-50	↓	↓	-	↓
	Livingston et al. (2019a,b)	Time-limited feeding program (8 h/day from 8 to 42 days)	42	↓	↓	-	↓
	Trocino et al. (2015)	Reduction of feed consumption to 80% of *ad libitum* basis from 13 days to 21 days of age	46	↑	=	-	↓
	Gratta et al. (2019)	Feed restriction (80% of *ad libitum* intake) during early (13–23 days) or late growth phases (27–37 days)	48	=	=	-	↓
Dietary nutrient density	Meloche et al. (2018b)	Reduction of digestible amino acid and metabolizable energy density (-10% and -15%) during starter phase (8–14 days)	35	=	↑	-	=
	Meloche et al. (2018b)	Reduction of nutrient density (-10%) for the entire rearing period (from 8 days to 49 days)	49	↓	↓	-	↓
	Bodle et al. (2018)	Reduction of digestible amino acids density (-15%) during the grower phase	45	=	↓	-	=
	Ahsan and Cengiz (2020)	Reduction of metabolizable energy and amino acid density (-5%) during grower or grower and finisher phases	49	=	-	-	= / (↓ feed efficiency)
Lysine	Cruz et al. (2017)	Increase of dig. Lys concentration during 12–28 days	35	↑	↑	-	↑
	Cruz et al. (2017)	Increase of dig. Lys concentration during 28–42 days	42	↑	↑	-	↑
	Meloche et al. (2018c)	Reduction of Lys concentrations at 75% of recommended dose from 12 days to 26 days of age	47	↓	↓	-	= / (↓breast yield)
	Meloche et al. (2018c)	Reduction of Lys concentrations at 75% of the recommended concentrations from 12 days to 40 days	61	↓	↓	-	=

	Soglia et al. (2021)	Lys restriction (-13%) either from 10 days to 20 days or 10 days to 34 days	49	=	=	=	=
Arginine	Bodle et al. (2018)	Increase of dig. Arg:Lys ratios during starter phase, grower phase, finisher phase and withdrawal phase	45	=	↓	-	=
	Zampiga et al.(2018)	Increase of dig. Arg:Lys ratios beyond current recommendations (+10%) in all feeding phases	42	=	=	=	↑
	Zampiga et al. (2019)	Increase of dig. Arg:Lys ratios beyond current recommendations (+30%) in all feeding phases	42	↓	-	↓	=
	Livingston et al. (2019b)	Dietary supplementation of Arg (0.25%) from 8 days to 44 days	45	=	=	-	↑
Glutamine	Livingston et al. (2019b)	Dietary supplementation of glutamine (1%) from 8 days to 44 days	45	↑	=	-	↑
Methionine	Sachs et al. (2019)	Alternative dietary regimen with natural methionine sources (i.e. sunflower and roasted cowpea)	42	↓	-	-	↓
Guanidinoacetic acid	Oviedo-Rondón and Córdova-Noboa (2020)	Dietary supplementation of guanidinoacetic acid (600 g/ton), a precursor of creatine	51-56	=/↓	↓	-	↑
Histidine	Lackner et al. (2021)	Higher dig. His:Lys ratio (0.54 and 0.64 vs. 0.44)	33	=	-	-	=
	Lackner et al. (2021)		53	↑	-	-	=
β-Alanine	Lackner et al. (2021)	Dietary supplementation of β-alanine (0.5%)	33	↑	-	-	=/(↓ breast yield)
	Lackner et al. (2021)		53	↑	-	-	=/(↓ breast yield)
Vitamin E	Kuttappan et al. (2012b)	Increasing dosage of dietary vitamin E (from 15 IU/kg to 400 IU/kg)	49	=	-	-	=
	Wang et al. (2020)	Vitamin E supplementation (10 IU/kg) in starter or grower diets	58	↓	↓	-	=
Vitamin C	Bodle et al. (2018)	Increase of dietary vitamin C throughout the rearing cycle	45	=	↓	-	↓
Vitamin premix	Bodle et al. (2018)	Doubling the vitamin premix supplementation	45	=	=	-	=/(↓ feed efficiency)

(Continued)

Table 1 (*Continued*)

Strategy	Reference	Features of the study	Slaughter age (days)	Effect on myopathy occurrence			Overall effect on growth and processing performance
				WS	WB	SM	
Ethoxyquin	Kuttappan et al. (2021)	Dietary administration of ethoxyquin in diets presenting oxidized fats	59	-	↓	-	=
Trace minerals	Sirri et al. (2016)	Dietary supplementation of zinc, manganese and copper (either organic or inorganic)	51	=	=	-	↑
	Kuttappan et al. (2021)	Dietary supplementation of methionine hydroxy-analog chelate of zinc, copper and manganese in diets with oxidized fats	59	-	=	-	=/↑
	Cemin et al. (2018)	High dietary dosages of organic Se (from 0.0 ppm to 1.0 ppm of Zinc-L-selenomethionine)	42	↑	↑	-	↑
	Sihvo et al. (2017)	Low selenium concentrations throughout the rearing cycle	10–42	-	=	-	=
ω-6/ω-3 ratio	Wang et al. (2020)	Reduction of ω-6/ω-3 ratio from 30:1 to 3:1 in starter or grower diets	58	↑	↑	-	↓

WS, white striping; WB, wooden breast; SM, spaghetti meat.
↓: reduction; ↑: increase; =: no significant effect; -: not evaluated.

breast myopathies. The rationale behind this approach lies in the modulation of growth pattern and nutrient intake during specific stages of the rearing cycle. In general, two different restriction strategies have been applied involving either quantitative approaches (i.e. reduction of the amount of feed available to the birds) or qualitative (i.e. reduction of the concentration of certain nutrients within the diet) approaches. Meloche et al. (2018a) observed a linear reduction in WS and WB severity scores at 33 days, 43 days and 50 days of age by providing 85% to 95% of feed than the amount assumed on *ad libitum* consumption, with only marginal improvements for reduction below 95%. Similarly, a time-limited feeding program (8 h/day from 8 days to 42 days) determined a significant reduction of both WS and WB scores (Livingston et al., 2019a). Conversely, Trocino et al. (2015), lowering feed consumption to 80% of *ad libitum* basis from 13 days to 21 days of age, detected a tendency for increased WS in 46-day-old broilers likely due to compensatory growth mechanisms occurring when birds had free access to the feed. On the other hand, neither early nor late feed restriction (80% of *ad libitum* intake during 13-23 days or 27-37 days, respectively) contributed to a significant reduction of WS and WB myopathies (Gratta et al., 2019). Lowering the digestible amino acid and metabolizable energy density during 8–14 days generated an increase of severe WB cases at 36 days in comparison to birds fed diets with adequate nutrient densities (Meloche et al., 2018b). However, the reduction of nutrient density (–10%) for the entire rearing period (from 8 days to 49 days) lessened the occurrence of severe WB and WS but compromised growth efficiency (Meloche et al., 2018b). Finally, Bodle et al. (2018) showed that reducing digestible amino acid density (–15%) during the grower phase lowered the average WB but not WS score. Recently, Ahsan and Cengiz (2020) pointed out that the reduction of metabolizable energy and amino acid density (-5%) during grower or grower and finisher phases did not affect the incidence and severity of WS. It is important to highlight that the lower occurrence of muscle abnormalities observed in some studies in response to the administration of reduced nutrient density diets was achieved at the expenses of growth rate, final body weight and/or breast yield, suggesting that this approach may not be completely sustainable and applicable in current commercial conditions.

Another approach that has been tested is the modulation of the dietary content of some specific amino acids. For instance, dietary Lys has a pivotal role on muscle development, protein turnover and meat quality, and thus some researchers argued that modifying the dietary concentrations of Lys could have consequences on myopathies occurrence. Firstly, Cruz et al. (2017) evaluated the effects of increasing dig. Lys concentration (0.77% to 1.17% during 12–28 days, and 0.68% to 1.07% during 28–42 days) on the incidence and severity of breast muscle myopathies. The results evidenced that higher Lys concentrations determine a greater incidence and severity of WS and WB, which could be likely associated with the improved growth performance. On the other hand, Meloche et al. (2018c) found lower occurrence of severe WS and WB cases in 48-day-old broilers by feeding diets with 75% of the recommended Lys concentrations from 12 days to 26 days of age. It is noteworthy to highlight that the dietary manipulation tested in that trial did not affect the growth performance in the whole rearing cycle but significantly reduced the breast meat yield. The administration of diets with dig. Lys at 85% of the recommended concentrations from 12 days to 40 days was able to decrease the incidence of severe WB cases compared to a control diet (100% of primary breeder recommendations) without negatively affecting live performance or processing yields at 61 days of age (Meloche et al., 2018c). Furthermore, the occurrence of WS at 49 days was decreased by limiting the dig. Lys concentration (-15%) during grower or grower and finisher phases, even though this strategy impaired growth performance and breast meat yield (Meloche et al., 2018c).In a recent study, Soglia et al. (2021) observed no significant effect of Lys restriction (-13% either from 10 days to 20 days or 10 days to 34 days) on the incidence and severity of WS, WB, and SM abnormalities at 49 days. Overall, these findings indicate that both the relative extent and the time duration of Lys depletion can affect the outcomes of this feeding strategy.

Arg is considered an essential amino acid in poultry nutrition due to the lack of a functional urea cycle in birds. In the context of meat quality, Arg has stimulated interest as nutritional intervention to limit the occurrence of muscle abnormalities because of its role in the synthesis of nitric oxide (Boerboom et al., 2018), a potent vasodilator compound that could alleviate the hypoxic condition observed in affected muscles. Zampiga et al. (2019) pointed out that increasing the dig. Arg:Lys ratios beyond current recommendations (dig Arg:Lys ratio = 1.35, 1.35, 1.36, 1.37, in 0–12 days, 13–22 days, 23–33 days and 34–43 days, respectively) could be considered a valuable strategy to mitigate the occurrence of WS and SM abnormalities while maintaining acceptable live performance, processing yields and breast meat quality traits. On the other hand, no significant effect was observed when the dig Arg:Lys ratio was increased by 10% (Zampiga et al., 2018). Bodle et al. (2018) detected positive effects on average WB scores in response to higher dig. Arg:Lys ratios (i.e. from 1.12 to 1.20 in starter 1.14 to 1.26 in grower 1.15 to 1.26 in finisher and 1.14 to 1.26 in withdrawal phases). However, the treatment had no significant effect on WS occurrence, growth performance and processing yields. Furthermore, it has been reported that the dietary supplementation of Arg (0.25% from 8 days to 44 days) had only a slight effect on the WB and WS scores (Livingston et al., 2019b). As concerns other amino acids, Sachs et al. (2019) showed that a dietary regimen with alternative Met sources (i.e. sunflower and roasted cowpea) reduced the WS defect even though this result was associated with a decrease in body weight at 42 days. Moreover, the dietary supplementation of glutamine (1% from 8 days to 44 days) seems not an effective strategy, as it contributed to a greater incidence and severity of the WS condition in 44-day-old broilers (Livingston et al., 2019b). Finally, recent reports have suggested that guanidinoacetic acid, a precursor of creatine, can exert some positive effects on the incidence and severity of the WB abnormality (reviewed in Oviedo-Rondón and Córdova-Noboa, 2020), whereas higher histidine:Lys ratios (0.54 and 0.64 vs. 0.44) or β-alanine supplementation (0.5%) increased the WS occurrence in broilers processed at 53 days of age (Lackner et al., 2021).

Other studies focused on the potential benefits that could be achieved on myopathy occurrence by elevating the vitamin and antioxidant content of broiler diets. Kuttappan et al. (2012b) tested different levels of vitamin E supplementation in broiler diets (15 IU, 50 IU, 100 IU, 200 IU, and 400 IU of vitamin E/kg of feed) and observed no significant effect on the occurrence of the WS defect. Recent findings of Wang et al. (2020) suggest that vitamin E supplementation (10 IU/kg), especially during the first 10 days of life, could be effective in reducing the incidence and severity of WB myopathy while exerting no adverse effects on growth performance and meat yield. The increase of dietary vitamin C dosages throughout the rearing cycle has been associated with a significant reduction of the WB score, even though doubling the vitamin premix supplementation did not provide substantial benefits on WB or WS occurrence (Bodle et al., 2018). As for other antioxidants, the dietary inclusion of ethoxyquin reduced the proportion of severe WB cases in birds fed diets including oxidized fats (Kuttappan et al., 2021). Furthermore, the authors observed a synergistic effect among ethoxyquin, organic trace minerals (i.e. Met hydroxy-analog chelate of zinc, copper and manganese) and organic Se on the incidence of fillets presenting no WB lesions. However, when considered separately, the impact of the organic trace minerals on WB occurrence was not statistically significant (Kuttappan et al., 2021). Concerning the role of Se, Cemin et al. (2018) reported a greater severity of WS and WB abnormalities in response to high dietary dosages of organic Se, which has been associated with the performance and breast yield improvements observed in birds fed Se-enriched diets. On the other hand, low Se concentrations were not effective in reducing the occurrence of WB (Sihvo et al., 2017). As for trace minerals, Sirri et al. (2016) showed that neither the dosage nor the source (inorganic vs. organic) of zinc, manganese and copper was able to determine significant improvements in the occurrence of WS and WB in broilers processed at 51 days of age. Finally, Wang et al. (2020) tested an alternative dietary strategy based on the reduction of the ω-6/ω-3 ratio (from 30:1 to 3:1) during the starter phase or grower phase. The results of this study indicate that lowering the ω-6/ω-3 ratio impaired growth and processing performance, while it increased the occurrence of WS and WB.

4 Consequences of heat stress on poultry meat quality and nutritional strategies to limit product alterations

The problems associated with global warming and climate change, such as the increased frequency and magnitude of extreme heat waves, have become more evident in the last decades, and this has stimulated the interest of the scientific community in characterizing the impact of heat stress on poultry production. The general consensus is that heat stress should be regarded as one of the most prominent environmental factors affecting quali-quantitative production traits of poultry, including productive and reproductive performance, health and welfare and product quality (Gonzalez-Rivas et al., 2020). In general, chickens are more susceptible to heat stress than other livestock species due to feather coverage and the absence of sweat glands (Renaudeau et al., 2012). Moreover, current fast-growing broiler lines have a lower capacity of controlling body core temperature under thermal stress conditions than unselected genotypes or ancestor lines, which could be ascribed to the high metabolic rate and growth potential of modern broiler hybrids (Yahav, 2009). When the ambient temperature exceeds the thermoneutral zone, chickens are no longer able to properly thermoregulate and, in order to deal with such conditions, they apply certain physiological, metabolic and behavioral adjustments such as increased respiratory rate and lower feed intake which could have relevant consequences on production traits including meat quality (Zhang et al., 2020).

Overall, there are different mechanisms by which heat stress can affect broiler meat quality. As previously mentioned, heat-stressed broilers increase the respiratory rate (i.e. panting) to enhance heat dissipation by means of evaporative cooling (Yahav, 2009). The hyperventilation associated with panting determines an acid–base disturbance known as respiratory alkalosis, which is characterized by decreased concentration of blood bicarbonate and higher blood pH (Gonzalez-Rivas et al., 2020). In this context, the anaerobic metabolism represents the main energy production pathway in the skeletal muscle, resulting in a greater conversion of pyruvate into lactate that can continue even during the early postmortem phases (Zaboli et al., 2019). Such increased acidification leads to a rapid change in muscle pH while the carcass temperature is still high, determining partial protein denaturation and loss of functionality, and meat with abnormal light or pale color, soft texture, reduced WHC and cooking yield. Overall, this phenotypical appearance of meat is similar to the widely known pale, soft and exudative (PSE)-like condition (Zaboli, 2019). The PSE-like condition represents a relevant concern considering either fresh retail or processed products due to the organoleptic alterations and the low technological properties of broiler (Petracci et al., 2015) and turkey meat (Zampiga et al., 2020). On the other hand, some studies (N'dri et al., 2007; Dai et al., 2012; Zeferino et al., 2016) demonstrated that the prolonged exposure to high environmental temperatures could result in a limited meat acidification (i.e. ultimate pH >6.0–6.2) due to the reduction of glycogen reserves *in vivo* as a consequence of the protracted stress. These opposite outcomes on meat quality could be ascribed to the duration of the thermal stress, which is generally classified as: (a) acute, which refers to a short and rapid rise of environmental temperature and (b) chronic, in which birds are exposed to high temperatures for a long period of time (e.g. days to weeks; Akbarian et al., 2016). Much of the evidence seems to indicate a greater risk of developing a PSE-like condition under acute heat stress, whereas long-term or prolonged heat stress could generate meat with high ultimate pH and darker color because of muscle glycogen depletion (Gregory, 2010).

Another relevant factor involved in the detriment of broiler meat quality under high environmental temperature is oxidative stress, which is known to occur when the amount of pro-oxidant compounds exceeds the antioxidant defenses (Fellenberg and Speisky, 2006). Typically, oxidation is triggered by reactive oxygen species (ROS), which are highly reactive compounds that can damage several cellular components, such as proteins, lipids and nucleic acids, thus leading to metabolic dysfunctions (Estévez, 2015). In normal physiological conditions, ROS production occurs as a result of the cellular metabolic activities, and the antioxidant defense systems can properly manage ROS concentrations. However, it has been widely demonstrated that heat stress leads to an overproduction of ROS (Lin et al., 2006; Mujahid et al., 2005, 2009; Azad et al., 2010a; Yang et al., 2010), which can induce relevant damages to cellular lipids or proteins. As the latter play important functional and structural roles in the cell, protein oxidation represents one of the most important factors involved in the worsening of meat quality during warm periods. For instance, oxidative processes associated with heat stress could target mitochondria proteins, leading to dysfunctions of mitochondria activity and metabolic alterations. The work of Lu et al. (2017) demonstrated that the oxidative stress induced by chronic heat exposure (14 days) causes irreversible mitochondrial damage, impairs aerobic metabolism of fat and glucose and increases glycolysis and intramuscular fat deposition resulting in meat with pale color and low WHC. It has been also reported that ROS-induced damages to sarcoendoplasmic reticulum calcium-ATPase can result in Ca^{2+} accumulation in the sarcoplasm and, in turn, in an uncontrolled muscle contraction as well as in reduction of muscle pH and meat WHC (Zaboli et al., 2019). Similarly, ROS can target proteolytic enzymes linked to meat tenderization or muscle proteins, mainly actin and myosin, leading to reduced proteolytic susceptibility (Zaboli et al., 2019) and higher drip and cooking losses because of the reduced solubility and water-binding capacity of oxidized proteins (Wang et al., 2009). Nonetheless, also lipid oxidation impairs meat quality by reducing its functional, sensorial and nutritional properties (Estévez, 2015). Heat-stress exposure (either acute or chronic) generally results in higher malondialdehyde levels in the skeletal muscle (Mujahid et al., 2009; Wang et al., 2009; Azad et al., 2010b; Yang et al., 2010; Tan et al., 2010; Sahin et al., 2016), suggesting that also lipid oxidation is involved in the worsening of meat quality under heat stress.

In addition, heat stress increases the levels of circulating corticosterone through the activation of the hypothalamic-pituitary-adrenal axis, which can modulate protein and lipid metabolism, body composition and meat quality (Scanes, 2016). Linking with the muscle receptors, catecholamines trigger a cascade reaction mediated by cyclic adenosine monophosphate which alters carbohydrate metabolism by accelerating muscle glycogenolysis and inhibiting glycogenosynthesis (Gonzalez-Rivas et al., 2020). In addition, corticosteroid hormones can further stimulate ROS formation exacerbating the scenario described previously. In their review paper, Zaboli et al. (2019) stated that the changes in body composition in response to the increased levels of corticosteroid hormones could affect meat quality traits, including higher drip loss and shear force as well as lower marinade uptake, protein solubility and cooking yields.

In this context, different nutritional strategies have been tested to mitigate the negative effects of heat stress on chicken meat quality. A widely known approach to alleviate heat-stress discomfort is limiting bird access to the feed during the warmest period of the day. This strategy was implemented with the aim of reducing metabolic heat production and the consequent increase of body temperature (Mujahid, 2011). However, Uzum and Toplu (2013) reported no significant effect of feed restriction on breast and thigh meat quality of broiler chickens exposed to chronic thermal stress (32–35°C, 8 h/day, from 21 days to 42 days of age). Also the increment of the dietary energy density through the addition of fats is thought to limit the negative effects of heat stress, compensating for the reduced feed intake while avoiding excessive heat production as lipid metabolism generates lower heat increment than either protein or carbohydrate (Daghir, 2009). However, when it comes to product quality, it should be considered that feeding diets with high metabolizable energy or fat content could exert negative effects on carcass characteristics, especially on the abdominal fat deposition (Ghazalah et al., 2007, 2008). Other dietary interventions concerned the modification of the dietary crude protein content, which is supported by the fact that heat-stressed broilers shown an altered protein metabolism likely aimed at mobilizing amino acids for energetic purposes (Jastrebski et al., 2017; Lu et al., 2018; Zampiga et al., 2021b). Two different approaches have been tested under heat stress, which involved either the reduction or the increase of dietary protein. Indeed, some studies advocated that lowering the dietary crude protein content could be beneficial to limit metabolic heat production (Furlan et al., 2004). However, such strategy could have detrimental effects on growth performance, processing yields and meat quality, likely due to the suboptimal provision of essential amino acids (Daghir, 2009; Houshmand et al., 2012; Zulkifli et al., 2018). In addition, the administration of low protein diets was associated with an increased fat deposition (Furlan et al., 2004). On the other hand, providing diets with higher protein levels could be justified by the reduced feed intake of heat-stressed broilers, yet this practice provided contrasting outcomes (Temim et al., 1999, 2000; de Faria Filho et al., 2007). Attia et al. (2018) reported that the administration of diets with high protein and metabolizable energy levels under heat-stress conditions increased meat dry matter, protein and lipid content compared to a standard-protein regimen, although no significant change was observed for technological traits such as pH, tenderness, color and WHC. In another study, the chemical composition and technological traits of breast meat were only slightly affected by the administration of diets with elevated protein and metabolizable energy content (Attia and Hassan, 2017).

Heat stress markedly alters the amino acid requirements and the ideal amino acid balance (Brake et al., 1998; Chamruspollert et al., 2004). The provision of diets with a balanced amino acid profile is paramount under heat-stress conditions to minimize the energy expenditure for excreting surplus nitrogen (Mujahid, 2011). In terms of meat quality, Attia et al. (2006) reported no ameliorative effects of the dietary Met or Arg supplementation on the chemical composition of breast meat in chickens exposed to chronic heat stress. Similarly, increasing the concentration of Met, Arg and Lys in broiler diets did not affect the proximate composition, pH value and color profile of breast meat from slow-growing broilers raised in thermal stress conditions (Attia et al., 2011). However, the dietary treatment increased meat tenderness and reduced WHC (Attia et al., 2011). Furthermore, Ayazi (2014) pointed out that glutamine supplementation was not effective in modifying meat quality traits of chronically heat-stressed broilers. One of the most interesting approaches seems to be the dietary supplementation of betaine, a naturally occurring amino acid derivative, which has exhibited some positive effects in attenuating the alterations induced by heat stress on chicken meat quality. In particular, the dietary supplementation of betaine has been associated with improvements in the drip loss of breast meat regardless of the dosage and the duration of heat stress (Attia et al., 2009; Wen et al., 2019; Shakeri et al., 2019, 2020a). Attia et al. (2009) demonstrated that dietary betaine could improve breast meat WHC and redness value even in slow-growing broilers. Several studies reported that dietary betaine can positively affect the oxidative status of chicken meat (Wen et al., 2019; Shakeri et al., 2019, 2020a; Nutautaitė et al., 2020), likely because of its antioxidant properties and ability to preserve the cellular antioxidant stores (Nutautaitė et al., 2020). Furthermore, betaine was reported to increase myofibrillar fragmentation and pH 24 postmortem (Shakeri et al., 2020a). Wen et al. (2019) stated that the administration of betaine-supplemented diets tends to enhance moisture and decrease crude protein content in the breast muscle. Conversely, Liu et al. (2019) found no significant difference in meat quality traits of yellow-feathered broilers exposed to long-term heat stress when diets with different concentrations of betaine were provided.

Adding natural or synthetic antioxidant compounds to broiler diets could represent a valuable strategy to counteract the oxidative stress induced by high environmental temperatures. Within this context, Zepeda-Velazquez et al. (2020) stated that Se supplementation could enhance the antioxidant status of broiler chickens under heat stress, although the response is affected by the dosage and the source of Se as well as by the magnitude of heat stress. Dietary supplementation with Se was associated with a reduction of malondialdehyde content and an increased activity of glutathione peroxidase in the breast muscle of cyclically heat-stressed broilers (Harsini et al., 2012). Interestingly, Harsini et al. (2012) detected a synergistic effect between Se and vitamin E on malondialdehyde content or activity of antioxidant enzymes (i.e. glutathione peroxidase and superoxide dismutase). Also, Habibian et al. (2016) stated that breast meat lipids can be preserved from heat-stress-induced oxidative stress by providing diets with supplemental doses of Se and vitamin E. The authors reported that the lowest malondialdehyde content was achieved with, respectively, 1 mg and 125 mg of Se and vitamin E per kg of feed. Positive effects of dietary vitamin E on lipid oxidation and antioxidant enzymes activity under heat stress were also demonstrated in other studies (Imik et al., 2012a). Concerning technological properties, dietary vitamin E has shown some beneficial effects on drip loss, WHC and fatty acid profile of breast meat in broilers exposed to chronic heat stress (Mazur-Kuśnirek et al., 2019). Vitamin C supplementation (250 mg/kg) in a diet containing chromium picolinate (200 μg/kg) was able to attenuate the effects of heat stress on total fat and ash content, color variables and pH value of breast meat without affecting sensory traits and consumer acceptability (Saracila et al., 2021). Similarly, Imik et al. (2012b) claimed that vitamin C and α-lipoic acid can improve pH and color parameters (particularly the b* value) and reduce microbial counts in meat of broilers raised under chronic heat-stress conditions. Adversely, other studies reported that the dietary supplementation with vitamin C and E or vitamin C alone did not substantially improve breast meat quality traits when broilers were subjected to thermal stress (Attia et al., 2009; Zeferino et al., 2016).

5 Effect of management practices on meat quality: the case of caponization

In the previous sections, large part of the discussion has been devoted to the negative effects played by certain environmental factors such as heat stress on meat quality traits. Notwithstanding, it should also be considered that some management practices can provide relevant benefits in terms of meat quality. In this context, the ancient practice known as 'caponization' (i.e. castration) should deserve a mention because of the relevant impact of gonadectomy on meat chemical composition and quality parameters. In detail, caponization refers to the surgical removal of testes in chickens before they have reached the sexual maturity (Winter and Funk, 1960; Symeon et al., 2010). The origins of this practice are dated back to more than 2000 years ago: capons were known in ancient China as well as in ancient Greece, even if the merit of having first obtained the capon has been generally attributed to the Romans (Toussaint-Samat, 2009). Indeed, when the *Lex Faunia* of 162 BC prohibited them from eating fattened hens to save grain, Roman chicken farmers found a way around this law. They castrated the cockerels, which then grew up more than their normal size and put on weight, like eunuchs do. Likewise, they achieved a similar result by sterilizing pullets (Toussaint-Samat, 2009). Currently, capon production has been displaced by modern chicken production, even though it is still quite common in small farms around the world. The commercialization of capon meat is, however, a peculiarity of only a few European countries (e.g. Italy, France, Spain, Greece and Portugal) where it has been established in traditional meals (Symeon et al. 2012). Nevertheless, capon production being a potential resourceful complement to the poultry meat market, it has gained interests also in other geographical areas such as the USA (Calik, 2014). Even if it represents a market niche, it encloses a great growth potential since capon meat has peculiar sensory attributes which are increasingly appreciated by consumers (Amorim et al., 2016a). While in the European Union and USA only capon meat is consumed, ovariectomized pullets are preferred by many consumers in China (Cui et al., 2017). In the European Union, capon production has been ruled with the EC Regulation 543/2008 (European Commission, 2008), establishing that only castrated male chickens slaughtered at a minimum age of 140 days and fattened after castration for at least 77 days can be considered as capon. In commercial practice, capons are generally slaughtered at about 6 months of age. In the past, heavy breeds were used for capon production, which were frequently raised in the farmyard to obtain birds of liveweights ranging from 4 kg to 5 kg with a high carcass fatness (Mast et al., 1981). Nowadays, slow-growing chicken genotypes are used instead of fast-growing ones, yielding leaner carcasses that better fit the current consumer attitudes. Moreover, in some geographical areas, local breeds are preferred for capon production, representing also a good practice for breed conservation and exploitation (Tor et al., 2002; Muriel Duran, 2004).

The surgical removal of testes produces considerable changes in chicken metabolism affecting growth (Symeon et al., 2012), behavior attitude (Schlinger et al., 2001), development of comb and wattles, vocalization, aggressiveness and activity (Mast et al., 1981; Calik et al., 2014). In addition, castration deeply alters tissue and carcass composition (Miguel et al., 2008; Symeon et al., 2010; Sinanoglou et al., 2011), as well as meat chemical characteristics, with particular regard to technological and sensory properties, lipid content and fatty acid profile (Sirri et al., 2009; Amorim et al., 2016a,b; U-Chupaj et al., 2017). One of the main physiological and metabolic alterations associated with castration is the reduction of the circulating concentration of testosterone (Scanes and Johnson, 1984; Symeon et al., 2010). As a consequence, body weight at slaughter could be either similar (Chen et al., 2005; Symeon et al., 2010, 2012; Paudel et al., 2018) or increased (Lin et al., 2002; Rahman et al., 2004; Chen et al., 2006) compared to that of intact males according to the chicken genotype investigated. On the contrary, caponization negatively affects carcass yield, with the abdominal fat pad being much greater than in intact males. In addition, capons generally show heavier breast and lower thigh and drumstick yields (Tor et al., 2002; Muriel Duran, 2004; Miguel et al., 2008; Chen et al., 2010; Symeon et al., 2012). According to these authors, the reduction of sexual hormones impacts on the early development of breast muscle, while it negatively affects thigh and drumstick muscle growth. Caponization was also associated with significant modifications of muscle fiber diameter, area and density, which could explain the higher breast yield generally observed in capons than intact males. Moreover, gonadectomy has important implications on meat texture, likely related to the decreased muscle fiber diameter and area and the increased fiber density (Lin and Hsu, 2002; Cui et al., 2017). As reported in several studies, lower shear force and energy as well as textural attributes such as hardness, cohesiveness, gumminess, chewiness and Young's modulus were detected either in raw or in cooked capon meat (Chen, 2006; Sirri et al., 2009; Symeon et al., 2010; Calik et al., 2015; Cui et al., 2017; U-Chupaj et al., 2017). In accordance with the instrumental textural profile analysis, also the sensory attributes are deeply affected by caponization. In particular, capon meat obtained lower scores for both firmness and number of chews than intact male meat and the highest score for juiciness. In addition, a strong correlation among instrumental texture parameters and sensory attributes were found (U-Chupaj et al., 2017). The greater tenderness observed in capon meat does not seem related to its collagen content, which was similar in capon and intact males (Amorim et al., 2016a). Other authors argued that the increased amount of intramuscular lipids in capons could play a positive role on meat tenderness (Cui et al., 2017). Among the other technological properties of capon meat, only the color profile is significantly affected (Fig. 1). In general, capon meat shows higher lightness (L*) and yellowness (b*) and lower redness (a*) values if compared to that of intact males (Sirri et al., 2009; Symeon et al., 2010, 2012; Cui et al., 2017). Moreover, capons exhibit lower content of heme pigments. The role exerted by these pigments on meat color measurements was confirmed, as meat redness (a*) and lightness (L*), respectively, showed positive and

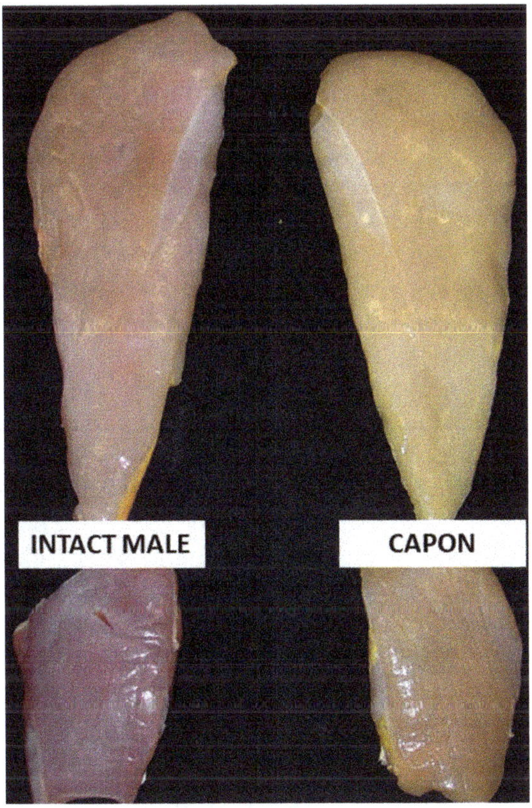

Figure 1 Visual appearance of breast meat from intact male and capon.

negative correlation with total heme pigments (Sirri et al., 2009). The higher yellowness (b*) value characterizing capon meat was associated with the greater percentage of intramuscular fat (Cui et al., 2017). Besides these effects on the technological properties, the most important metabolic effect of caponization is the increase in fat deposition capacity, considering abdominal (Cason et al., 1988; Symeon et al., 2012), subcutaneous and intramuscular fat, with respect to intact males at the same slaughter age and liveweight (Sirri et al., 2009; Symeon et al., 2010, 2012; Sinanoglou et al., 2011; Amorin et al., 2016; Cui et al., 2017). Although most of the abdominal fat is discarded during standard slaughtering conditions differently to other livestock species (e.g. pork), the intramuscular fat is of primary importance when it comes to poultry meat quality. The increased fat deposition capacity in response to castration has been related to different aspects. For instance, Chen et al. (2007, 2009) concluded that the decreased serum testosterone deriving from tests removal may positively regulate hepatic lipogenesis, as suggested by the upregulation of some genes involved in such metabolic function (Guo et al., 2015). In addition, also the modifications of behavioral traits in capons could play a role in fat deposition. Indeed, the energy expenditure for fighting and territorial protection is greatly reduced in capons than in roosters due to the lower levels of testosterone of the former, allowing a more efficient feed conversion and increased fat deposition (Calik, 2014). The modification of the growth hormone secretion due to caponization influences the energy metabolism as well (Pampori and Shapiro, 1994). Furthermore, the cholesterol content of breast and thigh meat is higher in capons than in cocks (Sirri et al., 2009). Chen et al. (2006) found higher concentrations of both blood cholesterol and triacylglycerol in capons than intact males and attributed this pattern to the increase of the hepatic lipogenic activity as a consequence of the surgical removal of their testicles.

The comparison of castrated with intact birds showed minor changes in the fatty acid profile of both breast and thigh meat. In particular, in some studies, the lipid profile of capon breast meat was characterized by higher saturated fatty acids (SFAs) (Sirri et al., 2009) and/or higher monounsaturated fatty acids (MUFAs) (Amorim et al., 2016b; Cui et al., 2017), whereas higher proportions of PUFA and MUFA and lower SFA content are commonly observed for thigh meat (Sirri et al., 2009; Sinanoglou et al., 2011). In general, the major changes in meat fatty acid composition in response to castration involve palmitic (C16:0), stearic (C18:0), oleic (C18:1) and arachidonic acid (C20:4). On the other hand, the moisture, protein and amino acid contents of the meat are only slightly influenced by the caponization (Sirri et al., 2009; Cui et al., 2017).

6 Conclusion and future trends in research

As summarized in the present work, nutrition and feeding strategies as well as heat stress and castration are important factors affecting meat quality traits in poultry species. Nutrition and feeding have shown great potential in modifying meat proximate composition, particularly the fatty acid profile and CLA content, improving the nutritional value of this product. Research in this field is currently focused on the use of more sustainable ω-3 PUFA sources such as microalgae and on the interpretation of the 'diet × genotype' interaction on fatty acid metabolism and deposition in chicken tissues. Although nutrition has proved to be an effective strategy for increasing the amount of antioxidant compounds in poultry meat (e.g. vitamin E and Se), oxidative stability still represents a great issue for the poultry meat industry. In this scenario, consumer's concerns regarding the use of synthetic compounds in animal nutrition is boosting the research interest on phytocompounds as potential substitutes for conventional antioxidants. Even if some beneficial effects of natural extracts have been delineated, their efficacy can be influenced by many factors (e.g. botanical species, method of extraction, storage conditions, dietary dosage and bird farming practices, etc.), and in general it seems to be lower than that of synthetic antioxidants. Moreover, the identification of nutritional strategies able to stimulate other endogenous antioxidant mechanisms of the muscle, such as those involving carnosine and anserine, can represent an alternative and valuable approach to mitigate oxidation problems. Future studies should be also aimed at unveiling the role of dietary amino acids as modulators of meat technological properties, including resistance to oxidation, and the optimal dietary amino acid profile that can guarantee high-quality standards for poultry meat.

The actual efficacy of nutritional strategies in mitigating the occurrence of breast abnormalities is questionable, and it seems to be more related to the modulation of growth pattern and/or slaughter yields rather than to a direct effect of the dietary treatment on muscle myopathies. These outcomes pose questions regarding the sustainability and viability of this type of approach, yet further stimulating researchers to identify novel interventions that can ameliorate myopathies occurrence without negatively affecting productive traits. Several nutritional approaches have also provided benefits with regard to the mitigation of heat-stress consequences on poultry meat quality. In this field, the adoption of more standardized experimental designs and procedures is advisable to ascertain the actual efficacy of the tested treatment and, consequently, to identify more practical, effective and sustainable solutions that could be put in place by poultry nutritionists during heat exposure. Finally, the physiological and metabolic alterations induced by caponization can deeply affect carcass and meat quality, with particular regard to lipid content, color profile, texture parameters and sensory attributes. Although it is considered a market niche, the valorization of capon production could represent an economic relevant complement for the poultry industry providing to the consumers a high-quality alternative to conventional chicken meat. In addition, capon production can be a potential strategy for the conservation of local chicken breeds and, from a scientific perspective, interesting insights regarding the muscle metabolism in poultry species could also be acquired evaluating the physiological responses triggered by caponization.

7 Where to look for further information

The following articles provide a good overview of the subject:

- Petracci, M., Mudalal, S., Soglia, F. and Cavani, C. (2015). 'Meat quality in fast-growing broiler chickens', *World Poult. Sci. J.*, 71(2), 363-374.
- Petracci, M. and Berri, C. (2017). *Poultry Quality Evaluation. Quality Attributes and Consumer Values*, Woodhead Publishing, Duxford, UK.

- Petracci, M., Soglia, F., Madruga, M., Carvalho, L., Ida, E. and Estévez, M. (2019). 'Wooden-breast, white striping, and spaghetti meat: causes, consequences and consumer perception of emerging broiler meat abnormalities', *Compr. Rev. Food Sci. Food Saf.*, 18(2), 565-583.

8 References

Abbasi, F., Samadi, F., Jafari, S. M., Ramezanpour, S. and Shams-Shargh, M. (2019). Production of omega-3 fatty acid-enriched broiler chicken meat by the application of nanoencapsultsed flaxseed oil prepared via ultrasonication, *J. Funct. Foods* 57, 373-381.

Abdullah, A. Y., Mahmoud, K. Z., Nusairat, B. M. and Qudsieh, R. I. (2010). Small intestinal histology, production parameters, and meat quality as influenced by dietary supplementation of garlic (Allium sativum) in broiler chicks, *Ital. J. Anim. Sci.* 9(4), e80.

Abidin, Z. and Khatoon, A. (2013). Heat stress in poultry and the beneficial effects of ascorbic acid (vitamin C) supplementation during periods of heat stress, *World Poult. Sci. J.* 69(1), 135-152.

Ahsan, U. and Cengiz, Ö. (2020). Restriction of dietary digestible lysine allowance in grower phase reduces the occurrence of white striping in broiler chickens, *Anim. Feed Sci. Technol.* 270, 114705.

Akbarian, A., Michiels, J., Degroote, J., Majdeddin, M., Golian, A. and De Smet, S. (2016). Association between heat stress and oxidative stress in poultry; mitochondrial dysfunction and dietary interventions with phytochemicals, *J. Anim. Sci. Biotechnol.* 7(1), 37.

Aksu, M. İ, İmik, H. and Karaoğlu, M. (2007). Influence of dietary sorghum (sorghum vulgare) and corn supplemented with methionine on cut-up pieces weights of broiler carcass and quality properties of breast and drumsticks meat, *Food Sci. Technol. Int.* 13(5), 361-367.

Albrecht, A., Herbert, U., Miskel, D., Heinemann, C., Braun, C., Dohlen, S., Zeitz, J. O., Eder, K., Saremi, B. and Kreyenschmidt, J. (2017). Effect of methionine supplementation in chicken feed on the quality and shelf life of fresh poultry meat, *Poult. Sci.* 96(8), 2853-2861.

Albrecht, A., Hebel, M., Heinemann, C., Herbert, U., Miskel, D., Saremi, B. and Kreyenschmidt, J. (2019). Assessment of meat quality and shelf life from broilers fed with different sources and concentrations of methionine, *J. Food Qual.* 2019, 1-10.

Alirezaei, M., Reza Gheisari, H., Reza Ranjbar, V. and Hajibemani, A. (2012). Betaine: a promising antioxidant agent for enhancement of broiler meat quality, *Br. Poult. Sci.* 53(5), 699-707.

Alnahhas, N., Berri, C., Chabault, M., Chartrin, P., Boulay, M., Bourin, M. C. and Le Bihan-Duval, E. (2016). Genetic parameters of white striping in relation to body weight, carcass composition, and meat quality traits in two broiler lines divergently selected for the ultimate pH of the pectoralis major muscle, *BMC Genet.* 17(1), 61.

Amorin, A., Rodrigues, S., Pereira, E., Valentim, R. and Teixeira, A. (2016). Effect of caponisation on physiochemical and sensory characteristics of chicken meat. *Animal* 10(6): 978-986.

Amorim, A., Rodrigues, S., Pereira, E., Valentim, R. and Teixeira, A. (2016a). Effect of caponisation on physicochemical and sensory characteristics of chickens, *Animal* 10(6), 978–986.

Amorim, A., Rodrigues, S., Pereira, E. and Teixeira, A. (2016b). Physicochemical composition and sensory quality evaluation of capon and rooster meat, *Poult. Sci.* 95(5), 1211–1219.

Attia, Y. A., Böhmer, B. M. and Roth-Maier, D. A. (2006). Responses of broiler chicks raised under constant relatively high am-bient temperature to enzymes, amino acid supplementations, or a high-nutrient diet, *Arch.Geflügelk.* 70(2), 80–91.

Attia, Y. A., Hassan, R. A. and Qota, E. M. A. (2009). Recovery from adverse effects of heat stress on slow-growing chicks in the tropics 1: effect of ascorbic acid and different levels of betaine, *Trop. Anim. Health Prod.* 41(5), 807–818.

Attia, Y. A., Hassan, R. A., Tag El-Din, A. E. and Abou-Shehema, B. M. (2011). Effect of ascorbic acid or increasing metabolizable energy level with or without supplementation of some essential amino acids on productive and physiological traits of slow-growing chicks exposed to chronic heat stress, *J. Anim. Physiol. Anim. Nutr. (Berl)* 95(6), 744–755.

Attia, Y. A., Al-Harthi, M. A. and Sh. Elnaggar, A. (2018). Productive, physiological and immunological responses of two broiler strains fed different dietary regimens and exposed to heat stress, *Ital. J. Anim. Sci.* 17(3), 686–697.

Attia, Y. A. and Hassan, S. S. (2017). Broiler tolerance to heat stress at various dietary protein/energy levels, *Europ. Poult. Sci.* 81. 10.1399/eps.2017.171.

Ayazi, M. (2014). The effect of dietary glutamine supplementation on performance and blood parameter, carcass characteristics, quality and characteristics meat of broiler chickens under continuous heat stress condition, *Int. J. Farming Allied Sci.* 3(12), 1234–1242.

Azad, M. A. K., Kikusato, M., Sudo, S., Amo, T. and Toyomizu, M. (2010a). Time course of ROS production in skeletal muscle mitochondria from chronic heat-exposed broiler chicken, *Comp. Biochem. Physiol. A Mol. Integr. Physiol.* 157(3), 266–271.

Azad, M. A. K., Kikusato, M., Maekawa, T., Shirakawa, H. and Toyomizu, M. (2010b). Metabolic characteristics and oxidative damage to skeletal muscle in broiler chickens exposed to chronic heat stress, *Comp. Biochem. Physiol. A Mol. Integr. Physiol.* 155(3), 401–406.

Baéza, E., Chartrin, P., Gigaud, V., Tauty, S., Meteau, K., Lessire, M. and Berri, C. (2013). Effects of dietary enrichment with n-3 fatty acids on the quality of raw and processed breast meat of high and low growth rate chickens, *Br. Poult. Sci.* 54(2), 190–198.

Baéza, E., Guillier, L. and Petracci, M. (2022). Review: production factors affecting poultry carcass and meat quality attributes, *Animal* 16(Suppl. 1), 100331. 10.1016/j.animal.2021.100331.

Baldi, G., Soglia, F. and Petracci, M. (2020). Valorization of meat by-products. In: Galanakis, C. (Ed.), *Food Waste Recovery. Processing Technologies, Industrial Techniques, and Applications*, Academic Press, Duxford, UK, pp. 419–443.

Baldi, G., Soglia, F. and Petracci, M. (2021). Spaghetti Meat abnormality in broilers: current understanding and future research directions, *Front. Physiol.* 12, 684497.

Baker, D. H. (2006). Comparative species utilization and toxicity of sulfur amino acids, *J. Nutr.* 136(6), 1670S–1675S.

Bami, M. K., Afsharmanesh, M., Salarmoini, M. and Ebrahimnejad, H. (2021). Effects of selenium-chitosan on growth performance, carcass traits, meat quality, and blood indices of broiler chickens, *Livest. Sci.* 250, 104562.

Belloir, P., Lessire, M., Lambert, W., Corrent, E., Berri, C. and Tesseraud, S. (2019). Changes in body composition and meat quality in response to dietary amino acid provision in finishing broilers, *Animal* 13(5), 1094–1102.

Belloir, P., Méda, B., Lambert, W., Corrent, E., Juin, H., Lessire, M. and Tesseraud, S. (2017). Reducing the CP content in broiler feeds: impact on animal performance, meat quality and nitrogen utilization, *Animal* 11(11), 1881–1889.

Berri, C., Besnard, J. and Relandeau, C. (2008). Increasing dietary lysine increases final pH and decreases drip loss of broiler breast meat, *Poult. Sci.* 87(3), 480–484.

Bodle, B. C., Alvarado, C., Shirley, R. B., Mercier, Y. and Lee, J. T. (2018). Evaluation of different dietary alterations in their ability to mitigate the incidence and severity of woody breast and white striping in commercial male broilers, *Poult. Sci.* 97(9), 3298–3310.

Boerboom, G., van Kempen, T., Navarro-Villa, A. and Pérez-Bonilla, A. (2018). Unraveling the cause of white striping in broilers using metabolomics, *Poult. Sci.* 97(11), 3977–3986.

Bölükbaşi, S. C. (2006). Effect of dietary conjugated linoleic acid (CLA) on broiler performance, serum lipoprotein content, muscle fatty acid composition and meat quality during refrigerated storage, *Br. Poult. Sci.* 47(4), 470–476.

Bonos, E., Kasapidou, E., Kargopoulos, A., Karampampas, A., Christaki, E., Florou-Paneri, P. and Nikolakakis, I. (2016). Spirulina as a functional ingredient in broiler chicken diets, *S. Afr. J. Anim. Sci.* 46(1), 94–102.

Boschetti, E., Bordoni, A., Meluzzi, A., Castellini, C., Dal Bosco, A. and Sirri, F. (2016). Fatty acid composition of chicken breast meat is dependent on genotype-related variation of FADS1 and FADS2 gene expression and desaturating activity, *Animal* 10(4), 700–708.

Botsoglou, N. A., Fletouris, D. J., Florou-Paneri, P., Christaki, E. and Spais, A. B. (2003). Inhibition of lipid oxidation in long-term frozen stored chicken meat by dietary oregano essential oil and α-tocopheryl acetate supplementation, *Food Res. Int.* 36(3), 207–213.

Bou, R., Guardiola, F., Barroeta, A. C. and Codony, R. (2005). Effect of dietary fat sources and zinc and selenium supplements on the composition and consumer acceptability of chicken meat, *Poult. Sci.* 84(7), 1129–1140.

Bou, R., Guardiola, F., Grau, A., Grimpa, S., Manich, A., Barroeta, A. and Codony, R. (2001). Influence of dietary fat source, α-tocopherol, and ascorbic acid supplementation on sensory quality of dark chicken meat, *Poult. Sci.* 80(6), 800–807.

Bou, R., Guardiola, F., Tres, A., Barroeta, A. C. and Codony, R. (2004). Effect of dietary fish oil, alpha-tocopheryl acetate, and zinc supplementation on the composition and consumer acceptability of chicken meat, *Poult. Sci.* 83(2), 282–292.

Brake, J., Balnave, D. and Dibner, J. J. (1998). Optimum dietary arginine: lysine ratio for broiler chickens is altered during heat stress in association with changes in intestinal uptake and dietary sodium chloride, *Br. Poult. Sci.* 39(5), 639–647.

Briens, M., Mercier, Y., Rouffineau, F., Mercerand, F. and Geraert, P. A. (2014). 2-Hydroxy-4-methylselenobutanoic acid induces additional tissue selenium enrichment in broiler chickens compared with other selenium sources, *Poult. Sci.* 93(1), 85–93.

Briens, M., Mercier, Y., Rouffineau, F., Vacchina, V. and Geraert, P. A. (2013). Comparative study of a new organic selenium source v. seleno-yeast and mineral selenium sources on muscle selenium enrichment and selenium digestibility in broiler chickens, *Br. J. Nutr.* 110(4), 617–624.

Calik, J. (2014). Capon production - breeding stock, rooster castration and rearing methods, and meat quality - a review, *Ann. Anim. Sci.* 14(4), 769–777.

Calik, J., Poltowicz, K., Świątkiewicz, S., Krawczyk, J. and Nowak, J. (2015). Effect of caponization on meat quality of Greenleg Partridge cockerels, *Ann. Anim. Sci.* 15(2), 541–553.

Cason, J. A., Fletcher, D. L. and Burke, W. H. (1988). Research note: effects of caponization on broiler growth, *Poult. Sci.* 67(6), 979–981.

Cavani, C., Petracci, M., Trocino, A. and Xiccato, G. (2009). Advances in research on poultry and rabbit meat quality, *Ital. J. Anim. Sci.* 8(sup2), 741–750.

Cemin, H. S., Vieira, S. L., Stefanello, C., Kindlein, L., Ferreira, T. Z. and Fireman, A. K. (2018). Broiler responses to increasing selenium supplementation using Zn-L-selenomethionine with special attention to breast myopathies, *Poult. Sci.* 97(5), 1832–1840.

Chamruspollert, M., Pesti, G. M. and Bakalli, R. I. (2004). Influence of temperature on the arginine and methionine requirements of young broiler chicks, *J. Appl. Poult. Res.* 13(4), 628–638.

Chen, K. L., Chi, W. T. and Chiou, P. W. S. (2005). Caponization and testosterone implantation effects on blood lipid and lipoprotein profile in male chickens, *Poult. Sci.* 84(4), 547–552.

Chen, K. L., Chi, W. T., Chu, C., Chen, R. S. and Chiou, P. W. (2007). Effect of caponization and testosterone implantation on hepatic lipids and lipogenic enzymes in male chickens, *Poult. Sci.* 86(8), 1754–1759.

Chen, K. L., Hsieh, T. Y. and Chiou, P. W. S. (2006). Caponization effects on growth performance and lipid metabolism in Taiwan country chicken cockerels nchu.edu.t w, *Asian-Australas. J. Anim. Sci.* 19(3), 438–443.

Chen, K. L., Lee, T. Y., Chen, T. W. and Chiou, P. W. (2009). Effect of caponization and different exogenous androgen on hepatic lipid and beta-oxidase of male chickens, *Poult. Sci.* 88(5), 1033–1039.

Chen, T. T., Huang, C. C., Lee, T. Y., Lin, K. J., Chang, C. C. and Chen, K. L. (2010). Effect of caponization and exogenous androgen implantation on muscle characteristics of male chickens, *Poult. Sci.* 89(3), 558–563.

Cheng, K., Niu, Y., Zheng, X. C., Zhang, H., Chen, Y. P., Zhang, M., Huang, X. X., Zhang, L. L., Zhou, Y. M. and Wang, T. (2016). A comparison of natural (D-α-tocopherol) and synthetic (DL-α-tocopherol acetate) vitamin E supplementation on the growth performance, meat quality and oxidative status of broilers, *Asian-Australas. J. Anim. Sci.* 29(5), 681–688.

Choi, I. H., Park, W. Y. and Kim, Y. J. (2010). Effects of dietary garlic powder and α-tocopherol supplementation on performance, serum cholesterol levels, and meat quality of chicken, *Poult. Sci.* 89(8), 1724–1731.

Cinar, M., Yildirim, E., Yigit, A. A., Yalcinkaya, I., Duru, O., Kisa, U. and Atmaca, N. (2014). Effects of dietary supplementation with vitamin C and vitamin E and their combination on growth performance, some biochemical parameters, and oxidative stress induced by copper toxicity in broilers, *Biol. Trace Elem. Res.* 158(2), 186–196.

Conde-Aguilera, J. A., Cholet, J. C. G., Lessire, M., Mercier, Y., Tesseraud, S. and van Milgen, J. (2016). The level and source of free-methionine affect body composition and breast muscle traits in growing broilers, *Poult. Sci.* 95(10), 2322–2331.

Corzo, A., Moran Jr., E. T. and Hoehler, D. (2003). Arginine need of heavy broiler males: applying the ideal protein concept, *Poult. Sci.* 82(3), 402–407.

Couloigner, F., Jlali, M., Briens, M., Rouffineau, F., Geraert, P. A. and Mercier, Y. (2015). Selenium deposition kinetics of different selenium sources in muscle and feathers of broilers, *Poult. Sci.* 94(11), 2708–2714.

Cruz, R. F. A., Vieira, S. L., Kindlein, L., Kipper, M., Cemin, H. S. and Rauber, S. M. (2017). Occurrence of white striping and wooden breast in broilers fed grower and finisher diets with increasing lysine levels, *Poult. Sci.* 96(2), 501–510.

Cui, X., Liu, R., Cui, H., Zhao, G., Zheng, M., Li, Q., Liu, J., Liu, Z. and Wen, J. (2017). Effects of caponization and ovariectomy on objective indices related to meat quality in chickens, *Poult. Sci.* 96(3), 770–777.

Daghir, N. J. (2009). Nutritional strategies to reduce heat stress in broilers and broiler breeders, *Lohmann Inf.* 44(1), 6–15.

Dai, S. F., Gao, F., Xu, X. L., Zhang, W. H., Song, S. X. and Zhou, G. H. (2012). Effects of dietary glutamine and gamma-aminobutyric acid on meat colour, pH, composition, and water-holding characteristic in broilers under cyclic heat stress, *Br. Poult. Sci.* 53(4), 471–481.

Dal Bosco, A., Mattioli, S., Cartoni Mancinelli, A., Cotozzolo, E. and Castellini, C. (2021). Extensive rearing systems in poultry production: the right chicken for the right farming system. A review of twenty years of scientific research in Perugia university, Italy, *Animals (Basel)* 11(5), 1281.

Dal Bosco, A., Mugnai, C., Ruggeri, S., Mattioli, S. and Castellini, C. (2012). Fatty acid composition of meat and estimated indices of lipid metabolism in different poultry genotypes reared under organic system, *Poult. Sci.* 91(8), 2039–2045.

Daneshyar, M., Ghandkanlo, M. A., Bayeghra, F. S., Farhangpajhoh, F. and Aghaei, M. (2011). Effects of dietary turmeric supplementation on plasma lipoproteins, meat quality and fatty acid composition in broilers, *S. Afr. J. Anim. Sci.* 41(4), 420–428.

de Faria, D. E., Borges Cam, D. M., Alfonso-To, K. A., Serpa Viei, B., Rosa, P. S., Vaz, A. M., Macari, M. and Furlan, R. L. (2007). Protein levels for heat-exposed broilers: performance, nutrients digestibility, and energy and protein metabolism, *Int. J. Poult. Sci.* 6(3), 187–194.

Du, M. and Ahn, D. U. (2002). Effect of dietary conjugated linoleic acid on the growth rate of live birds and on the abdominal fat content and quality of broiler meat, *Poult. Sci.* 81(3), 428–433.

Dukare, S., Mir, N. A., Mandal, A. B., Dev, K., Begum, J., Rokade, J. J., Biswas, A., Tyagi, P. K., Tyagi, P. K. and Bhanja, S. K. (2021). A comparative study on the antioxidant status, meat quality, and mineral deposition in broiler chicken fed dietary Nano zinc viz-a-viz inorganic zinc, *J. Food Sci. Technol.* 58(3), 834–843.

European Food Standards Agency (EFSA) (2012). Scientific opinion on the safety and efficacy of selenium as a feed additive for all animal species. *EFSA J.* 10(7): 2778.

European Food Standards Agency (EFSA) (2014). Scientific opinion on the safety and efficacy of DL-selenomethionine as a feed additive for all animal species. *EFSA J.* 12(2): 3567.

El-Senousey, H. K., Chen, B., Wang, J. Y., Atta, A. M., Mohamed, F. R. and Nie, Q. H. (2018). Effects of dietary vitamin C, vitamin E, and alpha-lipoic acid supplementation on the antioxidant defense system and immune-related gene expression in broilers exposed to oxidative stress by dexamethasone, *Poult. Sci.* 97(1), 30–38.

Estévez, M. (2015). Oxidative damage to poultry: from farm to fork, *Poult. Sci.* 94(6), 1368–1378.

Estévez, M., Geraert, P. A., Liu, R., Delgado, J., Mercier, Y. and Zhang, W. (2020). Sulphur amino acids, muscle redox status and meat quality: more than building blocks-Invited review, *Meat Sci.* 163, 108087.

European Commission (2008). Commission Regulation (EC) N°543/2008 of 16 June 2008 Laying down Detailed Rules for the Application of Council Regulation (EC) N° 1234/2007 as Regards the Marketing Standards for Poultry Meat, Off. J., p. L 157, 46-87.

Farhoomand, P. and Checaniazer, S. (2009). Effects of graded levels of dietary fish oil on the yield and fatty acid composition of breast meat in broiler chickens, *J. Appl. Poult. Res.* 18(3), 508–513.

Fellenberg, M. A. and Speisky, H. (2006). Antioxidants: their effects on broiler oxidative stress and its meat oxidative stability, *World Poult. Sci. J.* 62(1), 53–70.

Fernandes, J. I. M., Sakamoto, M. I., Peiter, D. C., Gottardo, E. T. and Tellini, C. (2013). Vitamin E: vitamin C relationship on meat quality of broiler chicken submitted to pre-slaughter stress, *Arq. Bras. Med. Vet. Zootec.* 65(1), 294–300.

Florou-Paneri, P., Giannenas, I., Christaki, E., Govaris, A. and Botsoglou, N. (2006). Performance of chickens and oxidative stability of the produced meat as affected by feed supplementation with oregano, vitamin C, vitamin E and their combinations, *Arch. Geflugelkunde* 70(5), 232–239.

Fu, Q., Leng, Z. X., Ding, L. R., Wang, T., Wen, C. and Zhou, Y. M. (2016). Complete replacement of supplemental DL-methionine by betaine affects meat quality and amino acid contents in broilers, *Anim. Feed Sci. Technol.* 212, 63–69.

Furlan, R. L., de Faria Filho, D. E., Rosa, P. S. and Macari, M. (2004). Does low-protein diet improve broiler performance under heat stress conditions?, *Braz. J. Poult. Sci.* 6(2), 71–79.

Gatrell, S. K., Kim, J., Derksen, T. J., O'Neil, E. V. and Lei, X. G. (2015). Creating ω-3 fatty-acid-enriched chicken using defatted green microalgal biomass, *J. Agric. Food Chem.* 63(42), 9315–9322.

Ghazalah, A. A., Abd-Elsamee, M. O. and Ali, A. M. (2008). Influence of dietary energy and poultry fat on the response of broiler chicks to heat therm, *Int. J. Poult. Sci.* 7(4), 355–359.

Ghazalah, A. A., Soliman, A. Z. M., Boulous, N. Z. Samia and Mobarez, M. (2007). Effect of using dry fat on performance, nutrients digestibility, carcass traits and blood constituents of broiler chicks, *Egypt Poult. Sci.* 27, 363–382.

Givens, D. I. and Gibbs, R. A. (2008). Current intakes of EPA and DHA in European populations and the potential of animal-derived foods to increase them, Symposium on 'How can the n-3 content of the diet be improved?, Proc. Nutr. Soc. 67(3), 273–280.

Gonzalez-Rivas, P. A., Chauhan, S. S., Ha, M., Fegan, N., Dunshea, F. R. and Warner, R. D. (2020). Effects of heat stress on animal physiology, metabolism, and meat quality: a review, *Meat Sci.* 162, 108025.

Gratta, F., Birolo, M., Sacchetto, R., Radaelli, G., Xiccato, G., Ballarin, C., Bertotto, D., Piccirillo, A., Petracci, M., Maertens, L. and Trocino, A. (2019). Effect of feed restriction timing on live performance, breast myopathy occurrence, and muscle fiber degeneration in 2 broiler chicken genetic lines, *Poult. Sci.* 98(11), 5465–5476.

Gregory, N. G. (2010). How climatic changes could affect meat quality, *Food Res. Int.* 43(7), 1866–1873.

Griffin, J. R., Moraes, L., Wick, M. and Lilburn, M. S. (2018). Onset of white striping and progression into wooden breast as defined by myopathic changes underlying pectoralis major growth. Estimation of growth parameters as predictors for stage of myopathy progression, *Avian Pathol.* 47(1), 2-13.

Guardia, S., Lessire, M., Corniaux, A., Métayer-Coustard, S., Mercerand, F., Tesseraud, S., Bouvarel, I. and Berri, C. (2014). Short-term nutritional strategies before slaughter are effective in modulating the final pH and color of broiler breast meat, *Poult. Sci.* 93(7), 1764-1773.

Guo, X., Nan, H., Shi, D., Zhou, J., Wan, Y., Zhou, B., Geng, Z., Chen, X. and Jiang, R. (2015). Effects of caponization on growth, carcass, and meat characteristics and the mRNA expression of genes related to lipid metabolism in roosters of a Chinese indigenous breed, *Czech J. Anim. Sci.* 60(7), 327-333.

Habibian, M., Ghazi, S. and Moeini, M. M. (2016). Effects of dietary selenium and vitamin E on growth performance, meat yield, and selenium content and lipid oxidation of breast meat of broilers reared under heat stress, *Biol. Trace Elem. Res.* 169(1), 142-152.

Harsini, S. G., Habibiyan, M., Moeini, M. M. and Abdolmohammadi, A. R. (2012). Effects of dietary selenium, vitamin E, and their combination on growth, serum metabolites, and antioxidant defense system in skeletal muscle of broilers under heat stress, *Biol. Trace Elem. Res.* 148(3), 322-330.

Haug, A., Eich-Greatorex, S., Bernhoft, A., Wold, J. P., Hetland, H., Christophersen, O. A. and Sogn, T. (2007). Effect of dietary selenium and omega-3 fatty acids on muscle composition and quality in broilers, *Lipids Health Dis.* 6(1), 29.

Herchi, W., Bahashwan, S., Trabelsi, H., Boukhchina, S., Kallel, H., Rochut, S. and Pepe, C. (2014). Changes in proximate composition and oil characteristics during flaxseed development, *Grasas Aceites* 65(2), e022. 10.3989/gya.097713.

Hou, L., Qiu, H., Sun, P., Zhu, L., Chen, F. and Qin, S. (2020). Selenium-enriched *Saccharomyces cerevisiae* improves the meat quality of broiler chickens via activation of the glutathione and thioredoxin systems, *Poult. Sci.* 99(11), 6045-6054.

Houshmand, M., Azhar, K., Zulkifli, I., Bejo, M. H. and Kamyab, A. (2012). Effects of prebiotic, protein level, and stocking density on performance, immunity, and stress indicators of broilers, *Poult. Sci.* 91(2), 393-401.

Imik, H., Atasever, M. A., Urcar, S., Ozlu, H., Gumus, R. and Atasever, M. (2012a). Meat quality of heat stress exposed broilers and effect of protein and vitamin E, *Br. Poult. Sci.* 53(5), 689-698.

Imik, H., Ozlu, H., Gumus, R. E. C. E. P., Atasever, M. A., Urcar, S. and Atasever, M. (2012b). Effects of ascorbic acid and α-lipoic acid on performance and meat quality of broilers subjected to heat stress, *Br. Poult. Sci.* 53(6), 800-808.

Jakubcova, Z., Mareš, P., Zeman, L., Horký, P., Jurí-ková, T., Mlček, J., Balla, Š, Kalhotka, L., Mrkvicová, E. and Sochor, J. (2014). Influence of garlic extract on antioxidant status of chicken, *Potr. S. J. F. Sci.* 8(1), 315-320.

Jankowski, J., Kubinska, M. and Zdunczyk, Z. (2014). Nutritional and immunomodulatory function of methionine in poultry diets-a review, *Ann. Anim. Sci.* 14(1), 17-32.

Jankowski, J., Zduńczyk, Z., Mikulski, D., Juśkiewicz, J., Pomianowski, J. F. and Zduńczyk, P. (2017). Fatty acid profile, oxidative stability and sensory quality of breast meat from turkeys fed diets with graded levels of flaxseed oil for different periods of time, *Anim. Prod. Sci.* 58(6), 1164-1174.

Jastrebski, S. F., Lamont, S. J. and Schmidt, C. J. (2017). Chicken hepatic response to chronic heat stress using integrated transcriptome and metabolome analysis, *PLoS ONE* 12(7), e0181900.

Jia, W., Rogiewicz, A., Bruce, H. L. and Slominski, B. A. (2010). Feeding flaxseed enhances deposition of omega-3 fatty acids in broiler meat portions in different manner, *Can. J. Anim. Sci.* 90(2), 203–206.

Jiang, W., Nie, S., Qu, Z., Bi, C. and Shan, A. (2014). The effects of conjugated linoleic acid on growth performance, carcass traits, meat quality, antioxidant capacity, and fatty acid composition of broilers fed corn dried distillers grains with solubles, *Poult. Sci.* 93(5), 1202–1210.

Jiao, P., Guo, Y., Yang, X. and Long, F. (2010). Effects of dietary arginine and methionine levels on broiler carcass traits and meat quality, *J. Anim. Vet. Adv.* 9(11), 1546–1551.

Jlali, M., Gigaud, V., Metayer-Coustard, S., Sellier, N., Tesseraud, S., Le Bihan-Duval, E. and Berri, C. (2012). Modulation of glycogen and breast meat processing ability by nutrition in chickens: effect of crude protein level in 2 chicken genotypes, *J. Anim. Sci.* 90(2), 447–455.

Juniper, D. T., Phipps, R. H. and Bertin, G. (2011). Effect of dietary supplementation with selenium-enriched yeast or sodium selenite on selenium tissue distribution and meat quality in commercial-line turkeys, *Animal* 5(11), 1751–1760.

Kakhki, R. A. M., Bakhshalinejad, R. and Shafiee, M. (2016). Effect of dietary zinc and α-tocopheryl acetate on broiler performance, immune responses, antioxidant enzyme activities, minerals and vitamin concentration in blood and tissues of broilers, *Anim. Feed Sci. Technol.* 221, 12–26.

Kennedy, O. B., Stewart-Knox, B. J., Mitchell, P. C. and Thurnham, D. I. (2005). Vitamin E supplementation, cereal feed type and consumer sensory perceptions of poultry meat quality, *Br. J. Nutr.* 93(3), 333–338.

Khan, A. Z., Kumbhar, S., Liu, Y., Hamid, M., Pan, C., Nido, S. A., Parveen, F. and Huang, K. (2018). Dietary supplementation of selenium-enriched probiotics enhances meat quality of broiler chickens (*Gallus gallus domesticus*) raised under high ambient temperature, *Biol. Trace Elem. Res.* 182(2), 328–338.

Kirkpinar, F. I. G. E. N., Ünlü, H. B., Serdaroğlu, M. and Turp, G. Y. (2014). Effects of dietary oregano and garlic essential oils on carcass characteristics, meat composition, colour, pH and sensory quality of broiler meat, *Br. Poult. Sci.* 55(2), 157–166.

Kiyma, Z., Küçükyilmaz, K., Çetinkaya, M. Atalay, H., Ateş, A., Gürsel, F. E. and Bozkurt, M. (2017). Effect of lavender (Lavandula stoechas) essential oil on growth performance, carcass characteristics, meat quality and antioxidant status of broilers, *S. Afr. J. Anim. Sci.* 47(2), 178–186.

Kumar, F., Tyagi, P. K., Mir, N. A., Dev, K., Begum, J., Biswas, A., Sheikh, S. A., Tyagi, P. K., Sharma, D., Sahu, B., Biswas, A. K., Deo, C. and Mandal, A. B. (2020). Dietary flaxseed and turmeric is a novel strategy to enrich chicken meat with long chain ω-3 polyunsaturated fatty acids with better oxidative stability and functional properties, *Food Chem.* 305, 125458.

Kuttappan, V. A., Brewer, V. B., Apple, J. K., Waldroup, P. W. and Owens, C. M. (2012a). Influence of growth rate on the occurrence of white striping in broiler breast fillets, *Poult. Sci.* 91(10), 2677–2685.

Kuttappan, V. A., Goodgame, S. D., Bradley, C. D., Mauromoustakos, A., Hargis, B. M., Waldroup, P. W. and Owens, C. M. (2012b). Effect of different levels of dietary vitamin E (DL-α-tocopherol acetate) on the occurrence of various degrees of white striping on broiler breast fillets, *Poult. Sci.* 91(12), 3230–3235.

Kuttappan, V. A., Manangi, M., Bekker, M., Chen, J. and Vazquez-Anon, M. (2021). Nutritional intervention strategies using dietary antioxidants and organic trace minerals to reduce the incidence of Wooden breast and other carcass quality defects in broiler birds, *Front. Physiol.* 12, 663409.

Lackner, J., Albrecht, A., Mittler, M., Marx, A., Kreyenschmidt, J., Hess, V. and Sauerwein, H. (2021). Effect of feeding histidine and β-alanine on carnosine concentration, growth performance, and meat quality of broiler chickens, *Poult. Sci.* 100(11), 101393.

Leroy, F. and Praet, I. (2015). Meat traditions: the co-evolution of humans and meat, *Appetite* 90, 200–211. 10.1016/j.appet.2015.03.014.

Leroy, F. and De Smet, S. (2019). Meat in the human diet: a biosocial perspective. In: Lorenzo, J., Munekata, P., Barba, F. and Toldrá, F. (Eds), *More than Beef, Pork and Chicken–the Production, Processing, and Quality Traits of Other Sources of Meat for Human Diet*, Springer, Basel, Switzerland, pp. 1–19.

Leskovec, J., Levart, A., Svete, A. N., Peric, L., Đukic Stojcic, M., Žikic, D., Salobir, J. and Rezar, V. (2018). Effects of supplementation with α-tocopherol, ascorbic acid, selenium, or their combination in linseed oil-enriched diets on the oxidative status in broilers, *Poult. Sci.* 97(5), 1641–1650.

Leskovec, J., Levart, A., Perić, L., Đukić Stojčić, M., Tomović, V., Pirman, T., Salobir, J. and Rezar, V. (2019). Antioxidative effects of supplementing linseed oil-enriched diets with α-tocopherol, ascorbic acid, selenium, or their combination on carcass and meat quality in broilers, *Poult. Sci.* 98(12), 6733–6741.

Li, J. L., Zhang, L., Yang, Z. Y., Zhang, Z. Y., Jiang, Y., Gao, F. and Zhou, G. H. (2018). Effects of different selenium sources on growth performance, antioxidant capacity and meat quality of local Chinese Subei chickens, *Biol. Trace Elem. Res.* 181(2), 340–346.

Liao, X., Lu, L., Li, S., Liu, S., Zhang, L., Wang, G., Li, A. and Luo, X. (2012). Effects of selenium source and level on growth performance, tissue selenium concentrations, antioxidation, and immune functions of heat-stressed broilers, *Biol. Trace Elem. Res.* 150(1–3), 158–165.

Lin, C. Y. and Hsu, J. C. (2002). Effects of surgical caponization on growth performance, fiber diameter and some physical properties of muscles in Taiwan country chicken cockerels, *Asian-Australas. J. Anim. Sci.* 15(3), 401–405.

Lin, H., Decuypere, E. and Buyse, J. (2006). Acute heat stress induces oxidative stress in broiler chickens, *Comp. Biochem. Physiol. A Mol. Integr. Physiol.* 144(1), 11–17.

Liu, G. Q., Zong, K., Zhang, L. L. and Cao, S. (2010). Dietary methionine affect meat quality and myostatin gene exon 1 region methylation in skeletal muscle tissues of broilers, *Agric. Sci. China* 9(9), 1338–1346.

Liu, W., Yuan, Y., Sun, C., Balasubramanian, B., Zhao, Z. and An, L. (2019). Effects of dietary betaine on growth performance, digestive function, carcass traits, and meat quality in indigenous yellow-feathered broilers under long-term heat stress, *Animals (Basel)* 9(8), 506.

Livingston, M. L., Landon, C., Barnes, H. J. and Brake, J. (2019a). White striping and wooden breast myopathies of broiler breast muscle is affected by time-limited feeding, genetic background, and egg storage, *Poult. Sci.* 98(1), 217–226.

Livingston, M. L., Ferket, P. R., Brake, J. and Livingston, K. A. (2019b). Dietary amino acids under hypoxic conditions exacerbates muscle myopathies including wooden breast and white stripping, *Poult. Sci.* 98(3), 1517–1527.

Loetscher, Y., Kreuzer, M. and Messikommer, R. E. (2013). Oxidative stability of the meat of broilers supplemented with rosemary leaves, rosehip fruits, chokeberry pomace, and entire nettle, and effects on performance and meat quality, *Poult. Sci.* 92(11), 2938–2948.

Long, S. F., Kang, S., Wang, Q. Q., Xu, Y. T., Pan, L., Hu, J. X., Li, M. and Piao, X. S. (2018). Dietary supplementation with DHA-rich microalgae improves performance, serum composition, carcass trait, antioxidant status, and fatty acid profile of broilers, *Poult. Sci.* 97(6), 1881–1890.

Lopez-Ferrer, S., Baucells, M. D., Barroeta, A. C., Galobart, J. and Grashorn, M. A. (2001a). n-3 enrichment of chicken meat. 2. Use of precursors of long-chain polyunsaturated fatty acids: linseed oil, *Poult. Sci.* 80(6), 753–761.

Lopez-Ferrer, S., Baucells, M. D., Barroeta, A. C. and Grashorn, M. A. (2001b). n-3 enrichment of chicken meat. 1. Use of very long-chain fatty acids in chicken diets and their influence on meat quality: fish oil, *Poult. Sci.* 80(6), 741–752.

Lu, Z., He, X., Ma, B., Zhang, L., Li, J., Jiang, Y., Zhou, G. and Gao, F. (2017). Chronic heat stress impairs the quality of breast-muscle meat in broilers by affecting redox status and energy-substance metabolism, *J. Agric. Food Chem.* 65(51), 11251–11258.

Lu, Z., He, X., Ma, B., Zhang, L., Li, J., Jiang, Y., Zhou, G. and Gao, F. (2018). Serum metabolomics study of nutrient metabolic variations in chronic heat-stressed broilers, *Br. J. Nutr.* 119(7), 771–781.

Lukanov, H., Genchev, A. and Ribarski, S. (2015). Effect of feed supplementation with garlic power on meat productivity and meat quality traits of classic Ross 308 male hybrid chickens, *Trakia J. Sci.* 13(1), 66–76.

Mack, S., Bercovici, D., De Groote, G., Leclercq, B., Lippens, M., Pack, M., Schutte, J. B. and Van Cauwenberghe, S. (1999). Ideal amino acid profile and dietary lysine specification for broiler chickens of 20 to 40 days of age, *Br. Poult. Sci.* 40(2), 257–265.

Marion, J. E. and Woodroof, J. G. (1963). The fatty acid composition of breast, thigh, and skin tissues of chicken broilers as influenced by dietary fats, *Poult. Sci.* 42(5), 1202–1207.

Marković, R., Glišić, M., Bošković, M. and Baltic, M. (2017). New scientific challenges-the possibilities of using selenium in poultry nutrition and impact on meat quality. In IOP Conference Series. Earth and Environmental Science (Vol. 85, No. 1, p. 012032), IOP Publishing.

Marković, R., Ćirić, J., Starčević, M., Šefer, D. and Baltić, M. Ž. (2018). Effects of selenium source and level in diet on glutathione peroxidase activity, tissue selenium distribution, and growth performance in poultry, *Anim. Health Res. Rev.* 19(2), 166–176.

Mast, M. G., Jordan, H. C. and Macneil, J. H. (1981). The effect of partial and complete caponization on growth rate, yield, and selected physical and sensory attributes of cockerels, *Poult. Sci.* 60(8), 1827–1833.

Mazur-Kuśnirek, M., Antoszkiewicz, Z., Lipiński, K., Kaliniewicz, J., Kotlarczyk, S. and Żukowski, P. (2019). The effect of polyphenols and vitamin E on the antioxidant status and meat quality of broiler chickens exposed to high temperature, *Arch. Anim. Nutr.* 73(2), 111–126.

Meloche, K. J., Fancher, B. I., Emmerson, D. A., Bilgili, S. F. and Dozier, W. A. (2018a). Effects of quantitative nutrient allocation on myopathies of the pectoralis major muscles in broiler chickens at 32, 43, and 50 days of age, *Poult. Sci.* 97(5), 1786–1793.

Meloche, K. J., Fancher, B. I., Emmerson, D. A., Bilgili, S. F. and Dozier, W. A. (2018b). Effects of reduced dietary energy and amino acid density on pectoralis major myopathies in broiler chickens at 36 and 49 days of age1, *Poult. Sci.* 97(5), 1794–1807.

Meloche, K. J., Fancher, B. I., Emmerson, D. A., Bilgili, S. F. and Dozier, W. A. (2018c). Effects of reduced digestible lysine density on myopathies of the pectoralis major muscles in broiler chickens at 48 and 62 days of age, *Poult. Sci.* 97(9), 3311-3324.

Mercier, Y., Gatellier, P., Viau, M., Remignon, H. and Renerre, M. (1998). Effect of dietary fat and vitamin E on colour stability and on lipid and protein oxidation in turkey meat during storage, *Meat Sci.* 48(3-4), 301-318.

Miguel, J. A., Ciria, J., Asenjo, B. and Calvo, J. L. (2008). Effect of caponisation on growth and on carcass and meat characteristics in Castellana Negra native Spanish chickens, *Animal* 2(2), 305-311.

Mikulski, D., Jankowski, J., Zdunczyk, Z., Wróblewska, M., Sartowska, K. and Majewska, T. (2009). The effect of selenium source on performance, carcass traits, oxidative status of the organism, and meat quality of turkeys, *J. Anim. Feed Sci.* 18(3), 518-530.

Mir, N. A., Rafiq, A., Kumar, F., Singh, V. and Shukla, V. (2017). Determinants of broiler chicken meat quality and factors affecting them: a review, *J. Food Sci. Technol.* 54(10), 2997-3009.

Mirghelenj, S. A., Golian, A. and Taghizadeh, V. (2009). Enrichment of chicken meat with long chain omega-3 fatty acids through dietary fish oil, *Res. J. Biol. Sci.* 4(5), 604-608.

Mohammadi, A. Ghazanfari, S. and Sharifi, S. D. (2019). Comparative effects of dietary organic, inorganic or nano-selenium complexes on meat quality and selenium deposition in muscles of broiler chickens. *Livest. Sci.* 226: 21-30.

Morales-Barrera, J. E., Gonzalez-Alcorta, M. J., Castillo-Dominguez, R. M., Prado-Rebolledo, O. F., Hernandez-Velasco, X., Menconi, A., Tellez, G., Hargis, B. M. and Carrillo-Dominguez, S. (2013). Fatty acid deposition on broiler meat in chickens supplemented with tuna oil, *Food Nutr. Sci.* 04(9), 16-20 10.4236/fns.2013.49A1003.

Mujahid, A. (2011). Nutritional strategies to maintain efficiency and production of chickens under high environmental temperature, *J. Poult. Sci.* 48(3), 145-154, 10.2141/jpsa.010115.

Mujahid, A., Yoshiki, Y., Akiba, Y. and Toyomizu, M. (2005). Superoxide radical production in chicken skeletal muscle induced by acute heat stress, *Poult. Sci.* 84(2), 307-314.

Mujahid, A., Akiba, Y. and Toyomizu, M. (2009). Olive oil-supplemented diet alleviates acute heat stress-induced mitochondrial ROS production in chicken skeletal muscle, *Am. J. Physiol. Regul. Integr. Comp. Physiol.* 297(3), R690-R698.

Muriel Duran, A. (2004). The effect of caponization on production indices and carcass and meat characteristics in free-range Extremena Azul chickens, *Span. J. Agric. Res.* 2(2), 211-216.

National Research Council (1994). *Nutrient Requirements of Poultry*. 9th rev ed., National Academies Press, Washington, DC.

Naylor, R. L., Hardy, R. W., Bureau, D. P., Chiu, A., Elliott, M., Farrell, A. P., Forster, I., Gatlin, D. M., Goldburg, R. J., Hua, K. and Nichols, P. D. (2009). Feeding aquaculture in an era of finite resources, *Proc. Natl. Acad. Sci. U. S. A.* 106(36), 15103-15110.

N'dri, A. L., Mignon-Grasteau, S., Sellier, N., Beaumont, C. and Tixierboichard, M. (2007). Interactions between the naked neck gene, sex, and fluctuating ambient temperature on heat tolerance, growth, body composition, meat quality, and sensory analysis of slow growing meat-type broilers, *Livest. Sci.* 110(1-2), 33-45.

Niu, Z. Y., Min, Y. N. and Liu, F. Z. (2018). Dietary vitamin E improves meat quality and antioxidant capacity in broilers by upregulating the expression of antioxidant enzyme genes, *J. Appl. Anim. Res.* 46(1), 397-401.

Nutautaitė, M., Alijošius, S., Bliznikas, S., Šašytė, V., Vilienė, V., Pockevičius, A. and Racevičiūtė-Stupelienė, A. (2020). Effect of betaine, a methyl group donor, on broiler chicken growth performance, breast muscle quality characteristics, oxidative status and amino acid content, *Ital. J. Anim. Sci.* 19(1), 621–629.

Oliveira, T. F. B., Rivera, D. F. R., Mesquita, F. R., Braga, H., Ramos, E. M. and Bertechini, A. G. (2014). Effect of different sources and levels of selenium on performance, meat quality, and tissue characteristics of broilers, *J. Appl. Poult. Res.* 23(1), 15–22.

Oviedo-Rondón, E. O. and Córdova-Noboa, H. A. (2020). The potential of guanidino acetic acid to reduce the occurrence and severity of broiler muscle myopathies, *Front. Physiol.* 11, 909.

Pampori, N. A. and Shapiro, B. H. (1994). Testicular regulation of sexual dimorphism in the ultradian profiles of circulating growth hormone in the chicken, *Eur. J. Endocrinol.* 131(3), 313–318.

Park, J. H. and Kim, I. H. (2019). The effects of betaine supplementation in diets containing different levels of crude protein and methionine on the growth performance, blood components, total tract nutrient digestibility, excreta noxious gas emission, and meat quality of the broiler chickens, *Poult. Sci.* 98(12), 6808–6815.

Paudel, T. P., Poudel, B. and Neupane, D. (2018). Evaluation of caponization on growth and meat quality parameters of dual-purpose chicken, *Int. J. Appl. Sci. Biotechnol.* 6(4), 339–343.

Petracci, M., Mudalal, S., Soglia, F. and Cavani, C. (2015). Meat quality in fast growing broiler chickens, *World Poult. Sci. J.* 71(2), 363–374.

Petracci, M., Soglia, F., Madruga, M., Carvalho, L., Ida, E. and Estévez, M. (2019). Wooden-breast, white striping, and spaghetti meat: causes, consequences and consumer perception of emerging broiler meat abnormalities, *Compr. Rev. Food Sci. Food Saf.* 18(2), 565–583.

Petrolli, T. G., Petrolli, O. J., Pereira, A. S. C., Zotti, C., Romani, J., Villani, R., Leite, F. and Zanandréa, F. (2019). Effects of the dietary supplementation with a microalga extract on broiler performance and fatty-acid meat profile, *Braz. J. Poult. Sci.* 21(3), 10.1590/1806-9061-2018-0958.

Pitargue, F. M., Kim, J. H., Goo, D., Delos Reyes, J. B. and Kil, D. Y. (2019). Effect of vitamin E sources and inclusion levels in diets on growth performance, meat quality, alpha-tocopherol retention, and intestinal inflammatory cytokine expression in broiler chickens, *Poult. Sci.* 98(10), 4584–4594.

Pompeu, M. A., Cavalcanti, L. F. L. and Toral, F. L. B. (2018). Effect of vitamin E supplementation on growth performance, meat quality, and immune response of male broiler chickens: a meta-analysis, *Livest. Sci.* 208, 5–13.

Rahman, M. M., Ali, Md. Y., Houssain, M. M., Islam, M. A. and Khondaker, M. E. A. (2004). Effect of caponization on body weight, hematological traits and blood cholesterol concentration of nara chicken, *Int. J. Poult. Sci.* 3(4), 284–286.

Renaudeau, D., Collin, A., Yahav, S., de Basilio, V., Gourdine, J. L. and Collier, R. J. (2012). Adaptation to hot climate and strategies to alleviate heat stress in livestock production, *Animal* 6(5), 707–728.

Ri, C. S., Jiang, X. R., Kim, M. H., Wang, J., Zhang, H., Wu, S., Bontempo, V. and Qi, G. (2017). Effects of dietary oregano powder supplementation on the growth performance, antioxidant status and meat quality of broiler chicks, *Ital. J. Anim. Sci.* 16(2), 246–252.

Rymer, C. and Givens, D. I. (2005). n− 3 fatty acid enrichment of edible tissue of poultry: a review, *Lipids* 40(2), 121–130.

Rymer, C., Gibbs, R. A. and Givens, D. I. (2010). Comparison of algal and fish sources on the oxidative stability of poultry meat and its enrichment with omega-3 polyunsaturated fatty acids, *Poult. Sci.* 89(1), 150–159.

Sachs, N. J., Hampton, A. R., Foster, K. K., Pechanec, M. Y., Henderson, J. D., King, A. J. and Mienaltowski, M. J. (2019). The effects of an alternative diet regimen with natural methionine ingredients on white striping breast myopathy in broiler chickens, *Poult. Sci.* 98(1), 413–421.

Sahin, K., Orhan, C., Tuzcu, M., Sahin, N., Hayirli, A., Bilgili, S. and Kucuk, O. (2016). Lycopene activates antioxidant enzymes and nuclear transcription factor systems in heat-stressed broilers, *Poult. Sci.* 95(5), 1088–1095.

Salas-Salvado, J., Marquez-Sandoval, F. and Bullo, M. (2006). Conjugated linoleic acid intake in humans: a systematic review focusing on its effect on body composition, glucose, and lipid metabolism, *Crit. Rev. Food Sci. Nutr.* 46(6), 479–488.

Saleh, A. A., Ragab, M. M., Ahmed, E. A. M., Abudabos, A. M. and Ebeid, T. A. (2018). Effect of dietary zinc-methionine supplementation on growth performance, nutrient utilization, antioxidative properties and immune response in broiler chickens under high ambient temperature, *J. Appl. Anim. Res.* 46(1), 820–827.

Salim, H. M., Lee, H. R., Jo, C. R., Lee, S. and Lee, B. (2011). Influence of various levels of organic zinc on the live performance, meat quality attributes, and sensory properties of broiler chickens, *Anim. Res.* 31(2), 207–214.

Saracila, M., Panaite, T. D., Mironeasa, S. and Untea, A. E. (2021). Dietary supplementation of some antioxidants as attenuators of heat stress on chicken meat characteristics, *Agriculture* 11(7), 638.

Scanes, C. G. (2016). Biology of stress in poultry with emphasis on glucocorticoids and the heterophil to lymphocyte ratio, *Poult. Sci.* 95(9), 2208–2215.

Scanes, C. G. and Johnson, A. L. (1984). Failure of castration to prevent the prepubescent decline in the circulating concentration of growth hormone in the domestic fowl, *Gen. Comp. Endocrinol.* 53(3), 398–401.

Schlinger, B. A., Soma, K. K. and Saldanha, C. (2001). Advances in avian behavioral endocrinology, *Auk* 118(2), 283–289.

Schmid, A., Collomb, M., Sieber, R. and Bee, G. (2006). Conjugated linoleic acid in meat and meat products: a review, *Meat Sci.* 73(1), 29–41.

Seidavi, A., Belali, M., Elghandour, M. M. Y., Adegbeye, M. J. and Salem, A. Z. M. (2020). Potential impacts of dietary inclusion of green tea (Camellia sinensis L.) in poultry feeding: a review, *Agrofor. Syst.* 94(4), 1161–1170.

Shakeri, M., Cottrell, J. J., Wilkinson, S., Le, H. H., Suleria, H. A. R., Warner, R. D. and Dunshea, F. R. (2019). Growth performance and characterization of meat quality of broiler chickens supplemented with betaine and antioxidants under cyclic heat stress, *Antioxidants (Basel)* 8(9), 336.

Shakeri, M., Cottrell, J. J., Wilkinson, S., Le, H. H., Suleria, H. A. R., Warner, R. D. and Dunshea, F. R. (2020a). Dietary betaine reduces the negative effects of cyclic heat exposure on growth performance, blood gas status and meat quality in broiler chickens, *Agriculture* 10(5), 176.

Shakeri, M., Oskoueian, E., Le, H. H. and Shakeri, M. (2020b). Strategies to combat heat stress in broiler chickens: unveiling the roles of selenium, vitamin E and vitamin C, *Vet. Sci.* 7(2), 71.

Sheldon, B. W., Curtis, P. A., Dawson, P. L. and Ferket, P. R. (1997). Effect of dietary vitamin E on the oxidative stability, flavor, color, and volatile profiles of refrigerated and frozen turkey breast meat, *Poult. Sci.* 76(4), 634–641.

Sihvo, H. K., Immonen, K. and Puolanne, E. (2014). Myodegeneration with fibrosis and regeneration in the pectoralis major muscle of broilers, *Vet. Pathol.* 51(3), 619–623.

Sihvo, H. K., Lindén, J., Airas, N., Immonen, K., Valaja, J. and Puolanne, E. (2017). Wooden breast myodegeneration of pectoralis major muscle over the growth period in broilers, *Vet. Pathol.* 54(1), 119–128.

Silva, V. A., Clemente, A. H. S., Nogueira, B. R. F., de Carvalho, A. C., de Freitas, L. F. V. B., Ramos, A. L. S. and Bertechini, A. G. (2019). Supplementation of selenomethionine at different ages and levels on meat quality, tissue deposition, and selenium retention in broiler chickens, *Poult. Sci.* 98(5), 2150–2159.

Simopoulos, A. P. (2002). The importance of the ratio of omega-6/omega-3 essential fatty acids, *Biomed. Pharmacother.* 56(8), 365–379.

Simopoulos, A. P. (2004). Omega-6/omega-3 essential fatty acid ratio and chronic diseases, *Food Rev. Int.* 20(1), 77–90.

Simopoulos, A. P. (2016). An increase in the omega-6/omega-3 fatty acid ratio increases the risk for obesity, *Nutrients* 8(3), 128.

Sinanoglou, V. J., Mantis, F., Miniadis-Meimaroglou, S., Symeon, G. K. and Bizelis, I. A. (2011). Effects of caponisation on lipid and fatty acid composition of intramuscular and abdominal fat of medium-growth broilers, *Br. Poult. Sci.* 52(3), 310–317.

Sirri, F., Tallarico, N., Meluzzi, A. and Franchini, A. (2003). Fatty acid composition and productive traits of broiler fed diets containing conjugated linoleic acid, *Poult. Sci.* 82(8), 1356–1361.

Sirri, F., Bianchi, M., Petracci, M. and Meluzzi, A. (2009). Influence of partial and complete caponization on chicken meat quality, *Poult. Sci.* 88(7), 1466–1473.

Sirri, F., Castellini, C., Bianchi, M., Petracci, M., Meluzzi, A. and Franchini, A. (2011). Effect of fast-, medium-and slow-growing strains on meat quality of chickens reared under the organic farming method, *Animal* 5(2), 312–319.

Sirri, F., Maiorano, G., Tavaniello, S., Chen, J., Petracci, M. and Meluzzi, A. (2016). Effect of different levels of dietary zinc, manganese, and copper from organic or inorganic sources on performance, bacterial chondronecrosis, intramuscular collagen characteristics, and occurrence of meat quality defects of broiler chickens, *Poult. Sci.* 95(8), 1813–1824.

Skřivan, M., Marounek, M., Englmaierová, M. and Skřivanová, E. (2012). Influence of dietary vitamin C and selenium, alone and in combination, on the composition and oxidative stability of meat of broilers, *Food Chem.* 130(3), 660–664.

Soglia, F., Zampiga, M., Baldi, G., Malila, Y., Thanatsang, K. V., Srimarut, Y., Tatiyaborworntham, N., Unger, O., Klamchuen, A., Laghi, L., Petracci, M. and Sirri, F. (2021). Lysine depletion during different feeding phases: effects on growth performances and meat quality of broiler chickens, *Animals (Basel)* 11(6), 1499.

Surai, P. F. (2002). Selenium in poultry nutrition 1. Antioxidant properties, deficiency and toxicity, *World Poult. Sci. J.* 58(3), 333–347.

Symeon, G. K., Mantis, F., Bizelis, I., Kominakis, A. and Rogdakis, E. (2010). Effects of caponization on growth performance, carcass composition, and meat quality of medium growth broilers, *Poult. Sci.* 89(7), 1481–1489.

Symeon, G. K., Mantis, F., Bizelis, I., Kominakis, A. and Rogdakis, E. (2012). Effects of caponization on growth performance, carcass composition and meat quality of males of a layer line, *Animal* 6(12), 2023–2030.

Szymczyk, B., Pisulewski, P. M., Szczurek, W. and Hanczakowski, P. (2001). Effects of conjugated linoleic acid on growth performance, feed conversion efficiency, and subsequent carcass quality in broiler chickens, *Br. J. Nutr.* 85(4), 465–473.

Tan, G. Y., Yang, L., Fu, Y. Q., Feng, J. H. and Zhang, M. H. (2010). Effects of different acute high ambient temperatures on function of hepatic mitochondrial respiration, antioxidative enzymes, and oxidative injury in broiler chickens, *Poult. Sci.* 89(1), 115–122.

Temim, S., Chagneau, A. M., Guillaumin, S., Michel, J., Peresson, R., Geraert, P. A. and Tesseraud, S. (1999). Effects of chronic heat exposure and protein intake on growth performance, nitrogen retention and muscle development in broiler chickens, *Reprod. Nutr. Dev.* 39(1), 145–156.

Temim, S., Chagneau, A. M., Guillaumin, S., Michel, J., Peresson, R. and Tesseraud, S. (2000). Does excess dietary protein improve growth performance and carcass characteristics in heat-exposed chickens?, *Poult. Sci.* 79(3), 312–317.

Tor, M., Estany, J., Villalba, D., Molina, E. and Cubiló, D. (2002). Comparison of carcass composition by parts and tissues between cocks and capons, *Anim. Res.* 51(5), 421–431.

Toussaint-Samat, M. (2009). *A History of Food*, Wiley and Sons Ltd, Chichester, UK.

Trocino, A., Piccirillo, A., Birolo, M., Radaelli, G., Bertotto, D., Filiou, E., Petracci, M. and Xiccato, G. (2015). Effect of genotype, gender and feed restriction on growth, meat quality and the occurrence of white striping and wooden breast in broiler chickens, *Poult. Sci.* 94(12), 2996–3004.

U-Chupaj, J., Malila, Y., Gamonpilas, C., Kijroongrojana, K., Petracci, M., Benjakul, S. and Visessanguan, W. (2017)Differences in textural properties of cooked caponized and broiler chicken breast meat, *Poult. Sci.* 96(7), 2491–2500.

Uzum, M. H. and Toplu, H. O. (2013). Effects of stocking density and feed restriction on performance, carcass, meat quality characteristics and some stress parameters in broilers under heat stress, *Rev. Méd Vét* 164(12), 546–554.

Wang, R. R., Pan, X. J. and Peng, Z. Q. (2009). Effects of heat exposure on muscle oxidation and protein functionalities of pectoralis majors in broilers, *Poult. Sci.* 88(5), 1078–1084.

Wang, J., Clark, D. L., Jacobi, S. K. and Velleman, S. G. (2020). Effect of vitamin E and omega-3 fatty acids early posthatch supplementation on reducing the severity of wooden breast myopathy in broilers, *Poult. Sci.* 99(4), 2108–2119.

Wang, D., Huang, H., Zhou, L., Li, W., Zhou, H., Hou, G., Liu, J. and Hu, L. (2015). Effects of dietary supplementation with turmeric rhizome extract on growth performance, carcass characteristics, antioxidant capability, and meat quality of Wenchang broiler chickens, *Ital. J. Anim. Sci.* 14(3), 3870.

Watanabe, G., Kobayashi, H., Shibata, M., Kubota, M., Kadowaki, M. and Fujimura, S. (2017). Reduction of dietary lysine increases free glutamate content in chicken meat and improves its taste, *Anim. Sci. J.* 88(2), 300–305.

Wen, C., Jiang, X. Y., Ding, L. R., Wang, T. and Zhou, Y. M. (2017). Effects of dietary methionine on growth performance, meat quality and oxidative status of breast muscle in fast-and slow-growing broilers, *Poult. Sci.* 96(6), 1707–1714.

Wen, C., Chen, Y., Leng, Z., Ding, L., Wang, T. and Zhou, Y. (2019). Dietary betaine improves meat quality and oxidative status of broilers under heat stress, *J. Sci. Food Agric.* 99(2), 620–623.

Whitehead, C. C. and Keller, T. (2003). An update on ascorbic acid in poultry, *World Poult. Sci. J.* 59(2), 161–184.

Windhorst, H. W. (2017). Dynamics and pattern of global poultry-meat production. In: Petracci, M. and Berri, C. (Eds), *Poultry Quality Evaluation*, Academic Press, Duxford, UK, pp. 1–25.

Winter, A. R. and Funk, E. M. (1960). *Poultry Science and Practice*, J.B. Lippincott Publishers Co., New York.

Yahav, S. (2009). Alleviating heat stress in domestic fowl: different strategies, *World Poult. Sci. J.* 65(4), 719–732.

Yang, L., Tan, G. Y., Fu, Y. Q., Feng, J. H. and Zhang, M. H. (2010). Effects of acute heat stress and subsequent stress removal on function of hepatic mitochondrial respiration, ROS production and lipid peroxidation in broiler chickens, *Comp. Biochem. Physiol. C Toxicol. Pharmacol.* 151(2), 204–208.

Yesilbag, D., Eren, M., Agel, H., Kovanlikaya, A. and Balci, F. (2011). Effects of dietary rosemary, rosemary volatile oil and vitamin E on broiler performance, meat quality and serum SOD activity, *Br. Poult. Sci.* 52(4), 472–482.

Yu, D. G., Namgung, N., Kim, J. H., Won, S. Y., Choi, W. J. and Kil, D. Y. (2021). Effects of stocking density and dietary vitamin C on performance, meat quality, intestinal permeability, and stress indicators in broiler chickens, *J. Anim. Sci. Technol.* 63(4), 815–826.

Zaboli, G., Huang, X., Feng, X. and Ahn, D. U. (2019). How can heat stress affect chicken meat quality?–A review, *Poult. Sci.* 98(3), 1551–1556.

Zampiga, M., Calini, F. and Sirri, F. (2021a). Importance of feed efficiency for sustainable intensification of chicken meat production: implications and role for amino acids, feed enzymes and organic trace minerals, *World Poult. Sci. J.* 77(3), 639–659.10.1080/00439339.2021.1959277.

Zampiga, M., Laghi, L., Zhu, C., Cartoni Mancinelli, A., Mattioli, S. and Sirri, F. (2021b). Breast muscle and plasma metabolomics profile of broiler chickens exposed to chronic heat stress conditions, *Animal* 15(7), 100275.

Zampiga, M., Laghi, L., Petracci, M., et al. (2018). Effect of dietary arginine to lysine ratios on productive performance, meat quality, plasma and muscle metabolomics profile in fast-growing broiler chickens, *J. Anim. Sci. Biotechnol.* 9(1), 1–14.

Zampiga, M., Soglia, F., Baldi, G., Petracci, M., Strasburg, G. M. and Sirri, F. (2020). Muscle abnormalities and meat quality consequences in modern turkey hybrids, *Front. Physiol.* 11, 554.

Zampiga, M., Soglia, F., Petracci, M., Meluzzi, A. and Sirri, F. (2019). Effect of different arginine-to-lysine ratios in broiler chicken diets on the occurrence of breast myopathies and meat quality attributes, *Poult. Sci.* 98(6), 2691–2697.

Zdanowska-Sąsiadek, Ż., Michalczuk, M., Damaziak, K., Niemiec, J., Poławska, E., Gozdowski, D. and Różańska, E. (2016). Effect of vitamin E supplementation on growth performance and chicken meat quality, *Eur. Poult. Sci.* 80(2), 1–14.

Zeferino, C. P., Komiyama, C. M., Pelícia, V. C., Fascina, V. B., Aoyagi, M. M., Coutinho, L. L., Sartori, J. R. and Moura, A. S. A. M. T. (2016). Carcass and meat quality traits of chickens fed diets concurrently supplemented with vitamins C and E under constant heat stress, *Animal* 10(1), 163–171.

Zepeda-Velazquez, A. P., Ramirez-Perez, A. H., Ambriz-Vilchis, V., Salinas-Martinez, J. A., Alvarez-Alonso, C., Plata-Perez, G., Ortega-Meneses, F. and Angeles-Hernandez, J. C. (2020). Effect of selenium supplementation on productive performance and antioxidant status of broilers under heat stress: a meta-analysis and a meta-regression, *Anim. Prod. Sci.* 61(3), 208–214.

Zhai, W., Schilling, M. W., Jackson, V., Peebles, E. D. and Mercier, Y. (2016). Effects of dietary lysine and methionine supplementation on Ross 708 male broilers from 21 to 42 days of age (II): breast meat quality, *J. Appl. Poult. Res.* 25(2), 212–222.

Zhang, M., Dunshea, F. R., Warner, R. D., DiGiacomo, K., Osei-Amponsah, R. and Chauhan, S. S. (2020). Impacts of heat stress on meat quality and strategies for amelioration: a review, *Int. J. Biometeorol.* 64(9), 1613–1628.

Zhao, J. P., Zhao, G. P., Jiang, R. R., Zheng, M. Q., Chen, J. L., Liu, R. R. and Wen, J. (2012). Effects of diet-induced differences in growth rate on metabolic, histological, and meat-quality properties of 2 muscles in male chickens of 2 distinct broiler breeds, *Poult. Sci.* 91(1), 237–247.

Zulkifli, I., Akmal, A. F., Soleimani, A. F., Hossain, M. A. and Awad, E. A. (2018). Effects of low-protein diets on acute phase proteins and heat shock protein 70 responses, and growth performance in broiler chickens under heat stress condition, *Poult. Sci.* 97(4), 1306–1314.

Part 2

Individual quality attributes: sensory, nutrition and health

Chapter 4

Advances in understanding color development in poultry meat

Ranjith Ramanathan and Frank Kiyimba, Oklahoma State University, USA; Surendranath Suman, University of Kentucky, USA; and Gretchen Mafi, Oklahoma State University, USA

1 Introduction

In the next decade, the demand for poultry meat will surpass other meats as an animal protein source (OECD-FAO, 2021). Meat quality plays an important role in consumer acceptance, especially in countries where meat is retailed in chilled conditions. Consumers use visual clues to make an assessment of quality. Beef is a red meat and color deviations leads to economic losses. For example, the US beef industry loses annually \$3.73 billion due to discoloration (Ramanathan et al., 2022). Although poultry is considered white meat due to lower myoglobin content, visual color impacts purchasing decisions. A recent report indicated that in 2020, the global loss due to chicken discoloration was \$7.4 billion (Maia Research Analysis, 2020). Poultry meat has more

http://dx.doi.org/10.19103/AS.2022.0112.04

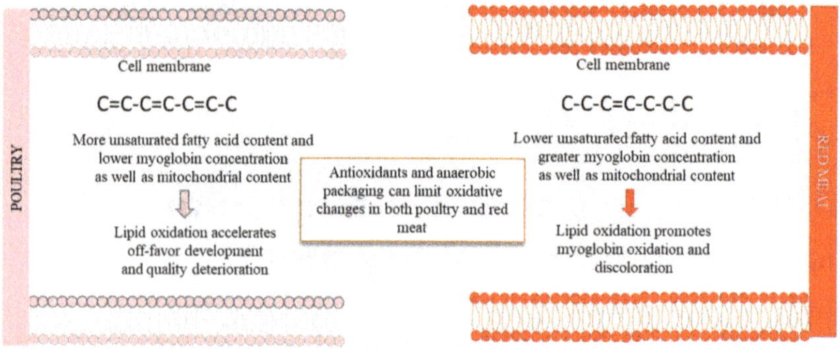

Figure 1 Comparison of quality changes in poultry and red meat.

unsaturated fatty acids than other species; hence oxidative changes and quality deterioration are faster (Figure 1). Various factors such as sex of the bird, age, breed, processing conditions, pH, and temperature can influence poultry meat color (Fletcher, 2002; Kuttappan et al., 2016; Hearn et al., 2021). The current chapter will discuss factors affecting poultry meat discoloration and recent advances in color development.

2 Myoglobin forms and poultry meat color

Myoglobin, hemoglobin, and cytochrome are the important heme proteins that contribute to meat color. In well-bled birds, the contribution of hemoglobin to color is less. Myoglobin is the primary heme protein that is responsible for meat color. Avians, in general, have a high metabolic rate and more glycolytic fiber type to provide energy during flight. Glycolytic muscles have lower myoglobin content than oxidative fibers. Hence, the concentration of myoglobin in poultry meat is less compared with beef. The myoglobin concentration varies depending on the anatomical location and the body's need for oxygen storage *in vivo*. For example, the thigh has more myoglobin than breast muscles (Nishida and Nishida, 1985). Hence, thigh muscles will have a darker color, while breast muscles will be pale. The 6-week chicken leg has 32.7 mg myoglobin/100 g tissue, and breast *pectoralis profundus* muscle has negligible to no myoglobin. Domesticated birds have lower myoglobin content than wild birds. Unlike beef, the color within each cut-up part of poultry is very inconsistent (Fletcher, 2002). For example, the lightness of breast fillets demonstrates a wide range with a package.

Depending on the ligand attached and the iron valence state, myoglobin can exist in three redox forms: deoxymyoglobin, oxymyoglobin, and metmyoglobin (AMSA, 2012). Predominant oxymyoglobin in poultry imparts a light pink color, while deoxymyoglobin gives a darker color. Oxidation of ferrous form leads to the formation of gray metmyoglobin. The use of modified atmospheric packaging has allowed the industry to modify gas compositions within packages such as high oxygen or carbon monoxide (McMillin, 2017). The binding of carbon monoxide to ferrous heme results in the formation of stable pink color. Both carboxymyoglobin and oxymyoglobin can impart a pinkish-red color in poultry, but carboxymyoglobin is very stable due to strong binding with heme.

3 Myoglobin primary structure and oxidation properties

Myoglobin is a single polypeptide consisting of 153 amino acids. The composition of amino acids and their arrangement around the heme ring influence myoglobin oxidation. Several studies have noted the importance of histidine number in oxymyoglobin oxidation. Bovine myoglobin has 13 histidines, while chicken and turkey have 9 histidines (Joseph et al., 2011). Auto-oxidation of chicken and turkey myoglobin is greater than bovine myoglobin (Yin et al., 2011). Similar to other myoglobin species, the secondary lipid oxidation products such as 4-hydroxy-2-nonenal and hexenal increased chicken and turkey myoglobin oxidation compared with control oxymyoglobin without lipid oxidation products (Maheswarappa et al., 2009; Yin et al., 2011). Research comparing lipid oxidation-induced oxidation in ratite myoglobins indicates that autoxidation and HNE-induced oxidation were greater in ostrich oxymyoglobin than in emu and beef oxymyoglobins (Nair et al., 2014). Covalent binding of HNE with histidine alters the tertiary structure of myoglobin, leading to heme release and oxymyoglobin oxidation.

4 Biochemical factors affecting poultry meat color

In the past, most of the research related to poultry meat color was related to pale soft exudative (PSE) or dark firm dry (DFD) conditions. Improvement in management practices, to a certain extent, has limited the occurrence of PSE and DFD conditions. The lightness of the breast muscle is an important factor that influences purchasing decisions and cooked meat color. Compared to red meat, limited studies have characterized the role of biochemical factors such as metmyoglobin reducing activity and oxygen consumption in poultry meat color.

4.1 Metmyoglobin reducing activity and oxygen consumption

In fresh meat, metmyoglobin reducing activity and oxygen consumption are two intrinsic biochemical processes that influence color development (Seyfert et al., 2007; Ramanathan and Mancini, 2018; Ramanathan et al., 2020).

Metmyoglobin reducing activity is the ability of the postmortem muscle to limit the accumulation of metmyoglobin by enzymatic, non-enzymatic, and mitochondria-mediated pathways (Arihara et al., 1995; Tang et al., 2005b; Denzer et al., 2020). In the enzymatic pathway, NADH-dependent cytochrome b_5 reductase transfers electron from NADH to an electron acceptor, while in non-enzymatic metmyoglobin reduction, electron carriers such as methylene blue transfer an reductase transfers an electron from NADH or ascorbate to metmyoglobin. In the mitochondria-mediated pathway, electrons available between complex III and IV are transferred to metmyoglobin by cytochrome b5. Although these pathways are well characterized in beef, limited studies have characterized their role in poultry meat color.

Oxygen consumption is the ability of the postmortem muscle to consume oxygen. Mitochondria, myoglobin, microbes, and protein/lipid oxidation contribute to oxygen consumption. In postmortem muscle, mitochondria and oxygen-consuming enzymes are the main competitor for available oxygen (Tang et al., 2005a; Ramanathan et al., 2013). A greater pH can promote mitochondrial activity and enzymes involved in oxygen consumption (English et al., 2016). In general, poultry meat has fewer type I fibers and has more glycolytic (less myoglobin and mitochondria) fibers. Hence the contribution of mitochondria to meat color may be less than in beef. In addition, a greater than normal pH of meat decreases light reflectance (Ramanathan et al., 2022). Water holding capacity will also be greater at higher pH, thus limiting the amount of free water on the surface of poultry meat (Swatland, 2008). Hence, a combination of biochemical (oxygen consumption) and physical (lack of muscle fiber shrinkage) factors promote darker color in high pH poultry meat.

4.2 Lipid oxidation

Lipid oxidation is the oxidation of unsaturated fatty acids in membranes and tissue (Faustman et al., 2010). Lipid oxidation generates primary and secondary oxidation products. Primary lipid oxidation products are short-lived, while secondary products such as aldehydes and ketones are long-lived, reactive, and bind with proteins. Both pre-and postharvest processes can increase lipid oxidation. Especially in ground poultry products, grinding can release prooxidants from cells and increase oxidative changes (Hearn et al., 2021). To impart a light pink color, ground chicken and turkey are often packaged in a high-oxygen environment to improve their appearance. Both oxygen-rich environment and grinding accelerate lipid oxidation. Hence, antioxidants are added to ground poultry products to improve their appearance and organoleptic qualities. Antioxidants are also included in the diet to promote oxidative stability of lipids (Leal et al., 2018). Turkey meat is more prone to lipid oxidation than other species because of its poor ability to store vitamin

E in membranes. Vitamin E is a phenolic fat-soluble antioxidant that limits free radicals such as singlet oxygen and peroxyl-induced lipid oxidation. The protective effect of vitamin E is attributed to its ability to stabilize the membrane. Oxidative changes also can decrease flavor profiles and consumer acceptance of cooked meat products.

4.3 Muscle myopathies and poultry color

Genetic selection to have a faster growth rate and larger breast muscles make birds very susceptible to oxidative stress (Carvalho et al., 2017). White striping and woody breast are two important myopathies that affect poultry quality. Although the mechanistic basis for white striping and woody breast conditions is not clear, the rapid growth rate is attributed to histological changes. White striations are observed parallel to muscle fibers of breast, thigh, and other tender meats in white striping conditions, while tougher breasts are observed in woody breast conditions (Kuttappan et al., 2016). Both conditions have more impact on processed meat characteristics and visual appearance than myoglobin redox changes.

5 Application of metabolomics and proteomics in poultry color research

Interaction of proteins and metabolites are involved in attaining ultimate pH and in biochemical processes such as lipid oxidation, metmyoglobin reducing activity, and oxygen consumption. Metabolomic is a study of small molecules such as sugars, amino acids, acids, and fatty acids. The application of the metabolomics approach helps to get a snapshot of molecular changes occurring in amino acids, tricarboxylic and glycolytic substrates (Abraham et al., 2017; Mitacek et al., 2019).

The use of nuclear magnetic resonance (NMR)-based metabolomics helped to identify metabolite differences in two chicken lines divergently selected on breast ultimate pH (Beauclercq et al., 2016). Their research noted darker breast color (*Pectoralis major*; lower L^* values, pH = 6.33) in high pH groups than lower pH (pH = 5.55) groups. Metabolomics analysis indicated lower glycogen, carbohydrates energy substrates, ketogenic amino acids, and lipid β-oxidation in high pH groups than low pH birds (Beauclercq et al., 2016).

The application of proteomics has also helped to understand the molecular basis of myopathics such as woody breast and white striping. Cai et al. (2018) reported that woody breast fillets were lighter, redder, and had more pH than normal pH breast fillets. Woody breast fillets had an overabundance of myosin regulatory light chain 2, 14-3-3 protein gamma, serum albumin, protein deglycase, phosphoglycerate mutase 1, and carbonic anhydrase III than normal breast fillets, indicating greater oxidative stress and lower ability to produce ATP (Cai et al., 2018; Zhang et al., 2021). In support, glycolysis and gluconeogenesis pathways were downregulated in woody breast meat than in normal meat (Kuttappan et al., 2017).

6 Cooked poultry meat color

Persistent pinkness is a defect in fully cooked, uncured poultry meat. Pink color leads to consumer rejection of fully cooked meat because of its undercooked appearance. Although the mechanistic basis for pink color is not clear, the greater molecular mass of poultry myoglobin (approximately 346 Da) compared with red meat myoglobin may provide a protective effect against heat-induced denaturation (Joseph et al., 2010). Feeding, pre-slaughter stress, the gases formed during combustion (carbon monoxide, nitric oxide), cytochrome *c*, nicotinamide binding as a ligand (Cornforth et al., 1986), and transportation (carbon monoxide) have also been attributed as causative factors for persistent pinking in poultry meat. The addition of ingredients such as tripolyphosphate, sodium citrate, calcium chloride, and non-fat dried milk have been shown to minimize the occurrence of pinking (Sammel and Claus, 2003, 2007).

7 Postharvest practices to improve meat color

In the United States, 94% of poultry meat is sold as case-ready meat. The introduction of a case-ready system has allowed processors to modify the gas composition within a meat package to improve appearance and shelf-life. Meat color, tenderness, and flavor are important sensory attributes that influence consumer purchasing and eating satisfaction. Immediately after harvest, prooxidant-antioxidant balance shifts, and cells are more prone to oxidative damage (Mancini et al., 2009; Yoder et al., 2021). Centralized packaging allows meat processors to adopt postharvest processes such as ingredient enhancement and modified atmospheric packaging. In ingredient enhancement, salt, phosphate, antioxidants, and antimicrobials are dissolved in water, and the solution is injected into the meat to increase organoleptic qualities and shelf-life. For example, the addition of phosphate helps to increase water holding capacity and promote juiciness and tenderness of cooked products. Case-ready meat allows modifying the gas composition of meat within the package. The formation of consumer-preferred light pink color depends on oxygen level within the package and oxymyoglobin formation. The use of high-oxygen modified atmospheric packaging (HiOx-MAP; 50–80% oxygen and remaining will be carbon dioxide) increases oxygen concentration within the package and formation of oxymyoglobin. Improved fresh meat color in HiOx-MAP is compromised by oxidative changes in meat. Antioxidants such as ascorbate and rosemary are added to limit oxidative changes in meat. For instance, the addition of rosemary to ground chicken thigh limited discoloration lowered lipid oxidation and raw meat sensory scores (Keokamnerd et al., 2008). In another research, chitosan and modified atmosphere (70% carbon dioxide and 30% nitrogen) increased the shelf-life of chicken breasts by 12 days compared with 5 days of shelf-life for aerobically packaged chicken breasts (Latou et al., 2014). Improved color by the addition of chitosan is also reported in beef (Suman et al., 2010).

The use of carbon monoxide within a package also creates a bright pink meat color. Both oxymyoglobin and carboxymyoglobin can impart pink color to poultry meat. The United States Food and Drug Administration has allowed carbon monoxide at a level of 0.4% in meat packaging. However, the European Union does not allow carbon monoxide to improve shelf-life in meat. In carboxymyoglobin modified atmospheric packaging (CO-MAP), 0.4% carbon monoxide is mixed with nitrogen and carbon dioxide. Carbon dioxide has a bacteriostatic effect, while nitrogen is used as a filler gas.

8 Case study 1: Impact of packaging on ground chicken color

Ground poultry products have gained popularity in several countries. As a result, retailing fresh ground poultry products has also increased in countries where the marketing of chilled meat is common and feasible. Meat grinding releases prooxidants from the inside of the cell to the outside. In addition, various oxidase enzymes and metals increase oxidative changes and decrease shelf-life. Myoglobin and lipid oxidation are interrelated. The introduction of case-ready packaging has allowed meat purveyors to have centralized processing and packaging. As a result, the benefit of improving color should not comprise flavor due to lipid oxidation.

- How does packaging influence color?
- What are the implications of packaging on lipid oxidation?
- Can anaerobic packaging improve color?

Chicken meat is more prone to lipid oxidation due to greater unsaturated lipid content within phospholipid membrane than beef and pork (Rhee et al., 1996). Therefore, low or no oxygen packages that limit lipid oxidation and maintain ground chicken's pink color are important to meet consumer expectations. The vacuum package removes all oxygen within the package and lipid oxidative changes. However, deformity of the product and lack of pink color limits its application. Recently brick-type vacuum packaging can retain either circular or square shape of patties. Turkey fillets stored at 0°C have a shelf-life of 25 days for turkey fillets under modified atmospheric packaging consisting of 0.5% CO, 80% CO_2, and 19.5% N_2 relative to 5 days shelf-life for aerobically packaged meat (Fraqueza et al., 2011). Inclusion of carbon monoxide along with anoxic gas mixtures for turkey meat under modified atmospheric packaging imparted bright pink color preferred by consumers without leading to the appearance of undercooked meat (Fraqueza and Barreto, 2011). The protective antioxidant effect of anaerobic modified atmospheric packaging with carbon monoxide was not observed in turkey breast muscles after 12 days of storage (Fraqueza and Barreto, 2011).

Research by Hearn et al. (2021) reported the beneficial effects of CO-MAP in enhancing the quality and shelf-life of ground chicken. Ground chicken patties were prepared, packaged in polyvinyl chloride packaging, HiOx-MAP, and CO-MAP, and stored for 4 days. Ground chicken patties packaged in both HiOx-MAP and CO-MAP had greater redness (a^* values) compared with polyvinyl packaging. Although HiOx-MAP improved redness than polyvinyl packaging, lipid oxidation was greater in high-oxygen conditions. CO-MAP creates an anaerobic system; hence lipid oxidation was lower compared with HiOx-MAP. The beneficial effect of CO-MAP is effectively utilized in the beef industry during transportation and retailing (larger bags that contain patties or steaks packaged in polyvinyl overwrap and kept inside mother bag flushed with 0.4% carbon monoxide, 69.6% nitrogen, and 30% carbon dioxide).

9 Case study 2: Molecular mechanisms regulating poultry color characteristics

With a shift toward retailing chilled poultry meat and further processed food products, there is increased demand to improve sensory and functional properties. However, the molecular and biochemical basis of poultry meat quality changes is still unclear. Over the years, several molecular techniques such as proteomics, metabolomics, transcriptomic, and nutrigenomics have been extended to poultry meat quality research to better understand relationships between protein, metabolite, and gene expression and poultry meat quality.

- How to identify proteins and metabolites in poultry meat of varying color characteristics?
- How to interpret data related to color changes?

The proteome is a set of all expressed proteins in a cell, tissue, or organism (Theodorescu and Mischak, 2007; Horgan and Kenny, 2011). Proteomics studies are broadly classified into qualitative, quantitative, and functional proteomics. In qualitative proteomics, global protein changes are compared using peptide fragmentation and peptide mass fingerprinting approaches (Pappin et al., 1993). Functional proteomics focuses on the functional relationships between proteins and other molecules such as metabolites and genes. Quantitative proteomics focuses on the differential abundance of proteins within compared samples. Proteomics data are generated using high-throughput platforms such as two-dimensional gel electrophoresis (2D) and liquid chromatography coupled with tandem liquid chromatography-mass spectrometry (LC-MS/MS).

The metabolome is defined as the state of global metabolites in a system of cells, tissues, and organisms within a given period (Horgan and Kenny, 2011). Based on the hypothesize of the experiment, metabolomics approaches can be classified as (i) targeted metabolomics profiling where the focus is on identifying or quantifying certain classes/groups of compounds such as polyphenols, amino acids, and lipids (Cavanna et al., 2018). While (ii) comprehensive (non-targeted) metabolomics profiling hypothesis-generating workflow, focusing on identifying and quantifying all detectable metabolites in a given sample extract (Naz et al., 2014). The analytical platforms available for metabolomics data acquisition include mass spectrometry methods coupled with gas or liquid chromatography (GC-MS and LC-MS), fluorescence spectroscopy, and NMR.

The advances in bioinformatics have enabled the analysis of omics datasets, enhancing understanding of the interactive nature of metabolites, proteins, and genes at a molecular level. Omics data is processed through the use of a number of bioinformatics platforms involving several data processing steps (Schneider and Orchard, 2011; Sugimoto et al., 2012; Tyanova et al., 2016). However, the capacity to integrate data from different analytical platforms is still limited due to the challenges of drawing different conclusions from identical starting datasets (Sauer et al., 2017). Therefore, to optimize research objectives, it is important to ensure that data quality is maintained. Cai et al. (2018) revealed that downregulation of glycolytic proteins and increased abundance of oxidative stress proteins modulates differences in the woody breast compared with normal breast poultry meat. Flavor precursor molecules such as low molecular weight peptides (LMWP) show differential abundance between the leg and chest muscles (Zhou et al., 2019). Identifying particular biomarkers enables the establishment of reference points that can assist in grading, detecting adulteration, and improving poultry meat quality properties. Furthermore, the proteomic, metabolomic, and transcriptomic differences provide additional information on the molecular and biochemical pathways regulating meat quality characteristics in poultry.

10 Tools to assess poultry quality

In recent years, real-time monitoring of color has been utilized in research and commercial settings. For example, the application of computer vision, near-infrared technique, hyperspectral imaging, and Raman spectroscopy has been reported in the literature (Xiong et al., 2015; Khatri et al., 2012). Surface reflectance/absorbance spectrophotometers are commonly used to determine meat's surface color and relative amounts of myoglobin forms (Krzywicki, 1979; Ramanathan et al., 2010; AMSA, 2012; Hernández et al., 2015). Differences in ultimate muscle pH can affect reflectance and absorbance properties. Recently, the computer vision system (CVS) has been utilized in color measurements (Tomasevic et al., 2019). In-line application of hyperspectral imaging to detect fecal matter on poultry carcasses (Seo et al., 2019). The use of in-line meat quality measurement requires developing algorithms and calibrations for accurate prediction.

11 Conclusion and future trends

The demand for safe, healthy, and consumer-acceptable poultry products is increasing. Any deviation from an acceptable consumer color leads to discarding meat products (Fig. 1). Selecting faster growth rates in poultry has led to the development of myopathies leading to lower meat quality. Several researchers have utilized omics research to elucidate the molecular basis of growth and development in live birds. In the last decade, meat quality research has also utilized genomics, proteomics, and metabolomics to understand how biomolecules change with storage time and influence meat quality. In-line application of near-infrared spectroscopy is utilized to get real-time segregation of meat based on quality characteristics. The application of machine learning will help to predict the real-time decision of quality and cooked product characteristics.

12 Where to look for further information

12.1 Journals

- *Meat Science:* https://www.journals.elsevier.com/meat-science.
- *Meat and Muscle Biology:* https://www.iastatedigitalpress.com/mmb/.
- *Poultry Science:* https://www.sciencedirect.com/journal/poultry-science.
- *British Poultry Science:* https://www.tandfonline.com/toc/cbps20/current.
- *Journal of Agricultural and Food Chemistry:* https://pubs.acs.org/journal/jafcau.
- *Journal of Food Science:* https://onlinelibrary.wiley.com/journal/17503841.
- *Journal of Animal Science:* https://academic.oup.com/jas.
- *Food Chemistry:* https://www.journals.elsevier.com/food-chemistry.

12.2 Recommended articles

- Beauclercq, S., Nadal-Desbarats, L., Hennequet-Antier, C., Collin, A., Tesseraud, S., Bourin, M., Le Bihan-Duval, E. and Berri, C. 2016. Serum and muscle metabolomics for the prediction of ultimate pH, a key factor for chicken-meat quality. *J. Proteome Res.* 15(4): 1168–1178. DOI: 10.1021/ACS.JPROTEOME.5B01050.

- Cai, K., Shao, W., Chen, X., Campbell, Y. L., Nair, M. N., Suman, S. P., Beach, C. M., Guyton, M. C. and Schilling, M. C. 2018. Meat quality traits and proteome profile of woody broiler breast (pectoralis major) meat. *Poult. Sci.* 97(1): 337–346. DOI: 10.3382/PS/PEX284.
- Carvalho, R., Shimokomaki, M. and Estévez, M. 2017. Poultry meat color and oxidation. *Poult. Qual. Eval. Qual. Attrib. Consum. Values:* 133–157. DOI: 10.1016/B978-0-08-100763-1.00006-4.
- Cornforth, D. P., Vahabzadeh, F., Carpenter, C. E. and Bartholomew, D. T. 1986. Role of Reduced Hemochromes in Pink Color Defect of Cooked Turkey Rolls. *J. Food Sci.* 51(5): 1132–1135. DOI: 10.1111/j.1365-2621.1986.tb13065.x.
- Fletcher, D. L. 2002. Poultry meat quality. *Worlds. Poult. Sci. J.* 58(2): 131–145. DOI: 10.1079/WPS20020013.
- Fraqueza, M. J., and Barreto, A. S. 2011. Gas mixtures approach to improve turkey meat shelf life under modified atmosphere packaging: The effect of carbon monoxide. *Poult. Sci.* 90(9): 2076–2084. DOI: 10.3382/ps.2011-01366.
- Mexis, S. F., Chouliara, E. and Kontominas, M. G. 2011. Shelf life extension of ground chicken meat using an oxygen absorber and a citrus extract. *LWT - Food Sci. Technol.* 49 (2012) 21-27. DOI: 10.1016/j.lwt.2012.04.012.
- Hearn, K., Denzer, M., Mitacek, R., Maheswarappa, N. B., McDaniel, C., Jadeja, R., Mafi, G., Beker, A., Pezeshki, A. and Ramanathan, R. 2021. Effects of Modified Atmospheric Packaging on Ground Chicken Color and Lipid Oxidation. *Meat Muscle Biol.* 5(1): 1–9. DOI: 10.22175/mmb.12599.
- Joseph, P., Suman, S. P., Li, S., Beach, C. M. and Claus, J. R. 2010. Mass spectrometric characterization and thermostability of turkey myoglobin. *LWT - Food Sci. Technol.* 43(2): 273–278. DOI: 10.1016/J.LWT.2009.08.019.
- Keokamnerd, T., Acton, J. C., Han, I. Y. and Dawson, P. L. 2008. Effect of commercial rosemary oleoresin preparations on ground chicken thigh meat quality packaged in a high-oxygen atmosphere. *Poult. Sci.* 87(1): 170–179. DOI: 10.3382/ps.2007-00066.
- Kuttappan, V. A., Bottje, W., Ramnathan, R., Hartson, S. D., Coon, C. N., Kong, B.-W., Owens, C. M., Vazquez-Añon, M. and Hargis, B. M. 2017. Proteomic analysis reveals changes in carbohydrate and protein metabolism associated with broiler breast myopathy. *Poult. Sci.* 96(8): 2992–2999. DOI: 10.3382/ps/pex069.
- Kuttappan, V. A., Hargis, B. M. and Owens, C. M. 2016. White striping and woody breast myopathies in the modern poultry industry: a review. *Poult. Sci.* 95(11): 2724–2733. DOI: 10.3382/PS/PEW216.

- Latou, E., Mexis, S. F., Badeka, A. V., Kontakos, S. and Kontominas, M. G. 2014. Combined effect of chitosan and modified atmosphere packaging for shelf life extension of chicken breast fillets. *LWT - Food Sci. Technol.* 55(1): 263–268. DOI: 10.1016/j.lwt.2013.09.010.
- McMillin, K. W. 2017. Advancements in meat packaging. *Meat Sci.* 132(April): 153–162. DOI: 10.1016/j.meatsci.2017.04.015.
- Nair, M. N., Suman, S. P., Li, S., Joseph, P. and Beach, C. M. 2014. Lipid oxidation-induced oxidation in emu and ostrich myoglobins. *Meat Sci.* 96(2): 984–993. DOI: 10.1016/J.MEATSCI.2013.08.029.
- Nishida, J. and Nishida, T. 1985. Relationship between the concentration of myoglobin and parvalbumin in various types of muscle tissues from chickens. *Br. Poult. Sci.* 26(1): 105–115. DOI: 10.1080/00071668508416792.
- Ramanathan, R., Hunt, M. C., Mancini, R. A., Nair, M. N., Denzer, M. L. 2020. Recent updates in meat color research: Integrating traditional and high-throughput approaches. *Meat Muscle Biol.* 4(2): 1–24. DOI: 10.22175/mmb.9598.
- Rhee, K. S., Anderson, L. M. and Sams, A. R. 1996. Lipid Oxidation Potential of Beef, Chicken, and Pork. *J. Food Sci.* 61(1): 8–12. DOI: 10.1111/j.1365-2621.1996.tb14714.x.
- Sammel, L. M. and Claus, J. R. 2003. Citric Acid and Sodium Citrate Effects on Reducing Pink Color Defect of Cooked Intact Turkey Breasts and Ground Turkey Rolls. *J. Food Sci.* 68(3): 874–878. DOI: 10.1111/j.1365-2621.2003.tb08259.x.
- Sammel, L. M. and Claus, J. R. 2007. Calcium chloride and tricalcium phosphate effects on the pink color defect in cooked ground and intact turkey breast. *Meat Sci.* 77(4): 492–498. DOI: 10.1016/J.MEATSCI.2007.04.025.
- Swatland, H. J. 2008. How pH causes paleness or darkness in chicken breast meat. *Meat Sci.* 80(2): 396–400. DOI: 10.1016/j.meatsci.2008.01.002.
- Yin, S., Faustman, C., Tatiyaborworntham, N., Ramanathan, R., Maheswarappa, N. B., Mancini, R. A., Joseph, P., Suman, S. P. and Sun, Q. 2011. Species-specific myoglobin oxidation. *J. Agric. Food Chem.* 59(22): 12198–12203. DOI: 10.1021/jf202844t.
- Zhang X., To, K. V., Jarvis, T. R., Campbell, Y. L., Hendrix, J. D., Suman, S. P., Li, S., Antonelo, D. S., Zhai, W., Chen, J., Zhu, H. and Schilling, M. W. 2021. Broiler genetics influences proteome profiles of normal and woody breast muscle. *Poult. Sci.* 100(4): 100994. DOI: 10.1016/j.psj.2021.01.017.

13 References

Abraham, A., Dillwith, J. W., Mafi, G. G., VanOverbeke, D. L. and Ramanathan, R. 2017. Metabolite profile differences between beef longissimus and psoas muscles during display. *Meat Muscle Biol.* 1(1): 18–27. DOI: 10.22175/mmb2016.12.0007.

AMSA. 2012. *AMSA Meat Color Measurement Guidelines* (2nd ed.). Champaign, IL: American Meat Science Association, pp. 1–135.

Arihara, K., Cassens, R. G., Greaser, M. L., Luchansky, J. B. and Mozdziak, P. E. 1995. Localization of metmyoglobin-reducing enzyme (NADH-cytochrome b5reductase) system components in bovine skeletal muscle. *Meat Sci.* 39(2): 205–213. DOI: 10.1016/0309-1740(94)P1821-C.

Beauclercq, S., Nadal-Desbarats, L., Hennequet-Antier, C., Collin, A., Tesseraud, S., Bourin, M., Le Bihan-Duval, E. and Berri, C. 2016. Serum and muscle metabolomics for the prediction of ultimate pH, a key factor for chicken-meat quality. *J. Proteome Res.* 15(4): 1168–1178. DOI: 10.1021/ACS.JPROTEOME.5B01050.

Cai, K., Shao, W., Chen, X., Campbell, Y. L., Nair, M. N., Suman, S. P., Beach, C. M., Guyton, M. C. and Schilling, M. W. 2018. Meat quality traits and proteome profile of woody broiler breast (pectoralis major) meat. *Poult. Sci.* 97(1): 337–346. DOI: 10.3382/PS/PEX284.

Carvalho, R., Shimokomaki, M. and Estévez, M. 2017. Poultry meat color and oxidation. *Poult. Qual. Eval. Qual. Attrib Consum. Values*: 133–157. DOI: 10.1016/B978-0-08-100763-1.00006-4.

Cavanna, D., Righetti, L., Elliott, C. and Suman, M. 2018. The scientific challenges in moving from targeted to non-targeted mass spectrometric methods for food fraud analysis: A proposed validation workflow to bring about a harmonized approach. *Trends Food Sci. Technol.* 80: 223–241. DOI: 10.1016/j.tifs.2018.08.007.

Cornforth, D. P., Vahabzadeh, F., Carpenter, C. E. and Bartholomew, D. T. 1986. Role of reduced hemochromes in pink color defect of cooked turkey rolls. *J. Food Sci.* 51(5): 1132–1135. DOI: 10.1111/j.1365-2621.1986.tb13065.x.

Denzer, M. L., Mowery, C., Comstock, H. A., Maheswarappa, N. B., Mafi, G., VanOverebeke, D. L. and Ramanathan, R. 2020. Characterization of the cofactors involved in non-enzymatic metmyoglobin/methemoglobin reduction in vitro. *Meat Muscle Biol.* 4(1): 1–10. DOI: 10.22175/mmb.9507.

English, A. R., Wills, K. M., Harsh, B. N., Mafi, G. G., VanOverbeke, D. L. and Ramanathan, R. 2016. Effects of aging on the fundamental color chemistry of dark-cutting beef. *J. Anim. Sci.* 94(9): 4040–4048. DOI: 10.2527/jas.2016-0561.

Faustman, C., Sun, Q., Mancini, R. and Suman, S. P. 2010. Myoglobin and lipid oxidation interactions: Mechanistic bases and control. *Meat Sci.* 86(1): 86–94. DOI: 10.1016/j.meatsci.2010.04.025.

Fletcher, D. L. 2002. Poultry meat quality. *Worlds Poult. Sci. J.* 58(2): 131–145. DOI: 10.1079/WPS20020013.

Fraqueza, M. J. and Barreto, A. S. 2011. Gas mixtures approach to improve turkey meat shelf life under modified atmosphere packaging: The effect of carbon monoxide. *Poult. Sci.* 90(9): 2076–2084. DOI: 10.3382/ps.2011-01366.

Fraqueza, M. J., Barreto, A. S., Keokamnerd, T., Acton, J. C. and Han, I. Y. 2011. Shelf life extension of ground chicken meat using an oxygen absorber and a citrus extract. *Poult. Sci.* 49(1): 503–510. DOI: 10.3382/ps.2011-01366.

Hearn, K., Denzer, M., Mitacek, R., Maheswarappa, N. B., McDaniel, C., Jadeja, R., Mafi, G., Beker, A., Pezeshki, A. and Ramanathan, R. 2021. Effects of modified atmospheric packaging on ground chicken color and lipid oxidation. *Meat Muscle Biol.* 5(1): 1–9. DOI: 10.22175/mmb.12599.

Hernández, B., Sáenz, C., Alberdi, C. and Diñeiro, J. M. 2015. Comparison between two different methods to obtain the proportions of myoglobin redox forms on fresh meat from reflectance measurements. *J. Food Sci. Technol.* 52:8212–8219. https://doi.org/10.1007/s13197-015-1917-x.

Horgan, R. P. and Kenny, L. C. 2011. 'Omic' technologies: Genomics, transcriptomics, proteomics and metabolomics. *Obstet. Gynaecol.* 13(3): 189–195. DOI: 10.1576/TOAG.13.3.189.27672.

Joseph, P., Suman, S. P., Li, S., Beach, C. M. and Claus, J. R. 2010. Mass spectrometric characterization and thermostability of turkey myoglobin. *LWT Food Sci. Technol.* 43(2): 273–278. DOI: 10.1016/J.LWT.2009.08.019.

Joseph, P., Suman, S. P., Li, S., Claus, J. R., Fontaine, M. and Steinke, L. 2011. Primary structure of turkey myoglobin. *Food Chem.* 129(1): 175–178. DOI: 10.1016/j.foodchem.2011.04.024.

Keokamnerd, T., Acton, J. C., Han, I. Y. and Dawson, P. L. 2008. Effect of commercial rosemary oleoresin preparations on ground chicken thigh meat quality packaged in a high-oxygen atmosphere. *Poult. Sci.* 87(1): 170–179. DOI: 10.3382/ps.2007-00066.

Khatri, M., Phung, V. T., Isaksson, T., Sørheim, O., Slinde, E. and Egelandsdal, B. 2012. New procedure for improving precision and accuracy of instrumental color measurements of beef. *Meat Sci.* 91:223–231. https://doi.org/10.1016/j.meatsci.2012.01.012.

Krzywicki, K. 1979. Assessment of relative content of myoglobin, oxymyoglobin and metmyoglobin at the surface of beef. Meat Sci. 3:1–10. https://doi.org/10.1016/0309-1740(79)90019-6.

Kuttappan, V. A., Bottje, W., Ramnathan, R., Hartson, S. D., Coon, C. N., Kong, B. W., Owens, C. M., Vazquez-Añon, M. and Hargis, B. M. 2017. Proteomic analysis reveals changes in carbohydrate and protein metabolism associated with broiler breast myopathy. *Poult. Sci.* 96(8): 2992–2999. DOI: 10.3382/ps/pex069.

Kuttappan, V. A., Hargis, B. M. and Owens, C. M. 2016. White striping and woody breast myopathies in the modern poultry industry: A review. *Poult. Sci.* 95(11): 2724–2733. DOI: 10.3382/PS/PEW216.

Latou, E., Mexis, S. F., Badeka, A. V., Kontakos, S. and Kontominas, M. G. 2014. Combined effect of chitosan and modified atmosphere packaging for shelf life extension of chicken breast fillets. *LWT Food Sci. Technol.* 55(1): 263–268. DOI: 10.1016/j.lwt.2013.09.010.

Leal, L. N., Beltrán, J. A., Alonso, V., Bello, J. M., den Hartog, L. A., Hendriks, W. H. and Martín-Tereso, J. 2018. Dietary vitamin E dosage and source affects meat quality parameters in light weight lambs. *J. Sci. Food Agric.* 98(4): 1606–1614 DOI: 10.1002/jsfa.8635.

Maheswarappa, N. B., Faustman, C., Tatiyaborworntham, N., Yin, S., Ramanathan, R. and Mancini, R. A. 2009. Mass spectrometric characterization and redox instability of Turkey and chicken myoglobins as induced by unsaturated aldehydes. *J. Agric. Food Chem.* 57(18): 8668–8676. DOI: 10.1021/jf902705q.

Maia Research Analysis. 2020. Global meat wastage at store or losses due to discoloration of meat 2015–2020. Available at: https://www.maiaresearch.com.

Mancini, R. A., Suman, S. P., Konda, M. K. R. and Ramanathan, R. 2009. Effect of carbon monoxide packaging and lactate enhancement on the color stability of beef steaks stored at 1°C for 9 days. *Meat Sci.* 81(1): 71–76. DOI: 10.1016/j.meatsci.2008.06.021.

McMillin, K. W. 2017. Advancements in meat packaging. *Meat Sci.* 132(April): 153–162. DOI: 10.1016/j.meatsci.2017.04.015.

Mitacek, R. M., Ke, Y., Prenni, J. E., Jadeja, R., VanOverbeke, D. L., Mafi, G. G. and Ramanathan, R. 2019. Mitochondrial degeneration, depletion of NADH, and oxidative stress decrease color stability of wet-aged beef longissimus steaks. *J. Food Sci.* 84(1): 38–50. DOI: 10.1111/1750-3841.14396.

Nair, M. N., Suman, S. P., Li, S., Joseph, P. and Beach, C. M. 2014. Lipid oxidation-induced oxidation in emu and ostrich myoglobins. *Meat Sci.* 96(2 Pt A): 984–993. DOI: 10.1016/J.MEATSCI.2013.08.029.

Naz, S., Vallejo, M., García, A. and Barbas, C. 2014. Method validation strategies involved in non-targeted metabolomics. *J. Chromatogr. A* 1353: 99–105. DOI: 10.1016/j.chroma.2014.04.071.

Nishida, J. and Nishida, T. 1985. Relationship between the concentration of myoglobin and parvalbumin in various types of muscle tissues from chickens. *Br. Poult. Sci.* 26(1): 105–115. DOI: 10.1080/00071668508416792.

OECD-FAO. 2021. *OECD-FAO Agricultural Outlook, OECD Agriculture Statistics (Database).*

Pappin, D. J. C., Hojrup, P. and Bleasby, A. J. 1993. Rapid identification of proteins by peptide-mass fingerprinting. *Curr. Biol.* 3(6): 327–332. DOI: 10.1016/0960-9822(93)90195-T.

Ramanathan, R., Hunt, M. C., Mancini, R. A., Nair, M. N., Denzer, M. L., Suman, S. P. and Mafi, G. G. 2020. Recent updates in meat color research: Integrating traditional and high-throughput approaches. *Meat Muscle Biol.* 4(2): 1–24. DOI: 10.22175/mmb.9598.

Ramanathan, R. and Mancini, R. A. 2018. Role of mitochondria in beef color: A review. *Meat Muscle Biol.* 2(1): 309–320. DOI: 10.22175/mmb2018.05.0013.

Ramanathan, R., Mancini, R. A., Joseph, P. and Suman, S. P. 2013. Bovine mitochondrial oxygen consumption effects on oxymyoglobin in the presence of lactate as a substrate for respiration. *Meat Sci.* 93(1): 893–897. DOI: 10.1016/j.meatsci.2012.12.005.

Ramanathan, R., Mancini, R. A., Naveena, B. M. and Konda. M. K. R. 2010. Effects of lactate-enhancement on surface reflectance and absorbance properties of beef longissimus steaks. *Meat Sci.* 84:219–226. https://doi.org/10.1016/j.meatsci.2009.08.027.

Ramanathan, R., Mancini, R. A., Konda, M. R., Bailey, K., More, S. and Mafi, G. G. 2022. Evaluating the failure to bloom in dark-cutting and lactate-enhanced beef longissimus steaks. *Meat Sci.* 184: 108684. DOI: 10.1016/J.MEATSCI.2021.108684.

Rhee, K. S., Anderson, L. M. and Sams, A. R. 1996. Lipid oxidation potential of beef, chicken, and pork. *J. Food Sci.* 61(1): 8–12. DOI: 10.1111/j.1365-2621.1996.tb14714.x.

Sammel, L. M. and Claus, J. R. 2003. Citric acid and sodium citrate effects on reducing pink color defect of cooked intact turkey breasts and ground turkey rolls. *J. Food Sci.* 68(3): 874–878. DOI: 10.1111/j.1365-2621.2003.tb08259.x.

Sammel, L. M. and Claus, J. R. 2007. Calcium chloride and tricalcium phosphate effects on the pink color defect in cooked ground and intact turkey breast. *Meat Sci.* 77(4): 492–498. DOI: 10.1016/J.MEATSCI.2007.04.025.

Sauer, U. G., Deferme, L., Gribaldo, L., Hackermüller, J., Tralau, T., van Ravenzwaay, B., Yauk, C., Poole, A., Tong, W. and Gant, T. W. 2017. The challenge of the application of 'omics technologies in chemicals risk assessment: Background and outlook. *Regul. Toxicol. Pharmacol.* 91 (Suppl. 1): S14–S26. DOI: 10.1016/J.YRTPH.2017.09.020.

Schneider, M. V. and Orchard, S. 2011. Omics technologies, data and bioinformatics principles. *Methods Mol. Biol.* 719: 3–30. DOI: 10.1007/978-1-61779-027-0_1.

Seo, Y., Lee, H., Mo, C., Kim, M.S., Baek, I. et al. 2019. Multispectral Fluorescence Imaging Technique for On-Line Inspection of Fecal Residues on Poultry Carcasses. *Sensors* 19(16): 3483. doi: 10.3390/s19163483.

Seyfert, M., Mancini, R. A., Hunt, M. C., Tang, J. and Faustman, C. 2007. Influence of carbon monoxide in package atmospheres containing oxygen on colour, reducing activity, and oxygen consumption of five bovine muscles. *Meat Sci.* 75(3): 432–442. DOI: 10.1016/j.meatsci.2006.08.007.

Sugimoto, M., Kawakami, M., Robert, M., Soga, T. and Tomita, M. 2012. Bioinformatics tools for mass spectroscopy-based metabolomic data processing and analysis. *Curr. Bioinform.* 7(1): 96–108. DOI: 10.2174/157489312799304431.

Suman, S. P., Mancini, R. A., Joseph, P., Ramanathan, R., Konda, M. K. R., Dady, G. and Yin, S. 2010. Packaging-specific influence of chitosan on color stability and lipid oxidation in refrigerated ground beef. *Meat Sci.* 86(4): 994–998. DOI: 10.1016/j.meatsci.2010.08.006.

Swatland, H. J. 2008. How pH causes paleness or darkness in chicken breast meat. *Meat Sci.* 80(2): 396–400. DOI: 10.1016/j.meatsci.2008.01.002.

Tang, J., Faustman, C., Hoagland, T. A., Mancini, R. A., Seyfert, M. and Hunt, M. C. 2005a. Postmortem oxygen consumption by mitochondria and its effects on myoglobin form and stability. *J. Agric. Food Chem.* 53(4): 1223–1230. DOI: 10.1021/jf048646o.

Tang, J., Faustman, C., Mancini, R. A., Seyfert, M. and Hunt, M. C. 2005b. Mitochondrial reduction of metmyoglobin: Dependence on the electron transport chain. *J. Agric. Food Chem.* 53(13): 5449–5455. DOI: 10.1021/jf050092h.

Theodorescu, D. and Mischak, H. 2007. Mass spectrometry based proteomics in urine biomarker discovery. *World J. Urol.* 25(5): 435–443. DOI: 10.1007/S00345-007-0206-3.

Tomasevic, I., Tomovic, V., Ikonic, P., Lorenzo Rodriguez, J. M., Barba, F. J., Djekic, I., Nastasijevic, I., Stajic, S. and Zivkovic. D. 2019. Evaluation of poultry meat colour using computer vision system and colourimeter: Is there a difference? *Br. Food J.* 121:1078–1087. https://doi.org/10.1108/BFJ-06-2018-0376.

Tyanova, S., Temu, T., Sinitcyn, P., Carlson, A., Hein, M. Y., Geiger, T., Mann, M. and Cox, J. 2016. The Perseus computational platform for comprehensive analysis of (prote) omics data. *Nat. Methods* 13(9): 731–740. DOI: 10.1038/nmeth.3901.

Xiong, Z., Sun, D.-W., Dai, Q., Han, Z., Zeng, X.-A. and Wang, L. 2015. Application of Visible Hyperspectral Imaging for Prediction of Springiness of Fresh Chicken Meat. *Food Anal. Methods* 8(2): 380–391. DOI: 10.1007/s12161-014-9853-3.

Yin, S., Faustman, C., Tatiyaborworntham, N., Ramanathan, R., Maheswarappa, N. B., Mancini, R. A., Joseph, P., Suman, S. P. and Sun, Q. 2011. Species-specific myoglobin oxidation. *J. Agric. Food Chem.* 59(22): 12198–12203. DOI: 10.1021/jf202844t.

Yoder, L., VanOverbeke, D. L., Ramanathan, R. and Mafi, G. G. 2021. Effects of rosemary and green tea antioxidants on ground beef patties in traditional and modified atmosphere packaging. *Meat Muscle Biol.* 5(1). DOI: 10.22175/mmb.11697.

Zhang, X., To, K. V., Jarvis, T. R., Campbell, Y. L., Hendrix, J. D., Suman, S. P., Li, S., Antonelo, D. S., Zhai, W., Chen, J., Zhu, H. and Schilling, M. W. 2021. Broiler genetics influences proteome profiles of normal and woody breast muscle. *Poult. Sci.* 100(4): 100994. DOI: 10.1016/j.psj.2021.01.017.

Zhou, R., Grant, J., Goldberg, E. M., Ryland, D. and Aliani, M. 2019. Investigation of low molecular weight peptides (<1 kDa) in chicken meat and their contribution to meat flavor formation. *J. Sci. Food Agric.* 99(4): 1728–1739. DOI: 10.1002/JSFA.9362.

Chapter 5

Understanding texture development in poultry meat

Clay J. Maynard and Casey M. Owens, University of Arkansas, USA

1 Introduction

This chapter is intended to inform the reader of advances and developmental strategies in poultry meat tenderness. As is common with all food products, texture or tenderness, as the terms are interchangeably used, is the final 'grade' that consumers place on a product. Meat tenderness is commonly known to be the hardest predicted trait for meat products as each consumer has their own level of acceptance. It should also be noted that texture development is a very intricate process that is multifaceted. The discussion to follow will focus on this multitude of factors that affect texture development and will express how complex final meat tenderness is in poultry products.

Poultry meat provides the necessary nutritional requirements for humans in modern diets as it is high in protein and low in fat, which is predominantly composed of unsaturated fatty acids. Furthermore, poultry meat is relatively low in sodium and has a tremendous reduction in total cholesterol when compared to beef or pork (John et al., 2016). Another reason for increasing demand for poultry is rising per capita availability in developing countries. As relative spending increases, the desire and ability to purchase variation in protein options becomes of utmost importance. Poultry provides a very economical option, due to high-efficiency production, and is therefore one of the most consumed animal proteins in the world. Texture development of the final product is considered important by the consumer.

http://dx.doi.org/10.19103/AS.2022.0112.05

2 Factors affecting poultry meat tenderness

Tenderness is one of the most important meat quality characteristics related to consumer acceptance. Deboning time, age, and cooking method are some of the major factors that can affect the tenderness of poultry (Lawrie, 1998; Northcutt et al., 2001; Cavitt et al., 2004). These factors can impact the fundamental mechanisms responsible for meat tenderness of poultry: the myofibrillar component including *rigor mortis*, connective tissue component, and juiciness.

2.1 Rigor mortis

Bechtel (1986) determined that postmortem conversion of muscle to meat can have a severe impact on quality attributes, the most important being tenderness. Biochemical transitions from antemortem to postmortem result in the conversion of muscle to meat, commonly known as *rigor mortis*. This process is important in the process of texture development as it can ultimately affect the contractile state, water holding capacity (WHC), and protein functionality.

The conversion of muscle to meat results in dramatic postmortem changes in the physical and biochemical aspects of muscle. *Rigor mortis* is a Latin term meaning 'stiffening of death.' It is characterized by stiffening and loss of extensibility of the muscle due to the formation of permanent actomyosin bonds. *Rigor mortis* occurs in a four-cycle pattern: delay, onset, completion, and resolution (Hedrick et al., 1994; Papinaho and Fletcher, 1996). During the delay stage, and post exsanguination, energy stores remaining in the muscle allow for contraction to continue. The presence of adenosine triphosphate (ATP) allows the muscle to remain flaccid and extensible (Hedrick et al., 1994). Aerobic metabolism is capable of producing high amounts of ATP in order to keep the muscle functioning. However, the loss of the circulatory system upon the slaughter of an animal ceases oxygen delivery to the muscle. This results in a shift from aerobic metabolism to anaerobic metabolism, which solely depends on glycolysis; it is relatively inefficient in producing ATP. The onset phase is the most critical for rigor development. ATP stores have been completely expended at this point, resulting in a reduction of cleaving actin-myosin bonds. Through the creation of a scaffolding network of permanent actomyosin bonds, the elasticity of the muscle slowly decreases (Lawrie, 1998). During the completion phase, all creatine phosphate is utilized to phosphorylate adenosine diphosphate so that no further ATP production can occur. A lack of ATP allows for complete muscle rigidity and non-breakable bonds between actin and myosin, without the induction of enzymes. The final phase, known as the resolution phase, begins as proteolytic enzymes from live tissue lysosomes are secreted to degrade actomyosin bonds to reduce stiffness

(Hedrick et al., 1994; Lawrie, 1998). The resolution phase has varying lengths in differing species, but poultry *rigor mortis* is complete between 2 h and 8 h postmortem (de Fremery and Pool, 1959; Wiskus et al., 1976; Addis, 1986; Etherington et al., 1987; Hedrick et al., 1994).

In addition, not only is there a depletion of ATP, but the pH of the muscle also declines as rigor develops. The end product of glycolysis is lactic acid. In the live animal, the circulatory system removes lactic acid from the muscle, but with the loss of the circulatory system after exsanguination, lactic acid accumulates in the muscle and lowers its pH. The pH decreases from 7.4 in living muscle to 5.5–5.7 after rigor development (Hedrick et al., 1994). The rate of decline in pH is important because it can affect many meat quality attributes including color, WHC, and texture.

Postmortem proteolysis is a significant factor in tenderness as the resolution of rigor completes through prolonged carcass aging (Lyon et al., 1985; Hedrick et al., 1994). Generally, postmortem proteolysis degrades proteins in and around the Z-line structure of the sarcomere through the calpain proteolytic system (Koohmaraie et al., 1988). In poultry, a general weakening of the myofibrillar lattice occurs in the first 24 h that results in an increase in tenderness (McKee et al., 1997) while the calpain system continues to degrade proteins at the Z-line beyond 24 h postmortem (Immonen et al., 2000). Increased fragmentation of the myofibril with longer periods of storage postmortem is evidence of this proteolytic degradation (McKee et al., 1997). However, Walker et al. (1995) reported that μ-calpain in avian breast muscle was inactive by 24-h PM which may explain the improvement in tenderness in the first 24 h. This also suggests that aging after 24 h is less important with poultry and differences exist between aging mechanisms between avian (e.g. chicken) and mammalian species (e.g. beef) which rely on extended aging for improved tenderness (Hedrick et al., 1994; Koohmaraie et al., 1988). Previous literature related to poultry meat tenderness has shown mixed results on tenderness improvement. Lyon et al. (1992) reported that holding fillets up to 24 h PM had very little effect on broiler breast fillet tenderness when fillets were deboned at 2 h PM and 3 h PM; however, McKee et al. (1997) showed improved tenderness after 23 h PM and 71 h PM, attributed to increased myofibrillar fragmentation and activity of the calpain proteolytic system. Lee et al. (2009) reported minimal improvement in tenderness (<10% change/decrease in shear value) over a 6-day period in fillets deboned early postmortem (1.5 h PM or 3 h PM) and no improvement in tenderness for fillets deboned post-rigor (6 h PM). This suggests that though proteolysis plays an important role in texture development, inducing sarcomere shortening, thus affecting the contractile state, can have greater effects on meat tenderness than the effects of rigor resolution and proteolysis. However, it is important to note that the 24 h of aging from the time of slaughter is important. In poultry processing, producing product is very streamlined, and it would be easy for processors to slaughter, chill, debone, and cook all in the same day. However, this would increase toughness because some proteolysis was not allowed. Furthermore, the term 'aging' used in poultry normally describes the period between slaughter and deboning rather than an extended period off the bone.

2.2 Contractile state of the muscle

The contractile state of the muscle, or more specifically the sarcomere length (basic contractile unit of muscle), is one of the primary factors that influence the tenderness of poultry meat (Cavitt et al., 2004). During normal rigor development, sarcomeres shorten due to increased number of actomyosin bonds in the muscle. However, induced shortening of sarcomeres can occur when muscle is stimulated by external stimuli when there are sufficient levels of ATP still present. Examples of external stimuli are the physical cutting of muscle and extreme cold temperatures. The common practice of deboning broiler breast meat can impact sarcomere shortening if practiced too early. This induced sarcomere shortening can result in tough meat when cooked due to increased density of muscle fibers (overlap of thick and thin filaments) (Locker, 1960; Herring et al., 1967; Papa and Fletcher, 1988), especially in animals that are slaughtered at young ages, such as commercial broilers (Nakamura et al, 1975). Cavitt et al. (2004) reported that sarcomere length along with razor blade shear energy (i.e., Meullenet Owens Razor Shear) were good predictors of descriptive sensory attributes such as initial hardness and chewdown hardness ($R^2 > 0.8$). Furthermore, they reported changes in tenderness and sarcomere length as debone time increased (Fig. 1). Cavitt et al. (2005) reported that the percentages of consumers tested who considered broiler breast fillets as 'Just About Right' for tenderness generally increased and the percentage for 'Too Tough' decreased as aging time increased. The percentage of consumers considering meat as 'Too Tough' was approximately 60% when deboned 2 h PM or earlier but decreased as more aging prior to deboning occurred; approximately 20% of consumers considered meat as 'Too Tough' when deboned for at least 4 h PM. The interaction between deboning time and rigor development plays a significant role in the contractile state of the muscle. Historically, it was recommended that broiler carcasses be aged under refrigeration for 4–6 h before deboning in order to avoid the toughening that accompanies pre-rigor harvesting of broiler breast meat (Stewart et al., 1984; Dawson et al., 1987; Cavitt et al., 2004). However, in the past couple of decades, processors have shortened aging periods to better streamline the processing of deboned breast fillets, increase production, and save on labor and utility costs associated with aging. Therefore, processes such as postmortem electrical stimulation were studied and implemented.

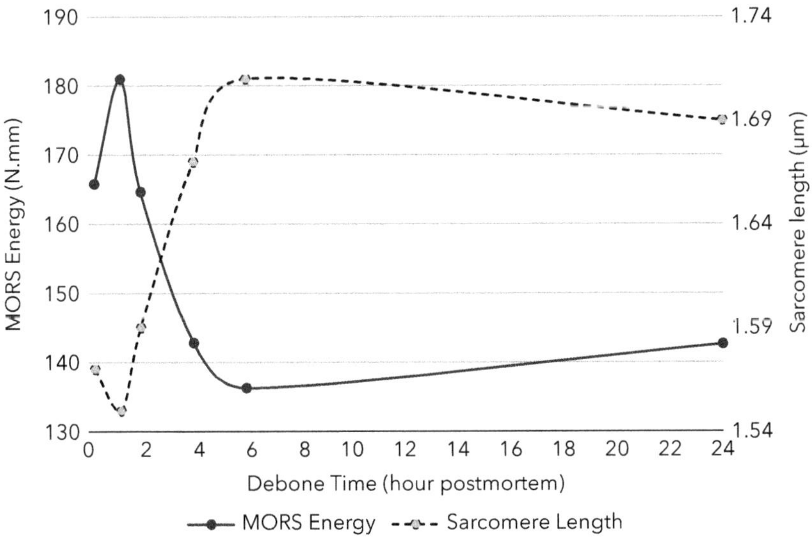

Figure 1 Meullenet Owens Razor Shear energy (MORSE) value and sarcomere length of boneless broiler breast meat deboned at various times postmortem. Fillets were aged on ice for 24 h postmortem before analysis. Data source: Cavitt et al. (2004).

A multitude of researchers speculate that electrical stimulation has allowed the poultry industry to make critical advances in deboning carcasses at earlier postmortem periods (Stewart et al., 1984; Dawson et al., 1987; Sams, 1990). The use of electrical stimulation has been assessed on tenderness properties and decreasing the presence of induced cold shortening. Birkhold and Sams (1993) reported that electrical stimulation aids in improving meat quality by accelerating ATP depletion, reducing pH, and disrupting muscle fibers for an overall improvement in tenderness (Castaneda et al., 2005a). Muscle fiber tearing has been found to be similar to the effects of prolonged carcass aging as Z-line degradation occurs (Birkhold et al., 1992). The combination of accelerated ATP depletion and muscle fiber disruption reduces prolonged exposure to cold environments to hasten rigor development and thus mitigates cold shortening (Lawrie, 1998). Improvements from electrical stimulation can then reduce aging times, energy consumption, and possible contamination (Li et al., 1993). Thus the use of electrical stimulation, post exsanguination, reduces chill times to 2.5 h (Li et al., 1993), improves tenderness (Sams, 1990; Pearson and Dutson, 1999; Zocchi and Sams, 1999), and allows for deboning of carcasses to occur earlier than non-stimulated poultry without negatively affecting tenderness (Clatfelter and Webb, 1987; Li et al., 1993; Pearson and Dutson, 1999).

Other practices such as carcass trimming early in processing can also have negative impacts on the contractile state. Breast blisters and damaged wings from over-scalding and improper picker placement are removed not only decreasing profit but decreasing tenderness through pre-rigor excision (Webb and Brunson, 1972). Severe transverse incisions on breast muscles and the removal of wings can decrease overall tenderness (Webb and Brunson, 1972). Castaneda et al. (2005b) further assessed the excision of breast blisters and determined that salvaged parts (such as breasts) produced a tougher product. Reducing the size of excision for breast trimming and limiting the removal of wings in the early postmortem stage directly increase tenderness and consumer acceptance (Webb and Brunson, 1972; Castaneda et al., 2005a)

2.3 Water-holding capacity

Juiciness has been related to tenderness, making WHC a key factor in customer acceptance (Pearson and Dutson, 1999). WHC has been shown to affect many meat quality attributes such as texture, juiciness, and color (Brewer et al., 2012). Water is present throughout the meat, constituting 75% of total muscle weight (Pearson and Dutson, 1999; Alvarado and Owens, 2005), with water present in three forms. Bound water constitutes approximately 1% of the total water. This water is tightly bound to protein spaces and can only be removed by ashing (Alvarado and Owens, 2005). Immobilized water comes as the intermediate state, contributing approximately 10–15% of the total water, and is held in place by ionic forces (Bechtel, 1986; Alvarado and Owens, 2005). Finally, free water constitutes the remaining 85–89% of water present in meat (Pearson and Dutson, 1999). This water is located extracellularly and can be easily extracted by applying external pressures.

WHC can be measured from different techniques including moisture content, drip loss, cook loss, and thaw loss (Allen et al., 1998; Woelfel et al., 2002; Lee et al., 2008). Differential methods of measurement are important for determinants that affect WHC, such as the ionic and steric effects. The ionic effect, related to muscle pH, is responsible for approximately one-third of the influence on WHC. This effect is a condition that refers to the isoelectric point. With an ultimate pH of around 5.1, the isoelectric point (pI) of meat produces the minimal WHC that can form in fresh meats (Alvarado and Owens, 2005; Alvarado and McKee, 2007). As the net charge of proteins nears zero, dipolar water molecules are no longer attracted to neutral proteins resulting in reduced WHC (Lawrie, 1998; Alvarado and Owens, 2005). The steric effect, or the spatial arrangement of ions, orients proteins to produce channels for water and nutrient incorporation. Spatial arrangement accounts for the remaining two-thirds influence on WHC, with sarcomere length development also contributing to spatial arrangement (Lawrie, 1998; Pearson and Dutson, 1999; Alvarado and Owens, 2005). As glycolysis completes during rigor development, and muscles begin to form permanent actomyosin bonds, sarcomere shortening forces water out as an exudate as myofibrils shrink laterally (Offer et al., 1989). Lawrie (1998) determined that as actomyosin bonds form, the resulting space between myofibrillar proteins is greatly diminished. Ionic and steric effects can both impact one another. As the pH value moves further away from the pI, the repulsion of similarly charged protein groups increases the distance between the protein network, allowing for a higher WHC (Bechtel, 1986).

Bechtel (1986) defines cook loss as the release of fluid after heating with or without the presence of external forces. This definition is beneficial in understanding cook loss' impact on meat quality. Releasing fluid or juice from a fillet will greatly diminish the eating quality of that product. Juiciness and tenderness are the first factors that consumers evaluate after visual confirmation. Pearson and Dutson (1999) describe cook loss to be affected by four main factors: size of the sample, temperature profile, final endpoint temperature, and environmental cooking type. Of these factors, Pearson and Dutson (1999) believe that environmental cooking type plays the largest role in meat quality. Differential cooking methods have been used to determine WHC and its impact on tenderness. Direct steam cooking and convection cooking in covered aluminum pans are two common techniques (Papa and Lyon, 1989; Cavitt et al., 2004; Mehaffey et al., 2006; Lee et al., 2008). Conditionally, the presence of a moist environment through direct steam cooking has led to increased tenderness when compared to dry heat methods such as convection (Mora et al., 2011). This can be expected as tenderness is highly affected by total juiciness (Pearson and Dutson, 1999). Generally, moisture content and shear value are negatively correlated (Lee et al., 2008) so that as moisture content decreases, shear values increase thus tenderness decreases.

In addition to environmental conditions, other factors also aid in total cook loss. The stage of rigor development and ultimate pH have shown to be highly correlated to cook loss (Offer et al., 1989; Yu et al., 2005). Cold-shortened muscle has shown to increase cook loss as more immobilized water is freely released (Bechtel, 1986). Bechtel (1986) also determined that by aging cold-shortened beef steaks for 7 days PM, the relative cook loss is improved and is similar to that of normal meat 24-h PM. Frozen samples are commonly utilized in the research field as time restraints are crucial for sample preparation in research settings. Previous studies have shown that freezing samples directly impacts meat quality (Farouk and Swan, 1998; Yu et al., 2005; Lee et al., 2008). Cooking from a frozen state has been shown to increase cook loss as opposed to thawed meat (Zhuang and Savage, 2013). Zhuang and Savage (2009) found that this increased cook loss can be accredited to varying degrees of intramuscular degradation of Z-lines and cytolysis of cellular structure releasing immobilized water. For this reason, samples that are fresh tend to have a lower cook loss than those stored under freezing conditions.

2.4 Connective tissue

Another key aspect of meat texture is the composition of collagen present within the meat. Collagen is a ubiquitous stromal protein found throughout live tissues. Moreover, collagen is known to be the most abundant protein in the body as the largest composition of connective tissues. Collagen is produced by three polypeptide chains that form a triple helix protein that provides the main skeletal support of animals. Stromal proteins are known to be very fibrous and highly insoluble in salt or water. The intrinsic development of stabilizing proteins reduces the tenderness of meat as they are not salt-soluble, and they become heat stable as mature collagen crosslinks are formed in the process of muscle growth and as the animal ages (Alvarado and Owens, 2005). Therefore, upon cooking of meat from a mature animal (older broilers/chickens), the mature heat stable crosslinks will not melt and the network of connective tissue remains intact resulting in decreased tenderness of the meat.

2.4.1 Woody breast

Modern, high-yielding broilers have seen an increased incidence of muscle myopathies as opposed to flocks from 50 years ago (Owens, 2014). The current increase in myopathies not only results in an economic loss to the industry but can cause issues for producers and consumers alike. Woody breast (WB) first appeared in the poultry industry in 2014 (Sihvo et al., 2014). This condition is accredited mostly to the stiff-like visual and touch-type detection measurements administered on individual breast fillets (Owens, 2014). Woody breast is characterized as having a natural stiffness to the muscle fibers from palpation techniques, as well as having noticeable 'ridges' throughout the breast fillet (Sihvo et al., 2014; Tijare et al., 2016). Woody fillets tend to show a greater stiffness or hard texture in the cranial region fading slowly to the middle and caudal regions respectively and have been graded by severity (Tijare et al., 2016). Woody breast fillets have connective tissue accumulation along with an increase in insoluble collagen which could explain the hard, rigid-like abnormality along with textural impacts (Owens, 2014; Sihvo et al., 2014; Tijare et al., 2016; Cai et al., 2018).

The WB condition can cause potential problems with meat quality. Research has shown primarily negative impacts on WHC attributes as indicated by increases in drip loss, decreases in marinade pickup and increases in cook loss (non-marinated or marinated), but little impact on shear value, an indirect measure of toughness (Mudalal et al., 2015; Tijare et al., 2016). However, the texture of WB is complex. Though shear values are not always impacted (Tijare et al., 2016), the texture can be affected, and it is related to other factors besides the contractile state. In fact, sarcomeres are longer in fillets with WB characteristics (Tijare et al., 2016). Chatterjee et al. (2016) found that affected fillets were harder and chewier than normal fillets using texture profile analysis (TPA). Other researchers have developed a new descriptor for WB meat. Crunchiness is a new descriptive texture attribute that has been considered to be prevalent with WB fillets, as a result of increased collagen in the muscle (Aguirre et al., 2018). This factor may have been directly influenced by the amount of compression force required to break the net surface tension of woody fillets and the layers of connective tissue and myofibrillar tissue. The crunchy textural aspect has a negative impact on consumer acceptability. Additionally, the lower WHC associated with WB has the potential to negatively impact the texture of WB meat.

2.5 Age and strain effects

Final tenderness and mouthfeel can be affected by a variety of factors such as bird age, sex, or even strain (common commercial broiler strains), but research has also been inconsistent at times. In poultry, a general trend for decreased tenderness is observed with increasing age; increasing age, advertently, increases the toughness of the meat (Musa et al., 2006). In terms of current market-ready broilers, an increase in flock age yields a tougher meat in some situations. Poole et al. (1999) conducted a study evaluating the texture of male and female commercial broilers slaughtered between 5 and 8 weeks of age and concluded that breast fillets from 5-week-old birds processed had lower shear force values than all other ages processed. In this study, fillets were deboned at 24 h PM, and therefore there would be little effect of contractile state or rate of rigor development on shear values. Additionally, Mehaffey et al. (2006) demonstrated that 7-week-old flocks had increased shear values compared to flocks that were 6 weeks in age. Collagen deposition increases as bird age increases which may have a direct effect on meat tenderness (Kuttappan et al., 2013) along with muscle fiber diameter (increases with age) (Velleman, 2007). Collagen deposits can also be impacted by sex differences (Hoffman et al., 2005). Northcutt et al. (2001) and Brewer et al. (2009) reported that fillets from females possess a higher degree of tenderness than fillets from males. Contrarily, Goodwin et al. (1969) and Poole et al. (1999) found that males generally have lower shear values than females when processed at the same market age. In addition, research has shown that differing breeds, combined with sex, can impact the overall tenderness of male chickens (Musa et al., 2006), but other studies have shown minimal impact (Mehaffey et al., 2006).

The effect of strain on meat tenderness of broilers has also been investigated. In the marketplace today, there are standard breast-yielding broilers and high breast-yielding broilers.

Brewer et al. (2012) reported that high-yielding broiler strains have tougher boneless breast fillets than standard breast-yielding broiler strains when breast fillets were deboned at 4 h PM. Mehaffey et al. (2006) also observed differences in yield between commercial broiler strains studied and when deboned at both 2 h PM and 6 h PM, differences in shear value among strains were noted. Brewer et al. (2012) reported a difference in tenderness at 2 h PM among strains reared to 40 days of age though the differences were not noted when fillets were deboned at 4 h PM or 6 h PM. Similarly, López et al. (2011) reported no difference in shear force between broiler breast fillets from different strains slaughtered at 42 days and deboned at 4 h PM. It is likely that strain itself has minimal impact on tenderness alone, and differences noted in studies may be affected by other factors.

3 Tenderness assessment

Evaluation of a product first comes from visual inspection, followed by tenderness evaluation. Since the early 1900s, tenderness has been evaluated through many methods of analysis. Instrumental and subjective-based analyses have been conducted to better understand the relationship between quantitative and qualitative tenderness, respectively. Although sensory characteristics give a more accurate result in consumer acceptability, instrumental methods utilize a quantitative method to determine values that can be correlated to perceived tenderness (Tornberg, 1996). To establish a comparative result, instrumental methods can utilize different force mediums to more accurately match consumer senses. Force application can be applied through compression, tension, or shear techniques, effectively producing quantitative values (Tornberg, 1996). In addition, force in the form of shear is generally applied perpendicular to the muscle fibers to create the greatest tensile stress (Tornberg, 1996). The Meullenet Owens Razor Shear, more commonly known as the MORS method, was developed to produce quantitative results in poultry (Cavitt et al., 2004, 2005). This method, along with others such as Allo-Kramer and Warner-Bratzler shear, has been correlated to sensory perception and thus used as an indirect measure of tenderness (Lyon and Lyon, 1990; Cavitt et al., 2004). The use of quantitative measures to predict tenderness and ultimately consumer acceptability has proven to be beneficial for researchers and integrators.

4 Conclusion

Though this chapter discussed the pertinent information on poultry meat tenderness, not all aspects were discussed. Therefore, the information here was discussed on a basic level and should be considered as a compliment to the research body of meat science. Texture development is an extensive process for meat science and will continually morph with time as new techniques are developed to assess product tenderness. Overall, the assessment of poultry meat texture is important not only to the consumer but also to integrators for the validity of production.

5 References

Addis, P. B. (1986). Poultry muscle as food. In: Schweigert, B. S. (Ed.), *Muscle as Food* 1st edn. Academic Press, Inc., Orlando, FL, pp. 372–405.

Aguirre, M. E., Owens, C. M., Miller, R. K. and Alvarado, C. Z. (2018). Descriptive sensory and instrumental texture profile analysis of woody breast in marinated chicken, *Poult. Sci.* 97(4), 1456–1461.

Allen, C. D., Fletcher, D. L., Northcutt, J. K. and Russell, S. M. (1998). The relationship of broiler breast color to meat quality and shelf-life, *Poult. Sci.* 76, 1042–1046.

Alvarado, C. Z. and McKee, S. R. (2007). Marination to improve functional properties and safety of poultry meat, *J. Appl. Poult. Res.* 16(1), 113–120.

Alvarado, C. Z. and Owens, C. M. (2005). Poultry: chemistry and biochemistry. In: Hui, Y. (Ed.), *Handbook of Food Science, Technology and Engineering (vol. I) - Food Science: Properties and Products*. CRC Press, Taylor and Francis Group, Boca Raton, FL.

Bechtel, P. J. (1986). *Muscle as Food* (1st edn.). Academic Press, Inc., Orlando, FL. p. 31.

Birkhold, S. G., Janky, D. M. and Sams, A. R. (1992). Tenderization of early-harvested broiler breast fillets by high-voltage post-mortem electrical stimulation and muscle tensioning, *Poult. Sci.* 71(12), 2106–2112.

Birkhold, S. G. and Sams, A. R. (1993). Fragmentation, tenderness, and post-mortem metabolism of early-harvested broiler breast fillets from carcasses treated with electrical stimulation and muscle tensioning, *Poult. Sci.* 72(3), 577–582.

Brewer, V. B., Emmert, J. L., Meullenet, J. F. and Owens, C. M. (2012). Small bird programs: impact of phase-feeding, strain, gender and debone time on meat quality of broilers, *Poult. Sci.* 91, 499–504.

Brewer, V. B., Kuttappan, V. A., Emmert, J. L., Meullenet, J. F. and Owens, C. M. (2009). Big bird program: impact of strain, gender, and debone time on meat quality of broilers, *Poult. Sci.* 91, 248–254 (Abstr.).

Cai, K., Shao, W., Chen, X., Campbell, Y. L., Nair, M. N., Suman, S. P., Beach, C. M., Guyton, M. C. and Schilling, M. W. (2018). Meat quality traits and proteome profile of woody broiler breast (pectoralis major) meat, *Poult. Sci.* 97(1), 337–346.

Castaneda, M. P., Hirschler, E. M. and Sams, A. R. (2005a). Early postmortem carcass trim effects on the tenderness of broiler breast fillets, *Poult. Sci.* 84(6), 951–954.

Castaneda, M. P., Hirschler, E. M. and Sams, A. R. (2005b). Functionality of electrically stimulated broiler breast meat, *Poult. Sci.* 84(3), 479–481.

Cavitt, L. C., Youm, G. W., Meullenet, J. F., Owens, C. M. and Xiong, R. (2004). Prediction of poultry meat tenderness using razor blade shear, Allo-Kramer Shear, and Sarcomere Length, *J. Food Sci.* 69(1) SNQ11-15.

Cavitt, L. C., Meullenet, J.-F. C., Xiong, R. and Owens, C. M. (2005). The correlation of Razor Blade shear, Allo-Kramer Shear, Warner-Bratzler Shear, and sensory tests to changes in tenderness of broiler breast fillets, *J. Muscle Foods* 16, 223–242.

Chatterjee, D., Zhuang, H., Bowker, B. C., Rincon, A. M. and Sanchez-Brambila, G. (2016). Instrumental texture characteristics of broiler pectoralis major with the wooden breast condition, *Poult. Sci.* 95(10), 2449–2454.

Clatfelter, K. A. and Webb, J. E. (1987). Method of eliminating aging step in poultry processing. U.S. Patent 4,675,947. July 9. 12 p. Int Cl. A22C 21/00.

Dawson, P. L., Janky, D. M., Dukes, M. G., Thompson, L. D. and Woodward, S. A. (1987). Effect of post-mortem boning time during simulated commercial processing on the tenderness of broiler breast meat, *Poult. Sci.* 66(8), 1331–1333.

de Fremery, D. and Pool, M. F. (1959). Biochemistry of chicken muscle as related to rigor mortis and tenderization, *J. Food Sci.* 38, 73–87.

Etherington, D. J., Taylor, A. J. and Dransfield, E. (1987). Conditioning of meat from different species, relationship between tenderising and the levels of cathepsin B, cathepsin L, calpain II and P-glucurinidase, *Meat Sci.* 20, 1–18.

Farouk, M. M. and Swan, J. E. (1998). Effect of rigor temperature and frozen storage on functional properties of hot-boned manufacturing beef, *Meat Sci.* 49(2), 233–247.

Fennema, O. (1996). *Food Chemistry: Third Edition*. Marcel Dekker, New York.

Goodwin, T. L., Andrews, L. D. and Webb, J. E. (1969). The influence of age, sex, and energy level on the tenderness of broilers, *Poult. Sci.* 48(2), 548–552.

Hedrick, H. B., Aberle, E. D., Forrest, J. C., Judge, M. D. and Merkel, R. A. (1994). *Principles of Meat Science* (3rd edn.). Kendall/Hunt Publishing, Dubuque, IA.

Herring, H. K., Cassens, R. G., Suess, G. G., Brungardt, V. H. and Briskey, E. J. (1967). Tenderness and associated characteristics of stretched and contracted bovine muscles, *J. Food Sci.* 32(3), 317–323.

Hoffman, L. C., Kritzinger, B. and Ferreira, A. V. (2005). The effects of region and gender on the fatty acid, amino acid, mineral, myoglobin and collagen contents of impala (Aepyceros melampus) meat, *Meat Sci.* 69(3), 551–558.

Immonen, K., Ruusunen, M. and Puolanne, E. (2000). Some effects of residual glycogen concentration on the physical and sensory quality of normal pH beef, *Meat Sci.* 55(1), 33–38.

John, K. A., Maalouf, J., Barsness, C. B., Yuan, K., Cogswell, M. E. and Gunn, J. P. (2016). Do lower calorie or lower fat foods have more sodium than their regular counterparts? *Nutrients* 8(8), 5–15.

Koohmaraie, M. A. S., Babiker, A. S., Merkel, R. A. and Dutson, T. R. (1988). Role of Ca ++ -dependent proteases and lysosomal enzymes in postmortem changes in bovine skeletal muscle, *J. Food Sci.* 53(5), 1253–1257.

Kuttappan, V. A., Brewer, V. B., Mauromoustakos, A., McKee, S. R., Emmert, J. L., Meullenet, J. F. and Owens, C. M. (2013). Estimation of factors associated with the occurrence of white striping in broiler breast fillets, *Poult. Sci.* 92(3), 811–819.

Lawrie, R. A. (1998). *Lawrie's Meat Science* (6th edn.). Woodhead Publishing Ltd, Cambridge, England.

Lee, Y. S., Saha, A., Xiong, R., Owens, C. M. and Meullenet, J. F. (2008). Changes in broiler breast fillet tenderness, water-holding capacity, and color attributes during long-term frozen storage, *J. Food Sci.* 73(4), E162–E168.

Lee, Y. S., Owens, C. M. and Meullenet, J. F. (2009). Tenderness perception of poultry major pectoralis muscle during mastication, *J. Food Sci.* 74(9), S413–S422.

Li, Y., Siebenmorgen, T. J. and Griffis, C. L. (1993). Electrical stimulation in poultry: a review and evaluation, *Poult. Sci.* 72(1), 7–22.

Locker, R. L. (1960). Degree of muscular contraction as a factor in tenderness of beef, *Food Res.* 25, 304-307.

López, K. P., Schilling, M. W. and Corzo, A. (2011). Broiler genetic strain and sex effects on meat characteristics, *Poult. Sci.* 90(5), 1105-1111.

Lyon, C. E., Hamm, D. and Thomson, J. E. (1985). pH and tenderness of broiler breast meat deboned various times after chilling, *Poult. Sci.* 64(2), 307-310.

Lyon, C. E. and Lyon, B. G. (1990). The relationship of objective shear values and sensory tests to changes in tenderness of broiler breast meat, *Poult. Sci.* 69(8), 1420-1427.

Lyon, C. E., Lyon, B. G., Papa, C. M. and Robach, M. C. (1992). Broiler tenderness: effects of post chill deboning time and fillet holding time, *J. Appl. Poult. Res.* 1(1), 27-32.

McKee, S. R., Hirschler, E. M. and Sams, A. R. (1997). Physical and biochemical effects of broiler breast tenderization by aging after pre-rigor deboning, *J. Food Sci.* 62(5), 959-962.

Mehaffey, J. M., Pradhan, S. P., Meullenet, J. F., Emmert, J. L., McKee, S. R. and Owens, C. M. (2006). Meat quality evaluation of minimally aged broiler breast fillets from five commercial genetic strains, *Poult. Sci.* 85(5), 902-908.

Mora, B., Curti, E., Vittadini, E. and Barbanti, D. (2011). Effect of different air/steam convection cooking methods on turkey breast meat: physical characterization, water status and sensory properties, *Meat Sci.* 88(3), 489-497.

Mudalal, S., Lorenzi, M., Soglia, F., Cavani, C. and Petracci, M. (2015). Implications of white striping and wooden breast abnormalities on quality traits of raw and marinated chicken meat, *Animal* 9(4), 728-734.

Musa, H. H., Chen, G. H., Cheng, J. H., Shuiep, E. S., and Bao, W. B. (2006). Breed and sex effect on meat quality of chicken, *Int. J. Poult. Sci.* 5(6), 566-568.

Nakamura, R., Sekoguchi, S. and Sato, Y. (1975). The contribution of intramuscular collagen to the tenderness of meat from chickens with different ages, *Poult. Sci.* 54(5), 1604-1612.

Northcutt, J. K., Buhr, R. J., Young, L. L., Lyon, C. E. and Ware, G. O. (2001). Influence of age and postchill carcass aging duration on chicken breast fillet quality, *Poult. Sci.* 80(6), 808-812.

Offer, G., Knight, P., Jeacocke, R., Almond, R., Cousins, T., Elsey, J., Parsons, N., Sharp, A., Starr, R. and Purslow, P. (1989). The structural basis of the water-holding, appearance and toughness of meat and meat products, *Food Microstruct.* 8, 151-170.

Owens, C. M. (2014). Identifying quality defects in poultry processing. *Watt Poult. U.S.A.*, 42-50.

Papa, C. M. and Fletcher, D. L. (1988). Pectoralis muscle shortening and rigor development at different locations within the broiler breast, *Poult. Sci.* 67(4), 635-640.

Papa, C. M. and Lyon, C. E. (1989). Shortening of the pectoralis muscle and meat tenderness of broiler chickens, *Poult. Sci.* 68(5), 663-669.

Papinaho, P. A. and Fletcher, D. L. (1996). The influence of temperature on broiler breast muscle shortening and extensibility, *Poult. Sci.* 75(6), 797-802.

Pearson, A. M. and Dutson, T. R. (1999). *Quality Attributes and Their Measurement in Meat, Poultry and Fish Products* (1st edn.). Aspen Publishers, Inc., Gaithersburg, MD.

Poole, G. H., Lyon, C. E., Buhr, R. J., Young, L. L., Alley, A., Hess, J. B., Bilgili, S. F. and Northcutt, J. K. (1999). Evaluation of age, gender, strain, and diet on the cooked yield and shear values of broiler breast fillets, *J. Appl. Poult. Res.* 8(2), 170-176.

Sams, A. R. (1990). Electrical stimulation and high temperature conditioning of broiler carcasses, *Poult. Sci.* 69(10), 1781-1786.

Sihvo, H. K., Immonen, K. and Puolanne, E. (2014). Myodegeneration With fibrosis and regeneration in the pectoralis major muscle of broilers, *Vet. Pathol.* 51(3), 619–623.

Stewart, M. K., Fletcher, D. L., Hamm, D. and Thomson, J. E. (1984). The effect of hot boning broiler breast meat muscle on postmortem pH decline, *Poult. Sci.* 63(11), 2181–2186.

Tijare, V. V., Yang, F. L., Kuttappan, V. A., Alvarado, C. Z., Coon, C. N. and Owens, C. M. (2016). Meat quality of broiler breast fillets with white striping and woody breast muscle myopathies, *Poult. Sci.* 95(9), 2167–2173.

Tornberg, E. (1996). Biophysical aspects of meat tenderness, *Meat Sci.* 43S1, 175–191.

Velleman, S. G. (2007). Muscle development in the embryo and hatchling, *Poult. Sci.* 86(5), 1050–1054.

Walker, L. T., Shackelford, S. D., Birkhold, S. G. and Sams, A. R. (1995). Biochemical and structural effects of rigor mortis-accelerating treatments in broiler *pectoralis*, *Poult. Sci.* 74(1), 176–186.

Webb, J. E. and Brunson, C. C. (1972). Effects of eviscerating line trimming on tenderness of broiler breast meat, *Poult. Sci.* 51(1), 200–203.

Wiskus, K. J., Addis, P. B. and Ma, R. T.-I. (1976). Distribution of βR, αR and αW fibers in turkey muscles, *Poult. Sci.* 55(2), 562–572.

Woelfel, R. L., Owens, C. M., Hirschler, E. M., Martinez-Dawson, R. and Sams, A. R. (2002). The characterization and incidence of pale, soft, and exudative broiler meat in a commercial processing plant, *Poult. Sci.* 81(4), 579–584.

Yu, L. H., Lee, E. S., Jeong, J. Y., Paik, H. D., Choi, J. H. and Kim, C. J. (2005). Effects of thawing temperature on the physicochemical properties of pre-rigor frozen chicken breast and leg muscles, *Meat Sci.* 71(2), 375–382.

Zhuang, H. and Savage, E. (2013). Comparison of cook loss, shear force and sensory descriptive profiles of meat cooked from a frozen or thawed state, *Poult. Sci.* 92(11), 3003–3009.

Zhuang, H. and Savage, E. M. (2009). Variation and Pearson correlation coefficients of Warner-Bratzler shear force measurements within broiler breast fillets, *Poult. Sci.* 88(1), 214–220.

Zocchi, C. and Sams, A. R. (1999). Tenderness of broiler breast fillets from carcasses treated with electrical stimulation and extended chilling times, *Poult. Sci.* 78(3), 495–498.

Chapter 6

Advances in understanding flavour development in poultry meat

Dinesh D. Jayasena, Uva Wellassa University, Sri Lanka; and Cheorun Jo, Seoul National University, Republic of Korea

1 Introduction

A tasty, nutritious and safe meat supply can ensure a continued consumption of meat and meat products by consumers (Joo and Kim, 2011). Global meat production is projected to reach 366 Mt by 2029, expanding by nearly 40 Mt, and consumption of poultry meat is forecast to increase globally to 145 Mt for the period 2020–2029 (OECD/FAO, 2020). Of all types of meat consumed in the world, chicken continues to be the cheapest commercially produced meat (Zhang et al., 2019) and the most commonly consumed meat source in many countries (Taşkıran et al., 2020). The global demand for chicken meat has been fulfilled over the years by fast-growing commercial broiler strains and a few native chicken breeds (Takahashi, 2018). A consumer's decision to purchase meat, however, is influenced by many criteria including appearance, taste, aroma, and texture. Therefore, quality and flavour enhancement of the poultry meat became the primary aim of many poultry breeders.

Meat flavour, one of the vital sensory attributes affecting consumers' meat-purchasing behaviour and preferences even before the meat is consumed, is comprised mainly of taste and aroma (Zhang et al., 2020; Sitz et al., 2005);

http://dx.doi.org/10.19103/AS.2022.0112.06

however, mouthfeel and juiciness of meat also affect the individual flavour perception (Khan et al., 2015). Raw meat has only a blood-like taste with little or no aroma. However, cooked meat flavour development is a complex scenario. Aromatic notes and most of the characteristic flavours responsible for the development of meat flavour are primarily contributed by the volatile compounds originated by the muscle matrix through thermally induced reactions between non-volatile components of lean and fatty tissues during cooking (Zhang et al., 2020; Mottram, 1998). Cooked meat flavour relies on several production and processing factors relating to the source animal, including age, breed, sex, nutritional status, chiller ageing, presence of free amino acids and nucleotides, antioxidants, pH, and method of cooking (Jayasena et al., 2013; Liu et al., 2012; Spanier et al., 1997). Thus, these pre- and post-mortem factors can change the degree of poultry meat flavour. Furthermore, off-flavour development in poultry meat through lipid oxidation is considered to be one of the main problems regarding the quality of poultry meat. As a result, the producers and processors in the poultry meat sector, and even consumers, try to avoid off-flavour development through various prevention mechanisms to preserve natural flavour and taste (Aymerich et al., 2008).

Understanding the chemistry and factors affecting poultry meat flavour is, therefore, vital in order to produce the most flavourful and consistent product possible by preserving and/or improving the flavour of poultry meat. The intent of this chapter is to highlight the flavour chemistry behind the development of flavour and off-flavour in poultry meat and to describe the main factors affecting flavour development in poultry meat, which can be manipulated to improve the flavour of poultry meat.

2 Flavour chemistry of poultry meat

Bloody, metallic, and salty taste with an aroma resembling blood serum is generally a unique characteristic of raw meat (Karabagias, 2018; Joo and Kim, 2011; Soncin et al., 2007). However, significant changes take place in the flavour of meat during cooking due to three main reactions responsible for flavour development, that is, Maillard reactions, thermal degradation of lipids, and Maillard–lipid interactions (Domínguez et al., 2019; Qi et al., 2017b; Jayasena et al., 2013). Flavour gets developed during cooking through complex reactions between components found in raw meat combining with heat. More than 1000 chemicals have so far been identified in the volatiles of different muscle foods (Shahidi, 1994). The majority of the volatile compounds identified in cooked poultry meat have been recognized in chicken (Brunton et al., 2002). However, many of these have little influence on flavour of meat and relatively few make a key contribution to the odour and flavour of cooked meat (Aliani and Farmer, 2005). The basic tastes of meat are sweet, sour, salty, bitter, and the 'umami' or savoury taste (Sabikun et al., 2021; Melton, 1999). Hydrocarbons, aldehydes, ketones, alcohols, furans, thiophenes, pyrrols, pyridines, pyrazines, oxazoles, thiazoles, sulphurous compounds, and many others have been identified as the flavour and aroma compounds found in meat (Kosowska et al., 2017; Ho et al., 1994; MacLeod, 1994). Sulphurous and carbonyl compounds are considered to be the principal contributors to meat flavour (Perez-Alvarez et al., 2010).

With respect to poultry meats, in particular chicken meat, many of their key flavour and odour compounds together with the mechanisms for the formation have been identified (Aliani and Farmer, 2005). According to Shi and Ho (1994), 16 primary odour components have been identified in chicken broth, of which 14 are structurally identified. It has been further demonstrated that 2-methyl-3-furanthiol, generated from the Maillard reaction and lipid oxidation, is the most vital chemical compound responsible for the meaty flavour of chicken broth. In addition, other volatile compounds originated from the above two reactions include 2-furfurylthiol, methionol, 2,4,5-trimethylthiazole, nonanol, 2-trans-nonenal, 2-formyl-5-methylthiophene, *p*-cresol, 2-trans-4-trans-nonadienal, 2-trans-4-trans-decadienal, 2-undecenal, β-ionone, γ-decalactone, and γ-dodecalactone. These compounds are obviously the major sources of chicken flavour (Varavinit et al., 2000; Shi and Ho, 1994). Sasaki et al. (2007) showed that components responsible for umami characteristics also contribute to the brothy taste of meat. With respect to the primary odourants of the broth, 2-trans-4-trans-decadienal and γ-dodecalactone predominated in chicken broth compared to that of beef (Table 1).

Recently, Fan et al. (2018) published an important study on chicken flavour where the unique flavour of Beijing Youji chicken broth compared with that of commercial broilers was investigated by solvent-assisted flavour evaporation combined with aroma extract dilution analysis of gas chromatography -olfactometry (AEDA/GC-O), quantitation, and aroma recombination. A total of 71 odourants were identified by GC-O in both broths. However, Beijing Youji chicken broth had 32 more extra odourants than commercial broiler broth, indicating the rich fragrance of the former. Aroma recombination and omission experiments demonstrated that 21 and 17 odourants with an odour activity value ≥ 1 contributed significantly to Beijing Youji chicken and commercial broiler broth aromas, respectively. Those key odourants mainly included sulphur-containing compounds and aliphatic aldehydes, such as 2-methyl-3-furanthiol, 3-(methylthio)propanal, (*E,E*)-2,4-decadienal, etc. (Table 2).

Furthermore, Takakura et al. (2014) reported methylpyrazine, 2-ethyl-4-methylthiazole, 3-(methylthio)propanal, and (*E,E*)-2,4-decadienal as primary aroma compounds of chicken soup stock. Recently, hexanal, nonanal, furfural, 5-methyl-2-furancarboxaldehyde, 2-acetyl-5-methylfuran, and 1-(2-furanyl)-ethanone were identified as the major volatiles of sugar-smoked chickens (Chang et al., 2021).

Table 1 Flavour dilution factors of odourants identified in broths from chicken and beef

	Flavour dilution factor		
Compounds	Chicken	Beef	Odour description
2-Methyl-3-furanthiol	1024	512	Meat-like, sweet
bis (2-Methyl-3-furyl)disulphide	<16	2048	Meat-like
2-Furfurylthiol	512	512	Roasty
2,5-Dimethyl-3-furanthiol	256	<16	Meaty
3-Mercapto-2-pentanone	128	32	Sulphurous
Methionol	128	512	Cooked potato
2,4,5-Trimethylthiazole	128	<16	Earthy
2-Formyl-5-methylthiophene	64	<16	Sulphurous
Phenylacetaldehyde	16	64	Honey-like
2-trans-4-trans-Decadienal	2048	64	Fatty
2-trans-4-cis-Decadienal	128	<16	Fatty, tallowy
2-Undecenal	256	<16	Tallowy, sweet
γ-Dodecalactone	512	<16	Tallowy, fruity
γ-Decalactone	64	<16	Peach-like
Nonanol	128	<16	Tallowy, green
2-trans-Nonenal	64	32	Tallowy, fatty
2-trans-4-trans-Nonadienal	64	<16	Fatty
β-Ionone	64	64	Violet-like
p-Cresol	64	<16	Phenolic

Source: Adapted from Gasser and Grosch (1990).

3 Flavour precursors of poultry meat

The major flavour precursors found in meat including poultry meat can be divided into two main groups: water-soluble components and lipids (Mottram, 1998). Differences in lipid-derived volatile components between species are mainly responsible for the species differences in flavour (Perez-Alvarez et al., 2010), whereas the precursors provided by lean tissues develop the meaty flavour common to all cooked meats (Mottram, 1998). Eicosapentaenoic acid (EPA) and docosahexaenoic acid (DHA) are responsible for improving sweetness and umami characteristics while suppressing sourness and bitterness (Koriyama et al., 2002). In addition, Takahashi (2018) stated that chicken meat containing higher levels of arachidonic acid is more palatable than that containing low arachidonic acid content. Takahashi (2018) has revealed that autoxidized arachidonic acid has a flavour like cooked chicken meat. It has been reported that arachidonic acid can serve as a flavour enhancer for type II receptor cells by activating the TRPM5 cation channel, which is considered an essential component of the sweet, bitter, and umami taste pathways of type II receptor cells and blocking the delayed rectifying K^+ channel (Takahashi, 2018; Takahashi et al., 2012; Oike et al., 2006).

Table 2 Concentrations and odour active values of key odourants identified in the chicken broths of Beijing Youji (BYJ) and commercial broilers (CB)

Compounds	Concentration (ng/g broth)[a,b]		Odour description	Odour activity value (OAV)		Thresholds (μg/kg)
	BYI	CB		BYI	CB	
3-Hydroxy-2-butanone	1592.27[a]	1078.49[b]	Buttery	2	1	800
Hexanal	107.39[b]	141.16[a]	Grass, green	23	30	4.5
Dimethyl disulphide	0.75[a]	0.62[b]	Onion, cabbage	12	9	0.06
2-Methyl-3-furanthiol	1.07[b]	1.48[a]	Meaty	2686	14 522	0.0004
Heptanal	6.55[a]	4.89[b]	Fatty, green	2	1	3
3-(Methylthio) propanal	158.15[a]	117.99[b]	Cooked potato	3953	2949	0.04
2-Furfurylthiol	2.71	-	Coffee, roasted	543	-	0.005
(E)-2-Heptenal	16.11	-	Green, fatty	1	-	13
1-Octen-3-ol	107.69[a]	92.13[b]	Mushroom	53	46	2
2-Methyl-3-(methylthio)furan	0.51	-	Meaty	102	-	0.005
Octanal	10.02[a]	3.68[b]	Citrus, green	14	5	0.7
(E)-2-Octenal	13.97[a]	14.47[b]	Green, fatty	5	5	3
4-Hydroxy-2,5-dimethyl-3(2*H*)-furanone	6.82[b]	13.65[a]	Caramel	1	3	5
2-Ethyl-3,5-dimethylpyrazine	1.63[b]	2.32[a]	Earthy, roasted	1	2	1
Nonanal	25.45[a]	26.99[b]	Green, fatty	25	26	1
(*E*,*Z*)-2,6-Nonadienal	0.47[a]	0.37[b]	Cucumber	23	18	0.02
(E)-2-Nonenal	2.9[b]	7.62[a]	Green, fatty	36	95	0.08
Decanal	1.94[b]	2.63[a]	Green, fatty	19	2	0.1
(*E*,*E*)-2,4-nonadienal	7.52[a]	5.72[b]	Fatty, green	125	95	0.06
(*E*,*E*)-2,4-decadienal	104.22[a]	71.37[b]	Deep-fried, fatty	2084	1427	0.05
(E)-2-undecenal	34.97[a]	14.78[b]	Fatty, metallic	11	4	3
Bis (2-methyl-3-furyl) disulphide	1.62	-	Meat broth	810	-	0.002

[a,b] Values within the same row with different letters indicate significant differences ($P < 0.05$).Source: Adapted from Fan et al. (2018).

Mottram (1998) identified free sugars, nucleotide-bound sugars, sugar phosphates, nucleotides, free amino acids, peptides, and other nitrogenous components such as thiamine as the main water-soluble flavour precursors. Sweet flavour in meat is derived from sugars, amino acids, and organic acids, whereas the sour flavours arise after the amino acids coupled with organic acids. Saltiness in meat is attributed to inorganic salts and sodium salts of glutamate and aspartate, and the bitterness is likely due to hypoxanthine, anserine, and carnosine as well as some amino acids (Ramalingam et al., 2019; MacLeod, 1994). Umami taste is a result of a number of non-volatile compounds in terms of amino acids predominantly glutamate and aspartic acid and 5′-nucleotides such as adenosine-5′-monophosphate (AMP), inosine-5′-monophosphate (IMP), and guanosine-5′-monophosphate (GMP) (Sabikun et al., 2021; Chen and Zhang, 2007).

The reaction of cysteine and sugar can lead to the characteristic meat flavour, especially for chicken and pork (Varavinit et al., 2000). This was further confirmed by research where it was demonstrated that the quantities of carbohydrates (reducing sugars) and amino acids, in particular ribose and cysteine, are reduced during heating (Jayasena et al., 2014b, 2015b; Meinert et al., 2009). The main carbohydrates with flavour-forming potential include ribose, ribose-5-phosphate, glucose, and glucose-6-phosphate (Meinert et al., 2009). Ribose which is associated with the ribonucleotides in the muscle is highly involved in flavour-producing reactions during heating of meat (Mottram, 1998). Aliani and Farmer (2005) demonstrated that ribose is more responsible than cysteine for chicken aroma since the increase of ribose can elevate the concentrations of sulphur-containing flavours, such as 2-furfurylthiol, 2-methyl-3-furanthiol, and 3-(methylthio)-propanal.

Many flavour compounds, however, may be formed by two or more possible mechanisms. For instance, principal precursors for 2-methyl-3-furanthiol, which is responsible for the meaty flavour of chicken broth (Aliani and Farmer, 2005; Shi and Ho, 1994), are pentose sugars and cysteine/cystine/glutathione or thiamine. As pentose sugars are mainly derived from ribonucleotides, in particular IMP, 2-methyl-3-furanthiol and its disulphide can be formed from reaction of IMP or ribose with cysteine or cystine or glutathione or by thiamine degradation (Melton, 1999; Shi and Ho, 1994). Therefore, the relative importance of each precursor for flavour generation in cooked chicken meat is still unclear (Aliani and Farmer, 2005).

3.1 Lipid-derived flavour compounds

The role of lipid-derived carbonyl compounds in poultry flavour has been assessed by many researchers over the years. Desirable as well as undesirable flavours can get generated in meat due to lipid oxidation as lean meat contains intramuscular triglycerides and structural phospholipids. Mild thermal oxidative changes of lipids lead to the generation of desirable flavour compounds and aromas in cooked meats (Jayasena et al., 2013; Shahidi, 2002).

Similar to other meats, the flavour development of poultry meat is partly attributed to its lipids (Perez-Alvarez et al., 2010). Lipid degradation, primarily the oxidation of the fatty acid components of lipids generates several hundred volatile compounds in cooked meat, including aliphatic hydrocarbons, aldehydes, alcohols, ketones, esters, carboxylic acids, some aromatic hydrocarbons, and oxygenated heterocyclic compounds such as lactones and alkylfurans (Fan et al., 2018; Mottram, 1998). Forty-one out of the 193 total compounds reported in the flavour of roasted chicken are lipid-derived aldehydes. Selected aldehydes identified in roasted and cooked chicken flavour are shown in Table 3 (Shi and Ho, 1994). According to Table 3, hexanal and 2,4-decadienal are the most abundant aldehydes identified in chicken flavour which are known to be the primary oxidation products of linoleic acid (Xu et al., 2017; Shi and Ho, 1994). However, 2,4-decadienal is considered a more important odourant for chicken flavour compared to hexanal due to its much lower odour threshold (Fan et al., 2018; Shi and Ho, 1994).

Phospholipids are much more important in the development of aroma volatiles during the cooking of meat than triglycerides (Jayasena et al., 2013; Soncin et al., 2007; Mottram, 1998). This is attributed to the presence of a much higher proportion of unsaturated fatty acids, including significant amounts of polyunsaturated fatty acids such as arachidonic acid (20:4) in phospholipids (Mottram and Edwards, 1983). Saturated and unsaturated aldehydes having green, fatty, and tallowy aroma play a vital role in all cooked meat aroma (Mottram, 1998). The higher levels of unsaturated fatty acids compared with red meat generate more unsaturated volatile aldehydes which are vital in the determination of specific aromas of chicken meat (Noleau and Toulemonde, 1987). Further, the aroma of fat-fried food is reported to be due to 2,4-decadienal (Mottram, 1998). Therefore, aliphatic aldehydes contribute to the fatty flavours of cooked meat including poultry meat.

A different composition of volatiles that is responsible for the desirable flavours is formed due to rapid reactions taking place in cooked meat (Mottram, 1998) compared to those volatiles formed during long-term storage that lead to rancid off-flavours. In chicken meat, lack of α-tocopherol is, however, considered as the main reason for meat flavour deterioration and formation of undesirable 'warmed over flavour (WOF)' in chicken meat products (Shahidi, 2002). Lipid-derived compounds in meat volatiles have greater odour threshold values as opposed to sulphur- and nitrogen-containing heterocyclic compounds which make them less significant as contributor to aroma. Even at relatively low concentrations, heterocyclic compounds possess a significant effect on aroma due to their low odour threshold values (Mottram, 1998).

Table 3 Selected aldehydes identified in roasted and cooked chicken flavour

	Concentration (mg/kg)	
Aldehyde	Roasted chicken[a]	Cooked chicken meat[b]
Butanal	0.133	
Pentanal	0.319	
Hexanal	1.804	25.6
Heptanal	0.212	2.1
Octanal	0.422	2.3
Nonanal	0.467	1.7
Decanal	0.052	0.3
Undecanal	0.058	
Dodecanal	0.022	
Tridecanal	0.151	
Tetradecanal	0.125	0.2
Pentadecanal	0.383	
Hexadecanal	19.788	1.4
Heptadecanal	0.276	0.1
Octadecanal	2.664	
trans-2-Butenal	tr	
cis-2-Pentenal	tr	
trans-2-Pentenal	0.085	1.1
cis-2-Hexenal	tr	
trans-2-Hexenal	0.060	0.3
trans-2-Heptenal	0.104	1.2
cis-2-Octenal	0.004	
trans-2-Octenal	0.195	3.7
trans-2-Nonenal	0.084	
cis-2-Decenal	0.003	
trans-2-Decenal	0.139	1.0
cis-2-Undecenal	0.002	
trans-2-Undecenal	0.139	0.4
trans-2-Dodecenal	0.002	0.3
trans,cis-2,4-Nonadienal	tr	
trans,trans-2,4-Nonadienal	tr	0.3
trans,cis-2,4-Decadienal	0.051	1.0
trans,trans-2,4-Decadienal	0.137	5.2
trans,trans-2,4-Undecadienal	0.001	0.2

tr = trace.
[a] Noleau and Toulemonde (1986).
[b] Schroll et al. (1988).

3.2 Flavour compounds from the Maillard reaction

The Maillard reaction is one of the main chemical reactions that take place during cooking of meat including poultry meat. This typically occurs between amino compounds and reducing sugars and eventually results in a large number of compounds responsible for the flavour of any meat (Fan et al., 2018; Mottram, 1994a). During the initial stages of this reaction, Amadori products are formed via glycosylamine as a result of condensation of the carbonyl group of a reducing sugar with amino compounds. Various sugar dehydration and degradation products such as furfural and furanone derivatives, hydroxyketones, and dicarbonyl compounds are formed by rearranging and dehydration of the resulting product via deoxyosones. Subsequently, these compounds interact with other reactive components such as amines, amino acids, aldehydes, hydrogen sulphide, and ammonia, and the corresponding aroma compounds are formed (Calkins and Hodgen, 2007; Mottram, 1998).

Strecker degradation of amino acids by Maillard reaction-derived dicarbonyl compounds is an essentially associated reaction, during which an aldehyde is formed due to the decarboxylation and deamination of an amino acid, and an α-aminoketone or amino alcohol results from the dicarbonyl compound as well. Hydrogen sulphide, ammonia, and acetaldehyde are also formed by Strecker degradation when cysteine is used as the amino acid (Mottram, 1998). Maillard reaction-derived carbonyl compounds interact with these compounds forming intermediates that further involve in flavour-forming reactions. This eventually produces many important classes of flavour compounds including furans, pyrazines, pyrroles, oxazoles, thiophenes, thiazoles, and other heterocyclic compounds (Melton, 1999). The sulphur-containing compounds which can be derived either during the reaction of reducing sugars (ribose) with cysteine/methionine or with the peptides containing cysteine/methionine are the principal contributors to meat flavour (Zhao et al., 2019a,b; Cao et al., 2017; Hou et al., 2017). Ribose is considered as the most important flavour precursor in chicken (Meinert et al., 2009). In addition, thiamine has also been proved to be an important precursor that provides a wide range of sulphur compounds (Fan et al., 2018; Cerny, 2007; Aliani and Farmer, 2005). Due to the degradation process of nucleotides such as IMP, ribose is formed and it is then involved in a number of secondary reactions yielding a large number of volatile compounds (Kavitha and Modi, 2007). Hence, IMP is generally considered as the major nucleotide in muscle that imparts flavour to the meat (Qi et al., 2017b; Jayasena et al., 2013; Yamaguchi, 1991). In general, the IMP content in chicken meat is 75–122 mg/100 g meat (Maga, 1983), but significant differences among poultry breeds exist (Jayasena et al., 2014b; Guan et al., 2013; Jung et al., 2011; Rikimaru and Takahashi, 2010; Tang et al., 2009).

Formation of 2-methyl-3-furanthiol in chicken broth via the Maillard reaction involves interaction between ribose and sulphur-containing amino acids (cysteine or cystine) or peptide (glutathione). Glutathione liberates hydrogen sulphide rapidly during the initial stages of cooking, whereas cysteine does upon prolonged heating. Another important odourant in chicken broth, 2-furfurylthiol, is formed from the reaction between furfural and cysteine (Shi and Ho, 1994). In addition, the Maillard reaction-derived volatiles are the major components in meat grilled under severe conditions (Mottram, 1998).

3.3 Flavour compounds from lipid-Maillard interactions

Interaction of lipid with the Maillard reaction leads to the formation of a number of volatiles that have been identified in meat (Fan et al., 2018; Mottram, 1994b; Whitfield, 1992). Several thiazoles with C_4-C_8 *n*-alkyl substituents in the 2-position and some other alkylthiazoles with much longer 2-alkyl substituents (C_{13}-C_{15}) have been reported in fried chicken and heated chicken, respectively (Farmer and Mottram, 1994; Tang et al., 1983). Melton (1999) reported that 2-octyl-4,5-dimethylthiazole found in chicken meat is also a product from the interaction of Maillard reactions and lipids.

3.4 Heterocyclic compounds in poultry meat flavour

Heterocyclic compounds such as pyrazines, thiazoles, and oxazoles are usually considered to be responsible for the roast flavours in foods including poultry meat. Melton (1999) reported that a large number of heterocyclic compounds are associated with roasted, grilled, fried, or pressure-cooked meats, but not boiled meat, due to higher temperatures used in those cooking methods. Different alkyl pyrazines and two classes of bicyclic compounds, 6,7-dihydro-5(H)-cyclopentapyrazines and pyrrolopyrazines, were found in meat volatiles (Mottram, 1998). Both classes of compounds increase greatly with the increasing severity of heat treatment. However, Mottram (1991) reported that boiled meat contained higher levels of sulphur-containing heterocyclic compounds such as thiophenes, trithiolanes, trithianes that have low odour thresholds with sulphurous, onion-like and, sometimes, meaty aromas (Fors, 1983). These compounds contributed to the overall flavour and aroma of boiled meat.

Thermal degradation of cysteine and glutathione results in two major volatile compounds of fried chicken. At frying temperature (180°C), cysteine produced 3,5-dimethyl-1,2,4-trithiolanes and 2,4,6-trimethyl perhydro-1,3,5-dithiazines (thialdine), whereas only the former compound was produced by glutathione under the same conditions. These volatiles were identified in cooked chicken as well (Shi and Ho, 1994). In addition to 3,5-dimethyl-1,2,4-trithiolanes, three other alkyl-substituted trithiolanes, 3,5-diisobutyl-1,2,4-trithiolane, 3-methyl-5-butyl-1,2,4-trithiolane, and 3-methyl-5-pentyl-1,2,4-trithiolane, were identified in fried chicken flavour. The formation of latter two volatiles involves thermal and oxidative degradation of lipids, pentanal and hexanal (Shi and Ho, 1994). According to the same authors, flavour of deep-fat-fried foods such as fried chicken is attributed to lipid-derived aldehyde, 2,4-decadienal. Being an oxidation product of 2,4-decadienal, trans-4,5-epoxy-trans-2-decenal having a low odour threshold plays a key role in the flavour of fried chicken.

Heterocyclic compounds, mainly pyrazines, pyridines, pyrroles, and thiazoles, found in fried and roasted chicken are listed in Table 4. Alkylpyrazines have roasted, nut-like, or toasted flavour notes and are present in the flavours of fried chicken and roasted chicken but not in chicken broth. This confirms that the generation of pyrazines requires high temperature and low moisture. Shi and Ho (1994) reported that fried chicken flavour was further intensified by 2-pentylpyridine (strong fatty and tallow-like odour), 2-isobutyl-3,5-diisopropylpyridine (roasted cocoa-like aroma), 2-pentyl-4-methyl-5-ethylthiazole (strong paprika pepper flavour), 2-heptyl-4,5-dimethylthiazole (strong spicy flavour), and 2-octyl-4,5-dimethylthiazole (sweet fatty aroma).

4 Factors influencing the flavour formation in poultry meat

Research findings have shown that several pre-slaughter and post-mortem factors influence the flavour formation in poultry meat. Hence, producers and processors in the poultry meat sector, and even consumers, should have at least a general understanding on factors influencing the flavour formation in poultry meat in order to preserve and/or enhance the flavour compounds in poultry meat during production and processing.

The main determinants influencing the flavour formation in poultry meat have been identified as breed/strain of the chicken, diet of the bird, production system, presence of free amino acids and nucleotides, irradiation, high-pressure treatment, cooking, antioxidants, pH, sex, and ageing (Jayasena et al., 2013; Farmer, 1999). Different breeds/strains contain different levels of flavour precursors leading to various types and concentrations of volatile compounds (Fan et al., 2018; Jayasena et al., 2013, 2014b, 2015a). Additionally, chicken results in more unsaturated volatile aldehydes as compared to beef or lamb because their muscle contains higher levels of polyunsaturated fatty acids in the triglycerides than beef and lamb (Calkins and Hodgen, 2007). These compounds contribute to the specific aromas of chicken (Mottram, 1991, Noleau and Toulemonde, 1987) which confirms the effect of lipids on

Table 4 Heterocyclic compounds identified in fried and roasted chicken flavours

	Compound	Fried chicken[a]	Roasted chicken[b]
Pyrazines	Pyrazine	+	
	2-Methylpyrazine	+	+
	2,3- Dimethylpyrazine	+	+
	2,5-Dimethylpyrazine	+	
	2,6-Dimethylpyrazine	+	+
	Trimethylpyrazine	+	+
	2-Isopropylpyrazine	+	
	2-Methyl-3-ethylpyrazine	+	+
	2-Methyl-6(5)-ethylpyrazine	+	+
	2-Butylpyrazine	+	
	2,3-Dimethyl-5-ethylpyrazine	+	+
	2,5-Dimethyl-3-ethylpyrazine	+	
	2,6-Dimethyl-3-ethylpyrazine		+
	2,6-Diethylpyrazine	+	+
	2-Methyl-5,6-diethylpyrazine	+	
	2-Methyl-3,5-diethylpyrazine	+	
	2-Methyl-3-butylpyrazine	+	
	2-Methyl-5-vinylpyrazine	+	
	2-Methyl-6-vinylpyrazine	+	
	2-Isopropenylpyrazine	+	
	6,7-Dihydro-5*H*-cyclopentapyrazine	+	
	2-Methyl-6,7-dihydro-5*H*-cyclopentapyrazine	+	
Pyridines	Pyridine	+	+
	2-Methylpyridine	+	+
	3-Ethylpyridine	+	+
	4-Ethylpyridine	+	
	2-Methyl-5-ethylpyridine	+	+
	2-Ethyl-3-methylpyridine	+	
	2-Butylpyridine	+	
	2-Pentylpyridine	+	
Pyrroles	Pyrrole	+	+
	N-methylpyrrole		+
	2-Methylpyrrole	+	
	2-Ethylpyrrole		+
	N-acetylpyrrole		+
	2-Acetylpyrrole	+	

	Compound	Fried chicken[a]	Roasted chicken[b]
	2-Isobutylpyrrole	+	
	N-isobutylpyrrole	+	
	N-(2-butanoyl)pyrrole	+	
	N-furfurylpyrrole		+
Thiazoles	Thiazole	+	+
	2-Methylthiazole	+	+
	2,4,5-Trimethylthiazole	+	
	2-Methyl-4-ethylthiazole	+	
	2-Methyl-5-ethylthiazole	+	+
	2,4-Dimethyl-5-ethylthiazole	+	
	2-Isopropyl-4,5-dimethylthiazole	+	
	2,5-Dimethyl-4-butylthiazole	+	
	2-Isopropyl-4-ethyl-5-methylthiazole	+	
	2-Butyl-4,5-dimethylthiazole	+	
	2-Butyl-4-methyl-5-ethylthiazole	+	
	2-Pentyl-4,5-dimethylthiazole	+	
	2-Hexyl-4,5-dimethylthiazole	+	
	2-Heptyl-4,5-dimethylthiazole	+	
	2-Heptyl-4-ethyl-5-methylthiazole	+	
	2-Octyl-4,5-dimethylthiazole	+	

+ Present.
[a]Tang et al. (1983).
[b]Noleau and Toulemonde (1986).

the flavour of chicken primarily via the differences in fatty acid composition and the resulting carbonyls (Perez-Alvarez et al., 2010). Furthermore, positive correlations between the flavour of chicken meat and the intramuscular contents of amino acids and nucleotides have been demonstrated (Takahashi et al., 2012; Rikimaru and Takahashi, 2010; Kurihara, 1987). Effect of sex on chicken meat flavour was demonstrated by many researchers, although the results were not consistent. Meat from male birds received higher scores for flavour as opposed to that from female birds (Farmer, 1999; Ramaswamy and Richards, 1982). However, it was also shown that the breast and leg meat of female birds were preferred to those of male birds. Meanwhile, freeze-dried broth made from female Korean native chickens had significantly higher IMP and arachidonic acid contents compared to that from male chickens. In contrast, freeze-dried broth prepared with male Korean native chickens contained higher levels of inosine, linoleic acid, glycine, alanine, lysine, and serine ($P < 0.05$). However, the sex had no effect on glutamic acid, oleic acid, and DHA contents in freeze-dried broth (Jayasena et al., 2014a). Many other researchers reported no significant relationship between the two parameters (Farmer, 1999).

Diet of the bird also plays a vital role towards the flavour of chicken meat (Perez-Alvarez et al., 2010; Fanatico et al., 2007) because the types of feed mainly affect the fatty acid profile of meat. In addition, production system has a more varying effect on taste and flavour of chicken meat (Farmer, 1999). Many research studies claim that chicken meat from alternative production systems such as free range, organic, or some combination is characterized by a better taste compared to conventional production systems (Lin et al., 2014; Horsted et al., 2012; Sogunle et al., 2012; Seabra et al., 2001; Culioli et al., 1990). Processing steps such as chiller ageing, cooking, irradiation, high pressure, and cold plasma treatment also affect the flavour of chicken meat. Post-mortem ageing results in the generation of many chemical compounds responsible for poultry meat flavour including sugars, organic acids, peptides, free amino acids (Spanier et al., 2004; Yano et al., 1995) and thereby leads to increased flavour. Cooking is important in flavour development, and it affects the acceptability and volatile flavour components of poultry meat (Sanudo et al., 2000). Cooking methods such as roasting, grilling, frying, and pressure cooking generates many pyrazines, pyridines, pyrroles, and thiazoles compared to boiling of chicken meat (Shi and Ho, 1994). Irradiation also affects flavour and aroma of chicken meat primarily through the production of free radicals. Aldehydes (hexanal, pentanal, heptanal, octanal, and nonanal) and sulphur volatiles mainly dimethyl trisulphide generated during irradiation result in the associated off-odour (Ahn et al., 2016a,b; Perez-Alvarez et al., 2010; Patterson and Stevenson, 1995). Effect of high-pressure treatment on flavour of meat including chicken meat had been variable over the years. Non-thermal plasma technology – an emerging technology for enhancing the microbiological safety of various products including poultry meat – can be considered a profitable tool in agri-food industries such as the poultry industry (Gavahian et al., 2019). Free radicals generated during plasma treatment can trigger lipid and/or protein oxidation and may negatively influence on sensory properties of chicken meat.

Lipid oxidation has been considered as the primary non-microbial cause of quality deterioration in meat and meat products (Domínguez et al., 2019), in particular flavour deterioration and development of off-flavour known as oxidized flavours or 'warmed-over flavour (WOF)' in poultry meat (Estévez, 2015; Shi and Ho, 1994). Generally, WOF is generated in precooked meat and meat products reheated after a short period of refrigerated storage and mainly attributed to the autoxidation of polyunsaturated fatty acids, mainly in the phospholipids (Byrne et al., 2002; Graf and Panter, 1991). It has been documented that lipid oxidation of muscle foods occurs in the order, fish > poultry (chicken and turkey) > pork > beef > lamb, which is attributed to the level of unsaturated fatty acids present in phospholipids of each of these muscle tissues (Byrne et al., 2002; Rhee et al., 1996; Tichivangana and Morrissey, 1985). WOF development is also related to rancid, and sulphur/rubber sensory notes and a parallel reduction in chicken 'meaty' flavour can also be observed (Byrne et al., 2002; Shi and Ho, 1994). A remarkable off-flavour problem in mechanically deboned chicken meat was reported, and it was attributed to rancidity of fat (Shi and Ho, 1994). Iron, being the major catalyst for the initiation of lipid peroxidation by generation of hydroxyl radicals, is therefore largely responsible for the off-flavour formation (Xiong et al., 2020; Buettner and Jurkiewicz, 1996). However, more recent studies indicated that raw poultry meats are highly resistant to oxidative changes due to various antioxidants present in chicken meat (Min et al., 2010). Meanwhile, it has also been reported that increasing cooking temperature is associated with increased roasted, toasted, and bitter sensory notes (Perez-Alvarez et al., 2010). Studies conducted to determine the critical changes in odour compounds of boiled chicken proved that refrigerated storage and reheating of boiled chicken leading to WOF was due to loss of meaty, chicken-like, and sweet odour notes, and the formation of green, cardboard-like, and metallic off-odours by the secondary by-products of lipid oxidation. These changes were caused primarily by an increase in hexanal (sevenfold) and sixfold decrease in both (*E,E*)-2,4-decadienal and 2-furfurylthiol (Kerler and Grosch, 1997). Rhee et al. (2005) reported that 'cardboard' flavour and 'sour' flavour intensity increased with storage time for chicken breasts, while 'cooked chicken' intensity decreased and 'cardboard' intensity increased for chicken thighs. Results further illustrated relative storage effects on intensity of the species-specific natural meat flavours and 'cardboard' flavour, with day-3 and day-6 scores expressed relative to day-0 scores. However, chicken exhibited the slowest rate of decrease in species-specific natural flavour intensity and the slowest rate of increase in 'cardboard' flavour intensity compared to pork and beef.

5 Advances in strategies for preserving or enhancing the flavour of poultry meat

Chemical compounds responsible for flavour of poultry meat can be preserved and/or enhanced by manipulating the factors discussed in the above section.

5.1 *Breed/strain of chicken*

The flavour of chicken meat can be improved by selecting the suitable breed/strain of chicken as different breeds/strains contain different levels of flavour precursors such as fat, IMP, amino acids, arachidonic acid, EPA, and DHA. Increased palatability of indigenous chicken compared to broilers is well documented by many authors. Superior flavour of Korean native/farm chicken,

Japanese Hinai-jidori chicken, Chinese Wenchang, Xianju, and Beijing Youji to broilers has been reported (Kim et al., 2021; Fan et al., 2018; Takahashi, 2018; Jung et al., 2011; Kiyohara et al., 2011; Tang et al., 2009). Further, it was reported that concentrations of IMP – the major nucleotide in muscle that imparts flavour to the cooked meat (Qi et al., 2017b; Jayasena et al., 2013) – in chicken meat from indigenous and slow-growing genotypes were higher than those in broiler chicken lines (Table 5). In addition, Korean native chicken contained significantly higher concentrations of taste-active compounds such as reducing sugars, glutamic acid, linoleic acid, arachidonic acid, and DHA in meat than commercial broilers did (Jayasena et al., 2014b; Lee et al., 2012). In addition, Jayasena et al. (2015a) revealed that defatted freeze-dried soup prepared from Korean native chicken was qualitatively superior due to enhanced nutritional and taste-related factors, that is, a higher content of IMP, inosine, cysteine, and polyunsaturated fatty acids (PUFAs) including essential fatty acids and DHA compared to that made with commercial broilers.

Wattanachant et al. (2004) showed that Thai indigenous chicken had significantly higher glutamic acid and DHA contents than the broilers. Similarly, chicken broth of Beijing Youji – an indigenous and rare chicken breed in China – contained higher contents of PUFAs (linoleic acid, arachidonic acid, EPA,

Table 5 Inosine-5′-monophosphate contents of different chicken breeds

Breed/strain	Inosine-5′-monophosphate (mg/g)
Broilers[a]	1.44
Hinai-jidori chicken[a]	1.57
Wenchang[b]	1.52
Xianju[b]	1.44
Avian[b]	0.95
Lingnanhuang[b]	0.97
Korean native chicken[c]	2.26–2.31
Broilers[c]	0.94–1.54
Ninghai chicken[d]	2.48
Frizzle chicken[d]	2.39
Ninghai xiang chicken[d]	2.51
Zhenning loquat chicken[d]	2.11
Arbor Acres plus[d]	1.90
Chinese yellow-feather broiler[e]	1.76

[a] Rikimaru and Takahashi (2010).
[b] Tang et al. (2009).
[c] Jayasena et al. (2014b) and Jung et al. (2011).
[d] Guan et al. (2013).
[e] Qi et al. (2017b).

and DHA) and water-soluble flavour precursors, including ribose, cysteine, methionine, and thiamine than that of commercial broilers (Fan et al., 2018). A recent study revealed that Village/Non-descriptive chicken strain had a significantly higher DHA compared with Naked Neck and Long Legged chicken strains. Among the same strains, the highest preference for flavour and overall acceptability was obtained by Village/Non-descriptive chicken strain during the sensory evaluation (Ranasinghe and Jayasena, 2019, 2020). Compared to broilers, Korean native chickens have greater levels of several unique proteins or peptides including phosphoglucomutase 1, NRXN3 96 kDa protein, HSPB1 22 kDa protein, TF 78 kDa protein, which were mostly attributed to degradation of muscle protein during post-mortem ageing period and thereby contributed to the characteristic flavour in the cooked meat (Liu et al., 2012).

5.2 Lipid class and fatty acid composition

Aldehydes produced during lipid degradation are probably characteristic features of certain species. Especially, the 2-alkenals such as hexenal, heptenal, octenal, nonenal, undecenal, and dodecenal, as well as other aldehydes, including octanal, nonanal, decanal, and decadienal, are related to both chicken-specific aroma and flavour (Ramarathnam et al., 1993). Greater levels of PUFAs in the triglycerides were reported in poultry and pork muscles as opposed to lamb or beef (Calkins and Hodgen, 2007) resulting in more unsaturated volatile aldehydes responsible for the specific aromas of chicken and pork (Mottram, 1991; Noleau and Toulemonde, 1987). Hence, better and characteristic flavour in poultry meat can be achieved through changing the fatty acid profile of meat, that is, increasing PUFA content such as linoleic acid, arachidonic acid, EPA, and DHA via genetic regulation and different feeds.

The possibility of genetic regulation of arachidonic acid content in chicken meat was recently tested by Rikimaru et al. (2016). In their study, it was supposed that fatty acid desaturase 1 (FADS1), fatty acid desaturase 2 (FADS2), and FADS1/2 clusters are the keys that regulate arachidonic acid and DHA content in chicken meat. The polymorphisms of FADS1 and FADS2 were genotyped and their association with the fatty acid profile in the Hinai-jidori meat was investigated. The arachidonic acid and DHA content of the G-G haplotype was significantly higher than that of the A-A haplotype. Based on the findings, Rikimaru et al. (2016) concluded that the single-nucleotide polymorphisms, in the FADS1 and FADS2 gene cluster, increase the content of arachidonic acid and DHA in Hinai-jidori chicken. At present, they are testing whether the flavour of Jidori meat can be improved by using molecular breeding and marker-assisted selection techniques (Takahashi, 2018). In a study conducted to investigate quantitative trait locus (QTL) analyses for fatty acid composition traits in Korean native chicken, Jin et al. (2018) detected 30 QTLs (12 in thigh and 18 in breast meat) by the half-sib analysis and 7 QTLs (3 in thigh and 4 in breast meat) by the full-sib analysis. With further verification of the QTL regions using additional markers and positional candidate gene studies, these results can provide valuable information for determining causative mutations affecting the fatty acid composition of Korean native chicken meat.

5.3 Intramuscular contents of amino acids and nucleotides

Positive correlations between the flavour of chicken meat and the intramuscular contents of amino acids, including glutamic acid and nucleotides, such as IMP have been demonstrated (Takahashi et al., 2012; Rikimaru and Takahashi, 2010; Kurihara, 1987). Glutamate and IMP are influential compounds for the taste of chicken meat, in particular umami taste (Qi et al., 2017b; Fujimura et al., 1996). In this regards, glutamic acid plays a key role singly or in combination with other taste-related compounds (Kurihara, 1987). Improvement detected in meat taste during storage was attributed to free amino acids (Rikimaru and Takahashi, 2010). Similarly, conditioning enhanced the free amino acids content and thereby affected the umami taste positively (Nishimura et al., 1988). In contrast, Fukunaga et al. (1989) has mentioned that phenylalanine content is closely associated with bitter taste in meat. In a series of sensory experiments in which isolated precursors were added to ground raw chicken before cooking, trained sensory panellists described the flavour of the cooked product as 'roasted' and 'vegetable soup' when thiamine precursor was added to ground raw chicken. On using cysteine and ribose, the same panel noted the flavours of samples as 'chicken' and 'savory' (Aliani and Farmer, 2005).

Different volatile compounds were, however, generated when different sugars were combined with amino groups under different conditions. Cysteine and glucose produced mainly sulphur compounds, whereas more pyrazines and furans were formed by the combination of cysteine with glucose under oxidation conditions (Tai and Ho, 1997). Sulphur-containing compounds such as thiophenes, thiazoles, and cyclic polysulphides were generated by Maillard volatile compounds from glutathione/glucose systems at pH 6 and 8, whereas a more acidic pH often resulted in furans as the end products. Glutathione sulphonic acid formed by oxidation of glutathione produces furans, carbonyls, pyrroles, and pyrazines with glucose. Sulphur-containing compounds are not generated when glutathione is oxidized (Tai and Ho, 1998).

IMP content in chicken meat is significantly varied with the breed/strain (Table 5), meat cut, age of the birds, storage condition, and cooking process. In general, breast meat contained higher IMP levels than leg meat (Zhang et al., 2020; Jayasena et al., 2015b). Furthermore, it has been shown that IMP content in meat increases with the age of the chicken (Rikimaru and Takahashi, 2010). The decreases in IMP content in meat during cooking could result from both their migration from the meat into the broth and the thermal degradation of IMP to inosine (Qi et al., 2017b; Jayasena et al., 2015b). Hence, above-mentioned factors should be taken into consideration in order to improve/preserve the flavour compounds with the aim of producing the most flavourful and consistent products.

Recently Zhang et al. (2020) elucidated the correlations between the IMP content in different parts of Jingyuan chicken and mRNA expression of adenylate kinase 1 (*AK1*) and verified the effect of *AK1* expression on the IMP content. The study revealed that the expression of *AK1* mRNA in the breast muscles of cocks and hens was significantly higher than that in the leg muscles. Moreover, *AK1* mRNA expression in cock breast muscles was negatively correlated with IMP content, whereas its expression in cock leg muscles was positively correlated with IMP content. In contrast, the expression of *AK1* mRNA in hen breast and leg muscles was significantly positively correlated with IMP content. These findings provide a scientific basis for enhancing the meat flavour of Jingyuan chicken and promoting the development and utilization of local variety resources, as well as constituting a basis for screening IMP-regulated genes.

5.4 Diet

Diet of the bird also plays a vital role towards the flavour of chicken meat (Perez-Alvarez et al., 2010; Fanatico et al., 2007). Although it was considered that poultry flavour could be improved by manipulating the diet (Fanatico et al., 2007), diet can either positively or negatively affect the flavour of chicken meat. Corn-fed chicken and arachidonic acid-enriched diet-fed chickens have produced tastier meat (Takahashi et al., 2012; Kiyohara et al., 2011; Lyon et al., 2004) while diets supplemented with fish meal have negatively affected the flavour of chicken meat (Poste, 1990). Although the off-flavour was less apparent after cooking, it increased when the samples were kept overnight under refrigerated condition and re-assessed 24 h later (Poste, 1990). However, chicken meat flavour was not adversely influenced when the diet was incorporated with 1.2% of formic acid or up to 7.5% of shrimp by-catch composed mainly of brush tooth lizardfish and herring (Poste, 1990). Broilers fed a dietary medicinal herb extract mix (mulberry leaf, Japanese honeysuckle, and goldthread; 48.5:48.5:3.0) at 0.3% level produced tastier breast meat than those fed control diet (Jang et al., 2008). Furthermore, Bou et al. (2001) demonstrated that the flavour of chicken meat was positively influenced by feed supplements. In this regards, variations in sensory scores (rancid aroma, rancid flavour, and acceptability) were attributed to dietary fat sources and α-tocopheryl acetate content. However, sensory panellists did not note any influence of ascorbic acid on the same sensory scores. In addition, Kim et al. (2009) demonstrated that supplementing broiler chicken diets with 4% garlic husk resulted in lower thiobarbituric acid reactive substance (TBARS) values and improved flavour scores in chicken meat.

Diet can also be used to increase the taste-active PUFA content of poultry meat. A diet incorporated with arachidonic acid-enriched oil contributed to better sensory scores for total taste intensity, umami, kokumi, aftertaste, and saltiness in both Hinai-jidori and broiler chicken meats compared with diets containing palm oil and corn oil (Takahashi et al., 2012; Kiyohara et al., 2011). In addition, sensory data and arachidonic acid content in thigh meat of broiler chickens were positively correlated ($P < 0.05$). Huang et al. (1990) demonstrated that DHA in thigh flesh could be increased without causing fishy flavour by feeding up to 3% fish oil to chickens. Additionally, Oh et al. (2012) demonstrated that the addition of local ingredients, such as perilla meal with conventional sources (flaxseed and fish oil), could be used to obtain value-enhanced meat by enhancing ω-3 PUFA content in Korean native chicken meat.

Administration of the bacterium, *Rhodopseudomonas palustris*, in drinking water significantly increased glutamic acid content and improved the sensory attributes including flavour and overall acceptability in broiler breast fillets (Xu et al., 2014). Effects of diets supplemented with exogenous IMP at 0.0%, 0.05%, 0.1%, 0.2%, or 0.3% on flavour compounds of chicken were investigated by Yan et al. (2018). Supplementation with 0.2–0.3% exogenous IMP promoted the deposition of IMP in the muscle tissues and increased free amino acids and delicious amino acids (aspartic acid, serine, glutamic acid, glycine, alanine, arginine) content in breast and thigh muscles. A recent work by Altmann et al. (2020) revealed that chicken fed a diet with microalga Spirulina (*Arthrospira platensis*) scored higher in terms of umami and chicken flavour than those fed a diet with soy meal. In addition, Spirulina-fed chicken meat had a reduced off-odour (barn odour). In contrast, comparable scores for umami, chicken flavour, and barn odour were observed between chicken fed a diet with partially de-fatted *Hermetia illucens* larval meal and soy meal-fed chicken. However, dietary incorporation of Spirulina, individually and in combination with exogenous enzymes, had no significant effect on sensorial characteristics such as flavour and off-flavour notes (Pestana et al., 2020).

Wang et al. (2017) suggested that probiotics can improve chicken meat flavour. *Pediococcus pentosaceus*, *Bacillus cereus* var. *albolactis*, *Bacillus macerans*, *Bacillus subtilis*, *Lactobacillus plantarum*, and *Issatchenkia orientalis* were isolated from the mucous membrane of various intestinal segments (jejunum, ileum, and caecum) of 154-day-old Caoke chickens and Qingjiama chickens raised in a free-range system and probiotic powders were prepared. Qingjiaoma chickens were then supplemented with the prepared probiotic powders in six different combinations. Compared to chickens without probiotic supplement, chickens supplemented with *P. pentosaceus* had more characteristic flavour compounds in the breast meat, in particular higher concentrations of (*E*)-2-heptenal, (*E,E*)-2,4-nonadienal, and certain C6-C9 unsaturated fatty acids. This resulted in a stronger chicken-fatty or fatty odour which directly improved the flavour.

5.5 Production system

Several studies have revealed that alternative production systems such as free-range, organic, or some combination positively affect the sensory quality of poultry meat including flavour. Free-range chicken received significantly higher taste panel scores for flavour than the cage group (Lin et al., 2014). In addition, the production system influenced the taste and flavour of the grilled chicken with birds on free-range having higher values compared to those managed on deep litter (Sogunle et al., 2012). There was a huge difference in the sensory profile between the conventional standard and the organic niche products. The positive aroma attributes 'chicken' and 'bouillon' were significantly more pronounced, and the negative attribute 'iron/liver' was significantly lower in the organic niche broilers than in the conventional standard broilers (Horsted et al., 2012). Similarly, the panellists classified meat from birds on free-range with the highest value for flavour (Seabra et al., 2001). Culioli et al. (1990) compared the meat characteristics of chicken raised according to the French 'label fermier' norms of production with those of standard type chickens raised intensively in confinement conditions. The sensory evaluation showed that the 'label' meat had a stronger flavour compared to standard meat.

Sogunle et al. (2012) stated that free-range production can reduce stress due to low bird density per unit area and increase comfort and bird welfare, thereby leading to products with better taste and flavour compared to conventionally produced broiler chicken (Fanatico et al., 2006). Birds are reared for a longer period of time in the unconventional rearing systems, which allows their meat to attain more intense meat flavours (Michalczuk et al., 2014; Horsted et al., 2012; Farmer, 1999). Furthermore, breast meat from outdoor-reared chicken has shown higher PUFA levels and ω-3 fatty acid levels compared to that from indoor-reared chicken (Michalczuk et al., 2014), which might contribute to the increased flavour of meat from outdoor-reared chicken.

Several prior studies, however, have demonstrated that production systems had no significant effect on the flavour of poultry meat. For instance, Cheng et al. (2008) and Husak et al. (2008) indicated that there were no significant differences in flavour and aroma of breast and thigh meats among conventionally reared, free-range, and organic chicken. Similarly, no difference was found for flavour attribute between conventional and Label Rouge-type chicken breast fillets (Smith et al., 2012). The production system alone did not have an effect on flavour of chicken meat reared in indoor or outdoor systems (Fanatico et al., 2007; Garcia et al., 1995). Furthermore, the intake of a subterranean clover pasture had no impact on the flavour of broiler meat. However, the consumer panel classified the meat from free-range broilers with greater scores for overall acceptance compared to indoor-reared broilers (Ponte et al., 2008). In addition, several prior studies have compared conventionally and alternatively reared broiler strains for meat sensory attributes. Due to many variables involved in those studies, no obvious trend for conventional or alternative strains for taste evaluation was detected (Smith et al., 2012).

Several factors such as breed, age at slaughter, diet, outdoor access, and housing conditions were very different between the alternative (free-range, organic, and Label Rouge-type) and standard broilers, which resulted in very different sensory profiles of these products as indicated above. Furthermore, Horsted et al. (2010) emphasized the significance of selecting appropriate birds to be suitable for alternative rearing systems as in such productions systems slow-growing birds have proven well due to their lower growth rate and lower body weight which facilitate better health status and lesser mortality (Mikulski et al., 2011; Fanatico et al., 2008).

5.6 Ageing

Post-mortem ageing generates many chemical flavour compounds including sugars, organic acids, peptides, free amino acids, and metabolites of adenine nucleotide metabolism and, therefore, is considered as one of the most important factors that determine the final flavour of meat (Liu et al., 2012). Amino acids and peptide levels are known to increase with post-mortem ageing in muscle (Spanier et al., 2004; Yano et al., 1995). Nishimura et al. (1988) also noted that an increase in free amino acids during conditioning is responsible for the meaty taste. These components serve either directly as flavour components or as a pool of reactive flavour intermediates that form many of the characteristic meat flavours after cooking (Spanier et al., 1997). Because the content of free amino acids is interrelated with the taste of chicken meat, ageing can improve the flavour of poultry meat.

A recent study by Li et al. (2018) revealed that storage temperature and storage time can significantly influence on the taste-active compounds and volatiles of pasteurized chicken soup. The results showed that each 5′-nucleotide and free amino acids, except lysine and aspartic acid, continuously decreased with the extent of storage time. However, super chilled storage (−2.5 ± 1°C) could better extend the flavour retention of chicken soup after 21 days compared with refrigerated storage (4 ± 1°C). Chicken soups from super chilled storage had significantly higher contents of free amino acids associated with sweet and umami notes and 5′-nucleotides such as IMP and AMP but lower contents of free amino acids responsible for bitter taste and hypoxanthine than those from refrigerated storage. Significantly lower values were found for ketones and hydrocarbons, whereas higher values were observed for aldehydes and alcohols during super chilled storage. The IMP content of broiler breast meat reached maximal level resulting in more pleasant flavour at 8 h after slaughter than that of the chicken immediately after slaughter (Terasaki et al., 1965) and then decreased gradually when the meat was stored at 4°C.

5.7 Cooking

Type and concentration of flavour compounds generated during thermal treatment depend on the method of cooking. For instance, cooking methods such as roasting, grilling, frying, and pressure cooking generates many more pyrazines, pyridines, pyrroles, and thiazoles compared to boiling of chicken meat (Shi and Ho, 1994). Perez-Alvarez et al. (2010) have revealed that many flavour compounds undergo alterations throughout storage and cooking. In addition, a relationship has been confirmed between raw and cooked meat in the contents of free amino acids, IMP, and peptides (Chikuni et al., 2002). Intramuscular content of taste-active components such as IMP, free amino acids, and reducing sugars decreased on cooking, and the decline may be caused by leaching of these substances from meat samples (Jayasena et al., 2014b, 2015b; Chikuni et al., 2002)

The flavour attributes of chicken meat were slightly, but positively, affected by the low temperature for a prolonged time (53°C and 58°C for 30 h) cooking (Christensen et al., 2012) where time had the main impact on the increasing intensity of cooked meat flavour. This confirms the fact that flavour compounds are mainly generated at higher temperatures and their intensities at low temperatures are low. Similarly, cooking of Indian country hen meat resulted in high levels of IMP degradation compounds at high temperatures leading to improved flavour (Vani et al., 2006).

Higher temperatures (above 100°C) involved in cooking methods such as roasting, grilling, frying, and pressure cooking compared to boiling favour the formation of a number of heterocyclic compounds found in the aroma of cooked meats (Melton, 1999). In particular, high temperature and low moisture conditions accelerate the generation of pyrazines. The occurrence of heterocyclic compounds due to the high temperature and low moisture combination is responsible for the flavours of chicken meat prepared by roasting, grilling, frying, or pressure cooking. It is further governed by the concentration and characteristics of the compounds formed via Maillard reaction, lipids degradation, and Maillard-lipid interactions discussed earlier

(Melton, 1999). Hence, the flavour of poultry meat can be improved by using the suitable cooking method with suitable conditions.

Recently, it has been reported that stewing time had significant effects on the taste-active and volatile compounds found in chicken soup (Qi et al., 2017a) and chicken meat (Qi et al., 2017b). Stewing improved the aroma profiles but decreased the content of taste components such as free amino acids and 5′-nucleotides in chicken meat. Based on the findings, stewing for 2 h was suggested as the appropriate cooking time to balance umami taste and aroma in stewed chicken meat (Qi et al., 2017b). The volatile compound concentrations in chicken soup were significantly different after different stewing times. In general, the contents of aldehydes, alcohols, ketones, furans, and esters increased significantly. However, there was a significant decrease in the aldehydes after stewing for 2 h. Sulphur-containing compounds and other compounds were only observed after stewing for 1–2 h (Qi et al., 2017a).

Jung et al. (2019) demonstrated that the broth quality of the chicken soup can be improved by heating with ultrasound treatment. The broth flavour of the chicken soup manufactured with the ultrasound treatment (45 kHz and 1.6 W cm^{-2}) in a water bath at 85°C received a higher score than that of the control. In contrast, no differences in the sensory properties of the chicken breast meat from the chicken soup were detected between the control and ultrasound treatment. Evidence has shown that sugar-smoking can successfully replace conventional wood smoking and significantly reduce the polycyclic aromatic hydrocarbon content in smoked foods (Chen et al., 2013). Further, sugar-smoking contributes to improving flavour attributes of meat products. The volatile fingerprinting results revealed heterocyclic, aldehydes, alcohol, and hydrocarbon compounds as the characteristic flavour compounds formed in sugar-smoked chicken. Furthermore, the flavour of sugar-smoked chicken was mainly derived from chicken skin, and sugar-smoked process improved the flavour of chicken skin (Chang et al., 2021)

5.8 Irradiation

It has been reported that irradiated raw chicken meat gave off a bloody and sweet aroma (Heath et al., 1990). Radiolytic degradation of sulphur-containing amino acids (Ahn, 2002) and lipid oxidation (Jo and Ahn, 2000) yield increased levels of sulphur-containing volatiles with cabbage-like or rotten vegetable-like putrid odours. In contrast, more intense 'fresh chickeny', bloody, and sweet aromatic odours were detected in irradiated raw chicken compared to non-irradiated samples (Hashim et al., 1995). It was further determined that cooked irradiated frozen dark chicken meat had a more intense chicken flavour than did non-irradiated samples. However, very little negative effect was expressed by electron beam irradiation on the flavour of preheated chicken breast meat

(Rababah et al., 2006). The mechanisms of volatile production from irradiated amino acids were well documented by Ahn et al. (2016a,b) and Feng et al. (2016). The sensory attributes of chicken meat treated with gamma irradiation decreased with the increase in irradiation dose and with the advancement in the duration of storage (Arshad et al., 2019). Further, irradiation-associated odours can be decreased through modified atmosphere packaging, freezing before irradiation, and addition of antioxidants (Jo et al., 1999, 2001). The detrimental effects of irradiation can be reduced or eliminated by ensuring oxygen-permeable packaging after irradiation. Vacuum packaging retains irradiation-generated sulphur-containing compounds, but re-packaging meat in oxygen-permeable materials dissipates these compounds (Nam and Ahn, 2003). Rababah et al. (2006) demonstrated that lipid oxidation in both irradiated and non-irradiated chicken meat could be decreased by infusing plant extracts and grape seed and green tea extracts into chicken meat.

5.9 High-pressure treatment

Sensory quality of various meat products was not affected by high-pressure treatment (Hayman et al., 2004). However, Kruk et al. (2011) demonstrated that high-pressure treatment at 300 MPa improved the flavour and taste of chicken meat as opposed to 450 MPa treatment. Pressure treatments at approximately 300 MPa at room temperature initiated the changes that eventually led to catalysis of lipid oxidation in pressure-processed meat (Cheah and Ledward, 1996). Additionally, high-pressure treatment of chicken breast at 400 MPa significantly changed the levels of some volatile compounds (Rivas-Canedo et al., 2008). The amount of alcohols and aldehydes decreased while 2,3-butanedione and 2-butanone were more abundant in high pressure-treated meats. This could have an impact on flavour, especially the aroma strength. High-pressure treatment may also accelerate denaturation of muscle proteins and pressure-induced increases in proteolytic enzyme activity in the muscle that have a bearing on food flavour.

5.10 Plasma treatment

Misra and Jo (2017) stated that plasma technologies could be used to at least partially replace conventional decontamination technologies in meat industry in the near future due to the rapid on-going developments in plasma science. However, sensory evaluation data on plasma-treated chicken meat are scarce: only few studies have been reported so far regarding the effect of plasma on sensory properties of chicken meat. For instance, jet plasma treatment did not adversely affect the sensory properties of sliced chicken (Aly, 2013). Similar appearance, flavour, and texture were observed between the jet plasma-treated samples and the control sample. On the other hand, it was reported that flexible thin-layer dielectric barrier discharge plasma treatment can induce unpleasant flavours in the chicken breast when plasma exposure time was increased from 2.5 min to 10 min ($P < 0.05$; Lee et al., 2016). The chicken breast treated with plasma had a less intensity of pleasant chicken flavour and a higher intensity of an unpleasant flavour.

Free radicals generated during plasma treatment can trigger lipid and/or protein oxidation and generate secondary oxidation products such as alkanes, alkenes, aldehydes, alcohols, ketones, and acids (Kim et al., 2013). These molecules produce fishy, metallic, rancid, and oxidized flavours (Kochhar, 1996). However, only a limited number of scientific studies have been published on the plasma-induced lipid oxidation in the poultry meat products. Lee et al. (2016) demonstrated that flexible thin-layer dielectric barrier discharge plasma treatment for 10 min did not induce lipid oxidation in the chicken breast. More importantly, the reported TBARS values were below the detectable threshold of 0.5–2.0 mg malondialdehyde (MDA)/kg (Kim et al., 2016) so that off-flavours were likely to be undetectable.

5.11 Additives

Synthetic antioxidants are being replaced with plant extracts in several meat products to control lipid oxidation mainly due to toxicological concerns over the former compounds (Ahn et al., 1998). Rababah et al. (2006) stated that plant extracts with high content of antioxidants such as grape seed extract, green tea extracts, and their combinations were effective in minimizing lipid oxidation products such as pentanal and hexanal from chicken meat. Thus, these additives may retard the formation of off-flavours in chicken meat. Jahan et al. (2005) revealed that PUFAs and α-tocopherol contents were strongly related ($r^2 = 0.96$ and 0.99, respectively) to total flavour component abundance in cooked chicken meat. In contrast, no relationship between PUFAs and abundance of nitrogen-rich components, such as pyrazines and pyridines, was reported. They further reported that two antioxidant enzymes, glutathione peroxidase and glutathione reductase, were related to chicken flavour components. Therefore, suitable antioxidants may retard the formation of off-flavours in poultry meat leading to preserving and/or improving the existing flavour.

The utilization of AMP is mainly proposed to enrich flavour and as additives for specific nutrition in food industry (Barido and Lee, 2021; Wang et al., 2013). Recently, spent hen meat was subjected to immersed marination for 24 h at 4 ± 2C in different concentrations of AMP solution (Barido and Lee, 2021). IMP content of spent hen breast meat was significantly elevated with the increased level of AMP in the marinade. The observed flavour enhancement was attributed to conversion of AMP to IMP, which intensifies the umami-like flavour. Addition of *Cordyceps militaris* mushrooms at 2% enriched the flavour and taste-related compounds, particularly 5′-AMP and umami-related free amino acid compounds (aspartic acid and glutamic acid) in retorted Samgyetang – a traditional Korean chicken soup (Barido et al., 2020).

6 Conclusion

The Maillard reaction, thermal degradation of lipids, and Maillard–lipid interactions are the main pathways by which a large number of flavour and aroma compounds responsible for poultry meat flavour are formed during cooking. Volatile compounds generated from the Maillard reaction and lipid oxidation including 2-methyl-3-furanthiol, 2-furfurylthiol, methionol, 2,4,5-trimethylthiazole, nonanol, 2-trans-nonenal, 2-formyl-5-methylthiophene, *p*-cresol, 2-trans-4-trans-nonadienal, 2-trans-4-trans-decadienal, 2-undecenal, β-ionone, γ-decalactone, and γ-dodecalactone are obviously the major sources of poultry meat flavour. However, 2-methyl-3-furanthiol that is produced from the reaction between ribose and cysteine or cystine and from degradation of thiamin is considered as the most important compound in chicken flavour. A large number of heterocyclic compounds are formed during roasting, grilling, frying, or pressure cooking of chicken meat due to higher temperature and low moisture conditions used in those cooking methods but not during the boiling of the meat. Flavour of cooked poultry meat is affected by the breed/strain, diet, intramuscular content of free amino acids and nucleotides, classes of the lipids and fatty acid composition, irradiation, cold plasma, high pressure, cooking, antioxidants, and ageing. In addition, the possibility of genetic regulation of taste-active compounds in chicken meat is being tested. Flavour of poultry meat can be preserved and/or improved by effective manipulation of the above-mentioned factors.

7 Future trends in research

Currently available data reveals that cold plasma technology could be utilized as a novel antimicrobial intervention for the inactivation of pathogens and improvement of safety of foods of animal origin (Fig. 1). However, literature reveals that sensory-evaluation data on cold plasma-treated poultry meat and meat products are scant. Therefore, there is a need for further evaluation of the sensory qualities of cold plasma-treated poultry meat and meat products. Furthermore, plasma-activated water is likely to become popular for meat applications, considering its simplicity of use and low cost of production (Misra and Jo, 2017). Additionally, the analysis of the marketability/consumer acceptance of plasma-treated foods of animal origin is crucial because the potential value of novel food technologies can only be realized if their application does not negatively affect the consumers' purchase decision making.

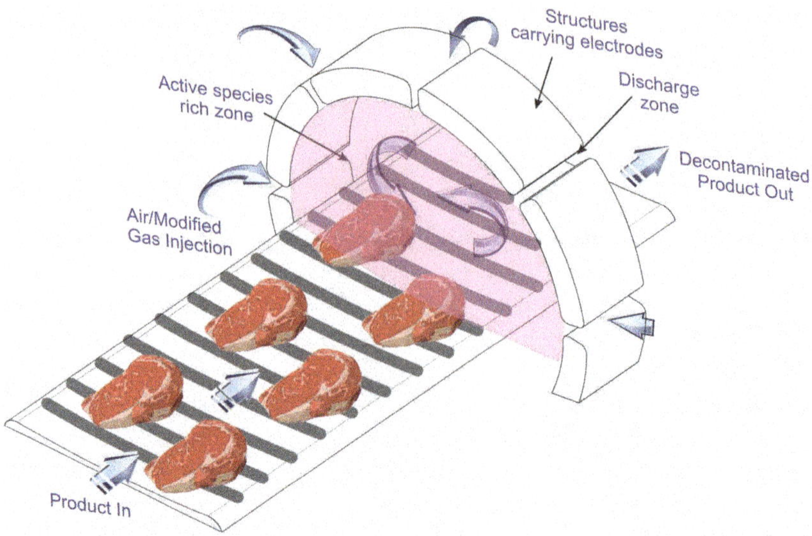

Figure 1 Conceptual design of an industrial-scale continuous cold plasma decontamination unit (Misra and Jo, 2017).

A recent study showed for the first time the application of supercritical carbon dioxide (SC-CO_2) drying process alone and in combination with high-power ultrasounds on raw chicken meat produced a dried microbiologically safe product with high nutritional value at 40°C (Morbiato et al., 2019). However, further studies need to be conducted to understand sensorial characteristics such as taste, texture, and flavour of the final product.

In addition, characterization of genetic profiles of indigenous/native chickens to identify, analyse, and interpret the nutritional virtues including fatty acid compositions, other flavour and aroma-related traits that define their unique and preferred features in meat quality needs to be elucidated adhering to their genetic information by advanced technologies.

8 Where to look for further information

1 Khan, M. I., Jo, C. and Tariq, M. R. (2015). Meat flavor precursors and factors influencing flavor precursors - a systematic review, *Meat Sci.* 110, 278–284.
2 Poultry Science (https://www.sciencedirect.com/journal/poultry-science) is an international journal publishing original papers, research notes, symposium papers and reviews of basic science as applied to poultry. World's Poultry Science Journal (https://www.tandfonline.com/toc/twps20/current) provides authoritative reviews in poultry science and an

international forum for the exchange and dissemination of information including research, education and industry structure. Both journals are of interest to academics, researchers, students, extension workers and commercial poultry producers.

3 World's Poultry Congress organized annually by World's Poultry Science Association.
4 Shahidi, F. (1994). *Flavour of Meat and Meat Products*. Springer Science+Business Media Dordrecht.
5 Jayasena, D. D., Ahn, D. U., Nam, K. C. and Jo, C. (2013). Flavour chemistry of chicken meat: a review, *Asian-Australas. J. Anim. Sci.* 26(5), 732–742.
6 Poulty Science Association (https://poultryscience.org/).

9 References

Ahn, D. U. (2002). Production of volatiles from amino acid homopolymers by irradiation, *J. Food Sci.* 67, 419–426.

Ahn, D. U., Lee, E. J., Feng, X., Zhang, W., Lee, J. H., Jo, C. and Nam, K. (2016a). Mechanisms of volatile production from non-sulfur amino acids by irradiation, *Radiat. Phys. Chem.* 119, 64–73.

Ahn, D. U., Joo Lee, E., Feng, X., Zhang, W., Lee, J. H., Jo, C. and Nam, K. (2016b). Mechanisms of volatile production from sulfur-containing amino acids by irradiation, *Radiat. Phys. Chem.* 119, 80–84.

Ahn, D. U., Olson, D. G., Jo, C., Chen, X., Wu, C. and Lee, J. I. (1998). Effect of muscle type, packaging, and irradiation on lipid oxidation, volatile production, and color in raw pork patties, *Meat Sci.* 49(1), 27–39.

Aliani, M. and Farmer, L. J. (2005). Precursors of chicken flavour II: identification of key flavour precursors using sensory methods, *J. Agric. Food Chem.* 53(16), 6455–6462.

Altmann, B. A., Wigger, R., Ciulu, M. and Mörlein, D. (2020). The effect of insect or microalga alternative protein feeds on broiler meat quality, *J. Sci. Food Agric.* 100(11), 4292–4302.

Aly, A. (2013). Comparison between gamma irradiation and plasma technology to improve the safety of cold sliced chicken, *Afr. J. Food Sci.* 7(12), 461–467.

Arshad, M. S., Amjad, Z., Yasin, M., Saeed, F., Imran, A., Sohaib, M., Anjum, F. M. and Hussain, S. (2019). Quality and stability evaluation of chicken meat treated with gamma irradiation and turmeric powder, *Int. J. Food Prop.* 22(1), 154–172.

Aymerich, T., Picouet, P. A. and Monfort, J. M. (2008). Decontamination technologies for meat products, *Meat Sci.* 78(1–2), 114–129.

Barido, F. H., Jang, A., Pak, J. I., Kim, D. Y. and Lee, S. K. (2020). Investigation of taste-related compounds and antioxidative profiles of retorted Samgyetang made from fresh and dried *Cordyceps militaris* mshrooms, *Food Sci. Anim. Resour.* 40(5), 772–784.

Barido, F. H. and Lee, S. K. (2021). Changes in proteolytic enzyme activities, tenderness-related traits, and quality properties of spent hen meat affected by adenosine 5′-monophosphate during cold storage, *Poult. Sci.* 100(5), 101056.

Bou, R., Guardiola, F., Grau, A., Grimpa, S., Manich, A., Barroeta, A. and Codony, R. (2001). Influence of dietary fat source, alpha-tocopherol, and ascorbic acid supplementation on sensory quality of dark chicken meat, *Poult. Sci.* 80(6), 800–807.

Brunton, N. P., Cronin, D. A. and Monahan, F. J. (2002). Volatile components associated with freshly cooked and oxidized off-flavours in turkey breast meat, *Flavour Frag. J.* 17(5), 327–334.

Buettner, G. R. and Jurkiewicz, B. A. (1996). Catalytic metals, ascorbate and free radicals: combinations to avoid, *Radiat. Res.* 145(5), 532–541.

Byrne, D. V., Bredie, W. L., Mottram, D. S. and Martens, M. (2002). Sensory and chemical investigations on the effect of oven cooking on warmed-over flavour development in chicken meat, *Meat Sci.* 61(2), 127–139.

Calkins, C. R. and Hodgen, J. M. (2007). A fresh look at meat flavour, *Meat Sci.* 77(1), 63–80.

Cao, C., Xie, J., Hou, L., Zhao, J., Chen, F., Xiao, Q., Zhao, M. and Fan, M. (2017). Effect of glycine on reaction of cysteine-xylose: insights on initial Maillard stage intermediates to develop meat flavor, *Food Res. Int.* 99(1), 444–453.

Cerny, C. (2007). Origin of carbons in sulfur-containing aroma compounds from the Maillard reaction of xylose, cysteine and thiamine, *LWT Food Sci. Technol.* 40(8), 1309–1315.

Chang, H., Wang, Y., Xia, Q., Pan, D., He, J., Zhang, H. and Cao, J. (2021). Characterization of the physicochemical changes and volatile compound fingerprinting during the chicken sugar-smoking process, *Poult. Sci.* 100(1), 377–387.

Cheah, P. B. and Ledward, D. A. (1996). High pressure effects on lipid oxidation in minced pork, *Meat Sci.* 43(2), 123–134.

Chen, D. W. and Zhang, M. (2007). Non-volatile taste active compounds in the meat of Chinese mitten crab (*Eriocheir sinensis*), *Food Chem.* 104(3), 1200–1205.

Chen, S., Kao, T. H., Chen, C. J., Huang, C. W. and Chen, B. H. (2013). Reduction of carcinogenic polycyclic aromatic hydrocarbons in meat by sugar-smoking and dietary exposure assessment in Taiwan, *J. Agric. Food Chem.* 61(31), 7645–7653.

Cheng, F. Y., Huang, C. W., Wan, T. C., Liu, Y. T., Lin, L. C. and Chyr, C. Y. L. (2008). Effect of free-range farming on carcass and meat qualities of black-feathered Taiwan native chicken, *Asian Australas. J. Anim. Sci.* 21(8), 1201–1206.

Chikuni, K., Sasaki, K., Emori, T., Iwaki, F., Tani, F., Nakajima, I., Muroya, S. and Mitsumoto, M. (2002). Effect of cooking on the taste- and flavor-related compounds in pork, *Jpn. J. Swine Sci.* 39(3), 191–199.

Christensen, L., Gunvig, A., Tørngren, M. A., Aaslyng, M. D., Knøchel, S. and Christensen, M. (2012). Sensory characteristics of meat cooked for prolonged times at low temperature, *Meat Sci.* 90(2), 485–489.

Culioli, J., Touraille, C., Bordes, P. and Girard, J. P. (1990). Caracteristiques des carcasses et de la viande du poulet label fermier, *Arch. Geflugelk* 53, 237–245.

Domínguez, R., Pateiro, M., Gagaoua, M., Barba, F. J., Zhang, W. and Lorenzo, J. M. (2019). A comprehensive review on lipid oxidation in meat and meat products, *Antioxidants (Basel)* 8(10), 429.

Estévez, M. (2015). Oxidative damage to poultry: from farm to fork, *Poult. Sci.* 94(6), 1368–1378.

Fan, M., Xiao, Q., Xie, J., Cheng, J., Sun, B., Du, W., Wang, Y. and Wang, T. (2018). Aroma compounds in chicken broths of Beijing Youji and commercial broilers, *J. Agric. Food Chem.* 66(39), 10242–10251.

Fanatico, A. C., Pillai, P. B., Cavitt, L. C., Emmert, J. L., Meullenet, J. F. and Owens, C. M. (2006). Evaluation of slow-growing broiler genotypes grown with and without outdoor access: sensory attributes, *Poult. Sci.* 85(2), 337–343.

Fanatico, A. C., Pillai, P. B., Emmert, J. L., Gbur, E. E., Meullenet, J. F. and Owens, C. M. (2007). Sensory attributes of slow- and fast-growing chicken genotypes raised indoors or with outdoor access, *Poult. Sci.* 86(11), 2441–2449.

Fanatico, A. C., Pillai, P. B., Hester, P. Y., Falcone, C., Mench, J. A., Owens, C. M. and Emmert, J. L. (2008). Performance, livability, and carcass yield of slow- and fast-growing chicken genotypes fed low-nutrient or standard diets and raised indoors or with outdoor access, *Poult. Sci.* 87(6), 1012–1021.

Farmer, L. J. (1999). Poultry meat flavour. In: Richardson, R. I. and Mead, G. C. (Eds), *Poultry Meat Science: Poultry Science Symposium Series Vol.* (vol. 25), CABI Publishing, Oxon, pp. 127–158.

Farmer, L. J. and Mottram, D. S. (1994). Lipid-Maillard interactions in the formation of volatile aroma compounds. In: Maarse, H. and vander Heij, D. G. (Eds), *Trends in Flavour Research*, Elsevier, Amsterdam, pp. 313–326.

Feng, X., Lee, E. J., Nam, K. C., Jo, C., Ko, K. and Ahn, D. U. (2016). Mechanisms of volatile production from amino acid esters by irradiation, *Food Res. Int.* 81, 100–107.

Fors, S. (1983). Sensory properties of volatile Maillard reaction products and related compounds. In: Waller, G. R. and Feather, M. S. (Eds), *The Maillard Reaction in Foods and Nutrition*, American Chemical Society, Washington, pp. 185–286.

Fujimura, S., Koga, H., Takeda, H., et al. (1996). Role of taste-active components, glutamic acid, 5′-inosinic acid and potassium ion in taste of chicken meat extract, *Anim. Sci. Technol.* 67, 423–429.

Fukunaga, T., Koga, K., Maita, Y. and Matsuoka, S. (1989). Free amino acid, carnosine and 5′-inosinic acid contents in the breast and leg meats from the cross and tripe-cross chickens of Satsuma native fowl, *Bull. Fac. Agric. Kagoshima Univ.* 39, 223–232.

Garcia, M., Cepero, R., Campo, M. M., Lafuente, R., Sanudo, C. and Canti, M. (1995). Effects of production system on the quality of label chickens and capons. *Proceedings of the 12th European Symposium on the Quality of Poultry Meat*. Zaragoza, Espana, pp. 207–212.

Gasser, U. and Grosch, W. (1990). Primary odorants of chicken broth, *Z. Lebensm. Unters Forch* 190(1), 3–8.

Gavahian, M., Chu, Y. H. and Jo, C. (2019). Prospective applications of cold plasma for processing poultry products: benefits, effects on quality attributes, and limitations, *Compr. Rev. Food Sci. Food Saf.* 18(4), 1292–1309.

Graf, E. and Panter, S. S. (1991). Inhibition of warmed-over flavour development by polyvalent cations, *J. Food Sci.* 56(4), 1055–1058.

Guan, R. F., Lyu, F., Chen, X. Q., Ma, J. Q., Jiang, H. and Xiao, C. G. (2013). Meat quality traits of four Chinese indigenous chicken breeds and one commercial broiler stock, *J. Zhejiang Univ. Sci. B* 14(10), 896–902.

Hashim, I. B., Resurrecci′on, A. V. A. and McWatters, K. H. (1995). Descriptive sensory analysis of irradiated frozen or refrigerated chicken, *J. Food Sci.* 60(4), 664–666.

Hayman, M. M., Baxter, I., O'riordan, P. J. and Stewart, C. M. (2004). Effects of high pressure processing on the safety, quality, and shelf life of ready-to-eat meats, *J. Food Prot.* 67(8), 1709–1718.

Heath, J. L., Owens, S. L., Tesch, S. and Hannah, K. W. (1990). Effect of high-energy electron irradiation of chicken on thiobarbituric acid values, shear values, odor, and cook yield, *Poult. Sci.* 69(2), 313–319.

Ho, C. T., Oh, Y. C. and Bae-Lee, M. (1994). The flavour of pork. In: Shahidi, F. (Ed.), *Flavour of Meat and Meat Products*, Blackie Academic and Professional, London, pp. 38–51.

Horsted, K., Allesen-Holm, B. H. and Hermansen, J. E. (2010). The effect of breed and feed-type on the sensory profile on breast meat in male broilers reared in an organic free-range system, *J. Sci. Food Agric.* 92, 258–265.

Horsted, K., Allesen-Holm, B. H., Hermansen, J. E. and Kongsted, A. G. (2012). Sensory profiles of breast meat from broilers reared in an organic niche production system and conventional standard broilers, *Brit. Poult. Sci.* 4, 515–524.

Hou, L., Xie, J., Zhao, J., Zhao, M., Fan, M., Xiao, Q., Liang, J. and Chen, F. (2017). Roles of different initial Maillard intermediates and pathways in meat flavor formation for cysteine-xylose-glycine model reaction systems, *Food Chem.* 232, 135–144.

Huang, Z. B., Leibovitz, H., Lee, C. M. and Millar, R. (1990). Effect of dietary fish oil on ω-3 fatty acid levels in chicken eggs and thigh flesh, *J. Agric. Food Chem.* 38(3), 743–747.

Husak, R. L., Sebranek, J. G. and Bregendahl, K. (2008). A survey of commercially available broilers marketed as organic, free-range, and conventional broilers for cooked meat yields, meat composition, and relative value, *Poult. Sci.* 87(11), 2367–2376.

Jahan, K., Paterson, A., Piggott, J. and Spickett, C. (2005). Chemometric modeling to relate antioxidants, neutral lipid fatty acids, and flavor components in chicken breasts, *Poult. Sci.* 84(1), 158–166.

Jang, A., Liu, X. D., Shin, M. H., Lee, B. D., Lee, S. K., Lee, J. H. and Jo, C. (2008). Antioxidative potential of raw breast meat from broiler chicks fed a dietary medicinal herb extract mix, *Poult. Sci.* 87(11), 2382–2389.

Jayasena, D. D., Ahn, D. U., Nam, K. C. and Jo, C. (2013). Factors affecting cooked chicken meat flavor: a review, *Worlds Poult. Sci. J.* 69(3), 515–526.

Jayasena, D. D., Jung, S., Alahakoon, A. U., Nam, K. C., Lee, J. H. and Jo, C. (2015a). Bioactive and taste-related compounds in defatted freeze-dried chicken soup made from two different chicken breeds obtained at retail, *Jpn Poult. Sci.* 52(2), 156–165.

Jayasena, D. D., Jung, S., Kim, H. J., Yong, H. I., Nam, K. C. and Jo, C. (2015b). Taste-active compound levels in Korean native chicken meat: the effects of bird age and the cooking process, *Poult. Sci.* 94(8), 1964–1972.

Jayasena, D. D., Jung, S., Kim, H. J., Alahakoon, A. U., Nam, K. C. and Jo, C. (2014a). Effect of sex on flavor-related and functional compounds in freeze-dried broth made from Korean native chicken, *Korean J. Food Sci. Anim. Resour.* 34(4), 448–456.

Jayasena, D. D., Kim, S. H., Lee, H. J., Jung, S., Lee, J. H., Park, H. B. and Jo, C. (2014b). Comparison of the amounts of taste-related compounds in raw and cooked meats from broilers and Korean native chickens, *Poult. Sci.* 93(12), 3163–3170.

Jin, S., Park, H. B., Seo, D., Choi, N. R., Manjula, P., Cahyadi, M., Jung, S., Jo, C., and Lee, J. H. (2018). Identification of quantitative trait loci for the fatty acid composition in Korean native chicken, *Asian Australas. J. Anim. Sci.*, 31(8), 1134–1140.

Jo, C. and Ahn, D. U. (2000). Production of volatile compounds from irradiated oil emulsion containing amino acids and proteins, *J. Food Sci.* 65(4), 612–616.

Jo, C., Ahn, D. U. and Byun, M. W. (2001). Lipid oxidation and volatile production in irradiated raw pork batters prepared with commercial soybean oil containing vitamin E, *Radiat. Phys. Chem.* 61(1), 41–47.

Jo, C., Lee, J. I. and Ahn, D.U. (1999). Lipid oxidation, color changes and volatiles production in irradiated pork sausage with different fat content and packaging during storage, *Meat Sci.* 51, 355–361.

Joo, S. T. and Kim, G. D. (2011). Meat quality traits and control technologies. In: Joo, S. T. (Ed.), *Control of Meat Quality*, Research Signpost, Kerala, pp. 1-20.

Jung, S., Jo, K., Lee, S. and Choi, Y.-S. (2019). Effect of ultrasound treatment on the quality properties of chicken breast meat and the broth from Korean chicken soup (Baeksuk), *Korean J. Agric. Sci.* 46(3), 539-548.

Jung, Y., Jeon, H. J., Jung, S., Choe, J., Lee, J., Heo, K., Kang, B. and Jo, C. (2011). Comparison of quality traits of thigh meat from Korean native chickens and broilers, *Korean J. Food Sci. Anim. Resour.* 31(5), 684-692.

Karabagias, I. K. (2018). Volatile profile of raw lamb meat stored at 4±1°C: the potential of specific aldehyde ratios as indicators of lamb meat quality, *Foods* 7(3), 40.

Kavitha, S. and Modi, V. K. (2007). Effect of water activity and temperature on degradation of 5′-inosine monophosphate in a meat model system, *LWT Food Sci. Technol.* 40(7), 1280-1286.

Kerler, J. and Grosch, W. (1997). Character impact odourants of boiled chicken: changes during refrigerated storage and reheating, *Food Res. Technol.* 205(3), 232-238.

Khan, M. I., Jo, C. and Tariq, M. R. (2015). Meat flavor precursors and factors influencing flavor precursors–a systematic review, *Meat Sci.* 110, 278-284.

Kim, H. C., Ko, Y. J. and Jo, C. (2021). Potential of 2D qNMR spectroscopy for distinguishing chicken breeds based on the metabolic differences, *Food Chem.* 342, 128316.

Kim, H. J., Jayasena, D. D., Yong, H. I. and Jo, C. (2016). Quality of cold plasma treated foods of animal origin. In: Misra, N. N., Schlüter, O. and Cullen, P. J. (Eds), *Cold Plasma in Food and Agriculture*, Academic Press, San Diego, pp. 273-291.

Kim, H. J., Yong, H. I., Park, S., Choe, W. and Jo, C. (2013). Effects of dielectric barrier discharge plasma on pathogen inactivation and the physicochemical and sensory characteristics of pork loin, *Curr. Appl. Phys.* 13(7), 1420-1425.

Kim, Y. J., Jin, S. K. and Yang, H. S. (2009). Effect of dietary garlic bulb and husk on the physicochemical properties of chicken meat, *Poult. Sci.* 88(2), 398-405.

Kiyohara, R., Yamaguchi, S., Rikimaru, K. and Takahashi, H. (2011). Supplemental arachidonic acid enriched oil improves the taste of thigh meat of Hinai-jidori chickens, *Poult. Sci.* 90(8), 1817-1822.

Kochhar, S. P. (1996). Oxidative pathways to the formation of off-flavors. In: Saxby, M. J. (Ed.), *Food Taints and Off-Flavours*, Springer, Boston, MA, pp. 168-225.

Koriyama, T., Wongso, S., Watanabe, K. and Abe, H. (2002). Fatty acid compositions of oil species affect the 5 basic taste perceptions, *J. Food Sci.* 67(2), 868-873.

Kosowska, M., Majcher, M. and Fortuna, T. (2017). Volatile compounds in meat and meat products, *Food Sci. Technol. (Campinas)* 37(1), 1-7.

Kruk, Z. A., Yun, H., Rutley, D. L., Lee, E. J., Kim, Y. J. and Jo, C. (2011). The effect of high pressure on microbial population, meat quality and sensory characteristics of chicken breast fillet, *Food Control* 22(1), 6-12.

Kurihara, K. (1987). Recent progress in the taste receptor. In: Kawamura, Y. and Kare, M. R. (Eds), *Umami: A Basic Taste*, Marcel Dekker, New York, pp. 3-39.

Lee, H., Yong, H. I., Kim, H. J., Choe, W., Yoo, S. J., Jang, E. J. and Jo, C. (2016). Evaluation of the microbiological safety, quality changes, and genotoxicity of chicken breast treated with flexible thin-layer dielectric barrier discharge plasma, *Food Sci. Biotechnol.* 25(4), 1189-1195.

Lee, K. H., Kim, H. J., Lee, H. J., Kang, M. and Jo, C. (2012). A study on components related to flavor and taste in commercial broiler and Korean native chicken meat, *Korean J. Food Preserv.* 19(3), 385-392.

Li, X., Zhu, J., Qi, J., Wang, P., Xu, X. and Zhou, G. (2018). Superchilled storage (−2.5 ± 1°C) extends the retention of taste-active and volatile compounds of yellow-feather chicken soup, *Anim. Sci. J.*, 89, 906–918.

Lin, C. Y., Kuo, H. Y. and Wan, T. C. (2014). Effect of free-range rearing on meat composition, physical properties and sensory evaluation in Taiwan game hens, *Asian Australas. J. Anim. Sci.* 27(6), 880–885.

Liu, X. D., Jayasena, D. D., Jung, Y., Jung, S., Kang, B. S., Heo, K. N., Lee, J. H. and Jo, C. (2012). Differential proteome analysis of breast and thigh muscles between Korean native chickens and commercial broilers, *Asian-Australas. J. Anim. Sci.* 25(6), 895–902.

Lyon, B. G., Smith, D. P., Lyon, C. E. and Savage, E. M. (2004). Effects of diet and feed withdrawal on the sensory descriptive and instrumental profiles of broiler breast fillets, *Poult. Sci.* 83(2), 275–281.

MacLeod, G. (1994). The flavour of beef. In: Shahidi, F. (Ed.), *Flavour of Meat and Meat Products*, Blackie Academic and Professional, Glasgow, pp. 4–37.

Maga, J. A. (1983). Flavour potentiator, *Crit. Rev. Food Sci. Nutr.* 18(3), 231–312.

Meinert, L., Schäfer, A., Bjergegaard, C., Aaslyng, M. D. and Bredie, W. L. (2009). Comparison of glucose, glucose 6-phosphate, ribose, and mannose as flavour precursors in pork; the effect of monosaccharide addition on flavour generation, *Meat Sci.* 81(3), 419–425.

Melton, S. L. (1999). Current status of meat flavour. In: Xiong, Y. L., Ho, C. T. and Shahidi, F. (Eds), *Quality Attributes of Muscle Foods*, Kluwer Academic/Plenum Publishers, New York, pp. 115–130.

Michalczuk, M., Lukasiewicz, M., Zdanowska-Sqsiadek, Ż and Niemiec, J. (2014). Comparison of selected quality attributes of chicken meat as affected by rearing systems, *Pol. J. Food Nutr. Sci.* 64(2), 121–126.

Mikulski, D., Celej, J., Jankowski, J., Majewska, T. and Mikulska, M. (2011). Growth performance, carcass traits and meat quality of slower-growing and fast-growing chickens reared with and without outdoor access, *Asian Australas. J. Anim. Sci.* 24(10), 1407–1416.

Min, B. R., Cordray, J. C. and Ahn, D. U. (2010). Effect of NaCl, myoglobin, Fe(II), and Fe(III) on lipid oxidation of raw and cooked chicken breast and beef loin, *J. Agric. Food Chem.* 58(1), 600–605.

Misra, N. N. and Jo, C. (2017). Applications of cold plasma technology for microbiological safety in meat industry, *Trends Food Sci. Technol.* 64, 74–86.

Morbiato, G., Zambon, A., Toffoletto, M., Poloniato, G., Dall'Acqua, S., de Bernard, M. and Spilimbergo, S. (2019). Supercritical carbon dioxide combined with high power ultrasound as innovate drying process for chicken breast, *J. Supercrit. Fluids* 147, 24–32.

Mottram, D. S. (1991). Meat. In: Maarse, H. (Ed.), *Volatile Compounds in Foods and Beverages*, Marcel Dekker, New York, pp. 107–177.

Mottram, D. S. (1994a). Flavour compounds formed during the Maillard reaction. In: Parliament, T. H., Morello, M. J. and McGorrin, R. J. (Eds), *Thermally Generated Flavours. Maillard, Microwave, and Extrusion Processes*, American Chemical Society, Washington, pp. 104–126.

Mottram, D. S. (1994b). Some aspects of the chemistry of meat flavour. In: Shahidi, F. (Ed.), *The Flavour of Meat and Meat Products*, Chapman and Hall, London, pp. 210–230.

Mottram, D. S. (1998). Flavour formation in meat and meat products: a review, *Food Chem.* 62(4), 415–424.
Mottram, D. S. and Edwards, R. A. (1983). The role of triglycerides and phospholipids in the aroma of cooked beef, *J. Sci. Food Agric.* 34(5), 517–522.
Nam, K. C. and Ahn, D. U. (2003). Use of double-packaging and antioxidant combinations to improve color, lipid oxidation, and volatiles of irradiated raw and cooked turkey breast patties, *Poult. Sci.* 82(5), 850–857.
Nishimura, T., Rhue, M. R., Okitani, A., et al. (1988). Components contributing to the improvement of meat taste during storage, *Agric. Biol. Chem.* 52, 2323–2330.
Noleau, I. and Toulemonde, B. (1986). Quantitative study of roasted chicken flavor, *Lebensm. Wiss. Technol.* 19, 122–125.
Noleau, I. and Toulemonde, B. (1987). Volatile components of roasted chicken fat, *Lebensm. Wiss. Technol.* 20, 37–41.
OECD/FAO (2020). *OECD-FAO Agricultural Outlook 2020–2029*, OECD Publishing, Paris, Rome.
Oh, S.-T., Jhun, H. K., Park, J.-M., Kim, J.-M., Kang, C.-W. and An, B.-K. (2012). Effects of dietary sources containing ω-3 fatty acids on the fatty acid composition of meats in Korean native chickens, *Food Sci. Anim. Resour.* 32(4), 476–482.
Oike, H., Wakamori, M., Mori, Y., Nakanishi, H., Taguchi, R., Misaka, T., Matsumoto, I. and Abe, K. (2006). Arachidonic acid can function as a signaling modulator by activating the TRPM5 cation channel in taste receptor cells, *Biochim. Biophys. Acta* 1761(9), 1078–1084.
Patterson, R. L. S. and Stevenson, M. H. (1995). Irradiation-induced off-odor in chicken and its possible control, *Br. Poult. Sci.* 36(3), 425–441.
Perez-Alvarez, J. A., Sendra-Nadal, E., Sanchez-Zapata, E. J., et al. (2010). Poultry flavour: general aspects and applications. In: Guerrero-Legarreta, I. and Hui, Y. H. (Eds), *Handbook of Poultry Science and Technology Volume 2: Secondary Processing*, John Wiley and Sons Inc., New Jersey, pp. 339–357.
Pestana, J. M., Puerta, B., Santos, H., Madeira, M. S., Alfaia, C. M., Lopes, P. A., Pinto, R. M. A., Lemos, J. P. C., Fontes, C. M. G. A., Lordelo, M. M. and Prates, J. A. M. (2020). Impact of dietary incorporation of Spirulina (*Arthrospira platensis*) and exogenous enzymes on broiler performance, carcass traits, and meat quality, *Poult. Sci.* 99(5), 2519–2532.
Ponte, P. I. P., Rosado, C. M. C., Crespo, J. P., Crespo, D. G., Mourão, J. L., Chaveiro-Soares, M. A., Brás, J. L., Mendes, I., Gama, L. T., Prates, J. A., Ferreira, L. M. and Fontes, C. M. (2008). Pasture intake improves the performance and meat sensory attributes of free-range broilers, *Poult. Sci.* 87(1), 71–79.
Poste, L. M. (1990). A sensory perspective of effect of feeds on flavor in meats: poultry meats, *J. Anim. Sci.* 68(12), 4414–4420.
Qi, J., Liu, D. Y., Zhou, G. H. and Xu, X. L. (2017a). Characteristic flavor of traditional soup made by stewing Chinese yellow-feather chickens, *J. Food Sci.* 82(9), 2031–2040.
Qi, J., Wang, H.-h., Zhou, G.-h., Xu, X., Li, X., Bai, Y. and Yu, X. (2017b). Evaluation of the taste-active and volatile compounds in stewed meat from the Chinese yellow-feather chicken breed, *Int. J. Food Prop.* 20(Suppl.3), S2579–S2595.
Rababah, T., Hettiarachchy, N. S., Horax, R., Cho, M. J., Davis, B. and Dickson, J. (2006). Thiobarbituric acid reactive substances and volatile compounds in chicken breast meat infused with plant extracts and subjected to electron beam irradiation, *Poult. Sci.* 85(6), 1107–1113.

Ramalingam, V., Song, Z. and Hwang, I. (2019). The potential role of secondary metabolites in modulating the flavor and taste of the meat, *Food Res. Int.* 122, 174–182.

Ramarathnam, N., Rubin, L. J. and Diosady, L. L. (1993). Studies on meat flavor: 4. Fractionation, characterization, and quantitation of volatiles from uncured and cured beef and chicken, *J. Agric. Food Chem.* 41(6), 939–945.

Ramaswamy, H. S. and Richards, J. F. (1982). Flavour of poultry meat- a review, *Can. Inst. Food Sci. Technol. J.* 15(1), 7–18.

Ranasinghe, M. and Jayasena, D. D. (2019). A study on sensory attributes of meat from three Sri Lankan indigenous chicken strains reared under semi-intensive systems, *Poult. Sci.* 98(E-Suppl. 1), 218.

Ranasinghe, M. and Jayasena, D. D. (2020). A comparative study on carcass and meat quality traits of three Sri Lankan indigenous chicken strains reared under semi-intensive system, Online, 145th Annual Meeting, Poultry Science Association, p. 109.

Rhee, K. S., Anderson, L. M. and Sams, A. R. (1996). Lipid oxidation potential of beef, chicken, and pork, *J. Food Sci.* 61(1), 8–12.

Rhee, K. S., Anderson, L. M. and Sams, A. R. (2005). Comparison of flavour changes in cooked-refrigerated beef, pork and chicken meat patties, *Meat Sci.* 71(2), 392–396.

Rikimaru, K. and Takahashi, H. (2010). Evaluation of the meat from Hinai-jidori chickens and broilers: analysis of general biochemical components, free amino acids, inosine 5′-monophosphate, and fatty acids, *J. Appl. Poult. Res.* 19(4), 327–333.

Rikimaru, K., Egawa, Y., Yamaguchi, S. and Takahashi, H. (2016). Association of chicken fatty acid desaturase 1 and 2 gene single-nucleotide polymorphisms with the fatty acid composition of thigh meat in Japanese Hinai-Dori crossbred chickens, *J. Fish. Livest. Prod.* 4, 202.

Rivas-Canedo, A., Fernandez-Garcia, E. and Nunez, M. (2008). Volatile compounds in fresh meats subjected to high pressure processing: effect of the packaging material, *Meat Sci.* 81(2), 321–328.

Sabikun, N., Bakhsh, A., Rahman, M. S., Hwang, Y. H. and Joo, S. T. (2021). Volatile and nonvolatile taste compounds and their correlation with umami and flavor characteristics of chicken nuggets added with milkfat and potato mash, *Food Chem.* 343, 128499.

Sanudo, C., Enser, M. E., Campo, M. M., Nute, G. R., María, G., Sierra, I. and Wood, J. D. (2000). Fatty acid composition and sensory characteristics of lamb carcasses from Britain and Spain, *Meat Sci.* 54(4), 339–346.

Sasaki, K., Motoyama, M. and Mitsumoto, M. (2007). Changes in the amounts of water-soluble umami-related substances in porcine longissimus and biceps femoris muscles during moist heat cooking, *Meat Sci.* 77(2), 167–172.

Schroll, W., Nitz, S. and Drawert, F. (1988). Determination of aroma compounds in cooked chicken meat and hydrolysate, *Z. Lebensm. Unters Forch* 187(6), 558–560.

Seabra, L. M., Zapata, J. F., Fuentes, M. F., Aguiar, C. M., Freitas, E. R. and Rodrigues, M. C. (2001). Effect of deboning time, muscle tensioning and calcium chloride marination on texture characteristics of chicken breast meat, *Poult. Sci.* 80(1), 109–112.

Shahidi, F. (1994). Flavour of meat and meat products-an overview. In: Shahidi, F. (Ed.), *Flavour of Meat and Meat Products*, Blackie Academic and Professional, Glasgow, pp. 1–3.

Shahidi, F. (2002). Lipid derived flavours in meat products. In: Kerry, J., Kerry, J. and Ledward, D. (Eds), *Meat Processing: Improving Quality*, Woodhead Publishing Ltd, Cambridge, pp. 105-121.

Shi, H. and Ho, C. T. (1994). The flavour of poultry meat. In: Shahidi, F. (Ed.), *Flavour of Meat and Meat Products*, Blackie Academic and Professional, Glasgow, pp. 52-69.

Sitz, B. M., Calkins, C. R., Feuz, D. M., Umberger, W. J. and Eskridge, K. M. (2005). Consumer sensory acceptance and value of domestic, Canadian, and Australian grass-fed beef steaks, *J. Anim. Sci.* 83(12), 2863-2868.

Smith, D. P., Northcutt, J. K. and Steinberg, E. L. (2012). Meat quality and sensory attributes of a conventional and a Label Rouge-type broiler strain obtained at retail, *Poult. Sci.* 91(6), 1489-1495.

Sogunle, O. M., Olaniyi, O. A., Shittu, T. A. and Abiola, S. S. (2012). Performance and meat attributes of chickens reared on deep litter and free-range, *Arch. Zootec.* 61(236), 569-576.

Soncin, S., Chiesa, L. M., Cantoni, C. and Biondi, P. A. (2007). Preliminary study of the volatile fraction in the raw meat of pork, duck and goose, *Journal of Food Composition and Analysis* 20(5), 436-439, Food Comp, J. Anal 20, 436-439.

Spanier, A. M., Flores, M., Mcmillin, K. W. and Bidner, T. D. (1997). The effect of postmortem aging on meat flavour quality. Correlation of treatment, sensory, instrumental, and chemical descriptors, *Food Chem.* 59(4), 531-538.

Spanier, A. M., Flores, M., Toldrá, F., Aristoy, M. C., Bett, K. L., Bystricky, P. and Bland, J. M. (2004). Meat flavor: contribution of proteins and peptides to the flavor of beef, *Adv. Exp. Med. Biol.* 542, 33-49.

Tai, C. Y. and Ho, C. T. (1997). Influence of cysteine oxidation on thermal formation of Maillard aromas, *J. Agric. Food Chem.* 45(9), 3586-3589.

Tai, C. Y. and Ho, C. T. (1998). Influence of glutathione oxidation and pH on thermal formation of Maillard-type volatile compounds, *J. Agric. Food Chem.* 46(6), 2260-2265.

Takahashi, H. (2018). Flavor of poultry meat: a new look at an old issue. In: Liu, X. (Ed.), *Application of Genetics and Genomics in Poultry Science*, IntechOpen Ltd., London, pp. 75-87.

Takahashi, H., Rikimaru, K., Kiyohara, R. and Yamaguchi, S. (2012). Effect of arachidonic acid-enriched oil diet supplementation on the taste of broiler meat, *Asian Australas. J. Anim. Sci.* 25(6), 845-851.

Takakura, Y., Mizushima, M., Hayashi, K., Masuzawa, T. and Nishimura, T. (2014). Characterization of the key aroma compounds in chicken soup stock using aroma extract dilution analysis, *Food Sci. Technol. Res.* 20(1), 109-113.

Tang, H., Gong, Y. Z., Wu, C. X., Jiang, J., Wang, Y. and Li, K. (2009). Variation of meat quality traits among five genotypes of chicken, *Poult. Sci.* 88(10), 2212-2218.

Tang, J., Jin, Q. Z., Shen, G. H., Ho, C. T. and Chang, S. S. (1983). Isolation and identification of volatile compounds from fried chicken, *J. Agric. Food Chem.* 31(6), 1287-1292.

Taşkıran, M., Olum, E. and Candoğan, K. (2020). Changes in chicken meat proteins during microwave and electric oven cooking, *J. Food Process. Preserv.* 44(2), e14324.

Terasaki, M., Kajikawa, M., Fujita, E. and Ishii, K. (1965). Studies on the flavor of meats, *Agric. Biol. Chem.* 29, 208-215.

Tichivangana, J. Z. and Morrissey, P. A. (1985). The influence of pH on lipid oxidation in cooked meats from several species, *Ir. J. Food Sci. Technol.* 9, 99-106.

Vani, N. D., Modi, V. K., Kavitha, S., Sachindra, N. M. and Mahendraker, N. S. (2006). Degradation of inosine-5′-monophosphate in aqueous and in laying chicken muscle fibre system: effect of pH and temperature, *LWT-J, Food Sci. Technol.* 39, 627–632.

Varavinit, S., Shobsngob, S., Bhidyachakorawat, M. and Suphantharika, M. (2000). Production of meat-like flavour, *Scienceasia* 26(4), 219–224.

Wang, D., Dong, H., Zhang, M., Liu, F., Bian, H., Zhu, Y. and Xu, W. (2013). Changes in actomyosin dissociation and endogenous enzyme activities during heating and their relationship with duck meat tenderness, *Food Chem.* 141(2), 675–679.

Wang, Y., Sun, J., Zhong, H., Li, N., Xu, H., Zhu, Q. and Liu, Y. (2017). Effect of probiotics on the meat flavour and gut microbiota of chicken, *Sci. Rep.* 7(1), 6400.

Wattanachant, S., Benjakul, S. and Ledward, D. A. (2004). Composition, color, and texture of Thai indigenous and broiler chicken muscles, *Poult. Sci.* 83(1), 123–128.

Whitfield, F. B. (1992). Volatiles from interactions of Maillard reactions and lipids, *Crit. Rev. Food Sci. Nutr.* 31(1–2), 1–58.

Xiong, Q., Zhang, M., Wang, T., Wang, D., Sun, C., Bian, H., Li, P., Zou, Y. and Xu, W. (2020). Lipid oxidation induced by heating in chicken meat and the relationship with oxidants and antioxidant enzymes activities, *Poult. Sci.* 99(3), 1761–1767.

Xu, L., Yu, X., Li, M., Chen, J. and Wang, X. (2017). Monitoring oxidative stability and changes in key volatile compounds in edible oils during ambient storage through HS-SPME/GC–MS, *Int. J. Food Prop.* 20(Suppl.3), S2926–S2938.

Xu, Q. Q., Yan, H., Liu, X. L., Lv, L., Yin, C. H. and Wang, P. (2014). Growth performance and meat quality of broiler chickens supplemented with *Rhodopseudomonas palustris* in drinking water, *Br. Poult. Sci.* 55(3), 360–366.

Yamaguchi, S. (1991). Roles and efficacy of sensory evaluation in studies of taste, *J. Japan. Soc. Food Sci. Technol.* 38(10), 972–978.

Yan, J., Liu, P., Xu, L., Huan, H., Zhou, W., Xu, X. and Shi, Z. (2018). Effects of exogenous inosine monophosphate on growth performance, flavor compounds, enzyme activity, and gene expression of muscle tissues in chicken, *Poult. Sci.* 97(4), 1229–1237.

Yano, Y., Kataho, N., Watanabe, M., Nakamura, T. and Asano, Y. (1995). Evaluation of beef aging by determination of hypoxanthine and xanthine contents: application of a xanthine sensor, *Food Chem.* 52(4), 439–445.

Zhang, J., Hu, H., Mu, T., Wang, W., Yu, B., Guo, J., Wang, Y., Zhou, Z., Gu, Y., Huang, Z., Cai, Z. and Xin, G. (2020). Correlation analysis between AK1 mRNA expression and inosine monophosphate deposition in Jingyuan chickens, *Animals (Basel)* 10(3), 439.

Zhang, Y., Wang, Y., Jiang, F. and Jin, H. (2019). Sensory characteristics of Maillard reaction products from chicken protein hydrolysates with different degrees of hydrolysis, *CyTA J. Food* 17(1), 221–227.

Zhao, J., Wang, T., Xie, J., Xiao, Q., Cheng, J., Chen, F., Wang, S. and Sun, B. (2019a). Formation mechanism of aroma compounds in a glutathione-glucose reaction with fat or oxidized fat, *Food Chem.* 270, 436–444.

Zhao, J., Wang, T., Xie, J., Xiao, Q., Du, W., Wang, Y., Cheng, J. and Wang, S. (2019b). Meat flavor generation from different composition patterns of initial Maillard stage intermediates formed in heated cysteine-xylose-glycine reaction systems, *Food Chem.* 274, 79–88.

Part 3

Poultry myopathies and shelf life

Chapter 7

Breast meat abnormalities associated with ischaemic necrosis: dorsal cranial myopathy and deep pectoral myopathy

Liris Kindlein, Federal University of Rio Grande do Sul, Brazil

1 Introduction

According to Shahbandeh (2022), the projected global consumption of poultry meat will amount to 151.83 metric kilotons by 2030 (carcass weight equivalent), with an average increase of 1.4% per year between 2021 and 2030. This shows that poultry meat will continue to be one of the most popular sources of animal protein for human consumption worldwide (OECD/FAO, 2021).

The intense selection of birds for increased growth rates and meat yields over the last six decades has resulted in modern birds with greatly altered morphologies which have been associated with muscle defects. The size of broilers showed an increase of 364% in the period from 1957 to 2005 (Zuidhof et al., 2014), and because of this evolution in the aviculture world, the body tissues of fast-growth broilers of modern lines require greater demands and, consequently, muscle abnormalities and meat quality problems become more evident.

These abnormalities lead to alterations in the physical structure, the metabolism and the repair mechanisms in poultry muscles. The effect of commercial selection on the growth, efficiency and yield of broilers has driven the emergence of unintended effects such as skeletal defects (Duggan et al., 2016; Lilburn, 1994; Rath et al., 2000) and metabolic disorders (Olkowski, 2007; Scheele, 1997).

http://dx.doi.org/10.19103/AS.2022.0112.08

Experiments conducted with broilers that evaluate specific crossing lines are explored in the scientific literature exhibiting gains in performance, although mostly in carcass and cut yields, which shows the concern of geneticists with the development of competitive lines to meet the consumer market requirements (Le Bihan-Duval et al., 1998). Genetic selection, added to nutrition and proper management, has achieved extremely productive results when it comes to increasing the development of carcass size, with the consequent intensification in meat production, particularly the *Pectoralis major* muscle (breast), which is the main target of genetic improvement programmes (Marcato, 2007). Genetic selection for increased production, fast-growth rate, proper slaughter weight and efficient feed conversion has been very successful in broilers; however, these implementations may be correlated to the increase of some negative effects on skeletal muscle and meat quality (MacRae et al., 2006, 2007).

One of the first myopathies reported in birds was deep pectoral myopathy (DPM), first identified in turkey breeder hens as a degenerative myopathy (Dickinson et al., 1968; Harper et al., 1975), but later observed in broilers (Richardson et al., 1980). Since then, it has become common in all fast-growing broilers (Grunder et al., 1984; Bilgili and Hess, 2002). This condition, also known as green muscle disease or Oregon disease, involves ischaemic necrosis of the *supracoracoideous* or *Pectoralis minor* muscles.

Additionally, another myopathy with similar characteristics of ischaemic necrosis was evidenced in the *latissimus dorsi* muscle and was named dorsal cranial myopathy (DCM). There are several factors related to the appearance of these two conditions, among them are: growth speed of the genetic lines, slaughter weight, age, sex, mobility of the birds, as well as rearing conditions and management (Kijowski and Kupińska, 2013).

Considering the economic impact of these myopathies in broilers, there is an increasing interest by the industry in estimating DPM and DCM incidences, meat quality issues and assessing alternative solutions. Thus, this chapter aims to provide an updated and critical overview of the knowledge concerning these abnormalities and their overall effect on poultry meat.

2 Deep pectoral myopathy

2.1 Morphological characteristics

DPM is defined as an ischaemic necrosis or haemorrhage occurring in the *Pectoralis minor* muscles, in which lesions can be detected in both unilateral (Fig. 1a) and bilateral (Fig. 1b) portions and can diversify in colour, progressing from a pale red appearance to a green discoloration (Bilgili and Hess, 2008), as displayed in Fig. 2. *Pectoralis major* is the main target of genetic improvement programmes and thus exhibits hypertrophy. Although DPM is distinctly visually perceived in *Pectoralis minor* muscles, degradative changes were also reported

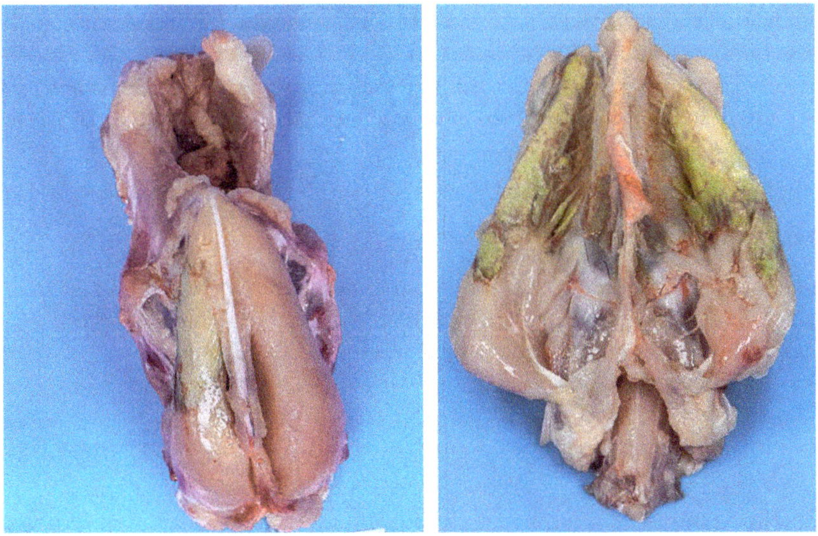

Figure 1 The *Pectoralis minor* muscle can exhibit unilateral lesions (Fig. 1a) or bilateral (Fig. 1b). Source: Images taken by the author.

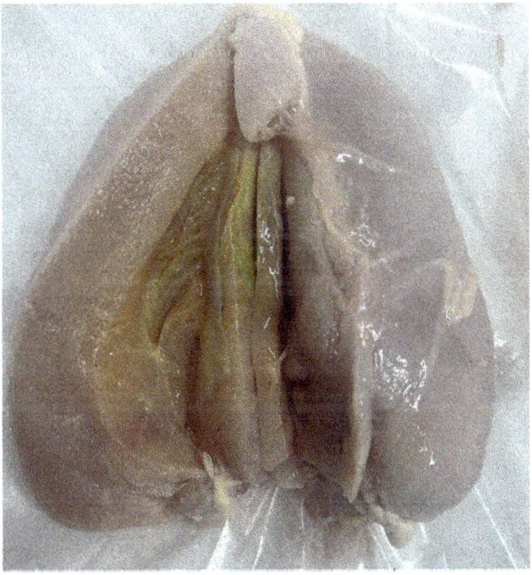

Figure 2 *Pectoralis major* and *minor* muscles from broilers. The *pectoralis minor* muscle exhibits a distinctive green colour that characterises deep pectoral myopathy, and these are internal to the hypertrophic *pectoralis major* muscle. Source: Images taken by the author.

for *Pectoralis major*. There are hypotheses that relate to the occurrence of this defect because *Pectoralis major* (breast fillet) and *Pectoralis minor* (tender) are the two principal muscles involved in wing movements. DPM prevalence has correlation to the genetic, anatomical and physiological background of birds and their management circumstances.

2.2 Microscopic characteristics

There are several studies exhibiting a broad range of muscle alterations caused by DPM. The macroscopic characteristics of the alterations caused by the myopathy are more commonly known and therefore significantly influence the categorisation stages since the visual progression of the myopathy allows the differentiation of these stages. In *Pectoralis minor*, the primarily affected muscle, DPM's most common macroscopic characteristics observed are alterations in colour, texture and meat composition (% of fibrous muscle fibres and adipose tissue) (Kijowski and Konstanczak, 2009). On the other hand, in the *Pectoralis major* muscle, other characteristics of the disease that are observed and analysed include alterations in pH, sarcomere length, water-holding capacity, shear force, protein and ion content (Giampietro-Ganeco et al., 2021; Traffano-Schiffo et al., 2020).

Giampietro-Ganeco et al. (2022) examined the quality parameters of colour and texture of the *Pectoralis major* muscle from turkeys affected by DPM. Alterations in the colour parameters of breast fillets due to the occurrence of the disease were verified and showed that L*, a* and b* were higher in samples with DPM in ventral and dorsal surfaces of turkey breeder hens' breast fillets. Additionally, water-holding capacity was higher ($P < 0.05$) for meat without myopathy (score 0) than for both degrees of DPM severity (scores 2 and 3).

Novel studies to characterise the different developmental stages of DPM have been carried out and provide further knowledge on the microscopic myopathy alteration characteristics. Since the traditional stage categorisation was based on visual appearance as the principal parameter to categorise the damage level, Traffano-Schiffo et al. (2020) studied the level of damage in poultry breast tenders by analysing biochemical and physicochemical parameters of muscle tissue metabolism. Microstructure, pH, protein and ion contents, together with colour, were determined allowing the identification of three DPM damage levels: normal, haemorrhagic samples (with angina) and necrotic tissues (with infarct) (Traffano-Schiffo et al., 2020).

This microstructural study allowed observation of the level of damage in the stages of DPM. At the microscopic level, it was possible to observe the following characteristics in the necrotic stage: excessive elongation of the myofibrils, separation areas developed in the sarcolemma, distinct presence of lesions in the cell membrane, loss of the cellular structure, endomysial inflammation and high level of collagen degradation (Traffano-Schiffo et al., 2020). The haemorrhagic category presented high levels of potassium and sodium ions.

As myopathies produce degradation in the structural proteins (myosin, actin, collagen and sarcoplasmic proteins), it was observed that the quantity of myosin decreased in *Pectoralis minor* depending on the level of damage. The necrotic samples showed a greater decrease of the collagen and sarcoplasmic protein contents compared with normal tissue in *Pectoralis minor*. Regarding *Pectoralis major*, there was a significant decrease in myosin content in necrotic tissue; collagen and sarcoplasmic proteins also showed a significant decrease compared to haemorrhagic samples. Additionally, actin degradation was significantly greater in DPM categories than in the normal samples in both *Pectoralis minor* and *major* (Traffano-Schiffo et al., 2020).

The sarcomere length of the breast fillets from carcasses with DPM scores 2 and 3 was significantly greater ($P < 0.05$) than fillets without myopathies. The authors suggested that the extended sarcomere length could be the result of dissociated actomyosin complexes of the myofibrils.

Furthermore, the percentages of total collagen, soluble and insoluble, were higher for the most severe category of myopathy (Giampietro-Ganeco et al., 2022). Other biochemical characteristics of muscle deterioration were detected in the *Pectoralis major* muscle with DPM: higher pH, drip loss, redness and lipid content but lower ash content (Yalcin et al., 2018).

2.3 Aetiology

The main hypothesis for DPM disease is that due to its confined position among the sternum and inelastic fascia, the *Pectoralis minor* muscle cannot expand in size during required activities, like wing flapping, leading to a cumulative pressure on the muscle, causing blood vessel constriction, oxygen insufficiency and subsequent ischaemia (Jordan and Pattison, 1998; Siller, 1985).

When birds are stressed, they are prone to sudden wing movements (Bilgili et al., 2000). In turkeys and broilers, the *Pectoralis minor* muscle lifts the wings expanding its volume by 20–25% to increase blood flow to the muscle. Analysing these muscle contractions, there is a differential response correlated to the bird weight: muscle pressure in lighter birds can rapidly return to the usual state; however, in heavy birds, the pressure may be maintained for prolonged periods of time, causing irreversible necrotic changes in the *Pectoralis minor* muscle (Bilgili and Hess, 2002).

DPM has been classified into three differential and successive stages, which reflect the progression of myopathy, within which the macroscopic and microscopic aspects are very characteristic. The known stages of DPM are divided into early, later and advanced according to the evolution of myopathy

The early stage shows acute muscle inflammation, oedema and haemorrhage. The later stage exhibits degenerative necrosis with muscle fibre atrophy, follicular, segmental and granular decomposition, macrophage infiltrations, connective tissue proliferation and eosinophil and phagocyte infiltration. Lastly, the advanced stage displays progressive lesions, fibrous muscle fibres, adipose tissue and a pink-to-green colour (Bianchi et al., 2006).

2.3.1 Associated factors

The incidence of DPM is correlated with several factors such as rearing conditions, sex, age, slaughter weight, bird mobility and genetic strain (Kijowski and Kupińska, 2013; Kijowski et al., 2014). Understanding the influence of environmental and/or management factors on the incidence of muscle defects is of fundamental importance. Bailey et al. (2015) found that such factors potentiate more than 90% of the incidence of DPM in broilers. There is evidence that the high growth rate of commercial broilers is associated with muscle disorders (Bianchi et al., 2006; Sihvo et al., 2014). Sex and increased weight in broilers are related to DPM incidence, as there are more DPM cases reported in higher-yielding strains and in males (Bianchi et al., 2006).

Increased bird activity (flock nervousness, flightiness, struggle and wing flapping) induced by factors such as feed or water outages, lighting programmes and intensity, human activity and excessive noise in and around chicken houses are also triggers for the development of myopathy (Bilgili et al., 2000).

Bailey et al. (2020) reported the genetic basis of breast muscle myopathies (BMM) and their relationship with bodyweight and breast yield in broiler chickens. The data presented by the authors showed a lower contribution of the BMM genetic variance to the phenotypic variance resulting in low heritability estimates and a large contribution of residual variance (Bailey et al., 2020). Thus, the authors showed that while there is a modest genetic component to the variation in myopathies, it is the non-genetic effects that are of greater importance, which agrees with previous results (Bailey et al., 2015) and questions the purely genetic hypothesis.

Previously, Bailey et al. (2015) demonstrated the polygenic nature of these myopathies and the low genetic relationships with body weight, processing body weight and breast yield highlighting the importance of understanding the environmental and/or management factors that contribute more than 90% of the variance of the incidence of wooden breast and DPM in broiler chickens. This indicates that selection for body weight and breast yield alone do not explain the manifestation of the myopathies and more research is needed to fully understand how the myopathies initiate and develop (Bailey et al., 2015, 2020).

There have been studies aiming to understand the pathogenesis of bird myopathies through the examination of gene expression, proteomic and metabolomic analyses (Boerboom et al., 2018; Cai et al., 2018; Kuttappan et al., 2017; Mutryn et al., 2015; Papah et al., 2018; Yalcin et al., 2019; Zambonelli et al., 2016). Generally, these studies show that there are differences in a range of metabolic (e.g. hypoxia, oxidative stress, calcium metabolism, fat metabolism, inflammation), anatomical and structural biological processes within the muscle of affected birds when compared to unaffected breast muscles. These results are important as they help to reveal the underlying biology and ongoing metabolic processes allowing for the development of biomarkers for further mapping of the pathophysiology of the bird myopathies.

For instance, Yalcin et al. (2019) evaluated the expression profiles of genes related to vascular development and muscle contraction regulation (vascular endothelial growth factor A (VEGFA), myosin light chain kinase 2 (MYLK2) and ATPase Ca+2 transporting gene 1 (ATP2A1)) and showed differences in the expression of those genes in the *Pectoralis major* muscle of broilers with or without induced DPM. The authors suggested that a regenerative process may be occurring in the *Pectoralis major* muscle of broilers after DPM induction since the higher capillary density and the expression of VEGFA in the *Pectoralis major* muscle of broilers with DPM may result in increased blood flow to increase oxygen availability (Yalcin et al., 2019).

2.4 Potential lines of work

DPM is currently a major challenge to the poultry industry because, although it does not have an impact on bird health, it is typically undetected until processing and deboning (Fig. 3) and, therefore, it becomes challenging to predict the frequency of DPM birds that will occur in a flock (Bianchi et al., 2006; Bilgili and Hess, 2008). Whereas the incidence of birds presenting with DPM remains small compared to the global number of processed birds, there are indications of its increasing incidence since breast meat yield is of fundamental importance to the industry. Additionally, the problem is rarely detected during processing if the birds are marketed as whole carcasses, resulting in consumer complaints and making the cause of the problem difficult to identify (Bilgili and Hess, 2008).

For these reasons, the improvement of management practices that lessen or minimise wing flapping in birds is a practical alternative to reduce DPM frequency until more advanced tools for predicting the disease can be developed. When birds are stressed, they are prone to sudden wing movements and as one of the triggers for this myopathy is excessive flapping or excitability within the flock, clear management recommendations must be available to

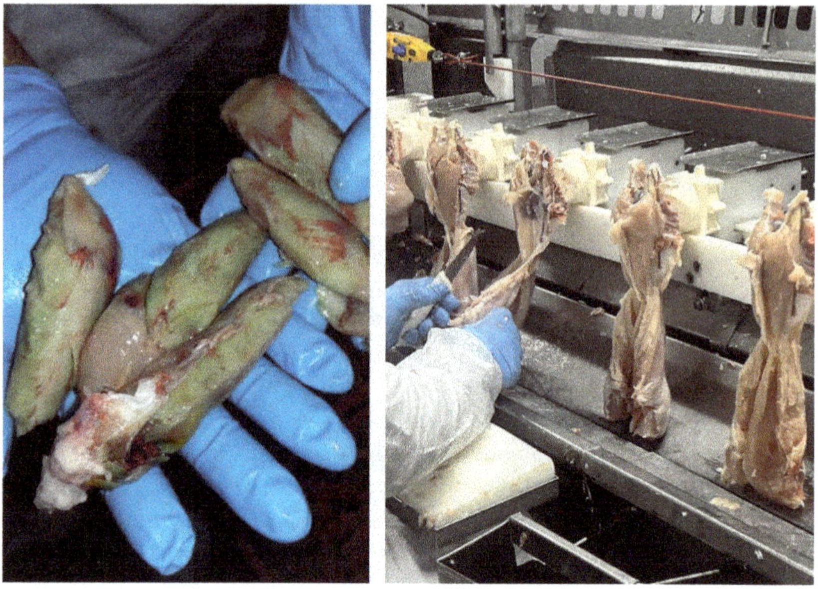

Figure 3 Deep pectoral myopathy is typically detected in deboning. Source: Images taken by the author.

control the field incidence of this condition (Siller et al., 1978; Bilgili and Hess, 2008; Lien et al., 2012).

Management practices that limit the amount of time spent in the broiler house to only that necessary may be advantageous, ensuring not to disregard good practices. Additionally, minimising excessive light exposure and excessive noise levels both inside and outside the house could prevent stressing the birds. The removal of mortalities and monitoring ventilation, air quality, temperature, ammonia levels and litter condition may also be beneficial to prevent excessive wing movement.

Furthermore, although not associated with infectious agents, DPM considerably affects appearance (Bilgili and Hess, 2008) and meat quality, as it affects both *Pectoralis minor* and *major* muscles and results in quality losses. Alterations in carcass-appreciated portions directly influence its sale as variations in the product's visual aspect deleteriously impact consumer acceptance. The partial condemnation of the pectoral muscle due to the injuries caused by myopathy is a major loss to the turkey meat market. Considering that difficulty, Cavalcanti et al. (2021) examined the quality of DPM breast meat from turkey breeder hens at disposal age and concluded that although the severe disorder caused variations in the quality of the *Pectoralis major* muscle, this type of raw meat has features (higher fat content and greater capacity for retaining intracellular water) which can be utilised in manufactured processed products. Thus, processing becomes an economically viable alternative to the commercialisation of breast meat from birds affected by myopathy (Cavalcanti et al., 2021).

3 Dorsal cranial myopathy

3.1 Morphological characteristics and incidence levels

DCM is a muscular abnormality that affects the anterior *latissimus dorsi* (ALD) muscle of fast-growing broilers and has been reported at processing plants all over the world. The characteristics of this type of lesion were first described by Coates (2003), who identified an uncommon degenerative muscle lesion in broilers of up to 5 weeks of age. In 2002, it was named DCM since it is positioned in the cranial dorsal portion of broilers occurring in the *latissimus dorsi* muscle. This muscle is superficially located in the dorsal region of birds, between the wings. It has a bilateral anatomy that can be divided into two independent portions: a cranial portion called *latissimus dorsi cranialis* or ALD and a caudal portion called *latissimus dorsi caudalis* or *posterior latissimus dorsi* (PLD). The PLD muscle provides energy to move the skeletal structure, coordinating the movements of the *humerus* during contraction of the flight muscles, also being involved in locomotion, posture, formation of the abdominal and thoracic cavities walls, support and activity of internal organs (Vanden Berge, 1975).

Both ALD and PLD are relatively thin and flat (especially the ALD), dorsally located on the bird's upper body and covered only by the skin, which allows the accumulation of a yellowish exudate in this region to be easily observed during slaughter (Fig. 4) (Zimermann, 2008). DCM can occur unilaterally or bilaterally in muscle. Broiler carcasses may have three different degrees of DCM: NORMAL

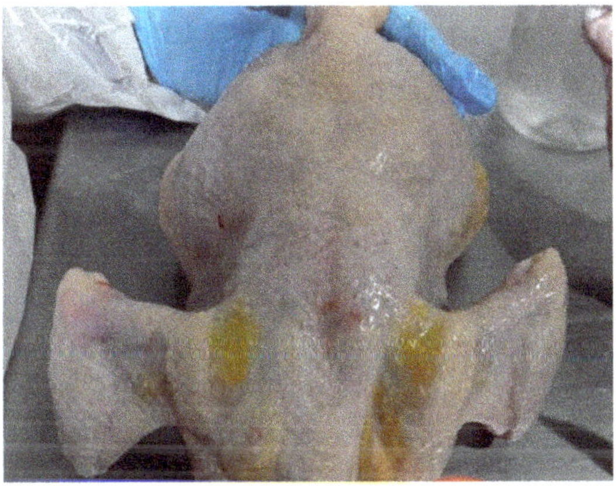

Figure 4 The dorsal cranial myopathy can be observed on the skin above the *anterior latissimus dorsi* muscles with a yellowish colour. Source: Images taken by the author.

(no lesion - Fig. 5a), MODERATE (muscles showing haemorrhage and petechiae - Fig. 5b) and SEVERE (subcutaneous volume increase and an oedematous gel-like yellow-citrine fluid - Fig. 5c). In more severe cases, the exudate extends to the wing region and an adjacent musculature greening can be observed, but still restricted in this area (Fig. 6). Thus, the literature suggested that DCM is a localised muscle lesion, with no consequences upon the systemic health of broilers (Amaral et al., 2017; Sesterhenn et al., 2017).

Sesterhenn et al. (2017) compared results when characterising *latissimus dorsi* muscle lesions to those of Sihvo et al. (2014), observing that the lesions are like those of woody breast, exhibiting whiteness, increased hardness and muscle thickness, in addition to the presence of a muscle surface exudate, characterised by an odourless and citrine-yellow gelatinous aspect. The macroscopic features mentioned above are directly related to the detrimental effects of the DCM condition on meat quality and are accountable for economic losses related to decreased yields during processing. It is also interesting to

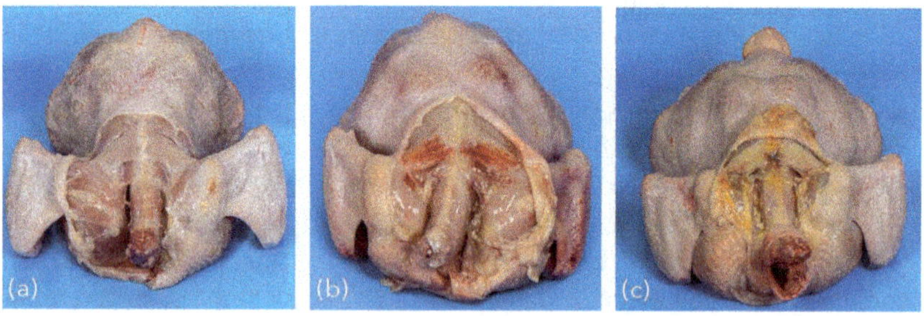

Figure 5 DCM in broiler, where (a) 0 = normal, (b) 1 = moderate and (c) 2 = severe. Source: Images taken by the author.

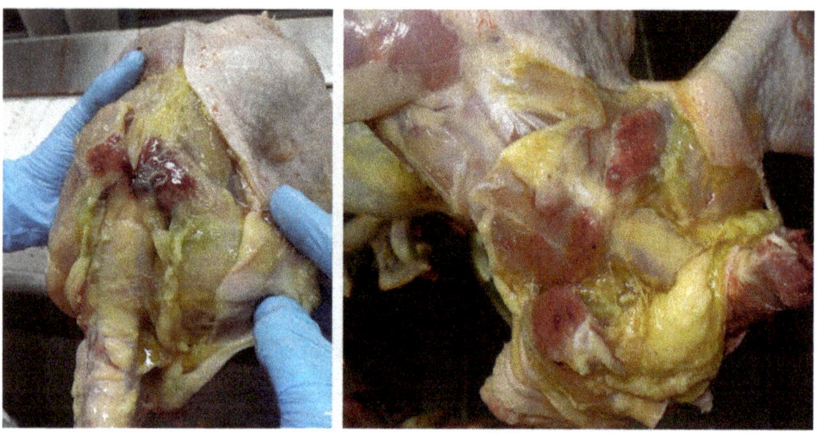

Figure 6 Severe DCM with extravasation of a greenish exudate from adjacent musculature. Source: Images taken by the author.

note that DCM-affected carcasses that had undergone the freezing stage, at the end of industrial processing, may exhibit an apparent macroscopic attenuation of the lesion by the temperature effect (Prado et al., 2021).

3.2 Microscopic characteristics

The histological characteristics of the ALD muscle of broilers presenting DCM showed the presence of fibres with moderate to severe floccular necrosis, hyalinisation, moderate to severe coagulative necrosis of muscle fibres, mild to moderate multifocal mononuclear infiltration and diffuse oedema. Sesterhenn et al. (2017) described the condition as chronic myopathic lesions showing polyphasic muscle degeneration, regeneration and extensive proliferation of fibrosis, floccular necrosis with a variable amount of interstitial connective tissue and the presence of adipocytes. Additionally, mononuclear inflammatory cell granulation tissue can be detected in DCM lesions. Furthermore, higher aspartate amino-transferase and creatine kinase serum concentrations were observed in affected broilers. Considering the microscopic alterations detected in affected muscles, the injury can be classified as multiphasic, suggesting that several aggressions to the muscle tissue occur in more than one moment.

3.3 Aetiology

Although still being considered a pathology of unknown causes, it is possible to find in the literature factors that may be related to its occurrence. Myopathic DCM lesions are non-specific and may result from several neuromuscular disorders. There are indications that the occurrence of DCM may be influenced by the seasons of the year, genetic factors, environmental factors, the incubation process, micronutrient deficiency (vitamin E, selenium) or be related to the energetic and vascular requirements during exercise (Bailey et al., 2015; Kuttappan et al., 2012, 2013; Lorenzi et al., 2014; Petracci et al., 2013; Roso and Dickel, 2010). However, it is a fact that the genetic selection pressure for a high carcass yield has led to an increase in the incidence of myopathies (Patricio et al., 2012).

Considering its location and ALD muscle functions, one of the hypotheses for the possible cause of DCM would be the birds' excessive wing flapping, possibly caused by an attempt to maintain corporal balance, given the increased development of the pectoral muscles and consequently greater weight in fast-growing strains. High growth rates induce morphological abnormality, a larger diameter of muscle fibres, a higher proportion of glycolytic fibres and less muscle proteolysis. Structural and metabolic muscle alterations modify meat quality; the decrease in muscle proteolysis increases the toughness of poultry meat (Dransfield and Sosnicki, 1999). This may explain why injuries are most present in male broilers of high-performance lineages and with the highest slaughter body weights (Zimermann et al., 2012). Another literature hypothesis is that DCM is an exercise-induced myopathy. Previous research analysing exercise inductions indicated that DPM (earlier described in this chapter) and DCM are not related because, in the exercise inductions executed, a high percentage of broilers had DPM when compared with the control group, while DCM was not detected in a high percentage, and there were no significant differences between the control group and the tested group (Giacomin et al., 2011).

Zimermann et al. (2012) performed macroscopic and microscopic evaluations of the ALD muscle, as well as evaluation of other muscles and viscera of carcasses with dorsal injuries in the search for information that could indicate an aetiological agent. The authors researched factors such as the bird age associated with the initiation of the lesion, the occurrence of potential nutritional and/or myotoxic complications, and the analysis of slaughterhouse carcass condemnation databases. It was verified that 89% of the cases presented bilateral lesions and that the correlation between DCM and DPM was 110:1.

Giacomin et al. (2011) performed an experiment aimed at inducing DCM by physical exercise, but the exercise protocol applied was unable to induce myopathy. One of the hypotheses for the non-development of DCM using the protocol is the possible occurrence of a muscular adaptation to the exercise, since it was induced weekly from the first to the last week of the birds' lives. The experimental reproduction of a lesion is a fundamental step towards understanding its aetiopathogenesis and could contribute to the development of strategies to prevent the occurrence and chronicity of the lesion.

In addition, there is a possibility that DCM is caused by a myotoxic attack due to the intense replacement of muscle fibres by fibrous connective tissue. This type of evolution occurs when there is sarcolemmal destruction and consequently loss of regenerative capacity (McGavin, 1998). Possessing a potential degenerative origin seems to characterise this myopathy, since the inflammatory changes were observed less frequently than degeneration and fibrosis. According to MacRae et al. (2006), the largest muscle fibres may be reaching the restriction of maximum functional size, since the oxygen diffusion increasing distances of the larger fibres can reduce the oxidative capacity and alter the mitochondrial distribution within the fibre.

3.3.1 Associated factors

3.3.1.1 Seasonal influence on the occurrence of dorsal cranial myopathy

Roso and Dickel (2010) suggested a relationship between the period of the year and the occurrence of DCM, since it was observed that in the winter months there were higher percentages of partial condemnations by DCM in Brazilian slaughterhouses. However, high incidences of DCM were also found in northeastern Brazilian regions where temperatures fluctuate between 20°C and 28°C in winters, reflecting a different reality than that observed in southern Brazil where negative temperatures can occur (De Oliveira Almeida et al., 2017). Further information and studies are required to assess the influence of environmental factors, such as ammonia levels, humidity, temperature, air velocity and ventilation, on the prevalence of DMC.

3.3.1.2 Genetic factors' influence on the occurrence of dorsal cranial myopathy

Zimermann (2008) observed that heavier strains of male broilers that exhibited higher weight and age averages during slaughter had the highest percentages of DCM present. The author also suggested that DCM may be a metabolic syndrome developed in broilers through the combination of musculature physiological characteristics and the oxygen deficit occasioned by genetic improvement. These propositions agree with that described by Bauermeister et al. (2009) and Kuttappan et al. (2013), who concluded that heavier birds with rapid growth rates are more susceptible to the development of myopathies. Thus, by applying genetic selection for rapid muscle development, cell hypertrophy can generate muscle injury, triggering cell lesions that produce free radicals leading to cell death by necrosis, as well as replacement of muscle cells with connective tissue in response to tissue repair.

3.3.1.3 Nutritional factors' influence on the occurrence of dorsal cranial myopathy

There are a few strategies to reduce the incidence of muscle defects such as myopathies. These are approaches largely related to bird growth rate and based on the use of antioxidants. Results from these methods vary and can occasionally affect performance factors such as growth rate, slaughter weight and carcass yield. However, an ideal solution is one that offers repeatable success in reducing the incidence of myopathies without sacrificing broiler performance.

Dietary restriction programmes have been shown to reduce myopathies (as well as other growth related metabolic diseases), either by limiting daily food allowances (amount) or by low-energy diets and low-protein foods (De Jong et al., 2012; Meloche et al., 2018a,b; Simões et al., 2020; Trocino et al., 2015). Raising broilers demonstrated the ability to achieve an acceptable final body weight and performance when initial feed restriction was applied to reduce the broilers' body fat content (Leeson et al., 1992). Therefore, broilers fed with diets that limit early growth may achieve zootechnic performance through compensatory gain in the growth phase and have shown lower incidences of myopathies such as DCM.

Nutrients that act on cellular repair and protection can prevent or minimise cellular oxidative stress. The body's antioxidant defence system has the main function of inhibiting or reducing the damage caused to cells by reactive oxygen species. There is a wide variety of antioxidant substances, which can be classified according to their origin and/or location as dietary antioxidants and intra- and extracellular antioxidants. The mechanism of action of antioxidants also allows them to be classified as preventive antioxidants (prevent the formation of free radicals), scavengers (prevent the attack of free radicals on cells) and repair (favour the removal of damaged DNA and the reconstitution of damaged cell membranes). Vitamin E acts as a cellular antioxidant, and one of its functions is to block the formation of toxic lipoperoxides preventing the formation of blood vessel lesions and alterations in capillary permeability, and this occurs through a mechanism of intervention in the stabilisation of polyunsaturated fatty acids from the lipid fraction of cell membranes (Barreto et al., 1999).

According to the poultry nutrient requirements of the National Research Council (NRC, 1994), the recommended dose of vitamin E for broilers is 30–50 IU/Kg; however, a footnote observation indicates the use of 60 IU/kg to avoid muscular dystrophy, encephalomalacia and exudative diathesis. The recommended dosage of vitamin E in the manuals of the main broiler lines supplying companies ranges from 50 IU/kg to 80 IU/Kg according to age group (Cobb, 2019). Consistent with Pompeu et al. (2015), to meet the requirements of male broilers, both in the initial and growth stages, the lowest level of vitamin E supplementation used in the study (10 mg/kg) is sufficient. This study also established a relationship with Nobakht's (2012) description, stating that the inclusion of 150 mg/kg of vitamin E in the diet of broilers resulted in negative effects on performance, including a reduction in weight gain and feed consumption, which could be explained by the interaction between vitamins A and E. Roman (2011) showed that DCM injured muscles had higher levels of selenium than control ones (normal muscles) and that the vitamin E levels at the time of slaughter did not influence the occurrence of the lesion. The literature research results involving vitamin E and selenium remain controversial, with the need for further analyses regarding the use of these additives in poultry feed and the suitable dosages to obtain positive results regarding DCM.

Another important aspect of nutritional studies is chromium supplementation. The involvement of chromium in metabolism is limited to an increase in the insulin sensitivity by the binding of four chromium atoms to a specific intracellular protein-denoted apochromodulin that, in turn, binds to the insulin receptor of peripheral tissue cells concomitantly with insulin located in the intracellular domain. This binding amplifies the cascade of intracellular signals responsible for stimulating the translocation of GLUT4 and increases glucose and amino acid uptake. Thus, the mechanism of action of chromium can assist in tissue adaptation and repair in situations of physiological or metabolic stress. Field studies have been carried out and are showing interesting results in reducing the incidence of myopathies.

Although there is little research literature on DCM to date, several studies revealed significant information and raised valuable new questions concerning DCM lesions. Further analyses are indispensable to reveal the cause of the injury and the necessary means to prevent it or minimise its chronicity and incidence.

4 Epidemiological data on muscle abnormalities in Latin American countries

Poultry production in Latin America has registered a steady growth over the last 10 years and, within this period, the volume of broiler meat produced increased by 29.3% (USDA, 2021). In this scenario, countries such as Brazil and Argentina are the two most competitive producers and exporters of animal protein, but others have also grown significantly, such as Peru, Chile, Colombia, Bolivia and Mexico. According to Rabobank International's global animal protein specialist, Nan-Dirk Mulder, the growth of broiler meat production in Latin America has been and will continue to be higher than the global annual average over the next decade.

In 2021, Brazil produced more than 14.329 million tons of chicken meat; 3.49% more than in 2020; of which 4.610 million tons (32.17%) were destined for exportation to 151 countries (ABPA, 2022). In Chile, expansion has been noted both in the volume of meat production and in meat commerce. Chicken is the most produced and consumed meat in Chile, with a per capita consumption of 31.7 kg. In 2020, it represented 44% of the total meat produced, reaching 690 647 tons, and statistics estimate a growing production and exportation.

The prevalence of myopathies in these countries significantly affects the economy of the food production chain, as these muscle defects are one of the main reasons for qualitative losses of meat. In a study with four Latin American companies producing broilers, in 2021, it was shown that DPM had the second highest prevalence (25.02%) among all studied myopathies and DCM was the least prevalent (0.19%) (Fig. 7). When stratifying the data, it was observed that both myopathies occurred in male, female and mixed lots (Fig. 8), yet they presented a higher prevalence in one of the two commercial genetic lineages of birds used (Fig. 9). However, Bailey et al. (2015) already indicated similar patterns of heritability and genetic correlations for two commercial lineages. Still, to support this conclusion it would be indispensable to also consider the weekly weight gain, body weight, processing body weight and age, as well as the management and environmental conditions used. The authors showed a

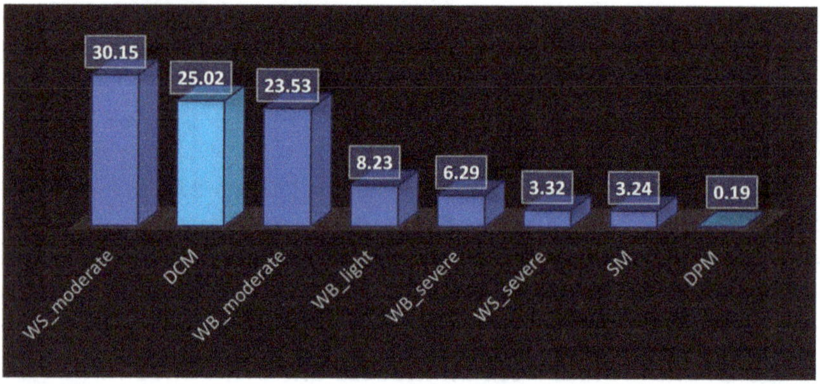

Figure 7 Prevalence of muscle defects in broiler carcasses in plants in Latin America (%). Abbreviations represent the types of myopathies (WS = white striping, DCM = dorsal cranial myopathy, WB = wooden breast, SM = spaghetti meat, DPM = deep pectoral myopathy) and their degree of severity (light, moderate or severe).

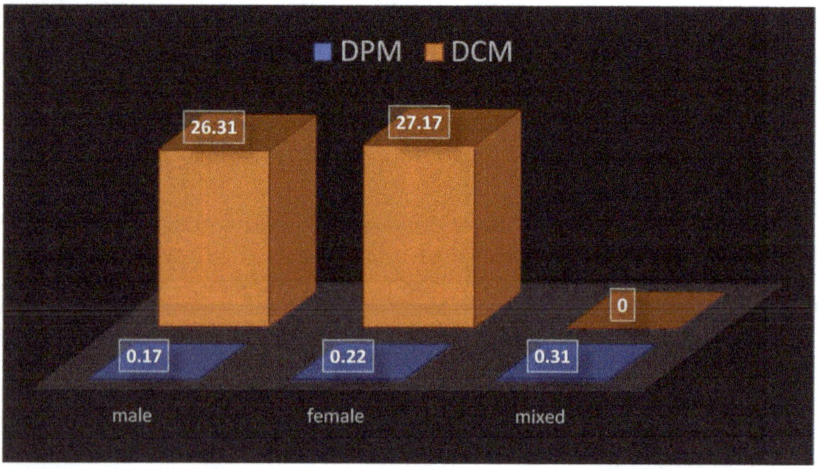

Figure 8 Prevalence of DPM and DCM myopathies considering male, female and mixed (male and female) broilers (%). DCM = dorsal cranial myopathy, DPM = deep pectoral myopathy.

smaller contribution of the genetic variance to the pectoral myopathies than the contribution of the phenotypic variance, resulting in low heritability estimates and a large contribution of residual variance. This means that while there is a modest genetic component to the variation in myopathies, it is the non-genetic effects that are of greatest importance. Subsequently, Bailey et al. (2020) cited the relationship between genetic and non-genetic factors that influence the prevalence of myopathies and highlighted those non-genetic factors as having a greater impact on the variance of myopathic traits than genetic factors.

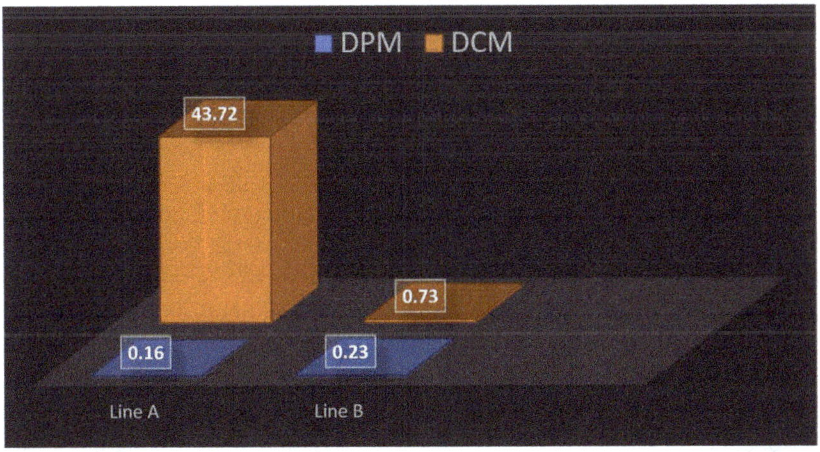

Figure 9 Prevalence of DPM and DCM myopathies considering the genetic lineage (lines A and B) of birds (%). DCM = dorsal cranial myopathy, DPM = deep pectoral myopathy.

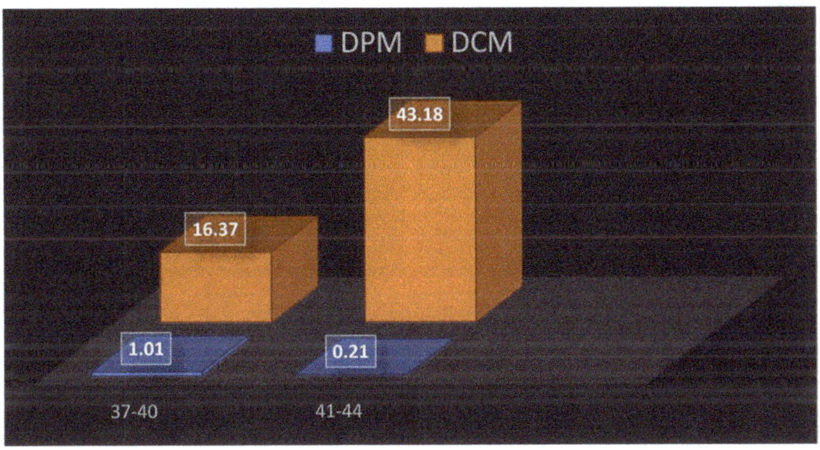

Figure 10 Prevalence (%) of DPM and DCM myopathies considering the age group of the birds (days). Birds were evaluated in two age groups, 37–40 and 42–44 days. DCM = dorsal cranial myopathy, DPM = deep pectoral myopathy.

The literature mentions that the weight and age of birds can influence the occurrence of muscle abnormalities; there are indications that heavy broilers are more likely to have such defects (Bauermeister et al., 2009; Kuttappan et al., 2009; Lorenzi et al., 2014), and other studies note that bird growth rate is a factor of great relevance for the appearance of Cobb lesions (Cruz et al., 2017; Kawasaki et al., 2018; Simões et al., 2020; Soglia et al., 2021). Figure 10 corroborates this information concerning DCM, which presented a prevalence of 43.18% against 16.35% in animals of 41–44 or 37–40 days of age, respectively.

5 Conclusion and future trends

DPM and DCM are associated with ischaemic necrosis, which may indicate a possible adaptive physiological response to injuries. Based on the literature presented, the occurrence, chronicity and degree of severity of muscle injuries have many associated factors. Studies are suggested to improve environmental and management conditions that guarantee good oxygenation to animals, animal welfare, good metabolism and animal health so that there is a physiological response to possible cellular damage, whether they are: (a) mechanisms of repair of the affected structure, (b) multiplication of initial cells (embryonic development) with losses occurring during production without compromising the tissue or (c) actions of elements that protect cells from free radicals (Fig. 11a,b). Elements like antioxidants can decrease oxidative stress by donating an electron to a rampaging free radical and neutralising it, thus reducing its capacity to damage. These antioxidants delay or inhibit cellular damage mainly through their free radical scavenging properties. Finally, we suggest that further research study possibilities for increasing the initial number of muscle cells (myocytes), improving the oxygenation and vascularisation of the affected muscle tissues, or preventing oxidative stress and free radical scavenging. Additionally, future research can evaluate the influence of animal welfare and the immune status on the occurrence and severity of these muscle defects.

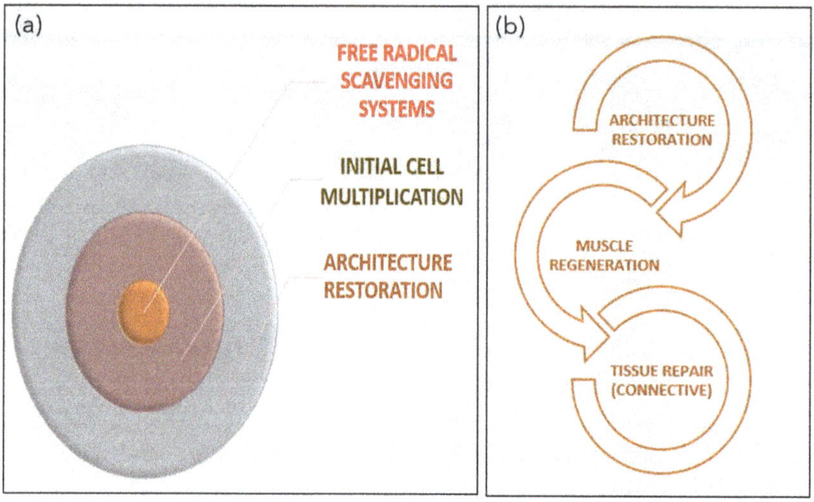

Figure 11 (a) Physiological response to possible cellular damage. (b) Tissue repair.

6 References

ABPA - Associação Brasileira de Proteína Animal. (2022), Relatório Anual 2022, Disponível em, https://abpa-br.org/mercados/#relatorios.

Amaral, P. C., Zimermann, C., Santos, L. R., Noro, M., Pra, M. D., Pilotto, F., Rodrigues, L. B. and Dickel, E. L. (2017), Evaluation of physiological parameters of broilers with dorsal cranial myopathy, *Rev. Bras. Ciênc. Avícola*, 19(1), 69–74.

Bailey, R. A., Watson, K. A., Bilgili, S. F. and Avendano, S. (2015), The genetic basis of pectoralis major myopathies in modern broiler chicken lines, *Poult. Sci.*, 94(12), 2870–2879. 10.3382/ps/pev304.

Bailey, R. A., Souza, E. and Avendano, S. (2020), Characterising the influence of genetics on breast muscle myopathies in broiler chickens, *Front. Physiol.*, 11, 1041. 10.3389/fphys.2020.01041.

Barreto, S. L. T., Ferreira, W. M. and Moraes, T. (1999), Efeito de níveis de vitamina E na dieta sobre o desempenho e concentração de alfa-tocoferol na carne de frangos de Corte, *Arquivo Brasileiro de Medicina Veterinária e Zootecnia*, 51(4), 387–392. Available at: doi.org/10.1590/S0102-09351999000400017.

Bauermeister, L. J., Morey, A. U., Moran, E. T., Singh, M., Owens, C. M. and McKee, S. R. (2009), Occurrence of white striping in chicken breast fillets in relation to broiler size, *Poult. Sci.*, 88, 33.

Bianchi, M., Petracci, M., Franchini, A. and Cavani, C. (2006), The occurrence of deep pectoral myopathy in roaster chickens, *Poult. Sci.*, 85(10), 1843–1846. 10.1093/ps/85.10.1843.

Bilgili, S. F. and Hess, J. B. (2002), Green muscle disease in broilers increasing, *Poult. Sci. J.*, 18, 42–43.

Bilgili, S. F. and Hess, J. B. (2008), *Green Muscle Disease: Reducing the Incidence in Broiler Flocks. March. Aviagen Brief*, Scotland: Aviagen Limited.

Bilgili, S. F., Hess, J. B., Lien, R. J. and Downs, K. M. (2000), Deep pectoral myopathy in broiler chickens. Proceedings of the XXI World's Poultry Congress, Montreal, Canada.

Boerboom, G., Van Kempen, T., Navarro-Villa, A. and Pérez-Bonilla, A. (2018), Unraveling the cause of white striping in broilers using metabolomics, *Poult. Sci.*, 97(11), 3977–3986. 10.3382/ps/pey266.

Cai, K., Shao, W., Chen, X., Campbell, Y. L., Nair, M. N., Suman, S. P., Beach, C. M., Guyton, M. C. and Schilling, M. W. (2018), Meat quality traits and proteome profile of woody broiler breast (pectoralis major) meat, *Poult. Sci.*, 97(1), 337–346. 10.3382/ps/pex284.

Cavalcanti, É. N. F., Giampietro-Ganeco, A., Mello, J. L. M., Fidelis, H. A., Oliveira, R. F., Pereira, M. R., Villegas-Cayllahua, E. A., Souza, R. A., Souza, P. A. and Borba, H. (2021), Breast meat quality of turkey breeder hens at disposal age affected by deep pectoral myopathy, *Poult. Sci.*, 100(8), 101259. 10.1016/j.psj.2021.101259.

Coates, J. (2003), An unusual degenerative muscle lesion (myopathy) in broilers, *Diagn. Diary*, 13(2), 3–4.

Cobb (2019), *Broiler Management Guide*, Siloam Springs, AR: Cobb - Vantress.

Cruz, R. F. A., Vieira, S. L., Kindlein, L., Kipper, M., Cemin, H. S. and Rauber, S. M. (2017), Occurrence of white striping and Wooden breast in broilers fed grower and finisher diets with increasing lysine levels, *Poult. Sci.*, 96(2), 501–510. 10.3382/ps/pew310.

De Jong, I., Berg, C., Butterworth, A. and Estevez, I. (2012), Scientific report updating the EFSA opinions on the welfare of broilers and broiler breeders, *Support. Publ. 2012EN-295*, 1-116.

De Oliveira Almeida, T. J., Sales de Assis, A., Mendonça, M. and de Queiroz Rolim, M. B. (2017), Causas de condenação de carcaças de Gallus gallus domesticus em abatedouros frigoríficos sob Inspeção Federal no Nordeste do Brasil, *Medicina Veterinária*, 11(4), 285. 10.26605/medvet-n4-1958.

Dickinson, E. M., Stevens, J. O. and Helfer, D. H. (1968), A degenerative myopathy in turkeys. Proceedings of the 17th Western Poultry Disease Conference. University of California, Davis, p. 6.

Dransfield, E. and Sosnicki, A. A. (1999), Relationship between muscle growth and poultry meat quality, *Poult. Sci. Savoy*, 78(5), 743-746.

Duggan, B. M., Hocking, P. M. and Clements, D. N. (2016), Gait in ducks (*Anas platyrhynchos*) and chickens (*Gallus gallus*) - similarities in adaptations to high growth rate, *Biol. Open*, 5(8), 1077-1085.

Giacomin, L., Tochetto, C., Roman, L. I., Testolin, G. and Barreta, M. H. (2011), Concórdia. Anais eletrônicos. O papel do exercício na indução da miopatia dorsal cranial em frangos de Corte. In: *MOSTRA DE INICIAÇÃO CIENTÍFICA* (vol. 1), p. 196, Concórdia: Instituto Federal Catarinense.

Giampietro-Ganeco, A., Owens, C. M., Borba, H., de Mello, J. L. M., de Souza, R. A., Ferrari, F. B., Cavalcanti, E. N., de Oliveira, R. F., Carvalho, L. T., Sun, X. and Trindade, M. A. (2021), Impact of deep pectoral myopathy on chemical composition and quality parameters of chicken breast fillet, *Poult. Sci.*, 100(9), 101377, 10.1016/j.psj.2021.101377.

Giampietro-Ganeco, A., Fidelis, H. A., Cavalcanti, E. N., Mello, J. L. M., Borba, H., de Souza, R. A., Carvalho, L. T., de Souza, P. A. and Trindade, M. A. (2022), Research note: quality parameters of turkey hens breast fillets detected in processing plant with deep pectoral myopathy and white striping anomaly, *Poult. Sci.*, 101(4), 101709, 10.1016/j.psj.2022.101709.

Grunder, A. A., Hollands, K. G., Gavora, J. S., Chambers, J. R. and Cave, N. A. G. (1984), Degenerative myopathy of the Musculus supracoracoideus and production traits in strains of meat-type chickens, *Poult. Sci.*, 63(4), 781-785, 10.3382/ps.0630781.

Harper, J. A., Bernier, P. E., Helfer, D. H. and Schmitz, J. A. (1975), Degenerative myopathy of the deep pectoral muscle in the turkey, *J. Hered.*, 66(6), 362-366.

Jordan, F. T. W. and Pattison, M. (1998), *Poultry Diseases*, London, UK: Saunders. Deep Pectoral Myopathy of Turkeys and Chickens, pp. 398-399.

Kawasaki, T., Iwasaki, T., Yamada, M., Yoshida, T. and Watanabe, T. (2018), Rapid growth rate results in remarkably hardened breast in broilers during the middle stage of rearing: A biochemical and histopathological study, *PLoS ONE*, 13(2), e0193307. 10.1371/journal.pone.0193307.

Kijowski, J. and Konstanczak, M. (2009), Deep pectoral myopathy in broiler chickens, *Bull. Vet. Inst. Pulawy.*, 53, 487-491.

Kijowski, J. and Kupińska, E. (2013), The evaluation of selected quality parameters of broiler chicken muscles with Deep Pectoral Myopathy (DPM) symptoms, *J. Eur. Poult. Sci.*, 77, 102-108.

Kijowski, J., Kupińska, E., Stangierski, J., Tomaszewska-Gras, J. and Szablewski, T. (2014), Paradigm of deep pectoral myopathy in broiler chickens, *Poult. Sci. J.*, 70(1), 125-138.

Kuttappan, V. A., Bottje, W., Ramnathan, R., Hartson, S. D., Coon, C. N., Kong, B. W., Owens, C. M., Vazquez-Añon, M. and Hargis, B. M. (2017), Proteomic analysis reveals changes in carbohydrate and protein metabolism associated with broiler breast myopathy, *Poult. Sci.*, 96(8), 2992–2999. 10.3382/ps/pex069.

Kuttappan, V. A., Brewer, V. B., Clark, F. D., McKee, S. R., Meullenet, J. F., Emmert, J. L. and Owens, C. M. (2009), Effect of white striping on the histological and meat quality characteristics of broiler fillets, *Poult. Sci.*, 88, 136–137.

Kuttappan, V. A., Brewer, V. B., Apple, J. K., Waldroup, P. W. and Owens, C. M. (2012), Influence of growth rate on the occurrence of white striping in broiler breast fillets, *Poult. Sci.*, 91(10), 2677–2685.

Kuttappan, V. A., Brewer, V. B., Mauromoustakos, A., McKee, S. R., Emmert, J. L., Meullenet, J. F. and Owens, C. M. (2013), Estimation of factors associated with the occurrence of White striping in broiler breast fillets, *Poult. Sci. Champ.*, 92(3), 811–819.

Le Bihan-Duval, E., Mignon-Grasteau, S., Millet, N. and Beaumont, C. (1998), Genetic analysis of a selection experiment on increased body weight and breast muscle weight as well as on limited abdominal fat weight, *Brit. Poult. Sci.*, 39(3), 346–353.

Leeson, S., Summers, J. D. and Caston, L. J. (1992), Response of broilers to feed restriction or diet dilution in the finisher period, *Poult. Sci.*, 71(12), 2056–2064.

Lien, R. J., Bilgili, S. F., Hess, J. B. and Joiner, K. S. (2012), Induction of deep pectoral myopathy in broiler chickens via encouraged wing flapping, *J. Appl. Poult. Res.*, 21(3), 556–562. 10.3382/japr.2011-00441.

Lilburn, M. S. (1994), Skeletal growth of commercial poultry species, *Poult. Sci.*, 73(6), 897–903, 10.3382/ps.0730897.

Lorenzi, M., Mudalal, S., Cavani, C. and Petracci, M. (2014), Incidence of white striping under commercial conditions in medium and heavy broiler chickens in Italy, *J. Appl. Poult. Res.*, 23(4), 754–758. 10.3382/japr.2014-00968.

MacRae, V. E., Mahon, M., Gilpin, S., Sandercock, D. A. and Mitchell, M. A. (2006), Skeletal muscle fibre growth and growth associated myopathy in the domestic chicken (*Gallus domesticus*), *Brit. Poult. Sci.*, 47(3), 264–272.

MacRae, V. E., Mahon, M., Gilpin, S., Sandercock, D. A., Hunter, R. R. and Mitchell, M. A. (2007), A comparison of breast muscle characteristics in three broiler great-grandparent lines, *Poult. Sci.*, 86(2), 382–385. 10.1093/ps/86.2.382.

Marcato, S. M. (2007), Características do crescimento corporal, dos órgãos e tecidos de duas linhagens comerciais de frangos de corte. Tese (doutorado) - Universidade Estadual Paulista, Faculdade de Ciências Agrárias e Veterinárias, Jaboticabal.

McGavin, R. (1998), Músculos. In: Carlton, W. W. and McGavin, R. (Eds), *Patologia Veterinária especial de Thompson* (2nd edn.), p. 426. Porto Alegre: ArtMed.

Meloche, K. J., Fancher, B. I., Emmerson, D. A., Bilgili, S. F. and Dozier, W. A. (2018a), Effects of reduced dietary energy and amino acid density on pectoralis major myopathies in broiler chickens at 36 and 49 days of Age 1, *Poult. Sci.*, 97(5), 1794–1807.

Meloche, K. J., Fancher, B. I., Emmerson, D. A., Bilgili, S. F. and Dozier, W. A. (2018b), Effects of quantitative nutrient allocation on myopathies of the pectoralis major muscles in broiler chickens at 32, 43, and 50 days of age, *Poult. Sci.*, 97(5), 1786–1793.

Mutryn, M. F., Brannick, E. M., Fu, W., Lee, W. R. and Abasht, B. (2015), Characterization of a novel chicken muscle disorder through differential gene expression and pathway analysis using RNA-sequencing, *BMC Genom.*, 16, 399. 10.1186/s12864-015-1623-0.

National Research Council (1994), *Nutrient Requirements of Poultry*, 9th rev. ed., Washington, DC: National Academy Press.

Nobakht, A. (2012), The effects of different levels of poultry fat with vitamin E on performance and carcass traits of broilers, *Afr. J. Agric. Res.*, 7, 1420–1424.

OECD/FAO (2021), *OECD-FAO Agricultural Outlook 2021-2030*, Paris: OECD Publishing. 10.1787/19428846-en.

Olkowski, A. A. (2007), Pathophysiology of heart failure in broiler chickens: structural, biochemical, and molecular characteristics, *Poult. Sci.*, 86(5), 999–1005.

Papah, M. B., Brannick, E. M., Schmidt, C. J. and Abasht, B. (2018), Gene expression profiling of the early pathogenesis of wooden breast disease in commercial broiler chickens using RNA-sequencing, *PLoS ONE*, 13(12), e0207346. 10.1371/journal.pone.0207346.

Patricio, I. S., Mendes, A. A., Ramos, A. A. and Pereira, D. F. (2012), Overview on the performance of Brazilian broilers (1990 to 2009), *Revista Brasileira de Ciência Avícola*, 14(4), 233–238.

Petracci, M., Sirri, F., Mazzon, M. and Meluzzi, A. (2013), Comparison of breast muscle traits and meat quality characteristics in 2 commercial chicken hybrids, *Poult. Sci.*, 92(9), 2438–2447.

Pompeu, M. A., Baião, N. C., Lara, L. J. C., Rocha, J. S. R., Cardeal, P. C., Baião, R. C., Pereira, L. F. P., Teixeira, M. P. F., Barbosa, V. M. and Cunha, C. E. (2015), Effect of vitamin E supplementation on the performance of broiler chickens, *Arquivos Brasileiro de Medicina Veterinária e Zootecnia*, 67(2), 506–510.

Prado, F., Orso, C., Ebbing, M. A., Kipper, M., Andretta, I. and Ribeiro, A. M. L. (2021), An in-situ assessment of Dorsal Cranial myopathy in broilers, approaching regarding meteorological influences in South Brazil, classification, and appearance of the lesions during industrial processing, *J. Appl. Poult. Res.*, 30(3), 100182, 10.1016/j.japr.2021.100182.

Rath, N. C., Huff, G. R., Huff, W. E. and Balog, J. M. (2000), Factors regulating bone maturity and strength in poultry, *Poult. Sci.*, 79(7), 1024–1032, 10.1093/ps/79.7.1024.

Richardson, J. A., Burgener, J., Winterfield, R. W. and Dhillon, A. S. (1980), Deep pectoral myopathy in seven-week-old broiler chickens, *Avian Dis.*, 24(4), 1054–1059.

Roman (2011), Quantificação dos níveis de selênio e vitamina E no músculo Anterior latíssimus dorsi de frangos de corte com miopatia dorsal cranial. In: *MOSTRA DE INICIAÇÃO CIENTÍFICA*, 1. 2011, Concórdia. Resumos. Instituto Federal Catarinense, Concórdia, 6p.

Roso, K. and Dickel, E. (2010), Study of the prevalence of cranial dorsal myopathy and economic losses occurred in a poultry slaughterhouse located in the north of the state of Rio Grande do Sul in the months of April to November 2010. 38th Brazilian Congress of Veterinary Medicine.

Scheele, C. W. (1997), Pathological changes in metabolism of poultry related to increasing production levels, *Vet. Q.*, 19(3), 127–130, 10.1080/01652176.1997.9694756.

Sesterhenn, R., Siqueira, F. M. A. C., Hamerski, A. C., Driemeier, D., Valle, S. F., Vieira, S. L., Kindlein, L. and Nascimento, V. P. (2017), Histomorphometric study of the anterior latissimus dorsi muscle and evaluation of enzymatic markers of broilers affected with dorsal cranial myopathy, *Poult. Sci.*, 96(12), 4217–4223.

Shahbandeh, M. (2022), Projected poultry meat consumption worldwide from 2021 to 2030. *Statista*, https://www.statista.com/statistics/739951/poultry-meat-consumption-worldwide/.

Sihvo, H. K., Immonen, K. and Puolanne, E. (2014), Myodegeneration with fibrosis and regeneration in the pectoralis major muscle of broilers, *Vet. Pathol.*, 51(3), 619–623.

Siller, W. G. (1985), Deep pectoral myopathy: a penalty of successful selection for muscle growth, *Poult. Sci.*, 64(8), 1591-1595.

Siller, W. G., Wight, P. A. L. and Martindale, L. (1978), Exercise-induced deep pectoral myopathy in broiler fowls and turkeys, *Vet. Sci. Commun.*, 2(1), 331-336. 10.1007/BF02291463.

Simões, C. T., Vieira, S. L., Stefanello, C., Kindlein, L., Ferreira, T., Favero, A. and Xavier, B. (2020), An in vivo evaluation of the effects of feed restriction regimens on wooden breast using ultrasound images as a predictive tool, *Brit. Poult. Sci.*, 61(5), 583-589. 10.1080/00071668.2020.1764909.

Soglia, F., Petracci, M., Davoli, R. and Zappaterra, M. (2021), A critical review of the mechanisms involved in the occurrence of growth-related abnormalities affecting broiler chicken breast muscles, *Poult. Sci.*, 100(6), 101180. 10.1016/j.psj.2021.101180.

Traffano-Schiffo, M. V., Chuquizuta, T., Castro-Giraldez, M. and Fito, P. J. (2020), Development of a methodology to categorize poultry meat affected by deep pectoral myopathy, *J. Food Proc. Preserv.*, 45(3), e15226. 10.1111/jfpp.15226.

Trocino, A., Piccirillo, A., Birolo, M., Radaelli, G., Bertotto, D., Filiou, E., Petracci, M. and Xiccato, G. (2015), Effect of genotype, gender and feed restriction on growth, meat quality and the occurrence of white striping and wooden breast in broiler chickens, *Poult. Sci.*, 94(12), 2996-3004.

USDA United States Department of Agriculture (2021), *In 2021, Production, Supply and Distribution*, Washington DC: United States Department of Agriculture.

Vanden Berge, J. C. (1975), Myology. In R. Getty, S. Sisson and J. D. Grossman (Eds), *The Anatomy of the Domestic Animals*, 5th edn., vol. 2, 1802-1848, Philadelphia: W.B. Saunders.

Yalcin, S., Şahin, K., Tuzcu, M., Bilgen, G., Özkan, S., Izzetoğlu, G. T. and Işik, R. (2019), Muscle structure and gene expression in pectoralis major muscle in response to deep pectoral myopathy induction in fast- and slow-growing commercial broilers, *Brit. Poult. Sci.*, 60(3), 195-201, 10.1080/00071668.2018.1430351.

Yalcin, S., Ozkan, S., Comert Acar, M. C. and Meral, O. (2018), The occurrence of deep pectoral myopathy in broilers and associated changes in breast meat quality, *Brit. Poult. Sci.*, 59(1), 55-62, 10.1080/00071668.2017.1401214.

Zambonelli, P., Zappaterra, M., Soglia, F., Petracci, M., Sirri, F., Cavani, C. and Davoli, R. (2016), Detection of differentially expressed genes in broiler pectoralis major muscle affected by White Striping - Wooden Breast myopathies, *Poult. Sci.*, 95(12), 2771-2785. 10.3382/ps/pew268.

Zimermann, F. C. (2008), Miopatia dorsal cranial em frangos de corte: caracterização anatomopatológica colheita e análise de dados. Porto Alegre: UFRGS, 2008. 83 p. Dissertação (mestrado) - Programa de Pós-Graduação em Ciências Veterinárias, Faculdade de Veterinária, Universidade Federal do Rio Grande do Sul, Porto Alegre.

Zimermann, F. C., Fallavena, L. C. B., Salle, C. T. P., Moraes, H. L. S., Soncini, R. A., Barreta, M. H. and Nascimento, V. P. (2012), Downgrading of heavy broiler chicken carcasses due to myodegeneration of the anterior latissimus dorsi: pathologic and epidemiologic studies, *Avian Dis.*, 56(2), 418-421.

Zuidhof, M. J., Schneider, B. L., Carney, V. L., Korver, D. R. and Robinson, F. E. (2014), Growth, efficiency, and yield of commercial broilers from 1957, 1978, and 2005, *Poult. Sci.*, 93(12), 2970-2982, 10.3382/ps.2014-04291.

Chapter 8

Quality defects associated with poultry muscle development: pale, soft and exudative meat

Giulia Baldi, Francesca Soglia and Massimiliano Petracci, Alma Mater Studiorum – Università di Bologna, Italy

1 Introduction

The constant pressure to improve profitability has led the poultry industry to prefer birds that are efficient feed converters, grow fast and have high lean

http://dx.doi.org/10.19103/AS.2022.0112.09

meat content with minimum production costs. This has been made possible thanks to improvements in the genetic selection of breeds. However, this has resulted in animals that are much more susceptible to environmental stressors and frequently develop muscular abnormalities that are responsible for great variations in meat quality (Petracci and Berri, 2017; Baldi et al., 2020). As meat moves along the distribution line from processors to the final consumer, quality becomes progressively more important. For this reason, intra- and inter-flock variations in poultry meat quality have been extensively investigated over the past 20 years (Dong et al., 2020a).

One condition known to be particularly problematic for the poultry industry is pale, soft and exudative (PSE) meat. PSE is a term formerly coined by the pork processing industry to define a meat quality defect resulting from an extremely rapid drop in muscular pH while the temperature of the carcass is still high (Briskey and Wismer-Pedersen, 1961). Although different factors are involved, this condition was later observed in turkeys and broilers (Van Hoof, 1979; Barbut, 1997). Initially, terms such as 'soft muscle tissue' and 'summer yield problem' were used (Sante-Lhoutellier et al., 1991), though terms like 'PSE-like condition' were later preferred (Sosnicki et al., 1998). In this chapter, the term 'PSE' will be used for convenience. PSE has now been found to occur in many species, with its distinctive characteristics observed e.g. in ostriches (Van Schalkwyk and Gertenbach, 2000), cattle (Aalhus et al., 1998) and even yellowfin tuna (i.e. 'burnt tuna') (Arroyo Mora et al., 2007).

Regardless of the species in which it occurs, PSE meat looks unattractive from a visual sensory standpoint, due to its pale color, flaccid texture and reduced water holding/binding capacity (Dong et al., 2020a; Woelfel et al., 2002). Since color, in particular, has a significant effect on consumer perceptions of quality, PSE affects poultry meat sales (Kim et al., 2014). Moreover, because of its poor processing characteristics, PSE meat seriously affects the functional quality of those processed products in which it is included (Barbut, 2009). This results in significant financial losses for both producers and retailers (Alvarado and Sams, 2002; Woelfel et al., 2002). Generally speaking, poultry meat is categorized as PSE if its lightness (L^*) is greater than 53 and/or pH is below 5.7 at 24-h post-mortem (Adzitey and Nurul, 2011; Dong et al., 2020a). However, this is just a rule of thumb, and values for determining PSE can vary among countries (Table 1).

PSE is a significant and growing concern for the poultry industry. However, significant recent progress has been made in understanding the factors underlying its occurrence and in developing methods to improve the functional properties of this meat. Within this context, this chapter aims to provide the readers with an overview of PSE in poultry and ways to reduce its occurrence or to improve the functional characteristics of affected meat.

Table 1 Incidence levels of PSE poultry meat observed in different countries.

Reference	Country	Species	L* cutoff[a]	$pH_{24\,h}$	$L^*_{24\,h}$	Incidence (%)
Barbut (1998)	Canada	Turkey	51	n.a.	n.a.	12.0
Owens et al. (2000a)	USA	Turkey	53	n.a.	54.7	40.0
Fraqueza et al. (2006)	Portugal	Turkey	50	5.76	51.2	8.0
Carvalho et al. (2014)	Brazil	Turkey	53	5.61	55.2	41.7
Barbut (1997)	Canada	Chicken	49	n.a.	n.a.	28.0
Woelfel et al. (2002)	USA	Chicken	54	5.76	59.8	47.0
Petracci et al. (2004)	Italy	Chicken	56	5.77	57.5	9.8
Lesiów et al. (2005)	Poland	Chicken	54	n.a.	n.a.	6.0
Zhu et al. (2012)	China	Chicken	53	5.73	54.7	23.0
Karunanayaka et al. (2016)	Sri Lanka	Chicken	58	5.83	61.8	70.0

n.a. = not available

[a] Type of colorimeter, illuminant and measuring conditions may vary among the studies.

2 Causes of the occurrence of pale, soft and exudative meat: genetic factors

Since its discovery in the 1960s, researchers and processors have endeavored to unravel the mechanisms underlying the development of PSE. A significant amount of data has been collected from experiments, ranging from live animal studies to highly controlled in-vitro trials. It is generally agreed that the occurrence of PSE particularly depends on genetic background, while ante-mortem stress and peri-mortem procedures are considered 'triggering factors' during poultry meat production (see Fig. 1), as will be further discussed in the following sections.

From the moment it was first observed in avian pectoral muscles in 1990s, meat scientists have studied the role of genetic factors in the development of PSE. So far the role of genetics appears to be indirectly related to birds' ability to adapt to stressors (i.e. acute stress) (Webb and Casey, 2010) rather than having direct involvement as a causative mechanism. Indeed, PSE seems to be the result of animals' reduced ability to cope with stress (Owens and Apple, 2010)

2.1 Genetic basis

Most of the studies carried out to define the molecular mechanisms and genetic factors leading to the development of PSE in poultry have focused on the

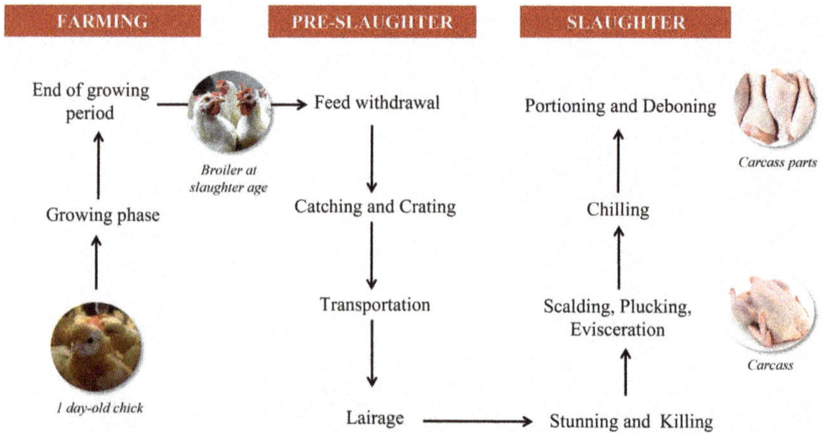

Figure 1 Schematic summary of farming, pre-slaughter and slaughtering phases of broilers.

ryanodine receptor (RYR) protein in light of its essential role in the regulation of calcium homeostasis in skeletal muscles. A point mutation (cytosine-to-thymine) with a single amino acid substitution in the *RYR* gene (involving the replacement of an arginine residue in wild-type at position 615 of *RYR1* with cysteine) has been demonstrated in swine (Fujii et al., 1991). This mutation, leading to the synthesis of a mutant and dysfunctional protein, is associated with a greater calcium release in muscle tissue (Paião et al., 2013; Strasburg and Chiang, 2009). The mutant protein was found to alter the functionality of the Ca^{2+}-releasing channels, which are sensitive to lower Ca^{2+} concentrations compared to the normal protein (i.e. the protein gate leaks or is locked open). This results in more intense muscle contraction and accelerated post-mortem glycolysis, triggering the development of PSE (Owens and Apple, 2010). Screening tests to identify animals' proneness to the development of PSE can be used in swine since PSE can be induced by exposure to anesthetics and depolarizing agents such as halothane and succinylcholine (Fujii et al., 1991).

Screening tests have been undertaken for poultry. However, exposing the animals to either halothane or succinylcholine was found to be a limited predictor of PSE in broilers and turkeys (Cavitt et al., 2004; Owens et al., 2000b; Wheeler et al., 1999). However, the similarity in the features associated with the development of PSE among different species suggests that the underlying mechanisms responsible for its onset may be similar and related to Ca^{2+} imbalances driven by genetic factors (England et al., 2017).

To complicate the matter, two isoforms of the RYR receptor in poultry, namely α and β, have been observed in defined locations in skeletal muscles (Airey et al., 1993; Chiang et al., 2007a; Paião et al., 2013). These exert different functions during the excitation-contraction mechanism (Paião et al., 2013).

Although a defect in the RYR receptor is hypothesized to cause PSE in poultry (Pietrzak et al., 1997), no differences in the expression of the *αRYR* gene have been found in chickens exhibiting PSE. However, a remarkable decrease in the expression of the gene encoding for the β isoform (*βRYR*) has been observed in muscular fibers of PSE broilers (Oda et al., 2009). These outcomes suggest the direct involvement of a differential expression of the RYR proteins in triggering the excess of cytosolic Ca^{2+} associated with the onset of PSE (Oda et al., 2009; Sporer et al., 2012).

None of the mutations in the *RYR1* found in pigs are observed in poultry (Chiang et al., 2004; Paião et al., 2013). A deletion mutation in *αRYR* leading to an alternative splicing (resulting in three different transcripts controlling the sensitivity of the RYR protein and Ca^{2+} release) was found in turkey (Chiang et al., 2004; Strasburg and Chiang, 2009). Substitutions (synonymous or leading to changes in protein structure) and deletion mutations have been detected in broiler chickens using cDNA sequencing (Oda et al., 2009; Sporer et al., 2012). A single nucleotide polymorphism in *αRYR* gene leading to a non-synonymous nucleotide substitution has been observed though it is not directly related to the development of PSE meat (Droval et al., 2012). Changes at mRNA level seem to be primarily associated with stress generated in birds from stressful conditions (i.e. heat stress) rather than being the primary cause of the defect itself. As an example, the expression level of *αRYR* and the occurrence of nonsense transcripts are modulated by exposure to heat stress and other stressors (Chiang et al., 2007b; Ziober et al., 2010). As a result, several genes and related proteins are differentially expressed and may contribute to the development of PSE (Malila et al., 2013, 2014; Sporer et al., 2012).

2.2 Growth rate

It has been suggested that there is a strong connection between the development of PSE and genetic selection for productive performances in poultry (Wang et al., 1999). Selection for growth, feed efficiency and meat yield has profoundly affected animal physiology (Velleman, 2019). Genetic selection has altered the ability of the animals to adapt to stressful conditions, leading to the development of stress-related syndromes (Gonzalez-Rivas et al., 2020; Solomon et al., 1998; Webb and Casey, 2010). Improved feed and management strategies may be additional contributing factors. This means that modern broiler chickens are generally more susceptible to stress and more prone to develop PSE (Cavitt et al., 2004). This susceptibility may be further exacerbated by the high body temperatures and metabolic heat production of fast-growing genotypes (Lu et al., 2007; Mills et al., 1999).

Functional differences along with a differential distribution of the proteins composing the sarcoplasmic reticulum are observed in turkey lines with different genetic backgrounds (i.e. fast-growing versus unimproved birds) (Wang et al., 1999). Upon being exposed to environmental stress, commercial poultry lines exhibit an altered thyroid response (i.e. increased plasma level of thyroid hormones) resulting in an increased incidence of PSE meat (Chiang et al., 2008; Everts, 1996; Gonzalez-Rivas et al., 2020).

3 Causes of the occurrence of pale, soft and exudative meat: ante-mortem stress

Aside from the genetic component, it is generally agreed that the most common factor leading to the development of PSE is pre-slaughter stress (Gonzalez-Rivas et al., 2020; Van Hoof, 1979; Vermeulen et al., 2015). As shown in Fig. 2, during the 24 h before slaughter birds are exposed to multiple stressors that make pre-slaughter procedures particularly critical for meat quality, since physiological stress can influence the metabolic state of the skeletal muscles at the time of slaughter (Bowker et al., 2017).

If animals are exposed to very stressful conditions before slaughter, the rate of glycogen breakdown into lactic acid (i.e. glycolysis) will be faster than normal. Rapid glycolysis also generates significant amounts of heat and increases carcass temperatures during the transformation of living tissue to meat (Matarneh et al., 2021). The combination of low pH and high carcass temperatures triggers the occurrence of meat quality alterations related to PSE (Scheffler and Gerrard, 2007). Adverse environmental temperatures (i.e. heat stress), as well as inadequate pre-slaughter procedures, have long been recognized as risk factors for the occurrence of PSE (Nicol and Scott, 1990).

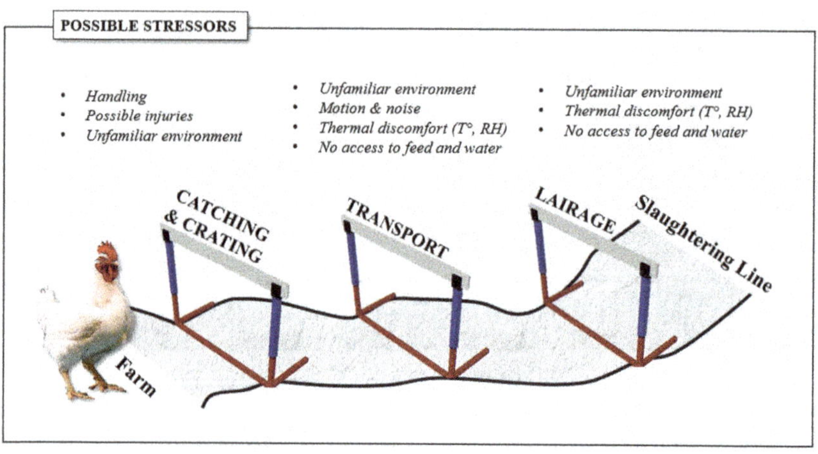

Figure 2 Possible stressors occurring during pre-slaughter phases.

3.1 Heat stress

The increase in average environmental temperatures and frequency of heat waves due to global warming have boosted the risk of heat stress in poultry (Gonzalez-Rivas et al., 2020). Heat stress is widely identified as one of the most stressful events in the life of birds, affecting their productivity and welfare in intensive and extensive systems worldwide (Lara and Rostagno, 2013). Moreover, due to their fast growth, rapid metabolic rate and increased basal metabolic heat production, modern poultry genotypes possess an altered thermoregulatory capacity, i.e. are more susceptible to suffer from heat stress than non-selected or indigenous breeds (Deeb et al., 2002).

Acute heat stress immediately before slaughter triggers the occurrence of PSE by accelerating muscle glycogenolysis and increasing the muscular concentration of lactic acid (Owens et al., 2009; Wang et al., 2017a). Fast-growing broilers show an increase in plasma thyroid hormone concentrations in response to heat stress (Bowen and Washburn, 1985). This alteration in thyroid responses seems to increase the incidence of PSE meat, since hyperthyroidism triggers the release of Ca^{2+} from the sarcoplasmic reticulum, stimulating energy metabolism and muscle contractility (Chiang et al., 2008; Everts, 1996).

When temperatures increase, birds breathe more rapidly to lose heat with carbon dioxide in the blood dramatically decreasing (Toyomizu et al., 2005) This condition is rapidly compensated for via urinary excretion of $HCO3^-$, triggering metabolic acidosis accompanied by a remarkable reduction in blood pH (Schneider et al., 1988). Under metabolic acidosis, the muscle generates energy through anaerobic metabolism by converting pyruvate into lactate to achieve a low muscular pH. If animals suffer heat stress before slaughter, metabolic acidosis significantly contributes to the development of PSE (Zaboli et al., 2019).

3.2 Feed withdrawal

Feed withdrawal refers to the total time that animals are deprived of food, including time at the farm, on the truck and at lairage (Barbut, 2015). This is a commonly used pre-slaughter procedure as it has a number of positive effects. One of these is to reduce the volume of stomach contents, thus limiting microbial contamination during transportation and evisceration (Eikelenboom et al., 1991). Fasting also affects *in vivo* energy metabolism and alters the levels of muscular glycogen concentration at slaughter (Sterten et al., 2010; Wang et al., 2013). If glycogen levels are reduced, the ultimate pH of meat will be higher than normal. Fasting is therefore a possible strategy to mitigate the incidence of PSE. On the other hand, prolonged on-farm fasting is not desirable because it excessively reduces glycogen reserves, increasing the risk of another quality defect: dark, firm and dry (DFD) meat (Adzitey and Nurul, 2011).

Feed withdrawals of 8–12 h for broilers and 6–10 h for turkeys are recommended, although there is no general consensus about the optimal duration of fasting (Barbut et al., 2008). However, if not conducted properly, fasting can become a potential stressor for birds. Several authors agree that fasting birds for more than 12 h is counterproductive since prolonged feed withdrawal not only reduces carcass weight at slaughter but also alters animals' hormonal responses (Barbut, 2015; Komiyama et al., 2008; Savenije et al., 2002). Wang et al. (2013) reported that prolonged fasting leads to higher plasma levels of corticosterone and altered thyroid response. The former is known to influence the onset of rigor mortis leading to paler meat, while the latter is responsible for an impaired regulation of calcium content in the skeletal muscle: both changes increase the risk of PSE (Gonzalez-Rivas et al., 2020; Kannan et al., 1998). It is important to highlight that, if done properly, fasting *per se* is not a stressor. However, if fasting is too protracted because of poor pre-slaughter handling practices (lairage, transportation, etc.), it becomes a risk factor for PSE (Costa et al., 2018). Proper communication between producer and slaughterhouse is essential when planning both fasting and lairage times to avoid the occurrence of meat quality defects.

3.3 Catching, crating and transportation

The period from when animals are harvested on the farm to slaughter at the processing facility is one of intense activity for birds. Catching, crating and transportation are generally considered the most stressful pre-slaughter procedures and may likely cause physical injuries (Ali et al., 2008; Schilling et al., 2008; Wang et al., 2017b). The stressors experienced by the animals could be further exacerbated by hot weather conditions, leading to metabolic alterations which ultimately affect muscular storage of glycogen (Kannan et al., 1997; Sandercock et al., 2001).

Prior to transportation to the slaughterhouse, birds are caught and placed into crates. Harvesting of birds is typically the least mechanized procedure and one of the most frightening and stressful for birds, since broilers are usually caught by hand and carried in an inverted position prior to placement into crates (Schilling et al., 2008). Handling chickens in an inverted position leads to prolonged tonic immobility and increases the production of glucocorticoids and epinephrine that may negatively affect meat quality, causing a rapid post-mortem pH decline (Kannan and Mench, 1996). However, current research suggests that the catching method (manual or mechanical) has only a limited impact on the incidence of PSE, though mechanical catching may be more advantageous under stressful conditions such as high summer temperatures (Bowker, 2017; Nijdam et al., 2005; Schilling et al., 2008).

Once in the crates, birds can only move in a half-erect posture. This means muscles maintain continuous tonicity, causing the breakdown of muscle glycogen into lactic acid (Kannan et al., 1997). After catching and crating, birds are loaded into trucks and transported to the abattoir. During transport, birds are exposed to a huge number of stressors, including the act of transporting (vibration, acceleration, noise, etc.), but also thermal stress, high relative humidity, crowding and poor airflow (Guàrdia et al., 2004; Knowles and Broom, 1990; Leheska et al., 2002). Transportation is thus generally associated with intense physiological stress, as shown by plasma corticosterone levels that increase proportionally with the duration of transport (Kannan et al., 1998; Nicol and Scott, 1990).

Prolonged transport quickly decreases the energy supplies of birds, making them progressively less able to cope with further stressors (Savenije et al., 2002). It has been shown that, after up to 6 h of transport, muscle glycogen and pH levels decrease dramatically, increasing the risk of PSE. However, it is important to highlight that the effects of transportation on the final quality of meat are tightly linked to the interacting influences of feed withdrawal, environmental stressors and subsequent lairage conditions (Bowker, 2017).

3.4 Lairage

After arrival at the slaughterhouse, there is usually a lairage period of 1–2 h before shackling the birds. A resting prior to slaughter both makes it easier to maintain continuous slaughter line operations and allows birds to recover from stress during transport (Costa et al., 2018; Warriss, 2000). During this rest period, the muscles can lose the acidic end-products derived from anaerobic metabolism triggered by stress and replenish depleted carbohydrate through mobilization of glycogen stored in the liver (Matarneh et al., 2017). However, since poultry are transported in plastic crates without access to food and water, lairage time should be kept as short as possible. A long lairage period exacerbates stress from transportation by coupling it with extensive fasting. Animals that face such stressful situations will inevitably have altered metabolism and likely produce PSE meat (Adzitey and Nurul, 2011; Van de Perre et al., 2010). After lairage, crates are usually moved to the shackling line where the slaughtering procedure begins.

4 Causes of the occurrence of pale, soft and exudative meat: slaughter and post-mortem factors

Although adverse environmental conditions and improper live bird management prior to slaughter play a key role in the occurrence of PSE, post-mortem processing of poultry carcasses can induce biochemical changes within the muscle tissue that may reduce meat quality. Indeed, PSE can also develop in non-stressed animals if they are improperly processed.

4.1 Stunning

Stunning is used to render animals unconscious prior to slaughter and to allow an easier, safer and efficient automated slaughtering process (Barbut, 2001). Stunning influences the onset of rigor mortis and, consequently, meat quality, depending on the type of stunning system as well as the degree of peri-mortem muscle activity of the animal entering the slaughtering process (Papinaho et al., 1995). Most slaughter plants utilize either low-voltage or high-current electrical stunning via an electrified water bath. Gas stunning is also used, although to a lesser extent (Bowker et al., 2017). Muscle contraction induced by electrical stunning increases the incidence of PSE meat in pigs (Channon et al., 2000). However, this has neither been observed in broilers (Mohan Raj et al., 1990) nor in turkeys (Northcutt et al., 1998, Alvarado and Sams, 2000). No differences have been found in the ultimate muscle pH of gas and electrically stunned birds, suggesting that stunning method may not reduce meat quality (Joseph et al., 2013; Papinaho et al., 1995).

However gas stunning makes it possible to stun animals inside the crates, avoiding the stressors related to pre-stunning activities such as manual removal of animals from crates and hanging on the shackle line (Barbut, 2009). Indeed, lower plasma corticosterone levels were found by Kannan and Mench (1996) in gas-stunned birds compared to those one subjected to electrical stunning, with decreased corticosterone levels potentially mitigating meat quality defects. A more recent study has highlighted a lower incidence (−10%) of PSE in broilers stunned by gas rather than electronarcosis (Da Silva-Buzanello et al., 2018).

4.2 Scalding

After being stunned, animals are killed. The resulting carcasses then need to be immersed in a hot water bath to soften the feather follicles and facilitate plucking. As a rule of thumb, the higher the scalding temperature, the easier the feather release from the follicles. In Europe, carcasses are scalded at about 52°C for 2–4 min while, in the USA, shorter times (1.5 min) and higher temperatures (approximately 58°C) are employed (Barbut, 2015).

Regardless of the scalding method, the immersion of carcasses in a hot water bath immediately after the death of the animals appears to accelerate post-mortem anaerobic glycolysis and triggers a rapid pH decline at elevated body temperatures, leading to protein denaturation and development of PSE (Lee and Choi, 1999). As scalding temperature increases from 66°C to 70°C, PSE incidence in pigs increased from 21.4% to 47.7% (Park, 1980). To the best of our knowledge, this trend has not been observed in birds, although higher scalding temperatures have been associated with paler appearance of meat (da Silva-Buzanello et al., 2019). However, although high scalding temperatures (>57°C) affected meat color, both ultimate pH and water-holding capacity (WHC) were not modified, suggesting that higher scalding temperatures did not trigger PSE in broilers.

4.3 Electrical stimulation

Electrostimulation is increasingly being used in commercial slaughter plants to prevent toughness problems associated with early deboning. The breast is brought into contact with a stainless-steel plate carrying an electric current. This hastens rigor mortis development by causing fast muscle contraction that facilitates consumption of glycogen and accelerates pH decline (Tornberg, 1996). Electrical stimulation reduces the time for meat to be ready to be sold. However, increasing the rate of early post-mortem changes while carcass temperature is still high may potentially cause protein denaturation (Bowker, 2017). However, in contrast to studies in pigs (Hammelman et al., 2003), experiments on chickens (Zhuang et al., 2010) and turkeys (Owens and Sams, 1997) did not show either a positive or negative effect of electrical stimulation on meat pH, color and WHC. Electrical stimulation *per se* does not promote the occurrence of PSE; however, since it stimulates post-mortem metabolism, its potentially negative effect might be a function of the state of muscle at the moment of slaughter (Craig et al., 1999).

4.4 Carcass chilling

Following evisceration, poultry carcasses are usually chilled to a temperature of 4°C over several hours, depending on the size of the carcass. Post-mortem carcass temperature is one of the most crucial processing factors affecting overall meat quality. It plays a significant role in the occurrence of PSE, to a degree that even normal glycolyzing carcasses can develop PSE meat if not properly chilled (Lee and Choi, 1999; Offer, 1991). This is particularly true for turkeys, as their large body size may cause slower chilling than for broilers (Alvarado and Sams, 2002). Improper and/or delayed carcass chilling can result in elevated muscle temperatures (>30°C) which, in turn, accelerate post-mortem pH decline and increase R-values (i.e. inosine:adenosine ratio) (Mckee and Sams, 1998; Zhu et al., 2011, 2013). Moreover, high carcass temperatures at early post-mortem may also hasten post-mortem metabolism through the activation of AMP-activated protein kinase, a key enzyme controlling early post-mortem pH decline, increasing the development of PSE meat (Shen et al., 2006; Zhu et al., 2013).

In order to reduce the risk of PSE meat characteristics, reaching a breast muscle temperature of 25°C or lower by 60 min post-mortem is recommended: if muscle temperature is still not reduced by 105 min post-mortem, meat quality could be compromised (Alvarado and Sams, 2002). For these reasons, accelerated carcass chilling (i.e. blast chilling) is nearly universal in the poultry industry (Matarneh et al., 2021).

5 Characteristics of pale, soft and exudative meat: physico-chemical and microbial properties

The development of PSE significantly affects the main quality parameters of meat, including color and texture. These color variations negatively affect consumer perception of fresh meat quality. The lower water-holding/binding capacity of PSE meat also affects its suitability for further processing.

Low ultimate pH (<5.7) and pale color are among the distinctive features of PSE broiler and turkey meat. The fall in pH following the animal's death is a key element determining meat quality as it can affect several meat quality attributes including color, WHC, texture and microbial shelf-life (Petracci et al., 2017). Post-mortem measurement of pH in poultry is typically performed 15 min (pH_{15}) and 24 h (pH_u) after slaughter (Petracci et al., 2017).

PSE meat develops as a consequence of an accelerated early post-mortem metabolism that, along with H^+ accumulation and heat production, ultimately result in rapid pH decline (pH < 6.0 within 1 h post-mortem; see Fig. 3) (England et al., 2017). Rapid acidification combined with high carcass temperatures immediately after post-mortem time result in a pale color (Kim et al., 2014). With rapid pH decline, extensive protein denaturation occurs which results in reduced solubility of the protein fraction composing the muscle tissue itself (Puolanne, 2017; Wilhelm et al., 2010).

Changes in protein functionality (due to denaturation occurring during the conversion of muscle to meat) can affect both myofibrillar and sarcoplasmic proteins. The latter is mainly responsible for the high light scattering and increased reflectance phenomena that ultimately result in a paler meat color (see Fig. 4) (Hughes et al., 2014; Kim et al., 2014; Puolanne, 2017; Swatland, 2008). In PSE meat, water migrates through the different muscular compartments and, accumulating on the meat surface, strongly affects light scattering. A higher proportion of incident light is reflected rather than absorbed, penetrates to a shallower depth and the color appears paler/lighter (Kim et al., 2014; Owens and Apple, 2010). Birefringence and light scattering is inversely proportional to pH (Zhang and Barbut, 2005a). Rapid post-mortem pH decline also affects the redox stability of myoglobin. By inducing the unfolding of the globin moiety, rapid pH fall allows a greater autooxidation of pigment and, by increasing susceptibility to metmyoglobin formation, leads to its denaturation (Zhu and Brewer, 2003).

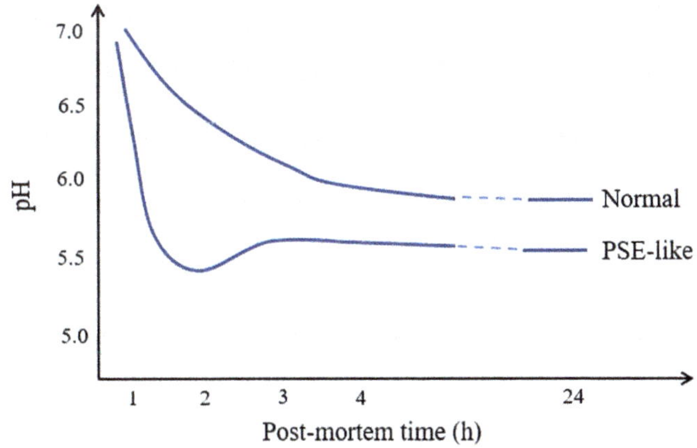

Figure 3 Schematic representation of pH decline in normal-acidifying and PSE poultry meat.

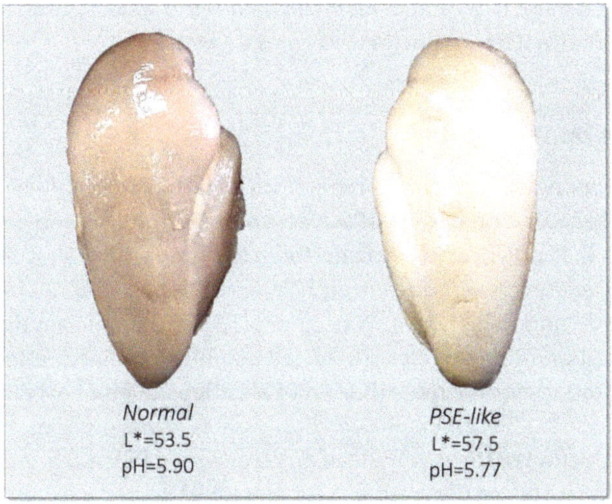

Figure 4 Representative images of normal and PSE chicken breast muscles. Data from Petracci et al. (2010).

5.1 Microbial properties

The physico-chemical properties of PSE poultry meat (i.e. low pH and impaired WHC) may also affect its microbial properties. The low pH of PSE meat inhibits microflora and could thus extend shelf-life (Allen et al., 1997, 1998). The significantly lower counts of aerobic mesophilic and psychrotropic microorganisms found in PSE poultry meat after 5 days and 12 days of storage seem to support this hypothesis (Fraqueza et al., 2008). On the other hand, the presence of drip and exudate on the meat surface (consisting of water but also containing proteins) is a potential source of nitrogen favoring the development of spoilage microorganisms (Lesiow and Kijowsi, 2003). However, the dominant flora in both turkey and broiler meat (including *Pseudomonas*, *Enterobacteriaceae* and *Brochotrix thermospacta*) do not seem to be influenced by ultimate pH (Fraqueza et al., 2008).

6 Characteristics of pale, soft and exudative meat: technological properties

Accelerated early post-mortem muscle pH decline results in denaturation of approximately 20% of muscular proteins, triggering lower WHC of PSE meat (Bowker, 2017; England et al., 2017; Liu et al., 2016). This phenomenon may thus account for the 'exudative' aspect of PSE meat which, if compared to normal meat, exhibit higher fluid loss following refrigerated storage (as drip) or exposure to thermal treatment (Petracci and Cavani, 2012).

6.1 Protein denaturation

Low pH conditions while muscle temperature is still high strongly affects myosin molecules. Denaturation of the ATPase active site is induced and conformation changes of the polypeptide chains (affecting their overall solubility and functionality) are observed (Bowker and Zhuang, 2015; Li et al., 2015b; Liu et al., 2016; Alvarado and Sams, 2004; Warner et al., 1997). Significant differences in the concentration and extractability of salt-soluble proteins are found in PSE meat compared to normal meat (Barbut et al., 2005; Chen et al., 2018; Li et al., 2015a; Zhao et al., 2017). Extensive shrinkage of myosin heads that, in turn, triggers contraction of the myofibrillar lattice occurs in PSE meat, pushing water into extracellular spaces (Bowker, 2017; Offer, 1991).

Muscle cell shrinkage, along with a higher number of intracellular spaces and a complete disorganization of the sarcomeric structure, is observed by comparing the microstructural features of PSE meat with normal meat (Barbut et al., 2005; Wilhelm et al., 2010). This phenomenon, along with the increased Ca^{2+} concentration associated with the onset of PSE, seems to promote either more intense activity or earlier activation (even before slaughter) of endogenous proteases (i.e. m-calpains) (Wilhelm et al., 2010). By affecting the integrity of the muscle structure, these further reduce functionality of protein.

Aside from denaturation, Carvalho et al. (2017) found higher protein carbonylation and more intense formation of protein cross-links in PSE meat compared to normal meat during refrigerated storage. Oxidation phenomena observed in PSE muscles may play a pivotal role in loss of protein functionality (de Avila Souza et al., 2022). Taken together, these features may help to explain the lower WHC (demonstrated by higher drip and cooking losses) observed in PSE meat compared with normal meat.

6.2 Water-holding capacity

Post-mortem changes in muscle tissue, including the rate and extent of pH decline, have the potential to influence WHC of the resulting meat (Bowker, 2017). PSE compromises the ability of breast meat to hold the water it contains as well as that added during processing (Barbut, 2015). WHC varies according to interactions between meat pH and the net electrical charges on muscle's proteins. As a rule of thumb, the further the ultimate pH of meat is from the isoelectric point of the myofibrillar proteins, the greater is the WHC (see Fig. 5). This is a direct consequence of the electrostatic repulsion phenomena existing among similar charges on the protein filaments (Bowker, 2017). When the final pH approximates the isoelectric point of the muscle proteins (i.e. the average pH value at which the number of positive and negative charges on the proteins' side groups are essentially equal), WHC is minimal since the overall net charge is zero. The attraction between these groups reduces the repulsive forces existing between the filaments and the space available to hold water (Huff-Lonergan and Lonergan, 2005).

The impaired water-holding and binding capacities observed in PSE meat can be explained by the shrinkage of both myofibrils and lattice space. These, along with the concurrent reduction in the repulsion phenomena among the proteins' side groups, ultimately result in a higher proportion of water migrating toward the extracellular compartments (Bowker, 2017; Offer and Knight, 1988). The longer relaxation time of water within extra-myofibrillar spaces suggest higher mobility of this water fraction that can thus be easily lost during processing and storage (Chen et al., 2018). Carvalho et al. (2017) reported that PSE muscle proteins are more susceptible to oxidative degradation. This is likely due to several mechanisms, including the pro-oxidant effect of low pH on the oxidizing cycle of iron, as well as the impact of fast pH decline on cell structure and compartmentalization that may enhance the interactions between pro-oxidants and amino acid residues. This is another plausible mechanism for the reduced WHC of PSE muscles, since it is widely known that protein functionality can be significantly affected by the occurrence of oxidation processes (Lund et al., 2011).

Although the mechanism is not fully understood, sarcoplasmic proteins may affect the WHC of PSE meat. Aside from scattering incident light, it has been suggested that sarcoplasmic protein denaturation may counteract the negative effects of lattice space shrinkage (Liu et al., 2016). The gel-like

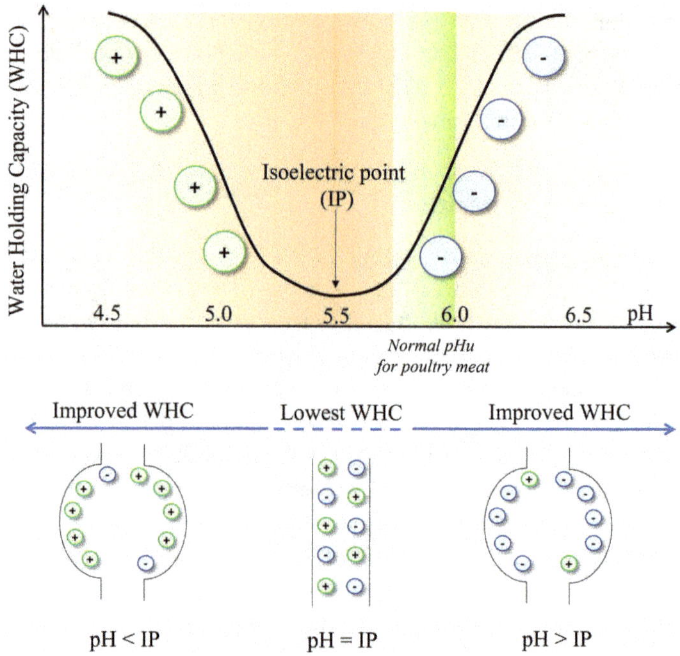

Figure 5 Influence of pH on meat water-holding capacity. The green frame indicates average ultimate pH values normally reached by poultry meat at 24 h post-mortem (5.8–6.0).

network formed upon coagulation of the denatured sarcoplasmic proteins in the extracellular spaces may trap the water migrated from the intracellular compartments, partially improving the WHC of PSE meat.

6.3 Textural and gelling properties

As the term 'soft' implies, PSE poultry meat has a flaccid consistency. The right degree of tenderness is one of the most important attributes used by consumers to assess the quality of poultry meat (Morey and Owens, 2017). PSE impairs the textural properties of poultry meat either when sold fresh (e.g. as cuts) or used for further processing. When compared to normal broiler and turkey breast meat, both whole muscle and processed PSE meat exhibit lower fracture forces and textural parameters (i.e. hardness, springiness, cohesiveness and chewiness) following texture profile analysis (Chan et al., 2011; Chen et al., 2018; Dong et al., 2020a; Li et al., 2015a, 2019; Zhang and Barbut, 2005b). These characteristics result from the altered functionality of salt-soluble proteins as well as the lower extractability and solubilization of myosin molecules in PSE poultry meat (Barbut et al., 2005; Li et al., 2019; Pietrzak et al.,

1997). This is due to the lower thermal gelling ability of the native proteins in PSE meat which results in a less effective trapping of water within muscular compartments (Li et al., 2019). Changes observed in the storage modulus (G′) of PSE meat batters (reflecting a less rigid, weaker and softer product) seem to be a direct consequence of a more intense protein denaturation affecting gel network formation (Li et al., 2015a, 2019; Somboonpanyakul et al., 2007; Zhang and Barbut, 2005b). It has been suggested that more protein interaction and aggregation occur in PSE meat at lower temperature (Li et al., 2019). Given their ability to stabilize protein conformation, the significant differences observed in the ionic and hydrogen bonds of PSE actomyosin may further alter gelling properties of these meats (Liu et al., 2011).

7 Reducing the incidence and mitigating the impact of pale, soft and exudative meat quality: genetic and epigenetic strategies

At the end of the growing phase, birds are caught, loaded into cages or coops and transported to a slaughterhouse that may be close to or far from the farm (see Fig. 1). The resulting stress (Fig. 2) stimulates the neuroendocrine system and affects immune and metabolic functions, leading to biochemical and metabolic changes in the early post-mortem period. These changes potentially increase the occurrence of PSE meat in poultry (Petracci et al., 2009). Glycolysis, protein modification, Ca^{2+} homeostasis, apoptosis and proteolysis have been identified as the main stress-related pathways or mediators that affect the main biochemical events occurring during the conversion of muscle to meat, possibly triggering PSE (Xing et al., 2019). Heat stress during pre-slaughter phases is considered the main stressor resulting in PSE meats, especially in fast-growing or heavier birds that are more sensitive to heat stress (Yunis and Cahaner, 1999). Sandercock et al. (2006) show that genetic selection for high muscle growth in broiler lines has compromised their thermoregulatory capacity because fast-growing broilers produce more heat than slower-growing chickens.

The negative correlation between heat tolerance and growth rate makes genetic selection to improve both traits challenging. El-Gendy and Washburn (1995) found that the heritability for heat tolerance was very low in fast-growing broilers. Substantial genetic variability has been observed in individual animal responses to increased heat loads, with a moderate degree of genotype-by-environment interaction, which implies that fast-growing animals in normal conditions may not adapt well to heat stress (Deeb et al., 2002). There have been a number of recent 'omics'-based studies on the underlying mechanisms of the heat stress (Nawaz et al., 2021). Many candidate genes and potential biomarkers have been proposed from DNA, RNA and metabolomics studies

(Goel et al., 2021; Kim et al., 2021; Sun et al., 2015). However, this research has still to be translated into selection tools to improve heat tolerance in breeding.

Thermal manipulation during embryogenesis may influence metabolic rate (Piestun et al., 2009; Tona et al., 2008) and heat production capacity during the growth and pre-slaughter phases (Loyau et al., 2014, 2016). Increasing the incubation temperature from 37.8°C to 39.5°C, and relative humidity from 56% to 65% from days 7 to 16 of embryogenesis for 12 h/day, changed body temperature, physiology and parameters of thyroid and adrenal axes, as well as respiratory parameters and stress levels of birds after hatching (Loyau et al., 2014; Piestun et al., 2008). This strategy can improve heat stress tolerance among birds particularly sensitive to heat (Piestun et al., 2008).

PSE is also directly related to muscle glycogen reserves at death, which is the main determinant of ultimate pH (pHu) in farmed animals. Higher glycogen storage in muscle predisposes breast meat to PSE in poultry. Manipulation of muscle glycogen could be used to change pHu and reduce levels of PSE in chickens and turkeys (Petracci et al., 2017). At the genetic level, the correlation between breast muscle glycolytic potential and pHu is close to -1 (Le Bihan-Duval et al., 2008). Several studies have shown that chicken breast meat pHu has a moderate to high heritability (as shown by heritability values of 0.3 to 0.5 found in different genetic lines), suggesting this trait could be modified by genetic selection (Gaya et al., 2011; Le Bihan-Duval et al., 2001, 2008). However, in contrast to pigs, where numerous quantitative trait loci (QTL) and a major gene have been identified for pHu (Milan et al., 2000), the genetic architecture of this trait remains poorly understood in chickens. A region of major interest with significant selection signatures associated with the pHu of Pectoralis major has been identified (Le Bihan-Duval et al., 2018). It has also been found that muscle glycogen content can be increased by exogenous supply of nutrients during the early post-hatching stage, suggesting nutrition may have a role in controlling muscle glycogen levels (Métayer-Coustard et al., 2021).

8 Reducing the incidence and mitigating the impact of pale, soft and exudative meat quality: pre-slaughter strategies

During the period before slaughtering, birds are exposed to different stressors, including on-farm feed withdrawal, catching, crating, transport and lairage at slaughterhouse (see Figs 1 and 2) (Saraiva et al., 2020; Xing et al., 2019). Responsibility for these stages is shared between the farmer, catching crew, lorry driver and the slaughter plant in modern integrated broiler and turkey production chains (Petracci et al., 2010). The farmer and the catching crew must first manage animals from feed withdrawal to catching and crating. The slaughter plant is then responsible for managing transport time and lairage conditions to maintain acceptable levels of welfare. The slaughter plant must also ensure proper processing conditions (i.e. stunning, scalding, chilling, etc.) to achieve optimal and uniform meat quality without defects such as PSE (Barbut et al., 2008; Petracci et al., 2010; Xing et al., 2019).

8.1 Feed withdrawal and dietary manipulation before slaughter

Feed withdrawal is a recommended on-farm practice in preparing birds before slaughter e.g. to reduce carcass contamination from spilling of gut contents during slaughtering (Xue et al., 2021). The amount of glycogen stored in skeletal muscles at the time of killing is influenced by pre-slaughter fasting times which can be used to control ultimate muscle pH and decrease the risk of PSE (England et al., 2013; Petracci et al., 2017). A short fasting (<8 h) could increase the prevalence of PSE meat, while prolonged feed withdrawal (>12 h) may prevent its occurrence, although resulting muscle glycogen exhaustion may increase incidence of DFD meat (Northcutt, 2000; Petracci et al., 2010). The best solution is to let sufficient time pass from the last on-farm feed intake to slaughter in order to promote a slight exhaustion of muscle glycogen.

Unlike large mammals, feed cannot be provided to poultry during transport and lairage because they are kept inside the crates without access to feed (Petracci et al., 2010). The addition of carbohydrates to drinking water during feed withdrawal may prevent muscle glycogen depletion during this period, especially if a long fasting time and cold conditions are expected during transport (Karacay et al., 2008; Kop-Bozbay and Ocak, 2015; Petrolli et al., 2016). The addition of electrolytes to broiler diets to promote greater retention of body fluids has also been investigated. This practice can help to reduce animal stress and body dehydration during transport and lairage at abattoirs, especially during hot weather season (Ahmad and Sarwar, 2019; Petrolli et al., 2016; Wasti et al., 2020).

Several dietary supplements have also been suggested to ameliorate the adverse effect of pre-slaughter stressors, such as guanidino acetic acid, betaine, *Forsythia suspensa* extract, resveratrol and L-threonine (Al-Abdullatif et al., 2021; Downing et al., 2017; Pan et al., 2018; Zhang et al., 2017, 2019a,b). Supplementation of gluconeogenic amino acids also seems to reduce tissue catabolism and maintain muscle glycogen (Schaefer et al., 2001), especially if birds are exposed to heat stress (Qaid and Al-Garadi, 2021).

8.2 Transport

The risk of PSE is increased by high temperatures and relative humidity that can develop rapidly inside transport vehicles (Aldridge et al., 2019; Nawaz et al., 2021; Xing et al., 2019). Trucks should have a covered deck and allow effective natural ventilation. However, passive ventilation may not allow sufficient air movement to prevent temperature extremes across both summer and winter conditions, particularly when vehicles are stationary (dos Santos et al., 2020; Mitchell and Kettlewell, 2009). During warm and dry conditions, maximum ventilation should be guaranteed by providing as many openings as possible (Cockram and Dulal, 2018). Innovative transport containers and crates have been developed to improve the microenvironment and reduce the risk of PSE (Carvalho et al., 2018). However, heat stress can still be a serious problem in open or semi-open vehicles, particularly when stationary in hot and humid weather conditions (Cockram and Dulal, 2018; Mitchell and Kettlewell, 2009). Mechanical ventilation to ensure adequate airflow is a potential solution (Mitchell and Kettlewell, 2009).

Stocking densities should also be adapted according to weather conditions and transport times. The space allowance per bird should be increased by 10–20% during hot seasons and in heavy traffic conditions where passive ventilation may be reduced (Delezie et al., 2007; Mitchell and Kettlewell, 2009). Delezie et al. (2007) have highlighted the importance of stocking density in transport crates compared to the effects of feed withdrawal and transport. Forced ventilation cooling systems during holding have also been suggested for broilers transported under heat stress (Zhang et al., 2020). The use of water-misting sprays with forced ventilation immediately after transport has been found to create a more comfortable environment to alleviate stress (Jiang et al., 2016).

8.3 Lairage

On arrival at the slaughter plant, crates are unloaded from the transport vehicle and kept in a protected area before slaughtering. Most abattoirs have a closed area equipped with forced ventilation and water-misting sprays, though areas are outside with a roof to protect the birds from sunlight and rain (Petracci et al., 2010). Holding birds at different temperatures can influence meat quality. PSE rates in chickens and turkeys may increase if they are kept in hot conditions before slaughter (Mckee and Sams, 1998; Petracci et al., 2001). Temperature and humidity should be monitored at the crate level to maintain values within the thermoneutrality zone of birds (15–25°C and 60–65% humidity) (Nielsen et al., 2019). Lairage areas in abattoirs can provide an opportunity for rest and recovery from stress during transport. A period of 2 h in a controlled lairage environment has been found to reduce the thermal load of broiler chickens (Vieira et al., 2011). However, prolonged lairage times at abattoirs should be avoided to reduce negative effects on animal welfare and meat quality (Saraiva et al., 2020). Blue light can also minimize stress and keep birds calm in lairage zones. Exposure to blue light just before slaughter provides a calming atmosphere which reduces the risk of defects such as PSE (Barbosa et al., 2013). Other studies have also focused on the combined effects of transport duration and lairage time and optimal holding times to alleviate stress and improve meat quality (Oba et al., 2009; Zhang et al., 2009).

9 Processing strategies to reduce the incidence and mitigating the impact of pale, soft and exudative meat quality: functional ingredients

Despite the efforts of farmers and producers, reducing the occurrence of PSE meat remains challenging, especially during the summer season. PSE can also develop after the death of the animal due to poor carcass processing. Since it is less attractive to consumers, PSE meat is often used in the manufacture of processed products. However, impaired water retention and reduced gelling properties complicate its inclusion in formulating meat products. Research has focused on ways to improve the sensorial and functional properties of PSE-based meat (Barbut, 2019).

When PSE meat is used in the formulation of processed products, its color and poor water-binding capacity need to be taken into account (Li et al., 2019). This has placed the focus on functional compounds able to enhance water retention and gelling properties of meat batters. Additives include carrageenan, collagen, protein isolates and modified starch (Dong et al., 2020a).

9.1 Starch

Starches are widely used in meat processing as thickeners to improve texture, water retention and gelling properties of processed products. Physical, enzymatic or chemical treatments can be used to modify the natural structure of starches to enhance their performance (e.g. lengthen/shorten gelatinization time, increase viscosity and stability, etc.) (Dong et al., 2020a). The effects of both regular and modified potato and tapioca starches on the functional properties of PSE chicken meat batters have been investigated by Zhang and Barbut (2005b). They found that, regardless of the type of starch, their addition efficiently improved cooking yield and texture of meat batters, thus partly compensating for protein functionality lost in PSE meat. Tapioca-modified starch was found to be the most efficient in enhancing the functional properties of meat batters. Native potato and tapioca starches are the most widely used in poultry meat products: the former is particularly used to increase viscosity, while the latter is employed to impart a shiny and smooth texture (Petracci et al., 2013). The gelatinization temperature of corn starch, being very close to that used for the cooking of meat products (62–72°C), makes this additive particularly efficient in improve the cooking yield of products formulated with PSE meat (Motzer et al., 1998).

9.2 Collagen

Collagen is an inexpensive meat by-product that is commonly used in restructured meat products. It increases cooking yield and tensile strength by improving protein functionality through immobilization of free water (Baldi et al., 2021b). Its incorporation in restructured boneless cured pork products increased WHC and protein functionality (Schilling et al., 2003). Incorporation of 1.5% collagen in chunked and formed deli rolls formulated with 100% PSE chicken meat decreased cooking loss and enhanced protein-protein binding (Schilling et al., 2005). The effects of including collagen in formulating restructured meat products vary according to the percentage and severity grade of PSE meat included and the amount of collagen added (up to a maximum of 3%) (Dong et al., 2020a).

9.3 Carrageenan

Carrageenan is a hydrophilic polysaccharide macromolecule extracted from red seaweed (Geonzon et al., 2019). This compound is widely used in the poultry processing industry because of its ability to react with proteins to improve yield and texture of processed meat products (BeMiller, 2018). Its effectiveness in increasing the usability of PSE meat was first described by Daigle et al. (2005), who reported that the inclusion of 0.3% of kappa and iota carrageenan improved the functionality of PSE turkey meat without affecting the sensory characteristics of the final product. Using this ingredient also increased cook yield and decreasing purge loss, though it can form a slimy precipitate on the surface of products that may be undesirable for consumers (Motzer et al., 1998).

9.4 Soy protein isolates

Protein isolates extracted from soybeans have good nutritional value and functional properties and therefore are commonly used in the formulation of processed product to increase protein gelation and water-binding capacity (Lusas and Rhee, 1995). Soy proteins can increase viscosity and create a stable gel matrix during heating due to their ability to form strong protein–water interactions (Petracci et al., 2013). The incorporation of 1.5% of soy protein isolate in the manufacture of deli rolls formulated with 100% PSE turkey meat significantly decreased cooking losses and improved overall water-binding capacity and total yield of the final product (Daigle et al., 2005).

9.5 Transglutaminase

Transglutaminase is an enzyme which has attracted attention for its ability to improve the WHC of restructured meat products (thus a reduction in salt content) (Petracci et al., 2013). Transglutaminase – extracted from microorganisms or from porcine and bovine blood – promotes cross-linking between myosin and other proteins. This creates a strong water-retaining gel network, improving texture and cooking yield of processed products (Baldi et al., 2021b; Park et al., 2017). Lesiow et al. (2017) showed that addition of transglutaminase improved the rheological and textural properties of sausage batters formulated from PSE pork meat, while its combination with sodium tripolyphosphate further enhanced meat microstructure, so much so that the texture of the sausages formulated with 100% PSE meat was judged equivalent to that made from normal, non-defective meat. Transglutaminase combined with different protein isolates significantly improved hardness values of chicken PSE meat (Barbut, 2009). Cegielka and Mroczek (2005) reported that adding 0.3% transglutaminase improved functional and sensory characteristics of low-salt chicken meat, suggesting its potential in commercial application of low-value meat in the food industry.

10 Processing strategies to reduce the incidence and mitigating the impact of pale, soft and exudative meat quality: technologies

The addition of functional ingredients can enhance the final quality of processed products formulated with PSE meat. On the other hand, many consumers prefer more 'natural' products with minimal addition of non-meat ingredients. This has focused attention on traditional and newer processing technologies to improve functional characteristics of PSE meat.

10.1 Marination

Marination is a traditional method widely used to tenderize and improve both sensory properties and protein functionality of meat and meat products. Marinades containing salt and phosphates are the most widely used in the meat industry to enhance meat WHC and total yield (Alvarado and McKee, 2007). Tumbling accelerates marinade solution absorption via extraction of salt-soluble proteins (Petracci et al., 2012). The combined effect of vacuum-tumbling and marination with phosphates and salt significantly improved the color and shelf-life of PSE chicken meat and reduced yield losses (Gorsuch and Alvarado, 2010). Alvarado and Sams (2003) tested pre-rigor injection of PSE chicken breast fillets with sodium phosphates and sodium bicarbonate and found that pre-rigor marination reduced cook loss by increasing meat pH.

However, given health concerns about excessive sodium salt intake, Haddad et al. (2018) found that the addition of <1% salt was sufficient to improve overall quality of a restructured meat product formulated with PSE meat.

10.2 High-pressure processing

High-pressure processing involves subjecting food to high isostatic pressure (ranging from 100 MPa to 900 MPa), improving shelf-life while preserving sensory and nutritional properties (Campus, 2010). Even though high-pressure processing is primarily used to reduce microbial spoilage, the technique also induces conformational changes in proteins (Messens et al., 1997). Chan et al. (2011) reported that, when subjected to 200 MPa pressure for 5 min, PSE meat improved water retention properties due to release of proteins' sulfhydryl groups and increased protein surface hydrophobicity. This produced higher total protein solubility and formation of a stronger gel network. However, high-pressure processing technology is constrained by factors such as cost and changes in meat color (i.e. whitening effect due to myoglobin denaturation), especially for PSE meat (Carlez et al., 1995; Dong et al., 2020a).

10.3 High-intensity ultrasound

High-intensity ultrasound is a nonthermal processing technique that generates pressure waves and cavitation bubbles whose collapse produces physical forces that change biological materials at both macro and microscopic levels (Al-Hilphy et al., 2020). This technique has been successfully used in the meat industry to enhance mass transfer and marination processes and particularly to improve meat tenderization and shorten aging time of red meats (Alarcon-Rojo et al., 2019). Ultrasound significantly affects the secondary structure of proteins and could thus enhance protein functionality of defective meat (Jayasooriya et al., 2004). PSE meat batter treated with ultrasound showed lower α-helical and higher β-sheet content compared to untreated batters, improving meat WHC and gel strength (Li et al., 2014). The technique has reduced drip losses and improved tenderness in PSE pork meat, suggesting its potential to improve technological properties of both fresh and processed meat products (Dolatowski et al., 2007).

10.4 Glycosylation

Glycosylation refers to the Maillard reaction in which a covalent connection between a carbonyl-containing moiety (usually a reducing sugar) and an available amino group occurs. This leads to the formation of glycoproteins, compounds that have both the macromolecular characteristics of proteins and the hydrophilic properties of carbohydrates (Dong et al., 2020a). This reaction can improve properties of food proteins such as gelling capacity, solubility and thermal stability (Liu et al., 2012).

Bian et al. (2018) have explored the effects of non-enzymatic glycosylation using glucosamine on the functionality of myofibrillar proteins extracted from PSE chicken breast meat. They reported that treatment modified the tertiary structure of proteins, increasing the density and strength of the protein gel network. However, since non-enzymatic glycosylation also alters meat color and produces potentially harmful compounds (e.g. acrylamide, furans and heterocyclic amines), enzymatic glycosylation has also been explored. Xu et al. (2018) evaluated the conjugation reaction between myofibrillar proteins of PSE meat and glucosamine using microbial transglutaminase. This process induced conformational modification of myofibrillar proteins and increased their solubility, thus improving the functionality of PSE chicken meat.

10.5 Pulsed electric fields

Pulsed electric field technology involves applying high-intensity electrical pulses for short periods to biological tissue placed between two electrodes (Bhat et al., 2019). This induces microstructural changes that can potentially improve poultry meat technological properties (Baldi et al., 2021a). Dong et al. (2020b) demonstrated that pulsed electric fields significantly changed the secondary structure of myofibrillar proteins extracted from PSE meat. This led to the release of hydrophobic and sulfhydryl groups and improved interaction between water and proteins, making them more soluble. The treatment also allowed formation of a strong, elastic protein network structure with the potential to improve the physico-chemical properties of PSE meat.

11 Detection of pale, soft and exudative meat

A key challenge for the meat industry is online detection of PSE using non-invasive, accurate and fast methods. Early detection of defective meat would prevent it from being supplied for fresh consumption so it could be diverted for use in processed products. Color lightness, pH and electrical conductivity were initially proposed as indicators to identify defective meat (Byrne et al., 2000; Garrido et al., 1995). However, these are hard to implement in industrial online systems. New techniques based e.g. on surface or low penetration depth measurements have been developed to detect defective meat more easily and objectively.

11.1 Computer vision system

Computer image analysis allows non-invasive and objective acquisition of data about color measured across the meat surface, avoiding more unreliable manual observation (Du and Sun, 2004). Chmiel et al. (2011) demonstrated that image analysis was effective in the detection of PSE condition occurring in pork *m. longissimus lumborum*, though it was less successful in identifying PSE in *m. semimembranosus* due to variability in color (Chmiel et al., 2016). Image analysis has been found to be effective in discriminating PSE from normal-colored meat in chicken fillets (Barbin et al., 2016).

11.2 Near-infrared spectroscopy

Near-infrared spectroscopy (NIR) is an efficient way to analyze the physical and chemical properties of a complex organic matrix and has been applied in the food processing industry to continuously monitor product quality (Alexandrakis et al., 2012). NIR can provide information concerning meat chemical and physical attributes through spectra arising from the different light absorption and/or reflection of chemical bonds (Jiang et al., 2019). Barbin et al. (2015) found distinctive spectral differences in wavelengths related to water distinguishing normal and PSE chicken breast meat (intact or minced). Jiang et al. (2017) demonstrated that spectra in the region of 400–1100 nm could be used to identify defective meat.

11.3 Dielectric spectroscopy

Dielectric spectroscopy involves measuring dielectric properties as a function of frequency. The technique is based on the electric interaction of an external electric field with the dipole moment of the sample (Pozar, 2011). Traffano-Schiffo et al. (2018) has explored dielectric spectroscopy to identify PSE chicken breast meat. PSE meat was easily discernible from normal meat thanks to its higher values of dielectric constant, likely due to increased protein degradation. As a direct consequence of protein degradation, higher amounts of polypeptides with smaller molecular weight produce a reduction in the relaxation frequency. The technique can detect PSE from the whole carcass, without needing to section muscles.

12 Conclusion

PSE is a condition that significantly affects meat quality and processing characteristics, causing significant financial losses to the poultry industry. However, remarkable progress has been made in understanding the circumstances underlying its occurrence and in developing methods to improve its functional properties. Particular attention should be taken when handling animals prior to slaughter and also at the post-slaughter stage to reduce the incidence of PSE meats. Despite efforts of farmers and producers, reducing the occurrence of PSE is still challenging, especially during the warmest periods of the year. Scientific research has made remarkable progress in developing innovative processing strategies to improve processing characteristics of PSE meat for use in processed products. Several online detection tools have been also developed to detect affected meat.

13 Where to look for further information

The following articles and book chapters provide a good overview of the subject:

- Solomon, M. B., Van Laack, R. L. J. M. and Eastridge, J. S. 1998. Biophysical basis of pale, soft, exudative (PSE) pork and poultry muscle: A review. J. Muscle Foods, 9, 1–11. https://doi.org/10.1111/j.1745-4573.1998.tb00639.x.
- Anthony, N. B. 1998. A review of genetic practices in poultry: efforts to improve meat quality. J. Muscle Foods 9, 25–33. https://doi.org/10.1111/j.1745-4573.1998.tb00641.x.
- Barbut, S. 2009. Pale, soft, and exudative poultry meat–Reviewing ways to manage at the processing plant. Poult. Sci. 88, 1506–1512. https://doi.org/10.3382/PS.2009-00118.
- Owens, C. M. and Apple, J. K. 2010. Pale, Soft and Exudative poultry meat, in: Guerrero-Legarreta, I. and Hui, Y. H. (Eds.), Handbook of Poultry Science and Technology, Volume 1: Primary Processing. John Wiley & Sons, Inc., Hoboken, USA, pp. 515–531.
- Barbut, S. 2015. The Science of Poultry and Meat Processing. Available online at: https://www.poultryandmeatprocessing.com/
- Petracci, M., Soglia, F. and Berri, C. 2017. Muscle metabolism and meat quality abnormalities, in: Petracci, M., Berri, C. (Eds.), Poultry Quality Evaluation: Quality Attributes and Consumer Values. 1st ed. Woodhead Publishing, pp. 51–75.

Key research in this area can be found at the following organizations:

- United States Department of Agriculture (USDA) (www.usda.gov).
- World's Poultry Science Association (WPSA) (www.wpsa.com).
- American Meat Science Association (www.meatscience.org).

14 References

Aalhus, J. L., Best, D. R., Murray, A. C. and Jones, S. D. M. 1998. A comparison of the quality characteristics of pale, soft and exudative beef and pork. *J. Muscle Foods* 9(3), 267–280. https://doi.org/10.1111/j.1745-4573.1998.tb00660.x.

Adzitey, F. and Nurul, H. 2011. Pale soft exudative (PSE) and dark firm dry (DFD) meats: causes and measures to reduce these incidences - a mini review. *Int. Food Res. J.* 18, 11-20.

Ahmad, T. and Sarwar, M. 2019. Dietary electrolyte balance: implications in heat stressed broilers. *Poult. Sci. J.* 62, 638–653. https://doi.org/10.1017/S0043933906001188.

Airey, J. A., Grinsell, M. M., Jones, L. R., Sutko, J. L. and Witcher, D. 1993. Three ryanodine receptor isoforms exist in avian striated muscles. *Biochemistry* 32(22), 5739–5745. https://doi.org/10.1021/bi00073a003.

Al-Abdullatif, A. A., Al-Sagan, A. A., Hussein, E. O. S., Saadeldin, I. M., Suliman, G. M., Azzam, M. M., Al-Mufarrej, S. I. and Alhotan, R. A. 2021. Betaine could help ameliorate transport associated water deprivation stress in broilers by reducing the expression of stress-related transcripts and modulating water channel activity. *Ital. J. Anim. Sci.* 20(1), 14–25. https://doi.org/10.1080/1828051X.2020.1865213.

Al-Hilphy, A. R., Al-Temimi, A. B., Al Rubaiy, H. H. M., Anand, U., Delgado-Pando, G. and Lakhssassi, N. 2020. Ultrasound applications in poultry meat processing: a systematic review. *J. Food Sci.* 85(5), 1386–1396. https://doi.org/10.1111/1750-3841.15135.

Alarcon-Rojo, A. D., Carrillo-Lopez, L. M., Reyes-Villagrana, R., Huerta-Jiménez, M. and Garcia-Galicia, I. A. 2019. Ultrasound and meat quality: a review. *Ultrason. Sonochem.* 55, 369–382 https://doi.org/10.1016/j.ultsonch.2018.09.016.

Aldridge, D. J., Luthra, K., Liang, Y., Christensen, K., Watkins, S. E. and Scanes, C. G. 2019. Thermal micro-environment during poultry transportation in South Central United States. *Animals (Basel)* 9(1), 31. https://doi.org/10.3390/ani9010031.

Alexandrakis, D., Downey, G. and Scannell, A. G. M. 2012. Rapid non-destructive detection of spoilage of intact chicken breast muscle using near-infrared and Fourier transform mid-infrared spectroscopy and multivariate statistics. *Food Bioprocess Technol.* 5(1), 338–347. https://doi.org/10.1007/S11947-009-0298-4.

Ali, M. S., Kang, G. H. and Joo, S. T. 2008. A review: influences of pre-slaughter stress on poultry meat quality. *Asian-Australas. J. Anim. Sci.* 21(6), 912–916. https://doi.org/10.5713/AJAS.2008.R.06.

Allen, C. D., Fletcher, D. L., Northcutt, J. K. and Russell, S. M. 1998. The relationship of broiler breast color to meat quality and shelf-life. *Poult. Sci.* 77(2), 361-366. https://doi.org/10.1093/ps/77.2.361.

Allen, C. D., Russell, S. M. and Fletcher, D. L. 1997. The relationship of broiler breast meat color and pH to shelf-life and odor development. *Poult. Sci.* 76(7), 1042–1046. https://doi.org/10.1093/ps/76.7.1042.

Alvarado, C. and McKee, S. 2007. Marination to improve functional properties and safety of poultry meat. *J. Appl. Poult. Res.* 16(1), 113–120. https://doi.org/10.1093/JAPR/16.1.113.

Alvarado, C. Z. and Sams, A. R. 2004. Turkey carcass chilling and protein denaturation in the development of pale, soft, and exudative meat. *Poult. Sci.* 83(6), 1039–1046. https://doi.org/10.1093/PS/83.6.1039.

Alvarado, C. Z. and Sams, A. R. 2003. Injection marination strategies for remediation of pale, exudative broiler breast meat. *Poult. Sci.* 82(8), 1332–1336. https://doi.org/10.1093/PS/82.8.1332.

Alvarado, C. Z. and Sams, A. R. 2002. The role of carcass chilling rate in the development of pale, exudative turkey pectoralis. *Poult. Sci.* 81(9), 1365-1370. https://doi.org/10.1093/ps/81.9.1365.

Alvarado, C. Z. and Sams, A. R. 2000. Rigor mortis development in turkey breast muscle and the effect of electrical stunning. *Poult. Sci.* 79(11), 1694–1698. https://doi.org/10.1093/PS/79.11.1694.

Arroyo Mora, D., Hamada, Y., Okamoto, A., Tateishi, A. and Tachibana, K. 2007. Characteristics of burnt meat in cultured yellowtail Seriola quinqueradiata. *Fisheries Sci.* 73(3), 651–659. https://doi.org/10.1111/j.1444-2906.2007.01378.x.

Alessandra de Avila Souza, M., Shimokomaki, M., Nascimento Terra, N. and Petracci, M. 2022. Oxidative changes in cooled and cooked pale, soft, exudative (PSE) chicken meat. *Food Chem.* 385, 132471.

Baldi, G., D'Elia, F., Soglia, F., Tappi, S., Petracci, M. and Rocculi, P. 2021a. Exploring the effect of pulsed electric fields on the technological properties of chicken meat. *Foods* 10(2), 241. https://doi.org/10.3390/foods10020241.

Baldi, G., Soglia, F. and Petracci, M. 2021b. Valorization of meat by-products. In: *Food Waste Recovery*. Academic Press, pp. 419–443. https://doi.org/10.1016/b978-0-12-820563-1.00017-2.

Baldi, G., Soglia, F. and Petracci, M. 2020. Current status of poultry meat abnormalities. *Meat Muscle Biol.* 4(2), 1–7. https://doi.org/10.22175/mmb.9503.

Barbin, D. F., Kaminishikawahara, C. M., Soares, A. L., Mizubuti, I. Y., Grespan, M., Shimokomaki, M. and Hirooka, E. Y. 2015. Prediction of chicken quality attributes by near infrared spectroscopy. *Food Chem.* 168, 554–560. https://doi.org/10.1016/J.FOODCHEM.2014.07.101.

Barbin, D. F., Mastelini, S. M., Barbon, S., Campos, G. F. C., Barbon, A. P. A. C. and Shimokomaki, M. 2016. Digital image analyses as an alternative tool for chicken quality assessment. *Biosyst. Eng.* 144, 85–93. https://doi.org/10.1016/J.BIOSYSTEMSENG.2016.01.015.

Barbosa, C. F., Carvalho, R. Hd de, Rossa, A., Soares, A. L., Coró, F. A. G., Shimokomaki, M. and Ida, E. I. 2013. Commercial preslaughter blue light ambience for controlling broiler stress and meat qualities. *Braz. Arch. Biol. Technol.* 56(5), 817–821. https://doi.org/10.1590/S1516-89132013000500013.

Barbut, S. 2019. Recent myopathies in broiler's breast meat fillets. *Worlds Poult. Sci. J.* 75(4), 559–582. https://doi.org/10.1017/S0043933919000436.

Barbut, S. 2009. Pale, soft, and exudative poultry meat—reviewing ways to manage at the processing plant. *Poult. Sci.* 88(7), 1506–1512. https://doi.org/10.3382/PS.2009-00118.

Barbut, S. 2001. *Primary Processing of Poultry, Poultry Products Processing*. https://doi.org/10.1201/9781420031744.ch4.

Barbut, S. 1998. Estimating the magnitude of the PSE problem in poultry. *J. Muscle Foods* 9(1), 35–49. https://doi.org/10.1111/j.1745-4573.1998.tb00642.x.

Barbut, S. 1997. Problem of pale soft exudative meat in broiler chickens. *Br. Poult. Sci.* 38(4), 355–358. https://doi.org/10.1080/00071669708418002.

Barbut, S., Sosnicki, A. A., Lonergan, S. M., Knapp, T., Ciobanu, D. C., Gatcliffe, L. J., Huff-Lonergan, E. and Wilson, E. W. 2008. Progress in reducing the pale, soft and exudative (PSE) problem in pork and poultry meat. *Meat Sci.* 79(1), 46–63. https://doi.org/10.1016/j.meatsci.2007.07.031.

Barbut, S., Zhang, L. and Marcone, M. 2005. Effects of pale, normal, and dark chicken breast meat on microstructure, extractable proteins, and cooking of marinated fillets. *Poult. Sci.* 84(5), 797–802. https://doi.org/10.1093/ps/84.5.797.

BeMiller, J. 2018. *Carbohydrate Chemistry for Food Scientists*. Woodhead Publishing, pp. 279–291.

Bhat, Z. F., Morton, J. D., Mason, S. L. and Bekhit, A. E. A. 2019. Current and future prospects for the use of pulsed electric field in the meat industry. *Crit. Rev. Food Sci. Nutr.* 59(10), 1660–1674. https://doi.org/10.1080/10408398.2018.1425825.

Bian, G., Xue, S., Xu, Y., Xu, X. and Han, M. 2018. Improved gelation functionalities of myofibrillar protein from pale, soft and exudative chicken breast meat by nonenzymatic glycation with glucosamine. *Int. J. Food Sci. Technol.* 53(8), 2006–2014. https://doi.org/10.1111/IJFS.13789.

Bowen, S. J. and Washburn, K. W. 1985. Thyroid and adrenal response to heat stress in chickens and quail differing in heat tolerance. *Poult. Sci.* 64(1), 149–154. https://doi.org/10.3382/PS.0640149.

Bowker, B. 2017. Developments in our understanding of water-holding capacity. In: Berri, C. and Petracci, M. (Eds), *Poultry Quality Evaluation : Quality Attributes and Consumer Values*. Woodhead Publishing, pp. 77–114.

Bowker, B. and Zhuang, H. 2015. Relationship between water-holding capacity and protein denaturation in broiler breast meat1. *Poult. Sci.* 94(7), 1657–1664. https://doi.org/10.3382/ps/pev120.

Bowker, B. C., Grant, A. L., Forrest, J. C. and Gerrard, D. E. 2017. Muscle metabolism and PSE pork. *J. Anim. Sci.* 79(E-Suppl_1), 1. https://doi.org/10.2527/jas.00.079ES1001c.

Briskey, E. J. and Wismer-Pedersen, J. 1961. Biochemistry of pork muscle structure. 1. Rate of Anaerobic Glycolysis and temperature Change versus the Apparent Structure of Muscle Tissue. *J. Food Sci.* 26(3), 297–305. https://doi.org/10.1111/j.1365-2621.1961.tb01658.x.

Byrne, C. E., Troy, D. J. and Buckley, D. J. 2000. Postmortem changes in muscle electrical properties of bovine M. longissimus dorsi and their relationship to meat quality attributes and pH fall. *Meat Sci.* 54(1), 23–34. https://doi.org/10.1016/S0309-1740(99)00055-8.

Campus, M. 2010. High pressure processing of meat, meat products and seafood. *Food Eng. Rev.* 2(4), 256–273. https://doi.org/10.1007/s12393-010-9028-y.

Carlez, A., Veciana-Nogues, T. and Cheftel, J.-C. 1995. Changes in colour and myoglobin of minced beef meat due to high pressure processing. *LWT Food Sci. Technol.* 28(5), 528–538. https://doi.org/10.1006/FSTL.1995.0088.

Carvalho, R. H., Honorato, D. C. B., Guarnieri, P. D., Soares, A. L., Pedrão, M. R., Oba, A., Paião, F. G., Ida, E. I. and Shimokomaki, M. 2018. Assessment of turkey vehicle container microclimate on transit during summer season conditions. *Int. J. Biometeorol.* 62(6), 961–970. https://doi.org/10.1007/s00484-018-1498-2.

Carvalho, R. H., Soares, A. L., Honorato, D. C. B., Guarnieri, P. D., Pedrão, M. R., Paião, F. G., Oba, A., Ida, E. I. and Shimokomaki, M. 2014. The incidence of pale, soft, and exudative (PSE) turkey meat at a Brazilian commercial plant and the functional properties in its meat product. *LWT Food Sci. Technol.* 59(2), 883–888. https://doi.org/10.1016/j.lwt.2014.07.019.

Carvalho, R. H., Ida, E. I., Madruga, M. S., Martínez, S. L., Shimokomaki, M. and Estévez, M. 2017. Underlying connections between the redox system imbalance, protein oxidation and impaired quality traits in pale, soft and exudative (PSE) poultry meat. *Food Chem.* 215, 129–137. https://doi.org/10.1016/j.foodchem.2016.07.182.

Cavitt, L. C., Hargis, B. M. and Owens, C. M. 2004. The use of halothane and succinylcholine to identify broilers prone to developing pale, soft, exudative meat. *Poult. Sci.* 83(8), 1440–1444. https://doi.org/10.1093/ps/83.8.1440.

Cegielka, A. and Mroczek, J. 2005. Effect of transglutaminase and phosphates on the quality of low-salt chicken meat ham. *51st Int. Congr. Meat Sci. Technol. Balt.* MD. pp. 832–836.

Chan, J. T. Y., Omana, D. A. and Betti, M. 2011. Application of high pressure processing to improve the functional properties of pale, soft, and exudative (PSE)-like turkey meat. *Innov. Food Sci. Emerg. Technol.* 12(3), 216–225. https://doi.org/10.1016/j.ifset.2011.03.004.

Channon, H. A., Payne, A. M. and Warner, R. D. 2000. Halothane genotype, pre-slaughter handling and stunning method all influence pork quality. *Meat Sci.* 56(3), 291–299. https://doi.org/10.1016/S0309-1740(00)00056-5.

Chen, H., Wang, H. H., Qi, J., Wang, M., Xu, X. and Zhou, G. 2018. Chicken breast quality - normal, pale, soft and exudative (PSE) and woody - influences the functional properties of meat batters. *Int. J. Food Sci. Technol.* 53(3), 654–664 https://doi.org/10.1111/ijfs.13640.

Chiang, W., Allison, C. P., Linz, J. E. and Strasburg, G. M. 2004. Identification of two αrYR alleles and characterization of αrYR transcript variants in turkey skeletal muscle. *Gene* 330, 177–184. https://doi.org/10.1016/j.gene.2004.01.023.

Chiang, W., Booren, A. and Strasburg, G. 2008. The effect of heat stress on thyroid hormone response and meat quality in turkeys of two genetic lines. *Meat Sci.* 80(3), 615–622. https://doi.org/10.1016/J.MEATSCI.2008.02.012.

Chiang, W., Byrem, T., Zhang, H. and Strasburg, G. 2007a. Binding property of avian skeletal muscle ryanodine receptor isoforms with dihydropyridine receptor and calmodulin. *J. Muscle Res. Cell Motil.* 28(1), 59–66. https://doi.org/10.1007/s10974-007-9106-9.

Chiang, W., Yoon, H. J., Linz, J. E., Airey, J. A. and Strasburg, G. M. 2007b. Divergent mechanisms in generating molecular variations of αRYR and βRYR in turkey skeletal muscle. *J. Muscle Res. Cell Motil.* 28(6), 343–354. https://doi.org/10.1007/s10974-008-9130-4.

Chmiel, M., Słowiński, M. and Dasiewicz, K. 2011. Lightness of the color measured by computer image analysis as a factor for assessing the quality of pork meat. *Meat Sci.* 88(3), 566–570. https://doi.org/10.1016/J.MEATSCI.2011.02.014.

Chmiel, M., Słowiński, M., Dasiewicz, K. and Florowski, T. 2016. Use of computer vision system (CVS) for detection of PSE pork meat obtained from m. semimembranosus. *LWT Food Sci. Technol.* 65, 532–536. https://doi.org/10.1016/J.LWT.2015.08.021.

Cockram, M. S. and Dulal, K. J. 2018. Injury and mortality in broilers during handling and transport to slaughter. *Can. J. Anim. Sci.* 98(3), 416–432 https://doi.org/10.1139/cjas-2017-0076.

Costa, F. A. D., Costa, O. A. D., Coldebella, A., de Lima, G. J. M. M. and Ferraudo, A. S. 2018. How do season, on-farm fasting interval and lairage period affect swine welfare, carcass and meat quality traits? *Int. J. Biometeorol.* 6311(63), 1497–1505. https://doi.org/10.1007/S00484-018-1527-1.

Craig, E. W., Fletcher, D. L. and Papinaho, P. A. 1999. The effects of antemortem electrical stunning and postmortem electrical stimulation on biochemical and textural properties of broiler breast meat. *Poult. Sci.* 78(3), 490–494. https://doi.org/10.1093/PS/78.3.490.

da Silva-Buzanello, R. A. D., Schuch, A. F., Gasparin, A. W., Torquato, A. S., Scremin, F. R., Canan, C. and Soares, A. L. 2019. Quality parameters of chicken breast meat affected by carcass scalding conditions. *Asian-Australas. J. Anim. Sci.* 32(8), 1186–1194. https://doi.org/10.5713/ajas.18.0692.

da Silva-Buzanello, R. A. D., Schuch, A. F., Nunes Nogues, D. R., Falcão de Melo, P., Gasparin, A. W., Sanches Torquato, A., Canan, C. and Lourenço Soares, A. 2018. Physicochemical and biochemical parameters of chicken breast meat influenced by stunning methods. *Poult. Sci.* 97(11), 3786–3792. https://doi.org/10.3382/PS/PEY281.

Daigle, S. P., Schilling, M. W., Marriott, N. G., Wang, H., Barbeau, W. E. and Williams, R. C. 2005. PSE-like turkey breast enhancement through adjunct incorporation in a chunked and formed deli roll. *Meat Sci.* 69(2), 319–324. https://doi.org/10.1016/J.MEATSCI.2004.08.001.

Deeb, N., Shlosberg, A. and Cahaner, A. 2002. Genotype-by-environment interaction with broiler genotypes differing in growth rate. 4. Association between responses to heat stress and to cold-induced ascites. *Poult. Sci.* 81(10), 1454–1462. https://doi.org/10.1093/PS/81.10.1454.

Delezie, E., Swennen, Q., Buyse, J. and Decuypere, E. 2007. The effect of feed withdrawal and crating density in transit on metabolism and meat quality of broilers at slaughter weight. *Poult. Sci.* 86(7), 1414–1423. https://doi.org/10.1093/ps/86.7.1414.

Dolatowski, Z. J., Stadnik, J. and Stasiak, D. 2007. Applications of ultrasound in food technology. *Acta Sci. Pol* 63, 89–99.

Dong, M., Chen, H., Zhang, Y., Xu, Y., Han, M., Xu, X. and Zhou, G. 2020a. Processing properties and improvement of pale, soft, and exudative-like chicken meat: a review. *Food Bioprocess Technol.* 13(8), 1280–1291. https://doi.org/10.1007/s11947-020-02464-3.

Dong, M., Xu, Y., Zhang, Y., Han, M., Wang, P., Xu, X. and Zhou, G. 2020b. Physicochemical and structural properties of myofibrillar proteins isolated from pale, soft, exudative (PSE)-like chicken breast meat: effects of pulsed electric field (PEF). *Innov. Food Sci. Emerg. Technol.* 59. https://doi.org/10.1016/j.ifset.2019.102277.

dos Santos, V. M., Dallago, B. S. L., Racanicci, A. M. C., Santana, Â. P., Cue, R. I. and Bernal, F. E. M. 2020. Effect of transportation distances, seasons and crate microclimate on broiler chicken production losses. *PLoS ONE* 15(4), e0232004. https://doi.org/10.1371/journal.pone.0232004.

Downing, J. A., Kerr, M. J. and Hopkins, D. L. 2017. The effects of pre-transport supplementation with electrolytes and betaine on performance, carcass yield and meat quality of broilers in summer and winter. *Livest. Sci.* 205, 16–23. https://doi.org/10.1016/J.LIVSCI.2017.09.006.

Droval, A. A., Binneck, E., Marin, S. R. R., Paião, F. G., Oba, A., Nepomuceno, A. L. and Shimokomaki, M. 2012. A new single nucleotide polymorphism in the ryanodine gene of chicken skeletal muscle. *Genet. Mol. Res.* 11(2), 821–829. https://doi.org/10.4238/2012.April.3.4.

Du, C. J. and Sun, D. W. 2004. Recent developments in the applications of image processing techniques for food quality evaluation. *Trends Food Sci. Technol.* 15(5), 230–249. https://doi.org/10.1016/J.TIFS.2003.10.006.

Eikelenboom, G., Bolink, A. H. and Sybesma, W. 1991. Effects of feed withdrawal before delivery on pork quality and carcass yield. *Meat Sci.* 29(1), 25–30. https://doi.org/10.1016/0309-1740(91)90020-Q.

el-Gendy, E. and Washburn, K. W. 1995. Genetic variation in body temperature and its response to short-term acute heat stress in broilers. *Poult. Sci.* 74(2), 225–230. https://doi.org/10.3382/ps.0740225.

England, E. M., Matarneh, S. K., Scheffler, T. L. and Gerrard, D. E. 2017. Perimortal muscle metabolism and its effects on meat quality. In: Purslow, P. P. (Ed.), *New Aspects of Meat Quality. From Genes to Ethics*. Woodhead Publishing, Elsevier Limited Company, Cambridge, pp. 63–89.

England, E. M., Scheffler, T. L., Kasten, S. C., Matarneh, S. K. and Gerrard, D. E. 2013. Exploring the unknowns involved in the transformation of muscle to meat. *Meat Sci.* 95(4), 837–843. https://doi.org/10.1016/j.meatsci.2013.04.031.

Everts, M. E. 1996. Effects of thyroid hormones on contractility and cation transport in skeletal muscle. In: *Acta Physiol. Scand.* John Wiley & Sons, Ltd 156(3), 325–333. https://doi.org/10.1046/j.1365-201X.1996.203000.x.

Fraqueza, M. J., Cardoso, A. S., Ferreira, M. C. and Barreto, A. S. 2006. Incidence of pectoralis major turkey muscles with light and dark color in a Portuguese slaughterhouse. *Poult. Sci.* 85(11), 1992–2000. https://doi.org/10.1093/ps/85.11.1992.

Fraqueza, M. J., Ferreira, M. C. and Barreto, A. S. 2008. Spoilage of light (PSE-like) and dark turkey meat under aerobic or modified atmosphere package: microbial indicators and their relationship with total volatile basic nitrogen. *Br. Poult. Sci.* 49(1), 12–20. https://doi.org/10.1080/00071660701821675.

Fujii, J., Otsu, K., Zorzato, F., De Leon, S., Khanna, V. K., Weiler, J. E., O'Brien, P. J. and Maclennan, D. H. 1991. Identification of a mutation in porcine ryanodine receptor associated with malignant hyperthermia. *Science* 253(5018), 448–451. https://doi.org/10.1126/science.1862346.

Garrido, M. D., Pedauyé, J., Bañon, S., López, M. B. and Laencina, J. 1995. On-line methods for pork quality detection. *Food Control* 6(2), 111–113. https://doi.org/10.1016/0956-7135(95)98915-N.

Gaya de, L. G., Mourão, G. B., Ferraz, J. B. S., de Mattos, E. C., da Costa, A. M. M. A., Michelan Filho, T., Rosa, A. F., Felício, A. M. and Eler, J. P. 2011. Estimates of heritability and genetic correlations for meat quality traits in broilers. *Sci. Agric.* 68, 620–625. https://doi.org/10.1590/S0103-90162011000600002.

Geonzon, L. C., Bacabac, R. G. and Matsukawa, S. 2019. Network structure and gelation mechanism of kappa and iota carrageenan elucidated by multiple particle tracking. *Food Hydrocoll.* 92, 173–180. https://doi.org/10.1016/J.FOODHYD.2019.01.062.

Goel, A., Ncho, C. M. and Choi, Y. H. 2021. Regulation of gene expression in chickens by heat stress. *J. Anim. Sci. Biotechnol.* 12(1), 11. https://doi.org/10.1186/S40104-020-00523-5.

Gonzalez-Rivas, P. A., Chauhan, S. S., Ha, M., Fegan, N., Dunshea, F. R. and Warner, R. D. 2020. Effects of heat stress on animal physiology, metabolism, and meat quality: a review. *Meat Sci.* 162, 108025 https://doi.org/10.1016/j.meatsci.2019.108025.

Gorsuch, V. and Alvarado, C. Z. 2010. Postrigor tumble marination strategies for improving color and water-holding capacity in normal and pale broiler breast fillets. *Poult. Sci.* 89(5), 1002–1008. https://doi.org/10.3382/PS.2009-00023.

Guàrdia, M. D., Estany, J., Balasch, S., Oliver, M. A., Gispert, M. and Diestre, A. 2004. Risk assessment of PSE condition due to pre-slaughter conditions and RYR1 gene in pigs. *Meat Sci.* 67(3), 471–478. https://doi.org/10.1016/J.MEATSCI.2003.11.020.

Haddad, G. B. S., Moura, A. P. R., Fontes, P. R., Cunha, S. F. V. D., Ramos, A. L. S. and Ramos, E. M. 2018. The effects of sodium chloride and PSE meat on restructured cured-smoked pork loin quality: a response surface methodology study. *Meat Sci.* 137, 191-200. https://doi.org/10.1016/J.MEATSCI.2017.11.030.

Hammelman, J. E., Bowker, B. C., Grant, A. L., Forrest, J. C., Schinckel, A. P. and Gerrard, D. E. 2003. Early postmortem electrical stimulation simulates PSE pork development. *Meat Sci.* 63(1), 69–77. https://doi.org/10.1016/S0309-1740(02)00057-8.

Huff-Lonergan, E. and Lonergan, S. M. 2005. Mechanisms of water-holding capacity of meat: the role of postmortem biochemical and structural changes. *Meat Sci.* 71(1), 194–204. https://doi.org/10.1016/J.MEATSCI.2005.04.022.

Hughes, J. M., Oiseth, S. K., Purslow, P. P. and Warner, R. D. 2014. A structural approach to understanding the interactions between colour, water-holding capacity and tenderness. *Meat Sci.* 98(3), 520–532. https://doi.org/10.1016/j.meatsci.2014.05.022.

Jayasooriya, S. D., Bhandari, B. R., Torley, P. and D'Arcy, B. R. 2004. Effect of high power ultrasound waves on properties of meat: a review. *Int. J. Food Prop.* 7(2), 301–319 https://doi.org/10.1081/JFP-120030039.

Jiang, H., Yoon, S. C., Zhuang, H., Wang, W., Li, Y. and Yang, Y. 2019. Integration of spectral and textural features of visible and near-infrared hyperspectral imaging for differentiating between normal and white striping broiler breast meat. *Spectrochim. Acta A Mol. Biomol. Spectrosc.* 213, 118–126. https://doi.org/10.1016/j.saa.2019.01.052.

Jiang, H., Yoon, S. C., Zhuang, H., Wang, W. and Yang, Y. 2017. Evaluation of factors in development of Vis/NIR spectroscopy models for discriminating PSE, DFD and normal broiler breast meat. *Br. Poult. Sci.* 58(6), 673–680. https://doi.org/10.1080/00071668.2017.1364350.

Jiang, N., Wang, P., Xing, T., Han, M. and Xu, X. 2016. An evaluation of the effect of water-misting sprays with forced ventilation on the occurrence of pale, soft, and exudative meat in transported broilers during summer: impact of the thermal microclimate. *J. Anim. Sci.* 94(5), 2218–2227. https://doi.org/10.2527/jas.2015-9823.

Joseph, P., Schilling, M. W., Williams, J. B., Radhakrishnan, V., Battula, V., Christensen, K., Vizzier-Thaxton, Y. and Schmidt, T. B. 2013. Broiler stunning methods and their effects on welfare, rigor mortis, and meat quality. *Worlds Poult. Sci. J.* 69(1), 99–112 https://doi.org/10.1017/S0043933913000093.

Kannan, G., Heath, J. L., Wabeck, C. J., Owens, S. L. and Mench, J. A. 1998. Elevated plasma corticosterone concentrations influence the onset of rigor mortis and meat color in broilers. *Poult. Sci.* 77(2), 322–328. https://doi.org/10.1093/PS/77.2.322.

Kannan, G., Heath, J. L., Wabeck, C. J., Souza, M. C. P., Howe, J. C. and Mench, J. A. 1997. Effects of crating and transport on stress and meat quality characteristics in broilers. *Poult. Sci.* 76(3), 523–529. https://doi.org/10.1093/PS/76.3.523.

Kannan, G. and Mench, J. A. 1996. Influence of different handling methods and crating periods on plasma corticosterone concentrations in broilers. *Br. Poult. Sci.* 37(1), 21–31. https://doi.org/10.1080/00071669608417833.

Karacay, N., Ocak, N., Sarica, M. and Erener, G. 2008. Effect of carbohydrate supplementation provided through drinking water during feed withdrawal on meat and liver colours in broilers. *J. Sci. Food Agric.* 88(3), 479–484. https://doi.org/10.1002/jsfa.3110.

Karunanayaka, D. S., Jayasena, D. D. and Jo, C. 2016. Prevalence of pale, soft, and exudative (PSE) condition in chicken meat used for commercial meat processing and its effect on roasted chicken breast. *J. Anim. Sci. Technol.* 58, 27. https://doi.org/10.1186/s40781-016-0110-8.

Kim, H., Kim, H., Seong, P., Arora, D., Shin, D., Park, W. and Park, J. E. 2021. Transcriptomic response under heat stress in chickens revealed the regulation of genes and alteration of metabolism to maintain homeostasis. *Animals (Basel)* 11(8), 2241. https://doi.org/10.3390/ANI11082241.

Kim, Y. H. B., Warner, R. D. and Rosenvold, K. 2014. Influence of high pre-rigor temperature and fast pH fall on muscle proteins and meat quality: a review. *Anim. Prod. Sci.* 54(4), 375–395. https://doi.org/10.1071/AN13329.

Knowles, T. G. and Broom, D. M. 1990. The handling and transport of broilers and spent hens. *Appl. Anim. Behav. Sci.* 28(1–2), 75–91. https://doi.org/10.1016/0168-1591(90)90047-H.

Komiyama, C. M., Mendes, A. A., Takahashi, S. E., Moreira, J., Garcia, R. G., Sanfelice, C., Borba, H. S., Leonel, F. R., Almeida Paz, I. C. L. and Balog, A. 2008. Chicken meat quality as a function of fasting period and water spray. *Rev. Bras. Cienc. Avic.* 10(3), 179–183. https://doi.org/10.1590/S1516-635X2008000300008.

Kop-Bozbay, C. and Ocak, N. 2015. Body weight, meat quality and blood metabolite responses to carbohydrate administration in the drinking water during pre-slaughter feed withdrawal in broilers. *J. Anim. Physiol. Anim. Nutr. (Berl)* 99(2), 290–298. https://doi.org/10.1111/jpn.12194.

Lara, L. J. and Rostagno, M. H. 2013. Impact of heat stress on poultry production. *Animals (Basel)* 3(2), 356–369. https://doi.org/10.3390/ani3020356.

Le Bihan-Duval, E., Berri, C., Baeza, E., Millet, N. and Beaumont, C. 2001. Estimation of the genetic parameters of meat characteristics and of their genetic correlations with growth and body composition in an experimental broiler line. *Poult. Sci.* 80(7), 839-843. https://doi.org/10.1093/ps/80.7.839.

Le Bihan-Duval, E., Debut, M., Berri, C. M., Sellier, N., Santé-Lhoutellier, V., Jégo, Y. and Beaumont, C. 2008. Chicken meat quality: genetic variability and relationship with growth and muscle characteristics. *BMC Genet.* 9, 53. https://doi.org/10.1186/1471-2156-9-53.

Le Bihan-Duval, E. L., Hennequet-Antier, C., Berri, C., Beauclercq, S. A., Bourin, M. C., Boulay, M., Demeure, O. and Boitard, S. 2018. Identification of genomic regions and candidate genes for chicken meat ultimate pH by combined detection of selection signatures and QTL. *BMC Genomics* 19(1), 294. https://doi.org/10.1186/s12864-018-4690-1.

Lee, Y. B. and Choi, Y. I. 1999. PSE (pale, soft, exudative) Pork : the causes and solutions - review -. *Asian-Australas. J. Anim. Sci.* 12(2), 244–252 https://doi.org/10.5713/ajas.1999.244.

Leheska, J. M., Wulf, D. M. and Maddock, R. J. 2002. Effects of fasting and transportation on pork quality development and extent of postmortem metabolism. *J. Anim. Sci.* 80(12), 3194–3202. https://doi.org/10.2527/2002.80123194x.

Lesiow, T. and Kijowsi, J. 2003. Impact of PSE and DFD meat on poultry processing: a review. *Pol. J. Food Nutr. Sci.* 12, 3–8.

Lesiów, T., Oziembłowski, M. and Wojciechowska, I. 2005. *PSE and DFD in Chicken Broiler Breast Muscle*, Available at: cabi.org.

Lesiow, T., Rentfrow, G. K. and Xiong, Y. L. 2017. Polyphosphate and myofibrillar protein extract promote transglutaminase-mediated enhancements of rheological and textural properties of PSE pork meat batters. *Meat Sci.* 128, 40–46. https://doi.org/10.1016/J.MEATSCI.2017.02.002.

Li, K., Chen, L., Zhao, Y. Y., Li, Y. P., Wu, N., Sun, H., Xu, X. L. and Zhou, G. H. 2015a. A comparative study of chemical composition, color, and thermal gelling properties of normal and PSE-like chicken breast meat. *CyTA J. Food* 13(2), 213–219. https://doi.org/10.1080/19476337.2014.941411.

Li, K., Zhao, Y. Y., Kang, Z. L., Wang, P., Han, M. Y., Xu, X. L. and Zhou, G. H. 2015b. Reduced functionality of PSE-like chicken breast meat batter resulting from alterations in protein conformation. *Poult. Sci.* 94(1), 111–122. https://doi.org/10.3382/ps/peu040.

Li, K., Kang, Z. L., Zhao, Y. Y., Xu, X. L. and Zhou, G. H. 2014. Use of high-intensity ultrasound to improve functional properties of batter suspensions prepared from PSE-like chicken breast meat. *Food Bioprocess Technol.* 7(12), 3466–3477. https://doi.org/10.1007/s11947-014-1358-y.

Li, K., Liu, J. Y., Fu, L., Zhao, Y. Y. and Bai, Y. H. 2019. Comparative study of thermal gelation properties and molecular forces of actomyosin extracted from normal and pale, soft and exudative-like chicken breast meat. *Asian-Australas. J. Anim. Sci.* 32(5), 721–733. https://doi.org/10.5713/ajas.18.0389.

Liu, J., Arner, A., Puolanne, E. and Ertbjerg, P. 2016. On the water-holding of myofibrils: effect of sarcoplasmic protein denaturation. *Meat Sci.* 119, 32–40. https://doi.org/10.1016/j.meatsci.2016.04.020.

Liu, J., Ru, Q. and Ding, Y. 2012. Glycation a promising method for food protein modification: physicochemical properties and structure, a review. *Food Res. Int.* 49(1), 170–183. https://doi.org/10.1016/J.FOODRES.2012.07.034.

Liu, R., Zhao, S. M., Xie, B. J. and Xiong, S. B. 2011. Contribution of protein conformation and intermolecular bonds to fish and pork gelation properties. *Food Hydrocoll.* 25(5), 898–906. https://doi.org/10.1016/j.foodhyd.2010.08.016.

Loyau, T., Hennequet-Antier, C., Coustham, V., Berri, C., Leduc, M., Crochet, S., Sannier, M., Duclos, M. J., Mignon-Grasteau, S., Tesseraud, S., Brionne, A., Métayer-Coustard, S., Moroldo, M., Lecardonnel, J., Martin, P., Lagarrigue, S., Yahav, S. and Collin, A. 2016. Thermal manipulation of the chicken embryo triggers differential gene expression in response to a later heat challenge. *BMC Genomics* 17, 329. https://doi.org/10.1186/s12864-016-2661-y.

Loyau, T., Métayer-Coustard, S., Berri, C., Crochet, S., Cailleau-Audouin, E., Sannier, M., Chartrin, P., Praud, C., Hennequet-Antier, C., Rideau, N., Couroussé, N., Mignon-Grasteau, S., Everaert, N., Duclos, M. J., Yahav, S., Tesseraud, S. and Collin, A. 2014. Thermal manipulation during embryogenesis has long-term effects on muscle and liver metabolism in fast-growing chickens. *PLoS ONE* 9(9), e105339. https://doi.org/10.1371/JOURNAL.PONE.0105339.

Lu, Q., Wen, J. and Zhang, H. 2007. Effect of chronic heat exposure on fat deposition and meat quality in two genetic types of chicken. *Poult. Sci.* 86(6), 1059–1064. https://doi.org/10.1093/ps/86.6.1059.

Lund, M. N., Heinonen, M., Baron, C. P. and Estévez, M. 2011. Protein oxidation in muscle foods: A review. *Mol. Nutr. food Res.* 55(1), 83–95. https://doi.org/10.1002/mnfr.201000453.

Lusas, E. W. and Rhee, K. C. 1995. Soy protein processing and utilization. In: *Practical Handbook of Soybean Processing and Utilization*, pp. 117–160. https://doi.org/10.1016/b978-0-935315-63-9.50012-7.

Malila, Y., Carr, K. M., Ernst, C. W., Velleman, S. G., Reed, K. M. and Strasburg, G. M. 2014. Deep transcriptome sequencing reveals differences in global gene expression between normal and pale, soft, and exudative Turkey meat. *J. Anim. Sci.* 92(3), 1250–1260. https://doi.org/10.2527/jas.2013-7293.

Malila, Y., Tempelman, R. J., Sporer, K. R. B., Ernst, C. W., Velleman, S. G., Reed, K. M. and Strasburg, G. M. 2013. Differential gene expression between normal and pale, soft, and exudative Turkey meat. *Poult. Sci.* 92(6), 1621–1633. https://doi.org/10.3382/ps.2012-02778.

Matarneh, S. K., England, E. M., Sheffler, T. L. and Gerrard, D. E. 2017. The conversion of muscle to meat. In: Lawrie, R. A. and Ledward, D. A. (Eds), *Lawrie's Meat Science*. Elsevier, pp. 159–184.

Matarneh, S. K., Silva, S. L. and Gerrard, D. E. 2021. New insights in muscle biology that alter meat quality. *Annu. Rev. Anim. Biosci.* 9, 355–377. https://doi.org/10.1146/annurev-animal-021419-083902.

McKee, S. R. and Sams, A. R. 1998. Rigor mortis development at elevated temperatures induces pale exudative turkey meat characteristics. *Poult. Sci.* 77(1), 169–174. https://doi.org/10.1093/ps/77.1.169.

Messens, W., Van Camp, J. and Huyghebaert, A. 1997. The use of high pressure to modify the functionality of food proteins. *Trends Food Sci. Technol.* 8(4), 107–112. https://doi.org/10.1016/S0924-2244(97)01015-7.

Métayer-Coustard, S., Tesseraud, S., Praud, C., Royer, D., Bordeau, T., Coudert, E., Cailleau-Audouin, E., Godet, E., Delaveau, J., Le Bihan-Duval, E. and Berri, C. 2021. Early growth and protein-energy metabolism in chicken lines divergently selected on ultimate pH. *Front. Physiol.* 0, 144. https://doi.org/10.3389/FPHYS.2021.643580.

Milan, D., Jeon, J. T., Looft, C., Amarger, V., Robic, A., Thelander, M., Rogel-Gaillard, C., Paul, S., Iannuccelli, N., Rask, L., Ronne, H., Lundström, K., Reinsch, N., Gellin, J., Kalm, E., Le Roy, P. L., Chardon, P. and Andersson, L. 2000. A mutation in PRKAG3 associated with excess glycogen content in pig skeletal muscle. *Science* 288(5469), 1248–1251. https://doi.org/10.1126/science.288.5469.1248.

Mills, L. J., Mitchell, M. A. and Mahon, M. 1999. Susceptibility to heat stress in fast and slow growing turkey lines. *Poult. Sci.* 78(Suppl 1, 247.

Mitchell, M. A. and Kettlewell, P. J. 2009. Welfare of poultry during transport - a review. In: Poultr. Welf. Symposium. Cervia, pp. 90–100.

Mohan Raj, A. B. M., Grey, T. C., Audsely, A. R. and Gregory, N. G. 1990. Effect of electrical and gaseous stunning on the carcase and meat quality of broilers. *Br. Poult. Sci.* 31(4), 725–733. https://doi.org/10.1080/00071669008417303.

Morey, A. and Owens, C. M. 2017. Methods for measuring meat texture. In: Petracci, M. and Berri, C. (Eds), *Poultry Quality Evaluation : Quality Attributes and Consumer Values*. Woodhead Publishing, pp. 115–132.

Motzer, E. A., Carpenter, J. A., Reynolds, A. E. and Lyon, C. E. 1998. Quality of restructured hams manufactured with PSE pork as affected by water binders. *J. Food Sci.* 63(6), 1007–1011. https://doi.org/10.1111/J.1365-2621.1998.TB15843.X.

Nawaz, A. H., Amoah, K., Leng, Q. Y., Zheng, J. H., Zhang, W. L. and Zhang, L. 2021. Poultry response to heat stress: its physiological, metabolic, and genetic implications on meat production and quality including strategies to improve broiler production in a warming world. *Front. Vet. Sci.* 0, 814. https://doi.org/10.3389/FVETS.2021.699081.

Nicol, C. J. and Scott, G. B. 1990. Pre-slaughter handling and transport of broiler chickens. *Appl. Anim. Behav. Sci.* 28(1–2), 57–73. https://doi.org/10.1016/0168-1591(90)90046-G.

Nielsen, S. S., Alvarez, J., Bicout, D. J., Calistri, P., Depner, K., Drewe, J. A., Garin-Bastuji, B., Gonzales Rojas, J. L., Gortázar Schmidt, C., Miranda Chueca, M. Á, Roberts, H. C., Sihvonen, L. H., Spoolder, H., Stahl, K., Velarde Calvo, A., Viltrop, A., Winckler, C., Candiani, D., Fabris, C., Van der Stede, Y., Michel, V. and EFSA Panel on Animal Health and Welfare (AHAW) 2019. Slaughter of animals: poultry. *EFSA J.* 17(11), e05849. https://doi.org/10.2903/j.efsa.2019.5849.

Nijdam, E., Delezie, E., Lambooij, E., Nabuurs, M. J. A., Decuypere, E. and Stegeman, J. A. 2005. Comparison of bruises and mortality, stress parameters, and meat quality in manually and mechanically caught broilers. *Poult. Sci.* 84(3), 467–474. https://doi.org/10.1093/PS/84.3.467.

Northcutt, J. K. 2000. Preslaughter factors affecting poultry meat quality. In: Sams, A. R. (Ed.), *Poultry Meat Processing*. CRC Press, Boca Raton, FL. https://doi.org/10.1201/9781420042177-5.

Northcutt, J. K., Buhr, R. J. and Young, L. L. 1998. Influence of preslaughter stunning on turkey breast muscle quality. *Poult. Sci.* 77(3), 487–492. https://doi.org/10.1093/PS/77.3.487.

Oba, A., Almeida, Md de, Pinheiro, J. W., Ida, E. I., Marchi, D. F., Soares, A. L. and Shimokomaki, M. 2009. The effect of management of transport and lairage conditions on broiler chicken breast meat quality and DOA (Death on Arrival). *Braz. Arch. Biol. Technol.* 52(spe), 205–211. https://doi.org/10.1590/S1516-89132009000700026.

Oda, S. H. I., Nepomuceno, A. L., Ledur, M. C., de Oliveira, M. C. Nd, Marin, S. R. R., Ida, E. I. and Shimokomaki, M. 2009. Quantitative differential expression of alpha and beta ryanodine receptor genes in PSE (pale, soft, exudative) meat from two chicken lines: broiler and layer. *Braz. Arch. Biol. Technol.* 52(6), 1519–1525. https://doi.org/10.1590/S1516-89132009000600024.

Offer, G. 1991. Modelling of the formation of pale, soft and exudative meat: effects of chilling regime and rate and extent of glycolysis. *Meat Sci.* 30(2), 157–184. https://doi.org/10.1016/0309-1740(91)90005-B.

Offer, G. and Knight, P. 1988. The structural basis of water-holding in meat. In: Lawrie, R. A. (Ed.), *Developments of Meat Science*. Elsevier Applied Science, London, New York, pp. 63–243.

Owens, C. M., Alvarado, C. Z. and Sams, A. R. 2009. Research developments in pale, soft, and exudative turkey meat in North America. *Poult. Sci.* 88(7), 1513–1517. https://doi.org/10.3382/PS.2009-00008.

Owens, C. M. and Apple, J. K. 2010. Pale, Soft and Exudative poultry meat. In: Guerrero-Legarreta, I. and Hui, Y. H. (Eds), *Handbook of Poultry Science and Technology* (vol. 1): Primary Processing. John Wiley & Sons, Inc., Hoboken, pp. 515–531.

Owens, C. M., Hirschler, E. M., McKee, S. R., Martinez-Dawson, R. and Sams, A. R. 2000a. The characterization and incidence of pale, soft, exudative Turkey meat in a commercial plant. *Poult. Sci.* 79(4), 553–558. https://doi.org/10.1093/ps/79.4.553.

Owens, C. M., Matthews, N. S. and Sams, A. R. 2000b. The use of halothane gas to identify Turkeys prone to developing pale, exudative meat when transported before slaughter. *Poult. Sci.* 79(5), 789–795. https://doi.org/10.1093/ps/79.5.789.

Owens, C. M. and Sams, A. R. 1997. Muscle metabolism and meat quality of pectoralis from turkeys treated with postmortem electrical stimulation. *Poult. Sci.* 76(7), 1047–1051. https://doi.org/10.1093/PS/76.7.1047.

Paião, F. G., Ferracin, L. M., Pedrão, M., Kato, T. and Shimokomaki, M. 2013. Skeletal muscle calcium channel ryanodine and the development of pale, soft, and exudative meat in poultry. *Genet. Mol. Res.* 12(3), 3017-3027. https://doi.org/10.4238/2013.August.20.3.

Pan, L., Ma, X. K., Zhao, P. F., Shang, Q. H., Long, S. F., Wu, Y. and Piao, X. S. 2018. Forsythia suspensa extract attenuates breast muscle oxidative injury induced by transport stress in broilers. *Poult. Sci.* 97(5), 1554–1563. https://doi.org/10.3382/ps/pey012.

Papinaho, P. A., Fletcher, D. L. and Buhr, R. J. 1995. Effect of electrical stunning amperage and peri-mortem struggle on broiler breast rigor development and meat quality. *Poult. Sci.* 74(9), 1533–1539. https://doi.org/10.3382/PS.0741533.

Park, H. K. 1980. Factors affecting the incidence of PSE muscle. Korean. *J. Anim. Sci.* 22, 447–452.

Park, Y. S., Choi, Y. S., Hwang, K. E., Kim, T. K., Lee, C. W., Shin, D. M. and Han, S. G. 2017. Physicochemical properties of meat batter added with edible silkworm pupae (Bombyx mori) and transglutaminase. *Korean J. Food Sci. Anim. Resour.* 37(3), 351-359. https://doi.org/10.5851/KOSFA.2017.37.3.351.

Petracci, M. and Berri, C. 2017. *Poultry Quality Evaluation : Quality Attributes and Consumer Values* (1st edn.). Woodhead Publishing.

Petracci, M., Betti, M., Bianchi, M. and Cavani, C. 2004. Color variation and characterization of broiler breast meat during processing in Italy. *Poult. Sci.* 83(12), 2086–2092. https://doi.org/10.1093/ps/83.12.2086.

Petracci, M., Bianchi, M. and Cavani, C. 2010. Pre-slaughter handling and slaughtering factors influencing poultry product quality. *Worlds Poult. Sci. J.* 66(1), 17–26 https://doi.org/10.1017/S0043933910000024.

Petracci, M., Bianchi, M. and Cavani, C. 2009. The European perspective on pale, soft, exudative conditions in poultry. *Poult. Sci.* 88(7), 1518–1523. https://doi.org/10.3382/PS.2008-00508.

Petracci, M., Bianchi, M., Mudalal, S. and Cavani, C. 2013. Functional ingredients for poultry meat products. *Trends Food Sci. Technol.* 33(1), 27–39. https://doi.org/10.1016/j.tifs.2013.06.004.

Petracci, M. and Cavani, C. 2012. Muscle growth and poultry meat quality issues. *Nutrients* 4(1), 1–12. https://doi.org/10.3390/nu4010001.

Petracci, M., Fletcher, D. L. and Northcutt, J. K. 2001. The effect of holding temperature on live shrink, processing yield, and breast meat quality of broiler chickens. *Poult. Sci.* 80(5), 670–675. https://doi.org/10.1093/PS/80.5.670.

Petracci, M., Laghi, L., Rocculi, P., Rimini, S., Panarese, V., Cremonini, M. A. and Cavani, C. 2012. The use of sodium bicarbonate for marination of broiler breast meat. *Poult. Sci.* 91(2), 526–534. https://doi.org/10.3382/PS.2011-01753.

Petracci, M., Soglia, F. and Berri, C. 2017. Muscle metabolism and meat quality abnormalities. In: Petracci, M. and Berri, C. (Eds), *Poultry Quality Evaluation. Quality Attributes and Consumer Values*. Woodhead Publishing, Elsevier limited company, Cambridge, pp. 51–76.

Petrolli, T. G., Junqueira, O. M., Domingues, C. H. F., Pereira, A. S. C., Santos, E. T. and Rocha, R. X. 2016. Effects of the addition of glucose, sodium bicarbonate, and vitamin E to the drinking water of pre-slaughter broiler chickens on carcass yield, gastric emptying and meat quality. *Rev. Bras. Cienc. Avic.* 18(1), 29–34. https://doi.org/10.1590/1516-635X1801029-034.

Piestun, Y., Halevy, O. and Yahav, S. 2009. Thermal manipulations of broiler embryos– the effect on thermoregulation and development during embryogenesis. *Poult. Sci.* 88(12), 2677–2688. https://doi.org/10.3382/PS.2009-00231.

Piestun, Y., Shinder, D., Ruzal, M., Halevy, O., Brake, J. and Yahav, S. 2008. Thermal manipulations during broiler embryogenesis: effect on the acquisition of thermotolerance. *Poult. Sci.* 87(8), 1516–1525. https://doi.org/10.3382/PS.2008-00030.

Pietrzak, M., Greaser, M. L. and Sosnicki, A. A. 1997. Effect of rapid rigor mortis processes on protein functionality in pectoralis major muscle of domestic turkeys. *J. Anim. Sci.* 75(8), 2106–2116. https://doi.org/10.2527/1997.7582106x.

Pozar, D. 2011. *Microwave Engineering*.

Puolanne, E. 2017. Developments in our understanding of water-holding capacity in meat. In: Purslow, P. P. (Ed.), *New Aspects of Meat Quality. From Genes to Ethics*. Elsevier Ltd., Amsterdam, Netherland, pp. 167–191.

Qaid, M. M. and Al-Garadi, M. A. 2021. Protein and amino acid metabolism in poultry during and after heat stress: a review. *Animals (Basel)* 11(4), 1167. https://doi.org/10.3390/ANI11041167.

Sandercock, D. A., Hunter, R. R., Mitchell, M. A. and Hocking, P. M. 2006. Thermoregulatory capacity and muscle membrane integrity are compromised in broilers compared with layers at the same age or body weight. *Br. Poult. Sci.* 47(3), 322–329. https://doi.org/10.1080/00071660600732346.

Sandercock, D. A., Hunter, R. R., Nute, G. R., Mitchell, M. A. and Hocking, P. M. 2001. Acute heat stress-induced alterations in blood acid-base status and skeletal muscle membrane integrity in broiler chickens at two ages: implications for meat quality. *Poult. Sci.* 80(4), 418–425. https://doi.org/10.1093/PS/80.4.418.

Sante-Lhoutellier, V., Bielicki, G., Renerre, M. and Lacourt, A. 1991. *Post Mortem Evolution in the Pectoralis Superficialis Muscle from Two Turkey Breeds: Relationship Between pH and Colour Changes. International Congress of Meat Science and Technology*, Kulmbach, Germany.

Saraiva, S., Esteves, A., Oliveira, I., Mitchell, M. and Stilwell, G. 2020. Impact of pre-slaughter factors on welfare of broilers. *Vet. Anim. Sci.* 10, 100146. https://doi.org/10.1016/J.VAS.2020.100146.

Savenije, B., Lambooij, E., Gerritzen, M. A., Venema, K. and Korf, J. 2002. Effects of feed deprivation and transport on preslaughter blood metabolites, early postmortem muscle metabolites, and meat quality. *Poult. Sci.* 81(5), 699–708. https://doi.org/10.1093/ps/81.5.699.

Schaefer, A. L., Dubeski, P. L., Aalhus, J. L. and Tong, A. K. W. 2001. Role of nutrition in reducing antemortem stress and meat quality aberrations. *J. Anim. Sci.* 79(E-Suppl), E91–E101. https://doi.org/10.2527/jas2001.79E-SupplE91x.

Scheffler, T. L. and Gerrard, D. E. 2007. Mechanisms controlling pork quality development: the biochemistry controlling postmortem energy metabolism. *Meat Sci.* 77(1), 7–16. https://doi.org/10.1016/j.meatsci.2007.04.024.

Schilling, M. W., Daigle, S. P., Alvarado, C. Z., Marriott, N. G. and Wang, H. 2005. Effects of collagen addition on the functionality of PSE-like and normal broiler breast meat in a chunked and formed deli roll. *J. Muscle Foods* 16(1), 46–53. https://doi.org/10.1111/j.1745-4573.2004.08104.x.

Schilling, M. W., Mink, L. E., Gochenour, P. S., Marriott, N. G. and Alvarado, C. Z. 2003. Utilization of pork collagen for functionality improvement of boneless cured ham manufactured from pale, soft, and exudative pork. *Meat Sci.* 65(1), 547–553. https://doi.org/10.1016/S0309-1740(02)00247-4.

Schilling, M. W., Radhakrishnan, V., Thaxton, Y. V., Christensen, K., Thaxton, J. P. and Jackson, V. 2008. The effects of broiler catching method on breast meat quality. *Meat Sci.* 79(1), 163–171. https://doi.org/10.1016/j.meatsci.2007.08.010.

Schneider, P. L., Beede, D. K. and Wilcox, C. J. 1988. Nycterohemeral patterns of acid-base status, mineral concentrations and digestive function of lactating cows in natural or chamber heat stress environments. *J. Anim. Sci.* 66(1), 112–125. https://doi.org/10.2527/JAS1988.661112X.

Shen, Q. W., Means, W. J., Thompson, S. A., Underwood, K. R., Zhu, M. J., McCormick, R. J., Ford, S. P. and Du, M. 2006. Pre-slaughter transport, AMP-activated protein kinase, glycolysis, and quality of pork loin. *Meat Sci.* 74(2), 388–395. https://doi.org/10.1016/j.meatsci.2006.04.007.

Solomon, M. B., Van Laack, R. L. J. M. and Eastridge, J. S. 1998. Biophysical basis of pale, soft, exudative (PSE) pork and poultry muscle: a review. *J. Muscle Foods* 9(1), 1–11. https://doi.org/10.1111/j.1745-4573.1998.tb00639.x.

Somboonpanyakul, P., Barbut, S., Jantawat, P. and Chinprahast, N. 2007. Textural and sensory quality of poultry meat batter containing malva nut gum, salt and phosphate. *LWT Food Sci. Technol.* 40(3), 498–505. https://doi.org/10.1016/j.lwt.2005.12.008.

Sosnicki, A. A., Greaser, M. L., Pietrzak, M., Pospiech, E. and Sante, V. 1998. PSE-like syndrome in breast muscle of domestic turkeys: a review. *J. Muscle Foods* 9(1), 13–23. https://doi.org/10.1111/j.1745-4573.1998.tb00640.x.

Sporer, K. R. B., Zhou, H. R., Linz, J. E., Booren, A. M. and Strasburg, G. M. 2012. Differential expression of calcium-regulating genes in heat-stressed turkey breast muscle is associated with meat quality. *Poult. Sci.* 91(6), 1418–1424. https://doi.org/10.3382/ps.2011-02039.

Sterten, H., Oksbjerg, N., Frøystein, T., Ekker, A. S. and Kjos, N. P. 2010. Effects of fasting prior to slaughter on pH development and energy metabolism post-mortem in M. longissimus dorsi of pigs. *Meat Sci.* 84(1), 93–100. https://doi.org/10.1016/j.meatsci.2009.08.019.

Strasburg, G. M. and Chiang, W. 2009. Pale, soft, exudative turkey-The role of ryanodine receptor variation in meat quality. *Poult. Sci.* 88(7), 1497–1505. https://doi.org/10.3382/ps.2009-00181.

Sun, H., Jiang, R., Xu, S., Zhang, Z., Xu, G., Zheng, J. and Qu, L. 2015. Transcriptome responses to heat stress in hypothalamus of a meat-type chicken. *J. Anim. Sci. Biotechnol.* 6(1), 6. https://doi.org/10.1186/S40104-015-0003-6.

Swatland, H. J. 2008. How pH causes paleness or darkness in chicken breast meat. *Meat Sci.* 80(2), 396–400. https://doi.org/10.1016/j.meatsci.2008.01.002.

Tona, K., Onagbesan, O., Bruggeman, V., Collin, A., Berri, C., Duclos, M. J., Tesseraud, S., Buyse, J., Decuypere, E. and Yahav, S. 2008. Effects of heat conditioning at d 16 to 18 of incubation or during early broiler rearing on embryo physiology, post-hatch growth performance and heat tolerance. *Arch. Geflügelkunde Sonderh.* 72, 75–83.

Tornberg, E. 1996. Biophysical aspects of meat tenderness. *Meat Sci.* 43S1, 175–191. https://doi.org/10.1016/0309-1740(96)00064-2.

Toyomizu, M., Tokuda, M., Mujahid, A. and Akiba, Y. 2005. Progressive alteration to core temperature, respiration and blood acid-base balance in broiler chickens exposed to acute heat stress. *J. Poult. Sci.* 42(2), 110–118. https://doi.org/10.2141/JPSA.42.110.

Traffano-Schiffo, M. V., Castro-Giraldez, M., Colom, R. J. and Fito, P. J. 2018. Innovative photonic system in radiofrequency and microwave range to determine chicken meat quality. *J. Food Eng.* 239, 1–7. https://doi.org/10.1016/j.jfoodeng.2018.06.029.

Van de Perre, V., Ceustermans, A., Leyten, J. and Geers, R. 2010. The prevalence of PSE characteristics in pork and cooked ham – effects of season and lairage time. *Meat Sci.* 86(2), 391–397. https://doi.org/10.1016/J.MEATSCI.2010.05.023.

Van Hoof, J. 1979. Influence of ante- and peri-mortem factors on biochemical and physical characteristics of turkey breast muscle. *Vet. Q.* 1(1), 29–36. https://doi.org/10.1080/01652176.1979.9693718.

Van Schalkwyk, A. P. and Gertenbach, W. D.. 2000. The effect of closing date on the performance of beef weaners grazing foggaged Digitaria eriantha and Acroceras macrum. *SA J. An. Sci.* 30(1). https://doi.org/10.10520/EJC94185.

Velleman, S. G. 2019. Recent developments in breast muscle myopathies associated with growth in poultry. *Annu. Rev. Anim. Biosci.* 7, 289–308. https://doi.org/10.1146/annurev-animal-020518-115311.

Vermeulen, L., Van de Perre, V., Permentier, L., De Bie, S., Verbeke, G. and Geers, R. 2015. Pre-slaughter handling and pork quality. *Meat Sci.* 100, 118–123. https://doi.org/10.1016/j.meatsci.2014.09.148.

Vieira, F. M. C., Silva, I. J. O., Barbosa Filho, J. A. D., Vieira, A. M. C. and Broom, D. M. 2011. Preslaughter mortality of broilers in relation to lairage and season in a subtropical climate. *Poult. Sci.* 90(10), 2127–2133. https://doi.org/10.3382/ps.2010-01170.

Wang, L. J., Byrem, T. M., Zarosley, J., Booren, A. M. and Strasburg, G. M. 1999. Skeletal muscle calcium channel ryanodine binding activity in genetically unimproved and commercial turkey populations. *Poult. Sci.* 78(5), 792–797. https://doi.org/10.1093/ps/78.5.792.

Wang, R. H., Liang, R. R., Lin, H., Zhu, L. X., Zhang, Y. M., Mao, Y. W., Dong, P. C., Niu, L. B., Zhang, M. H. and Luo, X. 2017a. Effect of acute heat stress and slaughter processing on poultry meat quality and postmortem carbohydrate metabolism. *Poult. Sci.* 96(3), 738–746. https://doi.org/10.3382/ps/pew329.

Wang, S., Li, C., Xu, X. and Zhou, G. 2013. Effect of fasting on energy metabolism and tenderizing enzymes in chicken breast muscle early postmortem. *Meat Sci.* 93(4), 865–872. https://doi.org/10.1016/j.meatsci.2012.11.053.

Wang, X., Li, J., Cong, J., Chen, X., Zhu, X., Zhang, L., Gao, F. and Zhou, G. 2017b. Preslaughter transport effect on broiler meat quality and post-mortem glycolysis metabolism of muscles with different fiber types. *J. Agric. Food Chem.* 65(47), 10310–10316. https://doi.org/10.1021/acs.jafc.7b04193.

Warner, R. D., Kauffman, R. G. and Greaser, M. L. 1997. Muscle protein changes post mortem in relation to pork quality traits. *Meat Sci.* 45(3), 339–352 https://doi.org/10.1016/S0309-1740(96)00116-7.

Warriss, P. D. 2000. *Meat Science: An Introductory Text*. CABI Publishing, New York.

Wasti, S., Sah, N. and Mishra, B. 2020. Impact of heat stress on poultry health and performances, and potential mitigation strategies. *Animals (Basel)* 10(8), 1266. https://doi.org/10.3390/ANI10081266.

Webb, E. C. and Casey, N. H. 2010. Physiological limits to growth and the related effects on meat quality. *Livest. Sci.* 130(1–3), 33–40. https://doi.org/10.1016/j.livsci.2010.02.008.

Wheeler, B. R., Mckee, S. R., Matthews, N. S., Miller, R. K. and Sams, A. R. 1999. A halothane test to detect turkeys prone to developing pale, soft, and exudative meat. *Poult. Sci.* 78(11), 1634–1638. https://doi.org/10.1093/ps/78.11.1634.

Wilhelm, A. E., Maganhini, M. B., Hernández-Blazquez, F. J., Ida, E. I. and Shimokomaki, M. 2010. Protease activity and the ultrastructure of broiler chicken PSE (pale, soft, exudative) meat. *Food Chem.* 119(3), 1201–1204. https://doi.org/10.1016/j.foodchem.2009.08.034.

Woelfel, R. L., Owens, C. M., Hirschler, E. M., Martinez-Dawson, R. and Sams, A. R. 2002. The characterization and incidence of pale, soft, and exudative broiler meat in a commercial processing plant. *Poult. Sci.* 81(4), 579–584. https://doi.org/10.1093/ps/81.4.579.

Xing, T., Gao, F., Tume, R. K., Zhou, G. and Xu, X. 2019. Stress effects on meat quality: a mechanistic perspective. *Compr. Rev. Food Sci. Food Saf.* 18(2), 380–401. https://doi.org/10.1111/1541-4337.12417.

Xu, Y., Zhao, X., Bian, G., Yang, L., Han, M., Xu, X. and Zhou, G. 2018. Structural and solubility properties of pale, soft and exudative (PSE)-like chicken breast myofibrillar protein: effect of glycosylation. *LWT* 95, 209–215. https://doi.org/10.1016/J.LWT.2018.04.051.

Xue, G., Cheng, S., Yin, J., Zhang, R., Su, Y., Li, X., Li, J. and Bao, J. 2021. Influence of pre-slaughter fasting time on weight loss, meat quality and carcass contamination in broilers. *Anim. Biosci.* 34(6), 1070–1077. https://doi.org/10.5713/ajas.20.0560.

Yunis, R. and Cahaner, A. 1999. The effects of the naked neck (Na) and frizzle (F) genes on growth and meat yield of broilers and their interactions with ambient temperatures and potential growth rate. *Poult. Sci.* 78(10), 1347–1352. https://doi.org/10.1093/ps/78.10.1347.

Zaboli, G., Huang, X., Feng, X. and Ahn, D. U. 2019. How can heat stress affect chicken meat quality? – a review. *Poult. Sci.* 98(3), 1551–1556. https://doi.org/10.3382/PS/PEY399.

Zhang, C., Geng, Z. Y., Chen, K. K., Zhao, X. H. and Wang, C. 2019a. L-theanine attenuates transport stress-induced impairment of meat quality of broilers through improving muscle antioxidant status. *Poult. Sci.* 98(10), 4648–4655. https://doi.org/10.3382/PS/PEZ164.

Zhang, L., Li, J. L., Wang, X. F., Zhu, X. D., Gao, F. and Zhou, G. H. 2019b. Attenuating effects of guanidinoacetic acid on preslaughter transport-induced muscle energy expenditure and rapid glycolysis of broilers. *Poult. Sci.* 98(8), 3223–3232. https://doi.org/10.3382/ps/pez052.

Zhang, C., Wang, L., Zhao, X. H., Chen, X. Y., Yang, L. and Geng, Z. Y. 2017. Dietary resveratrol supplementation prevents transport-stress-impaired meat quality of broilers through maintaining muscle energy metabolism and antioxidant status. *Poult. Sci.* 96(7), 2219–2225. https://doi.org/10.3382/PS/PEX004.

Zhang, L. and Barbut, S. 2005a. Rheological characteristics of fresh and frozen PSE, normal and DFD chicken breast meat. *Br. Poult. Sci.* 46(6), 687–693. https://doi.org/10.1080/00071660500391516.

Zhang, L. and Barbut, S. 2005b. Effects of regular and modified starches on cooked pale, soft, and exudative; normal; and dry, firm, and dark breast meat batters. *Poult. Sci.* 84(5), 789–796. https://doi.org/10.1093/ps/84.5.789.

Zhang, L., Huang, H., Wang, P., Xing, T. and Xu, X. 2020. Water-spraying forced ventilation during holding improves the water holding capacity, impedance, and microstructure of breast meat from summer-transported broiler chickens. *Poult. Sci.* 99(3), 1744–1749. https://doi.org/10.1016/j.psj.2019.10.077.

Zhang, L., Yue, H. Y., Zhang, H. J., Xu, L., Wu, S. G., Yan, H. J., Gong, Y. S. and Qi, G. H. 2009. Transport stress in broilers: I. Blood metabolism, glycolytic potential, and meat quality. *Poult. Sci.* 88(10), 2033–2041. https://doi.org/10.3382/PS.2009-00128.

Zhao, X., Xing, T., Chen, X., Han, M., Xu, X. and Zhou, G. 2017. Yield, thermal denaturation, and microstructure of proteins isolated from pale, soft, exudative chicken breast meat by using isoelectric solubilization/precipitation. *Process Biochem.* 58, 167–173. https://doi.org/10.1016/j.procbio.2017.04.035.

Zhu, L. G. and Brewer, M. S. 2003. Effects of urea denatureation and pH on the ability of porcine myoglobin to undergo reduction. *Meat Sci.* 63(4), 427–432. https://doi.org/10.1016/S0309-1740(02)00032-3.

Zhu, X., Ruusunen, M., Gusella, M., Ylä-Ajos, M., Xu, X., Zhou, G. and Puolanne, E. 2013. High early post-mortem temperature induces activation of AMP-activated protein kinase and development of pale, soft and exudative characteristics in turkey muscles. *Meat Sci.* 93(3), 600–606. https://doi.org/10.1016/j.meatsci.2012.11.011.

Zhu, X., Ruusunen, M., Gusella, M., Zhou, G. and Puolanne, E. 2011. High post-mortem temperature combined with rapid glycolysis induces phosphorylase denaturation and produces pale and exudative characteristics in broiler pectoralis major muscles. *Meat Sci.* 89(2), 181–188. https://doi.org/10.1016/j.meatsci.2011.04.015.

Zhu, X. S., Xu, X. L., Min, H. H. and Zhou, G. H. 2012. Occurrence and characterization of pale, soft, exudative-like broiler muscle commercially produced in China. *J. Integr. Agric.* 11(8), 1384–1390. https://doi.org/10.1016/S2095-3119(12)60137-3.

Zhuang, H., Savage, E. M. and Lawrence, K. 2010. Effect of 3 postmortem electrical stimulation treatments on the quality of early deboned broiler breast meat. *Poult. Sci.* 89(8), 1737–1743. https://doi.org/10.3382/PS.2009-00460.

Ziober, I. L., Paio, F. G., Marchi, D. F., Coutinho, L. L., Binneck, E., Nepomuceno, A. L. and Shimokomaki, M. 2010. Heat and chemical stress modulate the expression of the α-RYR gene in broiler chickens. *Genet. Mol. Res.* 9(2), 1258–1266. https://doi.org/10.4238/vol8-2gmr830.

Chapter 9

Quality defects associated with poultry muscle development: white striping

Yuwares Malila, Krittaporn V. Thanatsang and Yanee Srimarut, National Center for Genetic Engineering and Biotechnology, Thailand

1 Introduction

Consumer demand for poultry meat has been steadily rising over recent decades. There has been an increase in global poultry meat consumption from 11.1 kg per person in 2000 to 15.6 kg per person in 2018, with poultry expected to account for 50% of meat consumed globally by 2029 (USDA, 2021). To meet this high demand, chickens and turkeys have been intensively selected for high growth efficiency and heavy body mass. The success in breeding selection, however, has coincided with increasing challenges, including the development of the white striping (WS) abnormality among the birds.

WS was first reported in 2009 (Kuttappan et al., 2009) and has rapidly become one of the main growth-related issues globally within the poultry industry. The presence of WS is usually accompanied by wooden breast (WB), another emerging growth-related myopathy in broilers, but their macroscopic characteristics are distinctive. Development of WS is characterized by the appearance of white lines parallel to muscle fibers on the superficial pectoral muscle (Fig. 1). Research indicates that the white lines are the result of deposition of infiltrated fat and minerals on connective tissue of the muscle fibers (Barbut,

http://dx.doi.org/10.19103/AS.2022.0112.10

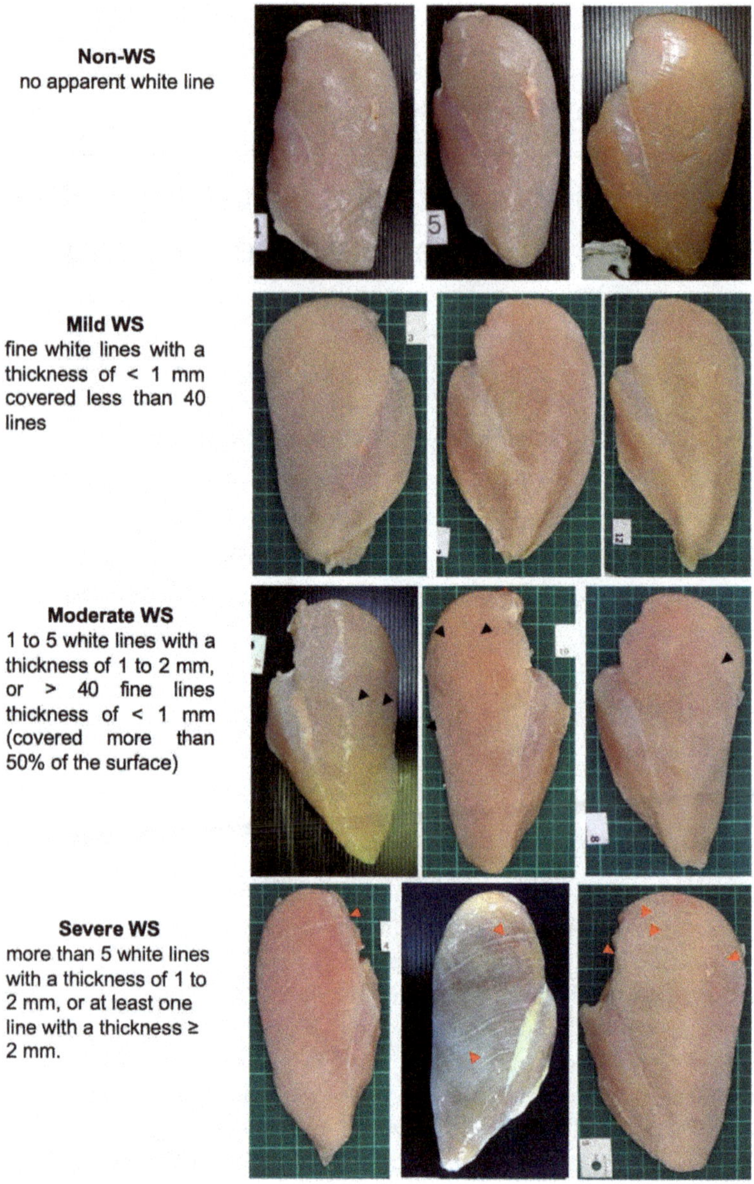

Figure 1 White striping (WS) abnormality. Broiler breasts in each row depict different WS severity levels, including non-WS, mild WS, moderate WS and severe WS. The classification is based on Malila et al. (2018). Black arrow heads indicate the lines with a thickness of 1 mm to 2 mm. Red arrowheads indicate the lines with thickness >2 mm.

2019). The lesions occur more toward the cranial end on the ventral surface of the poultry breast compared with other regions.

Based on their abnormal visual characteristics, breasts with severe WS have reduced appeal to consumers and are directed to other markets where the meat was not chosen based primarily on visual assessment (Owens and Vieira, 2012; Petracci et al., 2019). The severity of WS is associated with a decreased proportion of protein and increased fat, altering the nutritional profile of chicken breast meat. Severely affected meat also shows reduced protein functionality, particularly decreased water-holding capacity (WHC) (Kuttappan et al., 2012a, 2013a; Petracci et al., 2013, 2014; Mudalal et al., 2015; Tasoniero et al., 2016; Tijare et al., 2016). In the USA alone, growth-related myopathies such as WS have cost approximately US$200 million per year (Kuttappan et al., 2016). However, according to discussions with industry personnel, Barbut (2019) has suggested that this number accounts for only 1% of the actual economic losses from these myopathies within the US poultry industry.

The occurrence of WS has been extensively investigated among the world's largest broiler meat exporters, including Brazil (Ferreira et al., 2014), the USA (Kuttappan et al., 2012a), the European Union (Petracci et al., 2013; Sihvo et al., 2014; Tasoniero et al., 2016; Lucca et al., 2019) and Thailand (Malila et al., 2018). Initial studies suggested a prevalence of WS of approximately 10% of breast meat (Petracci et al., 2013). Further studies have suggested higher levels of WS, varying from 20% to as high as 98% (Kuttappan et al., 2013a; Lorenzi et al., 2014; Trocino et al., 2015; Russo et al., 2015; Tijare et al., 2016; Malila et al., 2018; Prisco et al., 2021). The lesions have begun to be detected in other cuts, e.g. the pectoralis minor muscle and thigh, of broilers (Petracci et al., 2019) and turkeys (Mudalal, 2019). During the preparation of this chapter, WS in commercial broilers has been less discussed in Eastern and Southeast Asia. This might be because most commercial broilers in this region are slaughtered at a younger age of 5–6 weeks and are sold with skin covering any undesirable appearances. Although the majority of WS-affected meats fall into the categories of 'mild' and 'moderate' categories of quality problems where properties are not drastically altered (Malila et al., 2018), the presence of a few severe cases of WS could still be responsible for technological challenges such as excessive water loss and inconsistent quality of processed products.

2 Impacts of white striping on quality traits of poultry meat

In 2012, when data were still limited, it appeared that general meat quality indices were not significantly influenced by WS (Owens and Vieira, 2012). However, as summarized in Table 1, cumulative evidence suggests that WS causes a wide range of physicochemical, technological and quality issues in

Table 1 Summary of previous studies indicating the effects of white striping (WS) abnormality on properties of poultry meat

Reference	Animal (*n*, age)	Live weight	Grading	Total incidence	Effects relative to normal meat		
					Increased	Decreased	Not different
Petracci et al. (2013)	(1) Commercial broilers (28 000, 45–54 days)	-	3-Scoring system[a]	12%	-	-	-
	(2) High-breast yield (HBY, Ross 708) vs standard-breast yield (SBY) (153, 7 weeks)	2.8 kg	3-Scoring system	HBY: 15.2%, SBY: 10%	• pH • a^* and b^*[b] • Purge loss • Cook loss	• Marinade uptake • Shear force	-
Mudalal et al. (2014)	Ross 708 broilers (12, 7 weeks)	2.8 kg	3-Scoring system	-	• Moisture • Fat • Collagen	• Protein • Ash	-
Petracci et al. (2014)	Ross 708 broilers (108, 7 weeks)	2.8 kg	3-Scoring system	-	• Fat • Collagen	• Protein	• Moisture
Mudalal et al. (2015)	Male Ross 708 broilers (192, 52 days)	3.7 kg	Normal or WS	-	**Non-marinated**		
					• Breast weight and thickness • pH • Cook loss	• Drip loss	• Breast length and width • L^*, a^* and b^* (raw and cooked)[b] • Shear force • Compression test
					Marinated		
						• Marinate uptake • Meat yield (cooked)	• L^*, a^* and b^* (cooked)
Trocino et al. (2015)	(1) HBY vs SBY (768, 46 days)	2.8–3.5 kg	3-Scoring system	HBY: 75% SBY: 74%	-	-	-

	(2) Broilers (256, 46 days)	-	3-Scoring system	-	• pH	• a^* and b^*	• L^* • Thaw loss • Cook loss • Shear force
Bowker and Zhuang (2016)	Mixed-sex Ross broilers (24, age not specified)	3.2–4.1 kg	3-Scoring system	-	• Breast weight • pH	• Thaw loss • SP3 solubility • MP3 emulsifying activity (only in severe WS)	• L^*, a^* and b^* • Cook loss • Salt-induced water uptake • MP3 solubility • SP3 emulsifying activity
Sanchez Brambila et al. (2016)	Broilers (42, 7–8 weeks)	-	3-Scoring system	-	• Breast weight • pH (only in moderate WS) • Hardness (only in severe WS)	-	• L^*, a^* and b^* • Sensory attributes: cohesiveness, chewiness and juiciness
Soglia et al. (2016)	Male high-breast-yield broilers (20, no information of age)	-	Normal or WS (Mudalal et al., 2015)	-	• Fat	• Protein • SFA4 • Hematin	• Moisture • Collagen • MUFA4 and PUFA4 • TBARS4 • Protein carbonyls
Tasoniero et al. (2016)	Ross 708 chickens (500, 51 days)	-	Normal or WS (Kuttappan et al., 2013)	-	• Breast weight • Fe • Overall intensity (sensory analysis score)	-	• pH • L^*, a^* and b^* • Ca, Na, K, P • Cook loss (48-h, 6-month storage) • Thaw loss (6-month storage) • Shear force

(Continued)

Table 1 (*Continued*)

Reference	Animal (n, age)	Live weight	Grading	Total incidence	Effects relative to normal meat		
					Increased	Decreased	Not different
Baldi et al. (2018)	Male Ross 308 broilers (96, 47 days)	2.8 kg	Normal or WS based on the presence of medium-to-thick white striations	-	• Breast weight • Breast length and height • pH • b^* (deep part) • Lipid	• Protein • Myofibril fragmentation index (superficial part)	• L^* and a^* • Moisture • Collagen • Ash • Carbonyls • Protein solubility
Malila et al. (2018)	Ross 308 broilers (100, 6 weeks)	-	Normal, mild, moderate or severe WS based on the amount and thickness of the white striations	97.8%	• Breast weight and yield • pH • Total collagen • Springiness (only in mild WS)	• Protein	• Moisture • Fat • Ash
Soglia et al. (2018)	Heavy male turkeys (BUT Big 6) (72, 137 days)	20 kg	Normal, moderate or severe WS based on the thickness of the white striations	-	• Breast weight • Lipid (only in severe WS)	• Ash • Drip loss (only in moderate WS)	• pH • L^*, a^* and b^* • Moisture • Protein • Collagen • Cook loss
Adabi and Soncu (2019)	(1) Ross 308 broilers (240, 28–39 days)	-	3-Scoring system	28 days: 44% 31 days: 50% 35 days: >60% 39 days: >70%	-	-	-

	(2) Ross 308 broilers (1000, 41 days)	-	3-Scoring system	73%	• Fat • PUFA	• L^* (dorsal) and a^* (ventral) • Protein • SFA • Tryptophan, histidine (only in severe WS) • Sulfhydryl	• a^* and b^* (dorsal) and, L^* and b^* (ventral) • MUFA, *n*-3 and *n*-6 • TBARS • Carbonyls
Baldi et al. (2019)	Ross 308 broilers (48, 46 days)	2.8 kg	Normal or WS based on visual appearance showing severe myopathic conditions	-	• Lipid (skin-side)	-	• pH • L^*, a^* and b^* • Moisture • Protein • Intramuscular collagen • Hydroxylysyl pyridinoline • Ash • Fatty acid composition
Gratta et al. (2019)	Ross 308 broilers (144, 49 days)	2.8 kg	3-Scoring system	-	• Weight • a^* • Cook loss • PUFA, *n*-3, *n*-6	• SFA • MUFA	• pH • L^* and b^* (cooked) • Shear force • Water • Protein • Ether extract • Ash
Mudalal (2019)	(1) Male turkeys (23 000, 20 weeks)	-	3-Scoring system	61.3%	-	-	-

(*Continued*)

Table 1 (*Continued*)

Reference	Animal (*n*, age)	Live weight	Grading	Total incidence	Effects relative to normal meat: Increased	Decreased	Not different
	(2) Male turkeys (60, 20 weeks)	-	3-Scoring system	-	**Raw, Non-marinated** • pH • a^* and b^* • Fat (only in severe WS) **Marinated** • b^* (cooked)	• Protein (only in severe WS) • L^*, a^* and b^* (marinated) • L^* and b^* (cooked) • Drip loss • Cook loss • Marinate uptake	• L^* • Moisture, ash • Calorie content • Freezing loss
Livingston et al. (2019c)	Male broiler (118, 42 days)	-	None, mild, moderate or severe WS based on the amount of striations	65.5%	• Body weight • Breast weight • pH • Drip loss	-	-
Salles et al. (2019)	Male Cobb 500 broiler (45, 42 days)	3.2 kg	3-Scoring system	-	• Moisture (only in severe WS) • Lipid (only in severe WS) • Collagen (only in severe WS) • ROS5 • AOPP5 level • TBARS (only in severe WS) • GPx5 activity (only in moderate WS) • GST5 activity	• Protein • Mineral matter	-

Soglia et al. (2019a)	Female Ross 308 (48, 48 days)	2.8 kg	Normal, or severe WS based on visual appearance according to the presence of muscular abnormality	-	• pH • Isoleucine • Leucine • Methionine • Phenylalanine • Threonine • Tryptophan • Valine	-	• L^*, a^* and b^* • TBARS • Carbonyls • Swelling
Campo et al. (2020)	Ross 308 and Cobb 500 broilers (300, 35 and 41 days)	-	3-Scoring system	41.7%	• Carcass weight • Breast yield • pH (only in severe WS) • L^*, a^* and b^* • Raw meat compression (C20 N/cm^2)	• Raw meat cohesiveness (only in severe WS)	• Cooked meat texture (shear force and TPA)
Carvalho et al. (2020)	(1) Cobb broilers (8959, 4–65 weeks)	-	Normal, moderate or severe WS based on the thickness of the white striations	5 weeks: 9.6% 7 weeks: 28.3% 9 weeks: 52% 65 weeks: 95%	-	-	-
	(2) Cobb broilers (32, 6–7 weeks)	-	-	-	• Lipid • Collagen	• Protein	• Moisture

(Continued)

Table 1 (*Continued*)

Reference	Animal (*n*, age)	Live weight	Grading	Total incidence	Effects relative to normal meat: Increased	Decreased	Not different
Carvalho et al. (2021a)	Nicholas breed male and female turkeys (87, 150–160 days)	18.5 kg	3-Scoring system	-	• L^* and b^* • Total and insoluble collagen • Ash (only in severe WS) • Capric acid (only in severe WS) • α-Linolenic acid (only in severe WS) • Eicosadienoic acid (only in severe WS) • Eicosatrienoic acid (only in severe WS) • Docosatetraenoic acid (only in severe WS)	• Protein • Ash (only in moderate WS) • Shear value	• pH • a^* • Moisture • Lipid • Cholesterol • Soluble collagen • SFA • MUFA and PUFA • Cook loss • Water-holding capacity
Carvalho et al. (2021b)	Chicken (60, age not specified)	-	Normal, moderate or severe WS based on the thickness of the white striations	-	• pH • a^* • Lipid • Collagen • Hardness	• L^* • Protein (only in severe WS)	• Protein (only in moderate WS)

AOPP = advanced oxygen protein products, GPx = glutathione peroxidase, GST = glutathione-S transferase, MP = myofibrillar protein, MUFA = monounsaturated fatty acid, PUFA = polyunsaturated fatty acid, ROS = reactive oxygen species, SFA = saturated fatty acid, SP = sarcoplasmic protein, TBARS = thiobarbituric acid reactive substances, TPA = texture profile analysis.

[a] 3-Scoring system indicates WS classification into normal (no lines), moderate (line thickness up to 1.0 mm) and severe (line thickness > 1.0 mm) according to Kuttappan et al. (2012b).

[b] L^* (lightness), a^* (redness) and b^* (yellowness) according to CIE system.

poultry meat. Impacts are in the broad spectrum due in part to factors such as differences in WS severity degree as well as breed, slaughter age, feed formulation and farming conditions. Table 1 lists mainly studies focusing on the breast samples collected from local commercial slaughterhouses or from the animals reared in large-scale industrial conditions.

2.1 Physicochemical properties

Chicken meat affected by WS abnormality shows an abnormally high final pH with a reduced content of protein (22.8–22.9% in normal, 18.7–20.9% in severe WS) and increased proportion of fat (0.8–1.0% in normal, 2.2–2.5% in severe WS) (Petracci et al., 2014; Mudalal et al., 2014). In moderate and severe cases, increases in moisture, total collagen and insoluble collagen have also been detected (Petracci et al., 2013; 2014; Mudalal et al., 2015; Trocino et al., 2015; Zambonelli et al., 2016; Baldi et al., 2018; Malila et al., 2018; Mudalal, 2019; Bordignon et al., 2020; Carvalho et al., 2020). WS did not alter the content of ash in broilers, but significant effects were found in turkey breasts (Soglia et al., 2018; Carvalho et al., 2021a).

Unlike WB, WS shows less impacts on oxidizing lipids and proteins in raw poultry meat (Soglia et al., 2016, 2019a; Baldi et al., 2018; Adabi and Soncu, 2019). Despite no significant differences in thiobarbituric acid reactive substances and protein carbonyls reported in some studies, increased oxidation of proteins and lipids was reported in severely affected WS raw chicken breast (Salles et al., 2019; Bordignon et al., 2020) and in chicken nuggets (Bordignon et al., 2020) in comparison to normal and moderate WS samples. However, when meats were subjected to freezing (−20°C for 45 days) followed by thawing (4°C for 16 h), protein carbonyls were drastically increased in the WS breasts compared to normal counterparts (Soglia et al., 2019a). Disruption of the membranes and cellular organelles during freezing and thawing seems to expose myofibrillar proteins to oxidative enzymes and pro-oxidant molecules, thus intensifying the effects of WS on protein oxidation.

2.2 Nutritional quality

The modified nutritional profile of WS-affected meat is a consequence of altered chemical composition. Consumers expect chicken meat to provide adequate amounts of essential amino acids with relatively low lipid and collagen content (Marangoni et al., 2015). However, the higher amount of fat accompanied by lower protein content in WS meat alters its energy profile (Petracci et al., 2014; Soglia et al., 2016, 2018; Baldi et al., 2018; Malila et al., 2018; Adabi and Soncu, 2019; Mudalal, 2019; Salles et al., 2019; Carvalho et al., 2021b).

Energy contribution from protein decreased, while energy contribution from fat increased (Petracci et al., 2014). The change was more pronounced as the WS severity increased.

WS condition also affects the profiles of fatty acids and amino acids in the meat. In the case of fatty acids, despite no changes in the amounts of monounsaturated fatty acids and PUFAs, total saturated fatty acids were significantly lower in WS meat compared to normal meat (Kuttappan et al., 2012a; Soglia et al., 2016; Adabi and Soncu, 2019; Salles et al., 2019). Individual fatty acids, particularly linoleic acid, linolenic acid, eicosapentaenoic acid and docosahexaenoic acid, were found to be affected by WS, though results were inconsistent among studies (Kuttappan et al., 2012a; Soglia et al., 2016; Adabi and Soncu, 2019; Carvalho et al., 2021a).

Essential amino acids were most affected by WS. Adabi and Soncu (2019) found significant decreases in histidine, arginine and tryptophan in cases of severe WS. More detrimental effects were observed when WS was accompanied by WB (Dalle Zotte et al., 2020). Concentrations of almost all free amino acids, except for beta-alanine, were significantly higher in WS meat (Soglia et al., 2019a). Myodegeneration in abnormal muscles might play a role in the breakdown of muscular proteins into free amino acids. The release of free amino acids may explain the loss of protein content in WS meat.

An increase in insoluble collagen in WS meat raises concern about the low susceptibility to proteolytic enzymes resulting in low protein digestibility. However, previous research suggests that, under optimum gastric conditions, high digestibility (95% of true digestibility) of insoluble collagens derived from bovine Achilles tendon can be obtained (Reuterswärd and Fabiansson, 1985). Further research in this area is required.

2.3 Technological properties

Differences in technological properties due to WS have been documented in several studies (Table 1). These include impaired WHC, indicated by high drip and cooking loss accompanied by low marinade uptake (Petracci et al., 2013; Mudalal et al., 2014; 2015; Tijare et al., 2016; Gratta et al., 2019; Bordignon et al., 2020; Yu et al., 2020). This corresponds to decreases in myofibrillar protein quantity and functionality (Mudalal et al., 2014; Bowker and Zhuang, 2016). Compared with non-WS meat, a greater proportion of water molecules in WS meat occur in extra-myofibrillar compartments with a lower intra-myofibrillar water fraction (Soglia et al., 2016, 2019a; Baldi et al., 2018). Those studies suggested that, in WS meat, water is less bound to the surrounding matrix in sarcoplasm and can be easily released by minor physical and mechanical forces, resulting in high fluid loss. The shift in muscular water distribution, due potentially to histopathological myodegeneration (Baldi et al., 2018), was more pronounced in the cranial-superficial section of abnormal fillets (Bowker and Zhuang, 2016; Baldi et al., 2018), corresponding to the number of degenerative lesions on the ventral surface (most affected) toward the dorsal side (least affected) (Bowker and Zhuang, 2016).

Alnahhas et al. (2016), Trocino et al. (2015) and Malila et al. (2018) observed no differences in texture profile between WS and normal samples. However, Petracci et al. (2013) and Zambonelli et al. (2016) reported greater toughness, in terms of increasing shear force, in WS cooked breasts relative to normal meat. The discrepancies with previous studies could be due in part to the heterogeneity of myopathic lesions and different sampling locations.

Sanchez Brambila et al. (2016) studied textural attributes of cooked WS breasts using a trained panel. Fillets affected with a severe WS degree received greater scores in cohesiveness, hardness and chewiness than those of non-defective and moderately affected breasts. An increased force required to shear through cooked WS breasts has been related to an accumulation of connective tissue, particularly insoluble collagen, in WS breasts (Kuttappan et al., 2013a; Petracci et al., 2014). Extensive fluid loss due to cooking may also intensify the toughness of cooked WS breast meat due to shrinkage and densely packed muscle fibers after cooking (Huff-Lonergan and Lonergan, 2005).

In turkey breast meat, no adverse effects of WS on technological parameters were detected, except for a slight decrease in drip loss from 1.1% in normal to 0.8% in moderate WS turkey meat with no change between normal and severe (1.0%) WS turkey samples (Soglia et al., 2018). Using nuclear magnetic resonance (NMR) imaging, Soglia et al. (2018) confirmed no differences in water mobility between normal and WS turkey samples. The less pronounced WS effects on the technological properties of turkey breast meat compared to chicken meat may be due to differences in breast muscle morphology between turkeys and chickens (Soglia et al., 2018).

2.4 Sensory quality and consumer acceptance

Sanchez Brambila et al. (2016) and Tasoniero et al. (2016) used sensory panels to assess the effects of WS on the texture and odor of cooked chicken breasts. Sanchez Brambila et al. (2016) found greater hardness and chewiness of cooked chicken breasts affected with severe WS compared to non-defective and moderately affected breasts. Tasoniero et al. (2016) found no differences in texture, but cooked WS breasts exhibited greater off odor intensity than normal samples.

Raw chicken breasts affected with WS received low hedonic scores from panels of consumers (Kuttappan et al., 2012c; De Carvalho et al., 2020). The higher the WS severity, the lower the score was acquired. Negative perceptions were due mainly to altered visual appearance due to white lines on the meat surface. Kuttappan et al. (2012c) found that over 50% of consumers refused to buy breasts affected by any degree of WS as they perceived the lines on the meat as a sign of high fat content, meat toughness and spoilage. Albrecht et al.

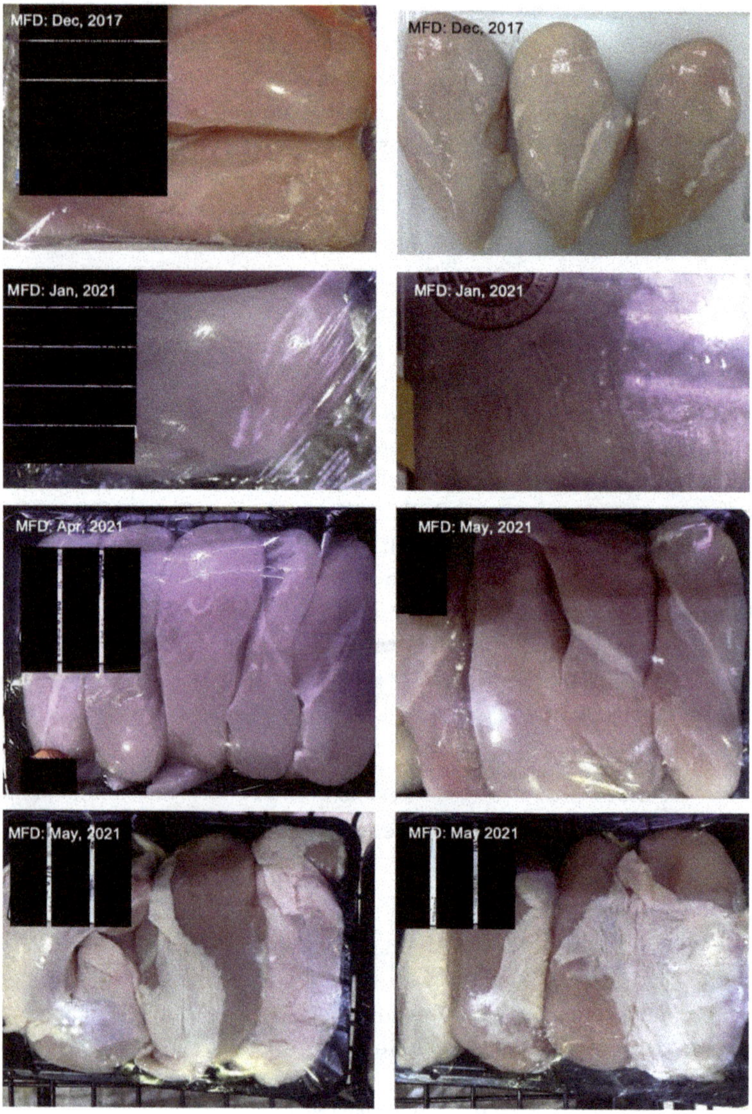

Figure 2 Chicken breast meat from local supermarkets in Thailand. The packaged meats were purchased during December 2017–May 2021. At least one of the meats showed white striping (WS) abnormality with varying severity levels, mostly mild or moderate WS.

(2019b) and De Carvalho et al. (2020) also found that WS reduced the intent to purchase.

Malila et al. (2018) addressed that the majority of Thai consumers could not distinguish between non-defective and WS samples even when packaged chicken fillets showed WS lesions, seeing lines as normal (Fig. 2). Despite the overall WS incidence of nearly 98%, only 3% of breasts with severe WS were rejected and used as raw materials for manufacturing other meat-processed products.

Consumers in Western countries have begun to relate WS with animal welfare. A leading welfare organization, The Humane League UK, investigated WS incidence in chicken fillets on supermarket shelves in 19 cities and towns across the UK during February and July to August 2020 (The Humane League UK, 2021). They found that 85% of packets examined had one breast or more affected with WS lesions, with over 90% of packets affected in some stores. They found WS incidence of only 11% in samples of chickens produced to higher welfare standards.

2.5 Shelf life

It has been assumed that, due to high pH (pH > 6.0), WS meat is more susceptible to microbial spoilage than normal meat since these conditions favor the growth of spoilage bacteria (Faucitano et al., 2010). However, Gratta et al. (2019) found that normal breasts from a commercial slaughterhouse had a greater initial total viable count of *Pseudomonas* spp. compared with WS samples. During 11-day storage at 4°C, normal samples showed a greater microbial count than WS samples. The authors suggested that the reduced glycolytic potential observed in WS breasts limited nutrients for microbial growth, prolonging the microbial shelf life of WS samples. In contrast, using a sensory panel to assess color, odor and texture, Albrecht et al. (2019a) found that the degree of WS was negatively correlated (Spearman's coefficient = -0.36) with the shelf life of chicken breasts. Meat with higher WS was rejected even if the threshold for microbial spoilage had not been reached.

3 Biological changes associated with the development of white striping

Research into growth-related myopathies such as WS suggests that a complex chain of linked biological events is involved. This research is discussed below.

Histological studies consistently show muscle fiber damage, as shown by accumulated macrophages, nuclei internalization, deposition of adipocytes, thickened endomysium and perimysium, inconsistent size of rounded myofibers, infiltration of lymphocytes and macrophages and necrosis in the affected muscle (Kuttappan et al., 2013a; Malila et al., 2018; Salles et al., 2019; Praud et al., 2020; Soglia et al., 2019b, 2021a). The increased occurrence of fiber necrosis, fibrosis and adipose tissue filtration has been correlated with increasing WS severity (Praud et al., 2020). It has been suggested that this is because hypertrophied muscle fibers have outgrown the capacity to support them (Mutryn et al., 2015).

Disrupted vascularization, often reported in high-breast-yield chickens, has been suggested as the main condition triggering the onset of WS (Ayansola et al., 2021). With increasing WS severity, the number of capillaries relative to muscle fibers is reduced while the distance between capillaries and myofibrils widens (Kindlein et al., 2017). This limits the supply of oxygenated blood, creating low oxygen availability, so-called hypoxia, and removal of metabolic wastes, e.g. reactive oxygen species (ROS) (Alnahhas et al., 2016; Kindlein et al., 2017; Malila et al., 2019; Salles et al., 2019; Praud et al., 2020; Soglia et al., 2021a).

In response to hypoxia, several biological processes are either activated or suppressed to maintain cellular homeostasis. Among the key stress-related transcription factors, an increased transcript abundance of hypoxia-inducible factor 1 (HIF-1), particularly alpha subunit (*HIF1A*), in WS muscle relative to non-defective samples has been consistently reported (Malila et al., 2019; Marchesi et al., 2019). Linked with increased *HIF1A*, differential expressions of several HIF-1-responsive genes involved in downstream biological pathways, such as carbohydrate metabolisms, energy production, angiogenesis and apoptosis, have been identified in myopathic muscles (Mutryn et al., 2015; Malila et al., 2019; Marchesi et al., 2019). Expression of myosin-heavy chain 6 is also elevated in WS muscle, suggesting replacement of fast-twitch IIb (glycolytic) with IIa (oxidative) fibers, potentially due to a need of oxidative respiration of muscle under hypoxia (Praud et al., 2020).

Another crucial role of HIF-1 in response to low oxygenation is to promote the switch of energy production from using oxygen to glycolysis. Oxygen is then reserved for essential cellular activities, and the formation of harmful metabolic products is suppressed. Lactate dehydrogenase isoform A (*LDHA*) is expected to increase together with high lactate content and low final pH in WS meat. However, in practice, the opposite appears to be true. Initial studies suggested the loss of lactate dehydrogenase (LDH) enzyme through leakage of muscle during myodegeneration as a possible cause of this discrepancy (Kuttappan et al., 2013b). Later, limited glycogen content and low glycolytic potential were reported in the skeletal muscle of WS-affected broilers, suggesting that dysregulated carbohydrate metabolism might be partly responsible for the reduced lactate level in the abnormal broilers (Alnahhas et al., 2016; Malila et al., 2019). Transcriptional and metabolomic analyses suggest that metabolic processes affecting macronutrients (i.e. carbohydrates, proteins and lipids), as well as other biochemical compounds, are significantly altered with WS (Boerboom et al., 2018; Malila et al., 2019, 2020; Marchesi et al., 2019).

Dysregulation of glycogen biosynthesis, glycolysis and gluconeogenesis has been consistently associated with growth-related myopathies (Zambonelli et al., 2016; Malila et al., 2019, 2020; Lake and Abasht, 2020; Bordini et al., 2021; Lake et al., 2021). Using droplet digital polymerase chain reaction, Malila et al. (2019) identified increased transcript abundance of genes encoding 6-phosphofructo-2-kinase/fructose-2,6-biphosphatase 4 (*PFKFB4*), *LDHA* and phosphorylase kinase beta subunit (*PHKB*) and decreased glyceraldehyde-3-phosphate dehydrogenase (*GAPDH*) in WS muscle, supporting disrupted glycogen breakdown and glycolysis.

Muscular proteins undergo greater breakdown in broilers with severe WS compared with non-defective birds (Vignale et al., 2017). This correlates with increased levels of short peptides in abnormal muscles (Boerboom et al., 2018). Metabolomic analysis by Boerboom et al. (2018) showed an accumulation of intermediates from the tricarboxylic acid cycle (i.e. fumarate and malate), suggesting the back flux of oxaloacetate converted into malate and fumarate when oxygen availability is low. In addition, increased levels of long- and medium-chain fatty acids (particularly MG18:4, MG18:2 and C16:0 and cis-5-tetradecanoylcarnitine), accompanied by decreased acylcarnitine esters, were found in the pectoralis major muscle of broilers affected with WS (Boerboom et al., 2018). Whether the accumulation of fatty acids and their derivatives are the results of defects of mitochondrial beta-oxidation or peroxisomal beta-oxidation, this phenomenon points to hypoxia-induced impairment of mitochondria in WS muscle (Prisco et al., 2021).

WS severity has been correlated with mitochondrial alterations, suggesting that mitochondrial dysfunction increases as WS progresses (Prisco et al., 2021). Increases in fatty acids and their derivatives in muscles can induce Ca^{2+} overload and are seen to cause a cascade of molecular processes (Lake and Abasht, 2020). Boerboom et al. (2018) pointed out that poultry breast muscle is mainly composed of the fast-twitch type IIb, which contains low numbers of mitochondria, making it more susceptible to damage from prolonged anaerobic metabolisms. This may explain the prevalence of WS incidence in the breast muscle.

Lake and Abasht (2020) have proposed fat and glucose metabolic syndromes as the etiological causes of growth-related myopathies. Accumulated lipids in the muscle (i.e. lipotoxicity) exert cellular stress on cells and further suppress glycolysis and gluconeogenesis in affected birds. Instead of an increased blood glucose level or increased stored glycogen in the liver, glucose molecules are re-routed through alternative metabolic pathways (e.g. the pentose phosphate and aldose reductase pathways) as well as advanced glycation of proteins and lipids. This aligns well with low glycogen content, reduced glycolytic potential and increased glycated proteins observed in broilers with myopathic lesions.

The metabolic products of these pathways can induce pathological effects, resulting in tissue injury with chronic inflammation and myopathies similar to muscles in diabetic mammalian species. Accumulated lipids can enhance oxidative stress through lipid peroxidation of fatty acids. Increased ROS and oxidation of proteins found in WS muscle (Praud et al., 2020) contribute to muscle damage. Differential expressions of several genes (e.g. muscle-specific glutathione-S transferase and nitric oxide synthase 3 (*NOS3*)) that play roles in antioxidant mechanisms are associated with increasing WS severity (Malila et al., 2019, 2020). In addition, while *NOS3* abundance increased, suggesting increasing blood flow in affected birds (Malila et al., 2020), the formation of nitric oxide (NO), known to raise vasodilation and blood flow, seems to be outpaced by the conversion of L-arginine to citrulline at a metabolic level (Boerboom et al., 2018). These findings support compromised antioxidant activity in the WS-affected muscle.

Muscle damage is followed by the onset of apoptosis aimed at removing damaged cells through caspase activity. However, programmed cell death is altered in severe WS muscle (Marchesi et al., 2019; Malila et al., 2020). Prolonged or incomplete containment and destruction of damaged cells during the apoptotic process can merge into secondary necrosis to trigger pro-inflammatory and pathological responses by phagocytes (Gordon and Plüddemann, 2018). Macrophages and cluster of differentiated (CD)8-positive T lymphocytes with a small number of Bu-1-positive B cells infiltrated in damaged cells of WS-affected birds (Prisco et al., 2021). This indicates an inflammatory response to hypoxia-induced muscle injury. In addition, Prisco et al. (2021) found that genes encoding cytokines, including interleukin-6 (IL-6) and IL-17, and lipopolysaccharide-induced tumor-necrosis factor alpha factor, were overexpressed in correlation with WS severity. Their findings indicate an enhanced response of T helper 17, potentially for angiogenesis, repair and remodeling after muscle ischemia (Prisco et al., 2021).

In response to muscle injury, pro-inflammatory cytokines recruit immune cells to the lesion site, thus promoting the migration and proliferation of satellite cells (Domingues-Faria et al., 2016). Satellite cells begin to fuse with existing myofibers or synthesize the new fibers to repair muscles. Sufficient vascularization is crucial during muscle regeneration for simultaneous formation of muscle and blood vessels. However, as capillary supply became limited in the muscle of birds with high-breast yield, satellite cell-mediated muscle regeneration is impaired. Satellite cells are converted into adipocytes accompanied by the replacement of muscle fibers with collagen tissue, resulting in the phenotypic abnormalities of WS meat (Biferali et al., 2019).

WS was suggested to be an early pathological stage of WB (Kuttappan et al., 2016) as it shares histological lesions with WB, with WS lesions detected as early as 16 days post-hatch (Griffin et al., 2018). However, some of the genes associated with glucose metabolisms, i.e. *LDHA, PFKFB4* and *PHKB*, and antioxidant mechanisms, i.e. glutathione-S transferase A4 (*GSTA4*), showed opposite changes in the muscle affected with WS compared to WB (Malila et al., 2019). Upregulation of genes encoding fatty acid-binding protein 4 (*FABP4*), a lipid transport in adipocytes, and cluster of differentiation 36 (*CD36*), a cell surface protein importing fatty acids inside cells, was identified only with WS, corresponding to high fluorescent signals of Bodipy 493, the marker of adiposis, in WS muscle samples (Praud et al., 2020), suggesting the molecular deviation between the two myopathies.

4 Factors contributing to the development of white striping

4.1 Genetic factors

There is growing evidence of an association between WS and artificial selection for modern meat type chickens and turkeys (Kuttappan et al., 2012a, 2013c; Lorenzi et al., 2014; Trocino et al., 2015; Alnahhas et al., 2016; Sanchez Brambila et al., 2016; Daughtry et al., 2017; Malila et al., 2018; Barbut, 2019; Petracci et al., 2019). Research suggests greater WS prevalence and severity in high-yielding strains compared to standard- or low-yielding strains (Kuttappan et al., 2012a; Lorenzi et al., 2014; Trocino et al., 2015).

While both genetic selection and nutrition have contributed to the productivity of modern broilers, 85–90% of high growth efficiency has been attributed to genetics (Zuidhof et al., 2014). From 1957 to 2005, growth of one commercial broiler (Ross 308) increased by over 400%, with a concurrent 50% reduction in feed conversion ratio, corresponding to a compound annual rate of increase in 42-day live body weight of 3.30% (Zuidhof et al., 2014). Over this period, the yield of the pectoralis major muscle, an economically important cut, increased in 42-day-old Ross 308 broilers by 79% in males and 85% in females. Average live weights of broilers have increased consistently to between 2.7 kg and 4.2 kg (Owens and Vieira, 2012).

However, a strong pressure on growth rate and muscle hypertrophy exerts challenges on birds. Hypertrophied muscle fibers are more prone to damage (Velleman, 2015), with impaired ability to regenerate due to exhaustion of satellite cell pools (Daughtry et al., 2017). Insufficient vascularization, as hypertrophied muscle fibers occupy spaces for capillaries, appears to trigger oxidative stress-induced myodogeneration with the onset of WS.

Alnahhas et al. (2016) has analyzed traits related to growth performance and meat quality indices, highlighting genetics as a major determinant of WS. WS is more closely related to muscle development than to overall growth. Kuttappan et al. (2013c), Bowker et al. (2019) and Malila et al. (2018) also found that WS was induced by disproportionately large breast muscle mass rather than absolute body weight. Alnahhas et al. (2016) also reported the positive genetic association between WS and ultimate meat pH, supporting a possible relationship between the ability of muscle to store energy as carbohydrates and its likelihood of developing WS. These authors have suggested breeding selection as an efficient means of reducing WS incidence in broiler chickens.

Following Alnahhas et al. (2016), Lake et al. (2021) estimated genetic parameters for growth-related myopathies using mixed male and female crossbred broiler population (Cobb 500, n = 1193) and reported heritability of 0.50 for WS. Based on a genome-wide association study, a total of 18 single-nucleotide polymorphisms (SNPs) were associated with WS, of which 17 were located on chicken chromosome 5 with one SNP on chromosome 11. The majority of the WS quality trait loci (QTL)-rich region of chicken chromosome 5 has highly conserved synteny with human chromosome 11. The area on human chromosome 11 is associated with a high number of growth and metabolic disorders, particularly insulin secretion. The findings of Lake et al. (2021) strengthen the association between WS and dysregulated metabolisms of macronutrients, lipids and glucose in particular, along with impaired muscle growth, development and repair (Abasht et al., 2016; Papah and Abasht, 2019; Malila et al., 2019; 2020; Lake and Abasht, 2020).

Using routinely recorded data from the Aviagen breeding program, Bailey et al. (2015, 2020) estimated the genetic basis for breast myopathies in pure commercial broiler lines. By accounting for the common maternal environment effect, they reported the estimated heritability of body weight, processing body weight and breast yield for WS at ≤0.33 and 0.25 in their first (Bailey et al., 2015) and second studies (Bailey et al., 2020), respectively. The authors concluded a low-to-moderate genetic influence for WS development. However, they still found heritability for WS to be greater than those for the other breast myopathies.

Variation in estimated heritability may be due to different populations and environments (Lake et al., 2021). Estimates for heritability of WS for purebred commercial broiler lines were lower in the studies of Bailey et al (2015, 2020) compared to other hybrid commercial broiler populations studied by Lake et al. (2021). Low-to-moderate heritability for growth-related myopathies indicates that both genetic and environmental factors are important. This may explain why post-hatch nutritional and management strategies implemented by industry alone have not completely eliminated the incidence of growth-related myopathies (Ferreira et al., 2020).

4.2 Nongenetic factors

Overnutrition has been proposed as a key factor triggering growth-related myopathies through lipotoxicity and glucotoxicity (Lake and Abasht, 2020). Negative dietary effects may be exacerbated by metabolic shifts through breeding selection. High-energy diets (Kuttappan et al., 2012a; Kindlein et al., 2017) have been shown to increase WS severity. Increasing digestible lysine, associated with muscle hypertrophy, during grower (from 12 days to 28 days of age) or finishing (from 28 days to 42 days of age) phases has been linked to higher WS prevalence in Cobb 500 × Cobb broilers (Cruz et al., 2017). The negative effects were more pronounced when digestible lysine was supplemented during the finishing period compared to the growing phase.

Suboptimal environmental conditions during early embryonic muscle development also affect the incidence and severity of growth-induced myopathies in poultry. Increasing WS incidence due to thermal stress during incubation was experimentally demonstrated either:

- continuously (37.8°C, embryonic day 1 to day 21) (Livingston et al., 2019c) or
- a short-period exposure (35.5°C to 37.8°C, embryonic day 12 to day 21) (Da Costa et al., 2016) or
- 39°C, embryonic day 24 to day 28 (Christensen et al., 2007).

However, Clark et al. (2017) found decreased numbers of degenerative muscle fibers when Ross 708 broilers were incubated under 39.5°C for 12 h from embryonic day 14 to day 18. Nyuiadzi et al. (2020) reported slightly less WS occurrence in 40-day-old Ross 308 broilers when embryos were exposed to a short cold treatment (15°C, 81% relative humidity (RH), 30 min) from embryonic day 18 to day 19, followed by optimal brooding conditions (32°C at day 1 post-hatch). No significant impacts on feed efficiency and total yield were observed. However, WS incidence in male broilers was tripled when cold-treated birds were exposed to initial low brooding temperatures (29°C). Increased WS severity accompanied by hypoxia in late incubation (the last 72–96 h) was reported in embryos of Ross 708 broilers from eggs subjected to prolonged storage (>5 days) (Livingston et al., 2019c). Oviedo-Rondon et al. (2020) identified overheating (>37°C) and reduced egg air exchange (oxygen concentration <18%) as the most common problems in industrial conditions. High temperatures and low oxygen concentrations during the last phase of incubation created hypoxia, which interfere with blood flow within the embryos, and disturbed metabolism, especially glucose and insulin balance, during the embryonic stage (Loyau et al., 2015). These conditions contribute to post-hatch muscle tissue architecture with limited vascularity and oxygenation capacity, creating a cellular environment that increases susceptibility to the development of WS in later life.

The influence of environmental conditions on WS may relate to their negative impact on poultry satellite cells. Satellite cells are the only source contributing to poultry post-hatch hypertrophic growth and muscle regeneration. Extreme environmental conditions, e.g. thermal manipulation, low oxygen availability and improper lighting, can change the quantity, ability to function and metabolic activities of satellite cells (Clark et al., 2017). Satellite cell numbers and activity are strongly influenced by incubation conditions and early life (Clark et al., 2016; Halevy, 2020). Thermal challenge during brooding and delayed access to feed, or inadequate amino acids at hatch, suppress proliferation of satellite cells and increase the level of fat deposition in muscles (Velleman et al., 2014; Clark et al., 2016; Patael et al., 2019; Xu et al., 2021), leading to characteristic signs of WS.

The effects of age on WS incidence have also been addressed (Kuttappan et al., 2017; Radaelli et al., 2017; Malila et al., 2018). However, it appears that age-related WS is due mainly to heavier weight and larger breast muscle size as birds age (Kuttappan et al., 2017). Clark et al. (2017) found that chickens that hatched early were associated with accelerated muscle development and thus showed a higher occurrence of moderate-to-severe degenerative changes in the breast muscle compared to those that hatched later.

5 Potential strategies for reducing white striping: feed modification

Many nutritional strategies have aimed at eliminating WS (Tables 2 and 3). Those strategies include restriction of feed intake or reduction of nutrient density to modify the growth curve of animals. There has been extensive discussion of supplementation with specific nutrients or additives either to modify biological characteristics associated with WS or to improve the overall health of birds.

5.1 Restriction of feed and nutrient density

Key findings of current research on nutritional modification to mitigate WS are summarized in Table 2. Meloche et al. (2018a) showed that decreasing feed intake (85%, 90% and 95% of ad libitum feed) throughout the life of broilers both decreased the growth and severity of WS in groups of broilers slaughtered at 33, 43 and 50 days of age. The most beneficial impacts were detected in 43-day broilers, suggesting that reducing the severity of WS depends on the timing of implementation and the age of the birds. As birds age, advanced myopathy may no longer be amenable to nutritional interventions. Quantitative feed restriction during the first growth period may increase necrosis and fat deposition in pectoralis major, which may increase the incidence of WS (Velleman et al., 2010). Trocino et al. (2015) reported a 10% increase in the occurrence of WS in

Table 2 Influences of feed control and restricted energy and amino acid (AA) density on white striping (WS) severity

Reference	Animal (slaughter age)	Total n (tested n for myopathies)	Modification and administration period	Treatment or dosage	Grading	Key findings associated with WS
Kuttappan et al. (2012a)	Cobb 500 (54 days)	280 (280)	Energy density Starter (0–18 days) Grower (18–32 days) Finisher (32–54 days)	Two treatments: (1) Low-energy diet = 0.5% supplemental poultry oil (2) High-energy diet (HED) = 6% supplemental poultry oil	Three-point scale: Normal = no lines MOD = line thickness up to 1 mm SEV = line thickness >1 mm	• HED-fed birds produced a greater ($P < 0.05$) percentage of SEV fillets. • Fillet weight and yield (percent of live weight) increased ($P < 0.05$) as the degree of white striping increased from normal to severe.
Meloche et al. (2018a)	Yield Plus × Ross 708, male (33, 43 and 50 days)	700 (112)	Quantitative nutrient allocation From day 2 onward	Four treatments: (1) Ad libitum (2) 95% of TRT1 intake (3) 90% of TRT1 intake (4) 85% of TRT1 intake	Normal = no lines Mild = visible striping <50% of the total fillet surface area or <2 mm wide on average Severe = the striping exceeded the limits of 'mild'	• Linear decreases ($P \leq 0.01$) in WS severity were observed at 33, 43 and 50 days with decreasing feed allocation. • WS severity at 43 and 50 days, decreased ($P \leq 0.05$) quadratically with decreasing feed allocation.

(*Continued*)

Table 2 *(Continued)*

Reference	Animal (slaughter age)	Total *n* (tested *n* for myopathies)	Modification and administration period	Treatment or dosage	Grading	Key findings associated with WS
Meloche et al. (2018b)	Yield Plus × Ross 708, male (36 days)	1386 (1134)	Dietary energy and AA density Starter (8–14 days) Grower (15–24 days) Finisher I (26 to 35 days)	Seven treatments: (1) 100% of recommended AA density and energy density throughout the experiment (2) 95% of TRT1 until 14 days of age, then as TRT1 (3) 95% of TRT1 until 24 days of age, then as TRT1 (4) 95% of TRT1 throughout the experiment (5) 90% of TRT1 until 14 d of age, then as TRT1 (6) 90% of TRT1 until 24 days of age, then as TRT1 (7) 90% of TRT1	Normal = no lines Mild = visible striping <50% of the total fillet surface area or <2 mm wide on average Severe = the striping exceeded the limits of 'mild'	• There was no reduction in the incidence of any given scoring category for WS relative to the control among the birds slaughtered at 36 days of age.
	Yield Plus × Ross 708, male (49 days)	1386 (1134)	Starter (8–14 days) Grower (15–24 days) Finisher I (26–42 days) Withdrawal (43–48 days)			• At 49 days, broilers fed at 90% of the recommended dietary density (TRT7) had lower incidence of severe WS, potentially due to their reduced growth efficiency.
Maynard et al. (2019)	Cobb MV × 700 broiler, male and female (47 days)	864 (288)	Metabolize energy and AA density Finisher (29–36 days) Withdrawal (37–46 days)	Three energy concentrations (low, medium and high) 2 AA density (medium and high)	Three-point scale: Normal = no lines MOD = line thickness up to 1 mm SEV = line thickness > 1 mm	• No significant effects of either factor on WS incidence.
Maynard et al. (2020)	Cobb MV × 700, male and female (47 days)	728 (288)	AA density Varied across feeding period Starter (0–14 days) Grower (15–28 days) Finisher (29–36 days) Withdrawal (37–46 days)	Six treatments: Combinations of low-, medium-, and high-AA density diets were varied across four feeding phases to evaluate the effect of six AA feeding regimens (HHHH, HHHM, HHMM, HMMM, MMMM and HHLL)	Three-point scale: Normal = no lines MOD = line thickness up to 1 mm SEV = line thickness > 1 mm	• No significant effects of AA density or interaction between AA and sex of the broilers on WS incidence ($P > 0.05$).

Pekel et al. (2020)	Ross 308 male (49 days)	288 (108)	AA density and stocking density (SD) Starter (0–15 days) Grower (16–30 days) Finisher (31–49 days)	3 × 2 factorial design AA densities -Normal, 10% or 20% lower than normal SD - high 35 kg/m^2 or low 26 kg/m^2	Three-point scale: Normal = no lines MOD = line thickness up to 1 mm SEV = line thickness > 1 mm	• Reducing dietary AA density linearly decreased WS severity as growth performance was decreased. • SD did not affect WS occurrence.
Butler et al. (2020)	Cobb MV × Cobb 700, male (56 days)	3600 (288)	Increasing digestible lysine (dLys) and metabolized energy (ME) Starter (0–12 days) Grower (12–26 days) Finisher (26–35 days) Withdrawal (35–55 days)	Six treatments with increasing dLys and ME Starter (1.2–1.38%, 2996–3084 kcal/kg) Grower (1.07–1.23%, 3084–3194 kcal/kg) Finisher (0.95–1.09%, 3172–3282 kcal/kg) Withdrawal (0.9–1.06%, 3172–3282 kcal/kg)	Four-point scale 0 = no incidence 1 = mild incidence 2 = moderate incidence 3 = severe incidence	• No significant effects on WS incidence.
Ahsan and Cengiz (2020)	Ross 308, male (49 days)	390 (390)	Lysine (Lys) ME and AA Grower (11–24 days) Grower and finisher (11–39 days)	Five treatments: (1) Control (2) 15% reduced Lys in grower (3) 15% reduced Lys in grower and finisher (4) 5% reduced ME and AA in grower (5) 5% reduced ME and AA in grower and finisher	Three-point scale: Normal = no lines MOD = line thickness up to 1 mm SEV = line thickness > 1 mm	• Feeding regimes based on treatment group 2 (reduced Lys in grower) and 3 (reduced Lys in grower and finisher) reduced the overall WS incidence (frequency of total lesions) compared with the other groups ($P < 0.001$).

(*Continued*)

Table 2 (*Continued*)

Reference	Animal (slaughter age)	Total *n* (tested *n* for myopathies)	Modification and administration period	Treatment or dosage	Grading	Key findings associated with WS
Cruz et al. (2017)	Slow-feathering Cobb 500 × Cobb broilers, male (35 days)	600 (96)	dLys (from 12–28 days)	Six treatments with increasing levels of dLys in 0.08% increments from 0.77% to 1.17%	Three-point scale: Normal = no lines MOD = line thickness up to 1 mm SEV = line thickness > 1 mm	• WS occurrence ranged from 18.8% to 56.3% among treatments. • The broilers fed a diet with 1.01% dLys developed WS severity at the greater extent than those in the control (0.77% dLys), while the birds fed the diets at other levels showed immediate WS scores between those two groups.
	Slow-feathering Cobb 500 × Cobb broilers, male (42 days)	600 (96)	From 28–42 days	Six treatments with increasing levels of dLys in 0.08% increments from 0.68% to 1.07%		• WS occurrence ranged from 58.1% (control, 0.77% dLys) to 100% (0.84% dLys) among treatments. • Increasing dLys level increased the occurrence and severity of WS lesions ($P < 0.01$), with the greatest WS score in the group of 0.92% dLys, corresponding to increased body weight and breast yield.
Meloche et al. (2018c)	Yield Plus × Ross 708, male (48 days)	1386 (1134)	Reduced dLys Grower I (12–18 days) Grower II (19–26 days)	Seven treatments: (1) 100% of dLys recommendations (2) 85% of TRT1 for grower I (3) 85% of TRT1 for grower II	Three-point scale: 0 = none 1 = mild (<50%, line thickness up to 2 mm) 2 = severe (>50%, line thickness > 2 mm)	• TRT7 (75%, grower I and grower II) reduced ($P < 0.05$) the incidence of severe WS relative to the control. • Broilers assigned to TRT7 resulted in reduced breast weights and yields, which likely resulted in the observed decrease in WS.

				(4) 85% of TRT1 for grower I and grower II (5) 75% of TRT1 for grower I (6) 75% of TRT1 for grower II (7) 75% of TRT1 for grower I and grower II		
	Yield Plus × Ross 708, male (62 days)	720 (480)	Reduced dLys Grower (12-28 days) Finisher I (29-40 days)	Four treatments: (1) 100% of dLys recommendations (2) 85% of TRT1 for grower (3) 85% of TRT1 for finisher I (4) 85% of TRT1 for grower and finisher I	Three-point scale: 0 = none 1 = mild (<50%, line thickness up to 2 mm) 2 = severe (>50%, line thickness > 2 mm)	• The severe WS incidence in TRT2 and TRT4 was 22.3% and 17.8%, respectively, which was significantly lower than the control (38.3%).
Soglia et al. (2021b)	Ross 308, male (49 days)	945 (840)	Depletion of synthetic lysine Starter (0-9 days) Grower I (10-20 days) Grower II (21-34 days) Finisher (35-49 days)	Three treatments: (1) CONT - dLys 1.15% and 1.05% in grower I and II, respectively (2) Grower I - dLys 1.00% and 1.05% in grower I and II, respectively (3) Grower I + II - dLys 1.00% and 0.91% in grower I and II, respectively	Three-point scale: Normal = no lines MOD = line thickness up to 1 mm SEV = line thickness > 1 mm	• No significant effects on WS incidence.

CONT = control, MOD = moderate, SEV = severe, TRT = treatment.

Table 3 Effects of supplements on white striping (WS) severity

Reference	Animal (slaughter age)	Total *n* (tested *n* for myopathies)	Modification and administration period	Treatment or dosage	Grading	Key findings associated with WS
Zampiga et al. (2018)	Ross 308 male (43 days)	1170	Crystalline L-arginine (purity 99%, Barentz, Hoofddorp, The Netherlands) Whole experimental period	Two treatments: (1) Control (2) Crystalline L-arginine at 1.20 g/kg, 1.15 g/kg, 1.10 g/kg and 0.95 g/kg in starter, grower I, grower II and finisher feeding phase, respectively	Three-point scale: Normal = no lines MOD = line thickness up to 1 mm SEV = line thickness > 1 mm	• No significant effects on WS incidence and severity.
Zampiga et al. (2019)	Ross 308 male (43 days)	1755 (450)	Digestible arginine-to-lysine ratio Whole experimental period	Three treatments: (1) Control (2) 20% increased L-arginine (3) 30% increased L-arginine	Three-point scale: Normal = no lines MOD = line thickness up to 1 mm SEV = line thickness > 1 mm	• Supplementation of arginine at either level showed a significantly higher percentage of breast without WS compared to control (28% and 41% vs 17%, respectively; $P < 0.001$). • TRT3 resulted in the lowest percentage of severe WS (11% vs 27% and 31%, respectively, for TRT3, TRT2 and control; $P < 0.001$).
Livingston et al. (2019a)	Male broilers (45 days)	512 (64)	Dietary glutamine and arginine Starter (8–14 days) Grower (15–44 days)	2 × 2 factorial of glutamine (Gln–, Gln+) and arginine (Arg–, Arg+) Gln+: added 1% glutamine Arg+: added 0.25% arginine	Four-point scale: 1 = no line 2 = line thickness < 1 mm 3 = line thickness 1–2 mm 4 = line thickness > 2 mm	• Supplementation of Gln increased WS incidence. • The change in WS scores over time demonstrated the greatest increase in WS severity (0.54, $P < 0.01$) occurred between 35 and 42 days.

Albrecht et al. (2019a)	Ross 308, male (35 days)	800 (210)	Type and concentration of supplemented methionine	Type of methionine: DL-methionine (DLM) or liquid DL-2-hydroxy-4-methylthiobutanoic acid (DL-HMTBA) Concentrations: Supplemented at 0.04%, 0.12% or 0.32% Seven treatments: (1) Control (basal diets) (2) DLM 0.04% (3) DLM 0.12% (4) DLM 0.32% (5) DL-HMTBA 0.04% (6) DL-HMTBA 0.12% (7) DL-HMTBA 0.32%	Three-point scale: Normal = no lines MOD = line thickness up to 1 mm SEV = line thickness > 1 mm	• No significant effects of methionine type. • Methionine concentration was positively but weakly (Spearman's coefficient = 0.159), correlated with occurrence and severity of WS abnormality.
Sachs et al. (2019)	Cobb 500 (46 days)	120 (30)	Natural methionine Whole experimental period	Two treatments: (1) Control (starter = 0.50%, grower = 0.47%, finisher = 0.46%) (2) Alternative starter = 0.49%, grower = 0.44%, finisher = 0.47%)	Three-point scale: 1 = Normal or mild (absent) 2 = Moderate (observed but <1 mm thick) 3 = Severe (>1 mm thick)	• An increase in mild WS and a decrease in severe WS was observed in treatment group compared to the group fed a control diet.
Khan et al. (2021)	Cornish cross broiler (43 days)	144 (54)	Omega-3 (docosahexaenoic acid (DHA)-rich microalgae) methionine (from 10 to 42 days)	• Three treatments:Control • 2% algae • 2% algae + 100% more methionine	Three-point scale: Normal = no lines MOD = line thickness up to 1 mm SEV = line thickness > 1 mm	• Methionine supplementation in DHA-rich microalgae-based diets enriches chicken tissues with ω-3 FA, and reduces breast muscle myopathic lesions and WS without affecting productivity.

(Continued)

Table 3 (*Continued*)

Reference	Animal (slaughter age)	Total *n* (tested *n* for myopathies)	Modification and administration period	Treatment or dosage	Grading	Key findings associated with WS
Córdova-Noboa et al. (2018)	Ross 708, male (56 days)	1280 (256)	Guanidinoacetic acid (GAA) and poultry by-product (PBP) Whole experimental period	2 × 2 factorial GAA - with or without GAA supplementation (0% or 0.06%) PBP - with or without the inclusion of PBP (0% or 5%)	Three-point scale: Normal = no lines MOD = line thickness up to 1 mm SEV = line thickness > 1 mm	• Overall average WS scores were not affected ($P > 0.05$) either by dietary PBP inclusion or by GAA supplementation.
Kuttappan et al. (2012b)	Male broilers (49 days)	2120 (200)	Vitamin E DL-alpha tocopherol acetate) Starter (1–18 days) Grower (19–32 days) Finisher (33–49 days)	15 IU/kg, 50 IU/kg, 100 IU/kg, 200 IU/kg and 400 IU/kg	Three-point scale: Normal = no lines MOD = line thickness up to 1 mm SEV = line thickness > 1 mm	• No significant effects on WS incidence.
Wang et al. (2020)	Ross 708, male (58 days)	210 (210)	Vitamin E and omega-3 Starter (0–10 days) or grower (11–24 days)	Seven treatments: (1) Control: VE 10 IU/kg and ω-6/ω-3 = 30:1 (2) Starter VE: VE 200 IU/kg (3) Starter ω-3: ω-6/ω-3 = 3:1 (4) Starter VE and ω-3: VE 200 IU/kg and ω-6/ω-3 = 3:1 (5) Grower VE: VE 200 IU/kg (6) Grower ω-3: ω-6/ω-3 = 3:1 (7) Grower VE and ω-3: VE 200 IU/kg and ω-6/ω-3 = 3:1	Four-point scale 0 = none 1 = mild 2 = moderate 3 = severe	• High concentration of VE (200 IU/kg) in the starter or grower diets tended to reduce moderate and severe WS but increase the proportion of normal and mild WS. • The starter VE group showed more of an effect on reducing the severity of WS compared with the grower VE group. • Omega-3 fatty acids increased the severity of WS in both the starter and grower diets.

Sirri et al. (2016)	Ross 308 male (51 days)	3600 (2160)	Trace minerals (TMs) (Zn, Mn and Cu) Whole experimental period	2 × 2 factorial TM source - organic TM vs inorganic TM Dosage - low (LO) vs high (HI) LO (starter-grower-finisher) Zn = 40-32-32 mg/kg Mn = 40-32-32mg/kg Cu = 10-8-8 mg/kg HO (starter-grower-finisher) Zn = 60-50-50 mg/kg Mn = 60-50-50 mg/kg Cu = 15-12.5-12.5 mg/kg	Three-point scale: Normal = no lines MOD = line thickness up to 1 mm SEV = line thickness > 1 mm	• No significant effects on WS incidence.
Cemin et al. (2018)	Slow-feathering Cobb × Cobb 500, male (42 days)	1872 (432)	Zn-L-selenomethionine (Availa-Se, Zinpro Corporation, Eden Prairie, MN) Whole experimental period	Six levels of supplementation of Se (0 ppm, 0.2 ppm, 0.4 ppm, 0.6 ppm, 0.8 ppm or 1.0 ppm)	Three-point scale: Normal = no lines MOD = line thickness up to 1 mm SEV = line thickness > 1 mm	• Broilers fed control showed a higher percentage of normal fillets compared to other Se supplementation levels. • Increasing Se supplementation to reach maximum growth performance led to higher degrees of WS severity.
Estevez and Fetracci (2019)	Ross 308, male and female (42 days)	360 (64)	Magnesium (Optibreast®) Grower (14–28 days) Finisher (28–42 days)	3000 mg/kg	Three-point scale: 1= normal 2 = line thickness up to 1 mm 3 = line thickness > 1 mm	• Mg supplementation reduced WS incidence of WS to almost 50% of the occurrence found in control. • Mg supplementation protected against protein oxidation in liver and plasma of broilers.

(Continued)

Table 3 (*Continued*)

Reference	Animal (slaughter age)	Total *n* (tested *n* for myopathies)	Modification and administration period	Treatment or dosage	Grading	Key findings associated with WS
Philpot et al. (2020)	Yield Plus × Ross 708, male (54 days)	1320 (840)	Copper Finisher I (29–40 days) Finisher II (41-53 days)	Five treatments: (1) Finisher I/finisher II = 0/0 (2) 135/0 mg/kg (3) 270/0 mg/kg (4) 135/135 mg/kg (5) 270/270 mg/kg	Four-point scale 0 = none 1 = mild 2 = moderate 3 = severe	• No significant effects on WS severity. • High copper concentrations may increase breast weight and yield of broilers processed at 54 days of age.
Ribeiro et al. (2020)	Cobb slow, male (42 days)	1800 (120)	Vitamin programs - commercial vitamin program and Optimum Vitamin Nutrition (OVN™, DSM) Mineral sources - Inorganic (sulfates) and carbo-amino-phospho-chelate (CAPC) (Tortuga Minerals) Whole experimental period	Four treatments: (1) Commercial vitamin programs + source of inorganic minerals (2) Optimum vitamin programs + source of inorganic minerals (3) Commercial vitamin programs + CAPC (4) Optimum vitamin programs + CAPC	Three-point scale: Normal = no lines MOD = line thickness up to 1 mm SEV = line thickness > 1 mm	• No significant effects on WS incidence.
Livingston et al. (2019b)	Ross YPM × Ross 708, male (35 and 43 days)	320 (48)	Dietary potassium (K) and available phosphorus (AvP) Starter (0–16 days) Grower (17–35 days) Finisher (36–43 days)	2 × 2 factorial AvP - fixed (0.45%) or declined (0.45%, 0.40% and 0.35% in the starter, grower and finisher diets, respectively). K - basal levels (K-) (0.86%, 0.77% and 0.68%) and increased levels (K+) (1.01%, 0.93% and 0.88%) in starter, grower and finisher diets, respectively	Four-point scale: 1 = normal (no line) 2 = mild (line thickness < 1 mm) 3 = moderate (line thickness 1-2 mm) 4 = severe (line thickness >2 mm)	• There was an interaction between K and AvP at 35 days on WS scores ($P < 0.01$), while no significant main effects were observed. • At 35 days, broilers fed with K-/declined AvP and K+/fixed AvP showed lower average WS severity scores than the other groups. • Among 45d broilers, no significant effects on WS incidence was found.

Joardar et al. (2020)	Ross 708, male (35 days)	384 (64)	Potassium (K) Starter (0-16 days) Grower (17-33 days)	2 × 2 factorial Limestone particle size - coarse (0.9 mm) and fine (0.2 mm) K - basal level (K-) and 0.2% added (K+)	Four-point scale: 1 = normal (no line) 2 = mild (line thickness < 1 mm) 3 = moderate (line thickness 1-2 mm) 4 = severe (line thickness > 2 mm)	• No significant effects on WS incidence.
Bodle et al. (2018)	Broiler, male (46 days)	1980 (462)	Dietary alteration Starter (1-12 days) Grower (13-24 days) Finisher (25-36 days) Withdrawal (37-45 days)	Six treatments: (1) Control (2) Increasing the ratio of digestible arginine: digestible lysine ranged from 113 to 126 (3) Supplementing TRT1 with 94.4 mg vitamin C/kg feed (4) Doubling the vitamin pack inclusion rate (5) Reducing the digestible amino acid density of only the grower phase by 15% and feeding the same TRT1 starter, finisher and withdraw diets (6) Combining the four strategies just mentioned	Three-point scale: Normal = no lines MOD = line thickness up to 1 mm SEV = line thickness > 1 mm	• No significant effects on WS incidence.

(Continued)

Table 3 (*Continued*)

Reference	Animal (slaughter age)	Total *n* (tested *n* for myopathies)	Modification and administration period	Treatment or dosage	Grading	Key findings associated with WS
Sobotik et al. (2018)	Cobb 500, male (49 days)	648 (288)	Phosphatidic acid (Mediator® PA, Chemi Nutra, Austin, TX, USA) Whole experimental period	Three treatments: (1) 5 mg/bird/day (2) 10 mg/bird/day (3) 15 mg/bird/day	Three-point scale: Normal = no lines MOD = line thickness up to 1 mm SEV = line thickness > 1 mm	• No significant effects on WS scores or in carcass or breast yield.
Pascual et al. (2020)	Ross 308, male and female (45 days)	768 (192)	Commercial microencapsulated sodium butyrate Whole experimental period	500 mg/kg; 30% Na-butyrate	Presence or absence (% occurrence)	• No significant effects on WS incidence.
Mudalal et al. (2021)	Ross 500 (34, 41 and 48 days)	504 (not specified)	Herb extract (HE) Whole experimental period	Three treatments: (1) Control (2) 0.2 ml/L in drinking water (3) 0.3 ml/L in drinking water	Four groups: Normal WS Wooden WS/wooden	• HE supplementation significantly reduced WS and WS/WB incidence in the broilers, particularly those slaughtered at 34 days of age.

FA = fatty acid, TRT = treatment, VE = Vitamin E.

broilers who received a restricted diet (80% of ad libitum) between 13 and 21 days of age as compared with birds fed ad libitum. During the refeeding period, Trocino et al. (2015) observed a compensatory growth in combination with poor muscle structure and more prevalent WS in the restricted birds.

In contrast, Radaelli et al. (2017) applied a similar feed restriction regime and reported an overall decline in muscle fiber degeneration in the feed-restricted birds compared to those fed ad libitum. The largest difference in fiber degeneration frequency between those two groups was at day 21 of age (6.3% vs 50%), the last day of feed restriction, but, from day 28 onward, this difference was reduced (62.5% vs 87.5%) and eventually disappeared on day 46 (96.9% vs 96.9%). This indicates the effectiveness of feeding restriction in controlling muscle degeneration only during the restriction period. Therefore, to maintain the benefit of the feed restriction in limiting the occurrence of growth-related myopathies, a refeeding program after feed restriction is crucial.

Apart from quantitative feed intake, restriction of energy and amino acid density intake has also been studied. Meloche et al. (2018b) reported that feeding 90% of recommended energy and amino acid density levels throughout life significantly decreased breast meat yield, hence reducing severe WS cases in Yield Plus × Ross 708 (49 days of age). Similar results were reported by Pekel et al. (2020). In contrast, Maynard et al. (2019; 2020) found that, when Cobb MV × Cobb 700 broilers (47 days of age) had dietary energy and amino acid density restriction either during the finishing and withdrawal periods (Maynard et al., 2019) or throughout their life (Maynard et al., 2020), there were no significant effects on the overall growth performance or reduction of WS. The authors suggested the differences in breeds to be the cause of this discrepancy. Aviagen conducted an internal trial along similar lines to Meloche et al. (2018b). This study found that the modern broilers were able to compensate for 5% to 10% reduction in energy by eating more feed and still achieved the same final body weight as the control group, with no apparent impact in reducing WS severity (Aviagen, 2019).

Restriction of digestible lysine is a possible approach since chickens are unable to recognize and therefore compensate for reduced digestible lysine in feed. Reducing digestible lysine levels tended to decrease WS incidence and severity in male Ross 308 (Ahsan and Cengiz, 2020) and male Ross 708 (Meloche et al., 2018c). The effective dosages and administration period varied between those studies. Meloche et al. (2018c) associated decreasing WS severity with lower breast muscle weight and yield related, in turn, to a 75% intake of dietary lysine (compared to normal intake). However, Soglia et al. (2021b) reported no significant effects of digestible lysine restriction (by 13%) during either grower I or grower I and II on decreasing the incidence of WS among male Ross 308 broilers slaughtered at the age of 49 days. The effects of dietary lysine restriction on the occurrence and severity of growth-related abnormalities depend greatly on both the extent of its reduction and the duration of such reduction within the overall rearing period. If the duration of reducing digestible lysine is too long, growth performance and breast meat yield will be adversely affected (Meloche et al., 2018c).

5.2 Supplementation of nutrients and functional ingredients

The roles of several additional nutrients and bioactive compounds against the incidence and severity of WS have been studied (Table 3). The main aims are to alter biological changes associated with growth-related myopathies. Among impaired biological pathways associated with severe WS is the arginine-nitric oxide pathway (Boerboom et al., 2018). L-arginine serves as a substrate for nitric oxide synthase to be converted into L-citrulline and nitric oxide, a critical regulator of vascular homeostasis (Jahangir et al., 2009). Supplying additional L-arginine may augment nitric oxide production and improve vasodilation and hence blood flow in the muscle. However, the results are inconsistent. Zampiga et al. (2018) found no significant effects of L-arginine supplementation (1.20 g/kg, 1.15 g/kg, 1.10 g/kg and 0.95 g/kg in starter, grower I, grower II and finisher feeding phases, respectively) on WS incidence among Ross 308 broilers, corresponding to findings by Bodle et al. (2018). In contrast, an increase in digestible arginine-to-lysine ratio at a high dosage (30% more than the basal diet) reduced the occurrence of growth-related myopathies among male Ross 308 broilers without exerting adverse effects on growth performance and meat quality traits (Zampiga et al., 2019).

A possible explanation for the inconsistent results of L-arginine supplementation may be the metabolic fate of L-arginine. The great proportion is attributable to the synthesis of creatine rather than conversion into nitric oxide. Formation of creatine consumes high methyl groups and reduces available methyl groups for other biological reactions (Jahangir et al., 2009). It has been suggested that the addition of methionine and guanidinoacetic acid (GAA), the methyl donor and metabolic methyl receiver, respectively, in the formation of creatine, might restore L-arginine flux for nitric oxide formation, therefore attenuating the severity of growth-related myopathies. Favorable effects of methionine in decreasing WS severity have been reported (Albrecht et al., 2019a; Sachs et al., 2019; Khan et al., 2021). No significant effects of GAA against WS severity were found (Córdova-Noboa et al., 2018).

Dietary administration of antioxidants has been suggested as playing a role in reducing inflammation and oxidative stress, thus preventing myodegeneration. Among feed ingredients, vitamin C and vitamin E are widely known for their antioxidant effects. Results on the use of vitamin E are not consistent. Supplemental vitamin E in the form of DL-alpha-tocopherol acetate at concentrations ranging from 15 IU/kg, 50 IU/kg, 100 IU/kg, 200 IU/kg and 400 IU/kg did not reduce WS incidence among commercial broilers slaughtered at 49 days (Kuttappan et al., 2012b). However, Wang et al. (2020) found that the addition of 200 IU/kg of DL-alpha-tocopherol acetate significantly decreased WS severity among 58-day-old Ross 708 broilers. Vitamin E supplementation in starter diets (day 0 to 10 post-hatch) showed the best effects compared with grower diets (day 11 to 24 post-hatch). Supplementation of vitamin C (94.4 mg vitamin C/kg feed) showed negligible effects on WS severity (Bodle et al., 2018).

The trace mineral selenium (Se) is vital in antioxidant systems and potentially provides protective effects against WS by maintaining the redox state in muscle. However, Cemin et al. (2018) found that inclusion of organic Se increased growth performance and yields but also WS severity. Estevez and Petracci et al. (2019) revealed the beneficial effects of magnesium (Mg, 0.3%) against protein oxidation in the liver and plasma of commercial broilers fed with oxidized oil. WS incidence in the Mg-supplemented group was reduced to almost half of the control group.

Other nutrients and feed additives have been studied as a way to regulate immune and metabolic functions, balance gut health and improve overall bird health. These include vitamins (Ribeiro et al., 2020), minerals (Sirri et al., 2016; Livingston et al., 2019b; Philpot et al., 2020; Joardar et al., 2020), lipids (Sobotik et al., 2018) and herb extracts (Mudalal et al., 2021). However, the protective effects of those additives against WS are inconsistent. Livingston et al. (2019b) and Joardar et al. (2020) supplemented dietary electrolytes, including potassium (K), available phosphorous and limestone, to counter ion imbalances. The results, however, were inconsistent. While Joardar et al. (2020) reported no effects of supplemental K^+ on WS incidence among 35-day-old Ross 708 broilers, Livingston et al. (2019b) found that the addition of 0.2% K^+ in combination with 0.45% available phosphorous significantly reduced WS severity scores in 35-day-old broilers but with no effects in 45-day-old broilers.

Overall, feed modification strategies to mitigate WS incidence and severity appear to be more complex than expected. Most of the studies that reported beneficial effects on WS were at the expense of diminishing growth performance and yield. Although there have been promising results, further studies, particularly on the optimization of those practices, are still required before actual implication.

6 Potential strategies for reducing white striping: farm management, processing and breeding

6.1 Farm management

Although the farm environment is a factor in WS development, research on this topic is still limited. It has been suggested that temperature control, gas ventilation and lighting, as well as access to feed early post-hatch, must be well controlled to assure adequate growth and high-quality meat. Sufficient ventilation is required to minimize heat accumulation and assure air distribution. Although thermal manipulation during incubation can reduce growth-related myopathies, it also limits growth (Clark et al., 2017). Verhoelst et al. (2011) found that increasing levels of carbon dioxide concentration during early prenatal development (embryonic day 4 to day 10, 1.5% increased CO_2) stimulated angiogenesis in chicken embryos. More data regarding optimum dose and administration period are needed before actual implementation.

6.2 Processing strategy

Processing steps can mitigate WS characteristics, particularly excessive moisture loss. U-Chupaj et al. (2021) developed a non-phosphate low-sodium (NPLS) marinade, containing 2.8% (w/v) potassium bicarbonate, 2.9% (w/v) potassium chloride and 1.5% (w/v) sorbitol and applied the marinade in combination with vacuum tumbling to improve the WHC of WS chicken meat. They found that marination can improve the total cooked yield of WS meat throughout 14 days of storage at 4°C. Time-domain NMR indicated that the water absorbed by the WS meat during marination mainly resided at the extra-myofibrillar compartment. Once cooked, the marinated WS meat was able to entrap more water in all compartments in comparison to non-marinated WS counterparts. NPLS marination exerted a mechanical action in conjunction with alkalinization that promoted solubility of myofibrillar and sarcoplasmic proteins from the marinated WS samples. The marination prolonged the duration of the lag phase of the psychrophiles from <1 to 3.5 days, which is desirable for chicken meat stability. The findings suggest the potential application of NPLS marination for improving the WHC of WS meat.

6.3 Breeding

Although many studies have suggested an association between WS and excessive broiler growth, reducing broiler growth rate may not be economically feasible and could make it harder to meet the growing meat demand (Barbut, 2019). The recent study of Bailey et al. (2020) has shown that, from the perspective of genetic selection, it is possible to continue to improve breast yield and body weight of the broilers while reducing the genetic propensity to express WB. A similar selection strategy could be used to reduce the severity of WS without compromising chicken growth performance.

7 Conclusion

WS has been observed worldwide, particularly during the past decade. WS may not have as negative impact as WB on chicken breast meat, but it certainly affects meat quality. Although the etiology of WS remains to be elucidated, an advanced understanding, particularly at molecular basis, shows the complex association between genetic and nongenetic factors, including nutrition and environmental conditions, in the development of WS. The knowledge lays a foundation for mitigating WS severity. However, more research is needed to ensure continuous and long-term reduction of WS in the field.

8 Where to look for further information

- Aviagen. (2019). Breast Muscle Myopathies (BMM) (http://en.aviagen.com/assets/Tech_Center/Broiler_Breeder_Tech_Articles/English/Breast-Muscle-Myopathies-2019-EN.pdf).
- Barbut, S. (2019). Recent myopathies in broiler's breast meat fillets. *World's Poultry Science Journal* 75(4), 559–582, DOI: 10.1017/S0043933919000436.
- Bordini, M., Zappaterra, M., Soglia, F., Petracci, M. and Davoli, R. (2021). Weighted gene co-expression network analysis identifies molecular pathways and hub genes involved in broiler White Striping and Wooden Breast myopathies. *Science Report* 11: 1776. https://doi.org/10.1038/s41598-021-81303-7.
- Velleman, S. G. and Petracci, M. (2020). Avian Muscle Development and Growth Mechanisms: Association with Muscle Myopathies and Meat Quality Vol I. Frontiers In Physiology (https://www.frontiersin.org/research-topics/10177/avian-muscle-development-and-growth-mechanisms-association-with-muscle-myopathies-and-meat-quality).

9 References

Abasht, B., Mutryn, M. F., Michalek, R. D. and Lee, W. R. (2016). Oxidative stress and metabolic perturbations in wooden breast disorder in chickens. *PLoS ONE* 11(4): e0153750.

Ahsan, U. and Cengiz, Ö. (2020). Restriction of dietary digestible lysine allowance in grower phase reduces the occurrence of white striping in broiler chickens. *Animal Feed Science and Technology* 270: 114705.

Albrecht, A., Hebel, M., Heinemann, C., Herbert, U., Miskel, D., Saremi, B. and Kreyenschmidt, J. (2019a). Assessment of meat quality and shelf life from broilers fed with different sources and concentrations of methionine. *Journal of Food Quality* 2019: 1–10.

Albrecht, A., Hebel, M., Mittler, M., Hurck, C., Kustwan, K., Heitkönig, B., Bitschinski, D. and Kreyenschmidt, J. (2019b). Influence of different production systems on the quality and shelf life of poultry meat: a case study in the German sector. *Journal of Food Quality* 2019: 1–11. DOI: 10.1155/2019/3718057.

Alnahhas, N., Berry, C., Chabault, M., Chartrin, P., Boulay, M., Bourin, M. C. and Le Bihan-Duval, E. (2016). Genetic parameters of white striping in relation to body weight, carcass composition, and meat quality traits in two broiler lines divergently selected for the ultimate pH of the pectoralis major muscle. *BMC Genetics* 17: 61.

Aviagen (2019). *Breast Muscles Myopathies*. Available at: http://en.aviagen.com/assets/Tech_Center/Broiler_Breeder_Tech_Articles/English/Breast-Muscle-Myopathies-2019-EN.pdf (Accessed April 13th, 2021).

Ayansola, H., Liao, C., Dong, Y., Yu, X., Zhang, B. and Wang, B. (2021). Prospect of early vascular tone and satellite cell modulations on white striping muscle myopathy. *Poultry Science* 100(3): 100945.

Bailey, R. A., Watson, K. A., Bilgili, S. F. and Avendano, S. (2015). The genetic basis of pectoralis major myopathies in modern broiler chicken lines. *Poultry Science* 94(12): 2870–2879.

Bailey, R. A., Souza, E. and Avendano, S. (2020). Characterising the influence of genetics on breast muscle myopathies in broiler chickens. *Frontiers in Physiology* 11: 1041.

Baldi, G., Soglia, F., Mazzoni, M., Sirri, F., Canonico, L., Babini, E., Laghi, L., Cavani, C. and Petracci, M. (2018). Implications of white striping and spaghetti meat abnormalities on meat quality and histological features in broilers. *Animal* 12(1): 164–173.

Baldi, G., Soglia, F., Laghi, L., Tappi, S., Rocculi, P., Tavaniello, S., Prioriello, D., Mucci, R., Maiorano, G. and Petracci, M. (2019). Comparison of quality traits among breast meat affected by current muscle abnormalities. *Food Research International* 115: 369–376.

Barbut, S. (2019). Recent myopathies in broiler's breast meat fillets. *World's Poultry Science Journal* 75(4): 559–582.

Biferali, B., Proietti, D., Mozzetta, C. and Madaro, L. (2019). Fibro-adipogenic progenitors cross-talk in skeletal muscle: the social network. *Frontiers in Physiology* 10: 1074.

Bodle, B. C., Alvarado, C., Shirley, R. B., Mercier, Y. and Lee, J. T. (2018). Evaluation of different dietary alterations in their ability to mitigate the incidence and severity of woody breast and white striping in commercial male broilers. *Poultry Science* 97(9): 3298–3310.

Boerboom, G., van Kempen, T., Navarro-Villa, A. and Pérez-Bonilla, A. (2018). Unraveling the cause of white striping in broilers using metabolomics. *Poultry Science* 97(11): 3977–3986.

Bordignon, S., Stefani, L. M. and Bolago, M. M. (2020). The use of white striped chicken breasts on the quality of nuggets and hamburgers. *Food Science and Technology* 41(3): 570–575. DOI: 10.1590/fst.16320.

Bordini, M., Zappaterra, M., Soglia, F., Petracci, M. and Davoli, R. (2021). Weighted gene co-expression network analysis identifies molecular pathways and hub genes involved in broiler white striping and wooden breast myopathies. *Scientific Reports* 11(1): 1776.

Bowker, B. and Zhuang, H. (2016). Impact of white striping on functionality attributes of broiler breast meat1. *Poultry Science* 95(8): 1957–1965.

Bowker, B., Zhuang, H., Yoon, S. C., Tasoniero, G. and Lawrence, K. (2019). Relationships between attributes of woody breast and white striping myopathies in commercially processed broiler breast meat. *Journal of Applied Poultry Research* 28(2): 490–496.

Butler, L. D., Scanes, C. G., Rochell, S. J., Mauromoustakos, A., Caldas, J. V., Keen, C. A., Maynard, C. W., Bolden, S. A., Brister, R. D., Smith, P. A., Latham, R. E., Owens, C. M. and Kidd, M. T. (2020). Cobb 700 response to increasing lysine by growth phase. *Journal of Applied Poultry Research* 29(2): 479–488.

Campo, M. D. M., Mur, L., Guerrero, A., Barahona, M., Resconi, V. C., Magalhaes, D. R., Lisbinski, E., Boito, B., Oliveira, I. M. and Olleta, J. L. (2020). Differentiating breast myopathies through color and texture analyses in broiler. *Foods* 9(6): 824 .

Carvalho, L. M., Madruga, M. S., Estevez, M., Badaró, A. T. and Barbin, D. F. (2020). Occurrence of wooden breast and white striping in Brazilian slaughtering plants and use of near-infrared spectroscopy and multivariate analysis to identify affected chicken breasts. *Journal of Food Science* 85(10): 3102–3112.

Carvalho, L. T., Owens, C. M., Giampietro-Ganeco, A., Malagoli de Mello, J. L., Ferrari, F. B., de Carvalho, F. A. L., Alves de Souza, R., Amoroso, L., Alves de Souza, P., Borba, H. and Trindade, M. A. (2021a). Quality of turkeys breast meat affected by white striping myopathy. *Poultry Science* 100(4): 101022.

Carvalho, L. M., Delgado, J., Madruga, M. S. and Estévez, M. (2021b). Pinpointing oxidative stress behind the white striping myopathy: depletion of antioxidant defenses, accretion of oxidized proteins and impaired proteostasis. *Journal of the Science of Food and Agriculture* 101(4): 1364–1371.

Cemin, H. S., Vieira, S. L., Stefanello, C., Kindlein, L., Ferreira, T. Z. and Fireman, A. K. (2018). Broiler responses to increasing selenium supplementation using Zn-L-selenomethionine with special attention to breast myopathies. *Poultry Science* 97(5): 1832–1840.

Christensen, V. L., J. Winelan, M., L. Grimes, J., O. Oviedo, E., Mozdziak, P. S., T. Ort, D. and M. Mann, K. (2007). Effect of incubator temperature and oxygen concentration at the plateau stage in oxygen consumption on turkey embryo muscle growth and development. *International Journal of Poultry Science* 6(6): 406–412.

Clark, D. L., Coy, C. S., Strasburg, G. M., Reed, K. M. and Velleman, S. G. (2016). Temperature effect on proliferation and differentiation of satellite cells from turkeys with different growth rates. *Poultry Science* 95(4): 934–947.

Clark, D. L., Walter, K. G. and Velleman, S. G. (2017). Incubation temperature and time of hatch impact broiler muscle growth and morphology. *Poultry Science* 96(11): 4085–4095.

Córdova-Noboa, H. A., Oviedo-Rondón, E. O., Sarsour, A. H., Barnes, J., Sapcota, D., López, D., Gross, L., Rademacher-Heilshorn, M. and Braun, U. (2018). Effect of guanidinoacetic acid supplementation on live performance, meat quality, pectoral myopathies and blood parameters of male broilers fed corn-based diets with or without poultry by-products. *Poultry Science* 97(7): 2494–2505.

Cruz, R. F. A., Vieira, S. L., Kindlein, L., Kipper, M., Cemin, H. S. and Rauber, S. M. (2017). Occurrence of white striping and wooden breast in broilers fed grower and finisher diets with increasing lysine levels. *Poultry Science* 96(2): 501–510.

Da Costa, M. J., Oviedo-Rondón, E. O., Wineland, M. and Jeffrey, D. (2016). Pathogeny of fatigued walking condition in Pekin ducks. *Avian Diseases* 60(4): 731–738.

Dalle Zotte, A., Ricci, R., Cullere, M., Serva, L., Tenti, S. and Marchesini, G. (2020). Research Note: effect of chicken genotype and white striping–wooden breast condition on breast meat proximate composition and amino acid profile. *Poultry Science* 99(3): 1797–1803.

Daughtry, M. R., Berio, E., Shen, Z., Suess, E. J. R., Shah, N., Geiger, A. E., Berguson, E. R., Dalloul, R. A., Persia, M. E., Shi, H. and Gerrard, D. E. (2017). Satellite cell-mediated breast muscle regeneration decreases with broiler size. *Poultry Science* 96(9): 3457–3464.

De Carvalho, L. M., Ventanas, S., Olegario, L. S., Madruga, M. S. and Estévez, M. (2020). Consumers awareness of white-striping as a chicken breast myopathy affects their purchasing decision and emotional responses. *LWT* 131: 109809.

Domingues-Faria, C., Vasson, M. P., Goncalves-Mendes, N., Boirie, Y. and Walrand, S. (2016). Skeletal muscle regeneration and impact of aging and nutrition. *Ageing Research Reviews* 26: 22-36.

Estevez, M. and Petracci, M. (2019). Benefits of magnesium supplementation to broiler subjected to dietary and heat stress: improved redox status, breast quality and decreased myopathy incidence. *Antioxidants* 8(10): 456.

Faucitano, L., Ielo, M. C., Ster, C., Lo Fiego, D. P., Methot, S. and Saucier, L. (2010). Shelf life of pork from five different quality classes. *Meat Science* 84(3): 466-469.

Ferreira, T. Z., Casagrande, R. A., Vieira, S. L., Driemeier, D. and Kindlein, L. (2014). An investigation of a reported case of white striping in broilers. *Journal of Applied Poultry Research* 23(4): 748-753.

Ferreira, T. Z., Kindlein, L., Flees, J. J., Shortnacy, L. K., Vieira, S. L., Nascimento, V. P., Meloche, K. J. and Starkey, J. D. (2020). Characterization of pectoralis major muscle satellite cell population heterogeneity, macrophage density, and collagen infiltration in broiler chickens affected by wooden breast. *Frontiers in Physiology* 11: 529.

Golzar Adabi, S. and Demirok Soncu, E. (2019). White striping prevalence and its effect on meat quality of broiler breast fillets under commercial conditions. *Journal of Animal Physiology and Animal Nutrition* 103(4): 1060-1069.

Gordon, S. and Plüddemann, A. (2018). Macrophage clearance of apoptotic cells: a critical assessment. *Frontiers in Immunology* 9: 127.

Gratta, F., Fasolato, L., Birolo, M., Zomeño, C., Novelli, E., Petracci, M., Pascual, A., Xiccato, G. and Trocino, A. (2019). Effect of breast myopathies on quality and microbial shelf life of broiler meat. *Poultry Science* 98(6): 2641-2651.

Griffin, J. R., Moraes, L., Wick, M. and Lilburn, M. S. (2018). Onset of white striping and progression into wooden breast as defined by myopathic changes underlying pectoralis major growth. Estimation of growth parameters as predictors for stage of myopathy progression. *Avian Pathology* 47(1): 2-13.

Halevy, O. (2020). Timing is everything- the high sensitivity of avian satellite cells to thermal conditions during embryonic and posthatch periods. *Frontiers in Physiology* 11: 235.

Huff-Lonergan, E. and Lonergan, S. M. (2005). Mechanisms of water-holding capacity of meat: the role of postmortem biochemical and structural changes. *Meat Science* 71(1): 194-204.

Jahangir, E., Vita, J. A., Handy, D., Holbrook, M., Palmisano, J., Beal, R., Loscalzo, J. and Eberhardt, R. T. (2009). The effect of L-arginine and creatine on vascular function and homocysteine metabolism. *Vascular Medicine* 14(3): 239-248.

Joardar, D., Livingston, K. A., Edens, F. W., Nusairat, B., Qudsieh, R., Livingston, M. L. and Brake, J. (2020). Effect of limestone particle size and potassium supplementation on growth performance, blood physiology, and breast muscle myopathy of male broiler chickens. *Frontiers in Veterinary Science* 7: 603284.

Khan, I. A., Parker, N. B., Löhr, C. V. and Cherian, G. (2021). Docosahexaenoic acid (22:6 n-3)-rich microalgae along with methionine supplementation in broiler chickens: effects on production performance, breast muscle quality attributes, lipid profile, and incidence of white striping and myopathy. *Poultry Science* 100(2): 865-874.

Kindlein, L., Ferreira, T. Z., Driemeier, D., Nascimento, V. P., Vieira, S. L., Moraes, L. E., King, A. and Sainz, R. D. (2017). Occurrence and severity of white striping in broilers until 50d of age fed with high and low-energy diets: body weight, histopathological changes and meat quality. *Journal of Veterinary Science and Technology* 8(6): 1000478.

Kuttappan, V. A., Brewer, V. B., Clark, F. D., et al. (2009). Effect of white striping on the histological and meat quality characteristics of broiler fillets. *Poultry Science* 88(E-SUPPL 1): 136–137.

Kuttappan, V. A., Brewer, V. B., Apple, J. K., Waldroup, P. W. and Owens, C. M. (2012a). Influence of growth rate on the occurrence of white striping in broiler breast fillets. *Poultry Science* 91(10): 2677–2685.

Kuttappan, V. A., Goodgame, S. D., Bradley, C. D., Mauromoustakos, A., Hargis, B. M., Waldroup, P. W. and Owens, C. M. (2012b). Effect of different levels of dietary vitamin E (dl-α-tocopherol acetate) on the occurrence of various degrees of white striping on broiler breast fillets. *Poultry Science* 91(12): 3230–3235.

Kuttappan, V. A., Lee, Y. S., Erf, G. F., Meullenet, J. F., McKee, S. R. and Owens, C. M. (2012c). Consumer acceptance of visual appearance of broiler breast meat with varying degrees of white striping. *Poultry Science* 91(5): 1240–1247.

Kuttappan, V. A., Shivaprasad, H. L., Shaw, D. P., Valentine, B. A., Hargis, B. M., Clark, F. D., McKee, S. R. and Owens, C. M. (2013a). Pathological changes associated with white striping in broiler breast muscles. *Poultry Science* 92(2): 331–338.

Kuttappan, V. A., Huff, G. R., Huff, W. E., Hargis, B. M., Apple, J. K., Coon, C. and Owens, C. M. (2013b). Comparison of hematologic and serologic profiles of broiler birds with normal and severe degrees of white striping in breast fillets. *Poultry Science* 92(2): 339–345.

Kuttappan, V. A., Brewer, V. B., Mauromoustakos, A., McKee, S. R., Emmert, J. L., Meullenet, J. F. and Owens, C. M. (2013c). Estimation of factors associated with the occurrence of white striping in broiler breast fillets. *Poultry Science* 92(3): 811–819.

Kuttappan, V. A., Hargis, B. M. and Owens, C. M. (2016). White striping and woody breast myopathies in the modern poultry industry: a review. *Poultry Science* 95(11): 2724–2733.

Kuttappan, V. A., Owens, C. M., Coon, C., Hargis, B. M. and Vazquez-Añon, M. (2017). Incidence of broiler breast myopathies at 2 different ages and its impact on selected raw meat quality parameters. *Poultry Science* 96(8): 3005–3009.

Lake, J. A. and Abasht, B. (2020). Glucolipotoxicity: a proposed etiology for wooden breast and related myopathies in commercial broiler chickens. *Frontiers in Physiology* 11: 169.

Lake, J. A., Dekkers, J. C. M. and Abasht, B. (2021). Genetic basis and identification of candidate genes for wooden breast and white striping in commercial broiler chickens. *Scientific Reports* 11(1): 6785.

Livingston, M. L., Ferket, P. R., Brake, J. and Livingston, K. A. (2019a). Dietary amino acids under hypoxic conditions exacerbates muscle myopathies including wooden breast and white stripping. *Poultry Science* 98(3): 1517–1527.

Livingston, M. L., Landon, C. D., Barnes, H. J., Brake, J. and Livingston, K. A. (2019b). Dietary potassium and available phosphorous on broiler growth performance, carcass characteristics, and wooden breast. *Poultry Science* 98(7): 2813–2822.

Livingston, M. L., Landon, C., Barnes, H. J. and Brake, J. (2019c). White striping and wooden breast myopathies of broiler breast muscle is affected by time-limited feeding, genetic background, and egg storage. *Poultry Science* 98(1): 217–226.

Lorenzi, M., Mudalal, S., Cavani, C. and Petracci, M. (2014). Incidence of white striping under commercial conditions in medium and heavy broiler chickens in Italy. *Journal of Applied Poultry Research* 23(4): 754–758.

Loyau, T., Bedrani, L., Berri, C., Métayer-Coustard, S., Praud, C., Coustham, V., Mignon-Grasteau, S., Duclos, M. J., Tesseraud, S., Rideau, N., Hennequet-Antier, C., Everaert, N., Yahav, S. and Collin, A. (2015). Cyclic variations in incubation conditions induce adaptive responses to later heat exposure in chickens: a review. *Animal* 9(1): 76–85.

Lucca, J. W., Zampar, A., Borba, H., Mello, J. L. Md, Barreta, M. and Boiago, M. M. (2019). The white striping myopathy in different broiler strains and its influence on composition and meat quality. *Arquivo Brasileiro De Medicina Veterinaria E Zootecnia* 71(4): 1331–1338.

Malila, Y., U-Chupaj, J., Srimarut, Y., Chaiwiwattrakul, P., Uengwetwanit, T., Arayamethakorn, S., Punyapornwithaya, V., Sansamur, C., Kirschke, C. P., Huang, L., Tepaamorndech, S., Petracci, M., Rungrassamee, W. and Visessanguan, W. (2018). Monitoring of white striping and wooden breast cases and impacts on quality of breast meat collected from commercial broilers (*Gallus gallus*). *Asian-Australasian Journal of Animal Sciences* 31(11): 1807–1817.

Malila, Y., Thanatsang, K., Arayamethakorn, S., Uengwetwanit, T., Srimarut, Y., Petracci, M., Strasburg, G. M., Rungrassamee, W. and Visessanguan, W. (2019). Absolute expressions of hypoxia-inducible factor-1 alpha (HIF1A) transcript and the associated genes in chicken skeletal muscle with white striping and wooden breast myopathies. *PLoS ONE* 14(8): e0220904.

Malila, Y., Uengwetwanit, T., Arayamethakorn, S., Srimarut, Y., Thanatsang, K. V., Soglia, F., Strasburg, G. M., Rungrassamee, W. and Visessanguan, W. (2020). Transcriptional profiles of skeletal muscle associated with increasing severity of white striping in commercial broilers. *Frontiers in Physiology* 11: 580.

Marangoni, F., Corsello, G., Cricelli, C., Ferrara, N., Ghiselli, A., Lucchin, L. and Poli, A. (2015). Role of poultry meat in a balanced diet aimed at maintaining health and wellbeing: an Italian consensus document. *Food and Nutrition Research* 59: 27606–27606.

Marchesi, J. A. P., Ibelli, A. M. G., Peixoto, J. O., Cantão, M. E., Pandolfi, J. R. C., Marciano, C. M. M., Zanella, R., Settles, M. L., Coutinho, L. L. and Ledur, M. C. (2019). Whole transcriptome analysis of the pectoralis major muscle reveals molecular mechanisms involved with white striping in broiler chickens. *Poultry Science* 98(2): 590–601.

Maynard, C. W., Latham, R. E., Brister, R., Owens, C. M. and Rochell, S. J. (2019). Effects of dietary energy and amino acid density during finisher and withdrawal phases on live performance and carcass characteristics of cobb MV × 700 broilers. *Journal of Applied Poultry Research* 28(3): 729–742.

Maynard, C. W., Latham, R. E., Brister, R., Owens, C. M. and Rochell, S. J. (2020). Effects of dietary amino acid regimens on live performance and processing characteristics of cobb MV × 700 male and female broilers. *Journal of Applied Poultry Research* 29(1): 64–76.

Meloche, K. J., Fancher, B. I., Emmerson, D. A., Bilgili, S. F. and Dozier, W. A. (2018a). Effects of quantitative nutrient allocation on myopathies of the pectoralis major muscles in broiler chickens at 32, 43, and 50 days of age. *Poultry Science* 97(5): 1786–1793.

Meloche, K. J., Fancher, B. I., Emmerson, D. A., Bilgili, S. F. and Dozier, W. A. (2018b). Effects of reduced dietary energy and amino acid density on pectoralis major myopathies in broiler chickens at 36 and 49 days of age1. *Poultry Science* 97(5): 1794–1807.

Meloche, K. J., Fancher, B. I., Emmerson, D. A., Bilgili, S. F. and Dozier, W. A. (2018c). Effects of reduced digestible lysine density on myopathies of the pectoralis major muscles in broiler chickens at 48 and 62 days of age. *Poultry Science* 97(9): 3311–3324.

Mudalal, S. (2019). Incidence of white striping and its effect on the quality traits of raw and processed turkey breast meat. *Food Science of Animal Resources* 39(3): 410–417.

Mudalal, S., Babini, E., Cavani, C. and Petracci, M. (2014). Quantity and functionality of protein fractions in chicken breast fillets affected by white striping. *Poultry Science* 93(8): 2108–2116.

Mudalal, S., Lorenzi, M., Soglia, F., Cavani, C. and Petracci, M. (2015). Implications of white striping and wooden breast abnormalities on quality traits of raw and marinated chicken meat. *Animal* 9(4): 728–734.

Mudalal, S., Zaazaa, A. and Omar, J. A. (2021). Effects of medicinal plants extract with antibiotic free diets on broilers growth performance and incidence of muscles abnormalities. *Brazilian Journal of Poultry Science* 23(1): eRBCA-2020-1342.

Mutryn, M. F., Brannick, E. M., Fu, W., Lee, W. R. and Abasht, B. (2015). Characterization of a novel chicken muscle disorder through differential gene expression and pathway analysis using RNA-sequencing. *BMC Genomics* 16(1): 399.

Nyuiadzi, D., Berri, C., Dusart, L., Travel, A., Méda, B., Bouvarel, I., Guilloteau, L. A., Chartrin, P., Coustham, V., Praud, C., Le Bihan-Duval, E., Tona, J. K. and Collin, A. (2020). Short cold exposures during incubation and postnatal cold temperature affect performance, breast meat quality, and welfare parameters in broiler chickens. *Poultry Science* 99(2): 857–868.

Oviedo-Rondon, E. O., Velleman, S. G. and Wineland, M. J. (2020). The role of incubation conditions in the onset of avian myopathies. *Frontiers in Physiology* 11: 545045.

Owens, C. M. and Vieira, S. L. (2012). White striping in broiler breast meat. In: *Broiler Carcass Quality-an Approach from the Production Sites*, S. L. Vieira (Ed.), Zinpro Corporation, São Paulo, SP, pp. 83–88.

Papah, M. B. and Abasht, B. (2019). Dysregulation of lipid metabolism and appearance of slow myofiber-specific isoforms accompany the development of wooden breast myopathy in modern broiler chickens. *Scientific Reports* 9(1): 17170.

Pascual, A., Trocino, A., Birolo, M., Cardazzo, B., Bordignon, F., Ballarin, C., Carraro, L. and Xiccato, G. (2020). Dietary supplementation with sodium butyrate: growth, gut response at different ages, and meat quality of female and male broiler chickens. *Italian Journal of Animal Science* 19(1): 1134–1145.

Patael, T., Piestun, Y., Soffer, A., Mordechay, S., Yahav, S., Velleman, S. G. and Halevy, O. (2019). Early posthatch thermal stress causes long-term adverse effects on pectoralis muscle development in broilers. *Poultry Science* 98(8): 3268–3277.

Pekel, A. Y., Tatlı, O., Sevim, Ö., Kuter, E., Ahsan, U., Khamseh, E. K., Atmaca, G., Köksal, B. H., Özsoy, B. and Cengiz, Ö (2020). Effects of reducing dietary amino acid density and stocking density on growth performance, carcass characteristics, meat quality, and occurrence of white striping in broiler chickens. *Poultry Science* 99(12): 7178–7191.

Petracci, M., Mudalal, S., Bonfiglio, A. and Cavani, C. (2013). Occurrence of white striping under commercial conditions and its impact on breast meat quality in broiler chickens. *Poultry Science* 92(6): 1670–1675.

Petracci, M., Mudalal, S., Babini, E. and Cavani, C. (2014). Effect of white striping on chemical composition and nutritional value of chicken breast meat. *Italian Journal of Animal Science* 13(1): 179–183.

Petracci, M., Soglia, F., Madruga, M., Carvalho, L., Ida, E. and Estévez, M. (2019). Wooden-breast, white striping, and spaghetti meat: causes, consequences and consumer perception of emerging broiler meat abnormalities. *Comprehensive Reviews in Food Science and Food Safety* 18(2): 565–583.

Philpot, S. C., Perryman, K. R. and Dozier, W. A. (2020). Growth performance and carcass characteristics of broilers fed diets varying in supplemental copper concentrations from 29 to 53 days of age. *Journal of Applied Poultry Research* 29(2): 289–300.

Praud, C., Jimenez, J., Pampouille, E., Couroussé, N., Godet, E., Le Bihan-Duval, E. and Berri, C. (2020). Molecular phenotyping of white striping and Wooden breast myopathies in chicken. *Frontiers in Physiology* 11: 633.

Prisco, F., De Biase, D., Piegari, G., d'Aquino, I., Lama, A., Comella, F., Mercogliano, R., Dipineto, L., Papparella, S. and Paciello, O. (2021). Pathologic characterization of white striping myopathy in broiler chickens. *Poultry Science* 100(7): 101150.

Radaelli, G., Piccirillo, A., Birolo, M., Bertotto, D., Gratta, F., Ballarin, C., Vascellari, M., Xiccato, G. and Trocino, A. (2017). Effect of age on the occurrence of muscle fiber degeneration associated with myopathies in broiler chickens submitted to feed restriction. *Poultry Science* 96(2): 309–319.

Reuterswärd, A. L. and Fabiansson, S. (1985). In vivo digestibility of insoluble collagen from bovine tendon as influenced by the inhibition of gastric acid secretion. *Journal of Food Science* 50(6): 1523–1525.

Ribeiro, M. V., Bittencourt, L. C., Hermes, R. G., Rönnau, M., Rorig, A., Lima, F. and Fernandes, J. (2020). Mineral source and vitamin level in broiler diets: effects on performance, yield, and meat quality. *Brazilian Journal of Poultry Science* 22(2): 1–14.

Russo, E., Drigo, M., Longoni, C., Pezzotti, R., Fasoli, P. and Recordati, C. (2015). Evaluation of white striping prevalence and predisposing factors in broilers at slaughter. *Poultry Science* 94(8): 1843–1848.

Sachs, N. J., Hampton, A. R., Foster, K. K., Pechanec, M. Y., Henderson, J. D., King, A. J. and Mienaltowski, M. J. (2019). The effects of an alternative diet regimen with natural methionine ingredients on white striping breast myopathy in broiler chickens. *Poultry Science* 98(1): 413–421.

Salles, G. B. C., Boiago, M. M., Silva, A. D., Morsch, V. M., Gris, A., Mendes, R. E., Baldissera, M. D. and da Silva, A. S. (2019). Lipid peroxidation and protein oxidation in broiler breast fillets with white striping myopathy. *Journal of Food Biochemistry* 43(4): e12792.

Sanchez Brambila, G., Bowker, B. C. and Zhuang, H. (2016). Comparison of sensory texture attributes of broiler breast fillets with different degrees of white striping. *Poultry Science* 95(10): 2472–2476.

Sihvo, H. K., Immonen, K. and Puolanne, E. (2014). Myodegeneration with fibrosis and regeneration in the pectoralis major muscle of broilers. *Veterinary Pathology* 51(3): 619–623.

Sirri, F., Maiorano, G., Tavaniello, S., Chen, J., Petracci, M. and Meluzzi, A. (2016). Effect of different levels of dietary zinc, manganese, and copper from organic or inorganic sources on performance, bacterial chondronecrosis, intramuscular collagen characteristics, and occurrence of meat quality defects of broiler chickens. *Poultry Science* 95(8): 1813–1824.

Sobotik, E. B., Lee, J. T., Hagerman, S. and Archer, G. S. (2018). Evaluation of the use of phosphatidic acid in the diet on growth performance and breast meat yield in broilers. *Animals: An Open Access Journal from MDPI* 8(6).

Soglia, F., Laghi, L., Canonico, L., Cavani, C. and Petracci, M. (2016). Functional property issues in broiler breast meat related to emerging muscle abnormalities. *Food Research International* 89: 1071–1076.

Soglia, F., Baldi, G., Laghi, L., Mudalal, S., Cavani, C. and Petracci, M. (2018). Effect of white striping on turkey breast meat quality. *Animal* 12(10): 2198–2204.

Soglia, F., Silva, A. K., Lião, L. M., Laghi, L. and Petracci, M. (2019a). Effect of broiler breast abnormality and freezing on meat quality and metabolites assessed by ^{1}H-NMR spectroscopy. *Poultry Science* 98(12): 7139–7150.

Soglia, F., Mazzoni, M., Zappaterra, M., Di Nunzio, M., Babini, E., Bordini, M., Sirri, F., Clavenzani, P., Davoli, R. and Petracci, M. (2019b). Distribution and expression of vimentin and desmin in broiler pectoralis major affected by the growth-related muscular abnormalities. *Frontiers in Physiology* 10: 1581.

Soglia, F., Petracci, M., Davoli, R. and Zappaterra, M. (2021a). A critical review of the mechanisms involved in the occurrence of growth-related abnormalities affecting broiler chickens breast muscle. *Poultry Science* 100(6): 101180.

Soglia, F., Zampiga, M., Baldi, G., Malila, Y., Thanatsang, K. V., Srimarut, Y., Tatiyaborworntham, N., Unger, O., Klamchuen, A., Laghi, L., Petracci, M. and Sirri, F. (2021b). Lysine depletion during different feeding phases: effects on growth performances and meat quality of broiler chickens. *Animals: An Open Access Journal from MDPI* 11(6): 1499.

Tasoniero, G., Cullere, M., Cecchinato, M., Puolanne, E. and Dalle Zotte, A. (2016). Technological quality, mineral profile, and sensory attributes of broiler chicken breasts affected by White Striping and Wooden Breast myopathies. *Poultry Science* 95(11): 2707–2714.

The Humane League UK (2021). *Investigative Report: White Striping Disease in Supermarket Chicken*. Available at: https://assets.ctfassets.net/ww1ie0z745y7/6c2m3SlCRioTbdMeBXpNnT/334cacbff0ce304740f3dd7ed17b0b28/White_Striping_in_Supermarket_Chicken.pdf.

Tijare, V. V., Yang, F. L., Kuttappan, V. A., Alvarado, C. Z., Coon, C. N. and Owens, C. M. (2016). Meat quality of broiler breast fillets with white striping and woody breast muscle myopathies. *Poultry Science* 95(9): 2167–2173.

Trocino, A., Piccirillo, A., Birolo, M., Radaelli, G., Bertotto, D., Filiou, E., Petracci, M. and Xiccato, G. (2015). Effect of genotype, gender and feed restriction on growth, meat quality and the occurrence of white striping and wooden breast in broiler chickens. *Poultry Science* 94(12): 2996–3004.

U-Chupaj, J., Malila, Y., Gozzi, G., Vannini, L., Dellarosa, N., Laghi, L., Petracci, M., Benjakul, S. and Visessanguan, W. (2021). Influence of non-phosphate and low-sodium salt marination in combination with tumbling process on properties of chicken breast meat affected by white striping abnormality. *Journal of Food Science* 86(2): 319–326.

USDA (2021). Livestock and poultry: world markets and trade (Accessed April 2021). Available at: https://apps.fas.usda.gov/psdonline/circulars/livestock_poultry.PDF?fbclid=IwAR1Hx_4LM3K6LWSgtFMY4eu8Ir8j4tgSZ7fvPgrtzxIPOViz9ro2d8taowM (Accessed April 13th, 2021).

Velleman, S. G. (2015). Relationship of skeletal muscle development and growth to breast muscle myopathies: a review. *Avian Diseases* 59(4): 525–531.

Velleman, S. G., Nestor, K. E., Coy, C. S., Harford, I. and Anthony, N. B. (2010). Effect of posthatch feed restriction on broiler breast muscle development and muscle transcriptional regulatory factor gene and heparan sulfate proteoglycan expression. *International Journal of Poultry Science* 9(5): 417–425.

Velleman, S. G., Coy, C. S. and Emmerson, D. A. (2014). Effect of the timing of posthatch feed restrictions on the deposition of fat during broiler breast muscle development. *Poultry Science* 93(10): 2622–2627.

Verhoelst, E., Ketelaere, B. D., Decuypere, E. and De Baerdemaeker, J. (2011). The effect of early prenatal hypercapnia on the vascular network in the chorioallantoic membrane of the chicken embryo. *Biotechnology Progress* 27(2): 562–570.

Vignale, K., Caldas, J. V., England, J. A., Boonsinchai, N., Magnuson, A., Pollock, E. D., Dridi, S., Owens, C. M. and Coon, C. N. (2017). Effect of white striping myopathy on breast muscle (pectoralis major) protein turnover and gene expression in broilers. *Poultry Science* 96(4): 886–893.

Wang, J., Clark, D. L., Jacobi, S. K. and Velleman, S. G. (2020). Effect of vitamin E and omega-3 fatty acids early posthatch supplementation on reducing the severity of wooden breast myopathy in broilers. *Poultry Science* 99(4): 2108–2119.

Xu, J., Strasburg, G. M., Reed, K. M. and Velleman, S. G. (2021). Effect of temperature and selection for growth on intracellular lipid accumulation and adipogenic gene expression in turkey pectoralis major muscle satellite cells. *Frontiers in Physiology* 12: 667814.

Yu, X., Feng, Y., Bowker, B. and Zhuang, H. (2020). Expressible fluid measurements of broiler breast meat affected by emerging muscle abnormalities. *LWT* 133: 110110.

Zambonelli, P., Zappaterra, M., Soglia, F., Petracci, M., Sirri, F., Cavani, C. and Davoli, R. (2016). Detection of differentially expressed genes in broiler pectoralis major muscle affected by White striping - wooden breast myopathies. *Poultry Science* 95(12): 2771–2785.

Zampiga, M., Laghi, L., Petracci, M., Zhu, C., Meluzzi, A., Dridi, S. and Sirri, F. (2018). Effect of dietary arginine to lysine ratios on productive performance, meat quality, plasma and muscle metabolomics profile in fast-growing broiler chickens. *Journal of Animal Science and Biotechnology* 9: 79.

Zampiga, M., Soglia, F., Petracci, M., Meluzzi, A. and Sirri, F. (2019). Effect of different arginine-to-lysine ratios in broiler chicken diets on the occurrence of breast myopathies and meat quality attributes. *Poultry Science* 98(6): 2691–2697.

Zuidhof, M. J., Schneider, B. L., Carney, V. L., Korver, D. R. and Robinson, F. E. (2014). Growth, efficiency, and yield of commercial broilers from 1957, 1978, and 2005. *Poultry Science* 93(12): 2970–2982.

Chapter 10

Quality defects associated with poultry muscle development: wooden breast

Martina Bordini, Francesca Soglia, Martina Zappaterra, Adele Meluzzi and Massimiliano Petracci, Alma Mater Studiorum – Università di Bologna, Italy

1 Introduction

Over the next decade, global meat supply will be mainly from poultry production (Whitton et al., 2021). Global consumption of meat proteins is expected to increase by 14% by 2030, with poultry meat accounting for about 41% of all protein sources by that point (OECD/FAO, 2021). To meet growing consumer demand, breeding companies have put great effort into improving the productive performance of modern broiler chickens (Havenstein et al., 2003). However, these improvements have indirectly led to the development of several muscular defects mainly affecting modern chicken hybrids: so-called 'growth-related abnormalities' (Petracci et al., 2019). Among them, wooden breast (WB) is a globally challenging meat quality issue (Hubert and Athrey, 2020), which mainly affects the *Pectoralis major* (PM) muscle of broilers selected for rapid growth and high breast yield (Griffin et al., 2018; Petracci et al., 2015; Caldas-Cueva and Owens, 2020).

The defect was first described by Sihvo et al. (2014) as a muscular abnormality manifested by the presence of hardened, pale and bulging areas with scattered hemorrhagic petechiae and viscous materials at the surface level of the pectoral muscle. WB has been demonstrated to cause considerable

http://dx.doi.org/10.19103/AS.2022.0112.11

economic losses for the poultry industry (Kuttappan et al., 2016; Zanetti et al., 2018; Barbut, 2019), due to its adverse impact on meat quality and consumer acceptability (Kuttappan et al., 2012; Oliveira et al., 2021; da Rocha et al., 2022). Severe WB also reduces the technological and nutritional properties of chicken meat, which can be ascribed to the proliferation of connective tissue (up to fibrosis) as well as to the occurrence of protein degradation characterizing WB fillets (Tasoniero et al., 2016; Baldi et al., 2018).

Various authors have estimated the incidence of WB in different countries. Trocino et al. (2015) estimated an incidence level of WB in Italy of 8–16%, whereas Dalle Zotte et al. (2017) reported a level of 53% WB among chicken fillets in the same country. A study of a Chinese commercial broiler processing plant found that 61.9% of birds were affected by WB (Xing et al., 2020). Studies carried out in Brazil and the United States reported higher levels of WB, with an incidence of up to about 85% and 96%, respectively (Tijare et al., 2016; Cruz et al., 2017).

Considering the worldwide distribution, high incidence and economic impact of WB (Kuttappan et al., 2016; Barbut, 2019), significant efforts have been made to understand the characteristics of and genetic and/or environmental factors potentially involved in the occurrence of WB (Mudalal et al., 2015; Kuttappan et al., 2016; Soglia et al., 2016; Tasoniero et al., 2016; Clark and Velleman, 2017; Sihvo et al., 2017). There is now a detailed understanding of the phenotypic and molecular changes characterizing this abnormality and that may underlie its occurrence in broiler chickens (Kuttappan et al., 2013; Trocino et al., 2015; Soglia et al., 2016; Papah et al., 2017; Pampouille et al., 2018; Lake and Abasht, 2020; Praud et al., 2020). A few papers have been published on WB in turkeys, even though none evaluates its effects on meat quality. Overall, although a similar abnormality has been observed in turkeys selected for high growth (i.e. toughening of the breast muscle; Soglia et al., 2018), its occurrence seems to be negligible (Zampiga et al., 2020).

This chapter aims to provide an overview of the current state of knowledge regarding WB, focusing on: (a) the possible factors underlying the occurrence of WB, (b) the main features of this abnormality and its effects on key poultry meat quality traits and (c) potential strategies to mitigate the detrimental effects of WB.

2 Possible factors underlying the occurrence of wooden breast

There have been numerous studies to identify factors potentially involved in the development of WB. Many studies have considered the following aspects:

- chicken genotype (hybrids selected for meat production vs. unselected breeds) (Petracci et al., 2013b; Trocino et al., 2015; Santos et al., 2021);

- productive performance (high or standard breast yield) (Trocino et al., 2015; Alnahhas et al., 2016; Santos et al., 2021);
- slaughter weight (heavier vs. lighter broilers) (Cruz et al., 2017; Papah et al., 2017); and
- gender (male vs. female) (Brothers et al., 2019).

Other studies have focused on nutritional and environmental factors as potentially involved in manifestation of WB (Radaelli et al., 2017; Livingston et al., 2019; Che et al., 2022).

Most studies have identified a greater prevalence of WB in broilers selected for rapid growth compared to unselected breeds (Petracci et al., 2019; Santos et al., 2021), as well as an increase in the incidence and severity of this defect as slaughter weight increases (Cruz et al., 2017; Papah et al., 2017). Given the high incidence of WB in modern fast-growing broilers, there is a consensus among researchers that its major causes are linked to the genetic selection of modern chicken hybrids (Trocino et al., 2015; Soglia et al., 2019; Petracci et al., 2019). Moreover, a strong genetic determinism has been hypothesized given the moderate (h^2 = 0.38) to high (h^2 = 0.65) hereditability values in high-yielding chickens (Bailey et al., 2015; Alnahhas et al., 2016). Several studies have tried to identify candidate genes possibly underlying WB (Hubert et al., 2018; Lilburn et al., 2019; Pampouille et al., 2018; Papah et al., 2018; Zambonelli et al., 2016). These suggest a polygenic inheritance of the defect rather than the existence of a single causative gene responsible for its development.

Looking at the environmental and nutritional factors potentially involved in WB, several studies have reported that feeding restriction strategies to reduce broiler weight lead to a reduced incidence and/or a less severe degree of WB (Livingston et al., 2019). Iwasaki et al. (2021) showed that inadequate nutrition at the early stage of bird development (1–12 days of age) may increase incidence of WB, highlighting the importance of an appropriate starter diet to ensure poultry meat quality.

Given the importance of gut health for the growth and function of skeletal muscles (Lahiri et al., 2019), recent studies have investigated the gut microbiota of chickens affected by WB (Maharjan et al., 2020a; Zhang et al., 2021; Kang et al., 2022). These studies highlighted a reduction in the intestinal microflora diversity of chicken characterized by severe WB, along with differences in gut microbial metabolites (e.g. anserine and choline increase with the severity of WB), which may have implications in skeletal muscle development and could be used as biomarkers for tracking the severity of the defect (Maharjan et al., 2020a; Kang et al., 2022).

In assessing environmental factors associated with the development of WB, Che et al. (2022) stress the multi-factorial nature of the defect. They showed a positive association between WB and the average live weight of broilers at slaughter, as well as association with high temperatures during the growth period (Che et al., 2022). They also found that birds that are not vaccinated against *coccidiosis* seem to have a higher risk of developing WB. However, they also pointed out the need to determine biological and molecular mechanisms underlying the association between these environmental factors and WB.

Even though the precise etiology of WB is not fully understood, remarkable progress has been made by the scientific community to understand and describe the metabolic pathways and molecular processes involved in the onset and progression of WB.

2.1 Main hypotheses regarding causes of wooden breast

As recently reviewed by Soglia et al. (2021a), one of the major hypotheses concerning the underlying mechanisms responsible for the cascade of events leading to WB focuses on the decreased capillarization of modern broiler PM muscle (Hoving-Bolink et al., 2000). The reduced vascularization/capillary-to-fiber ratio may determine the accumulation of metabolic waste products and an impaired oxygen supply to the breast muscle, which then leads to oxidative and hypoxic conditions (Kuttappan et al., 2013; Soglia et al., 2019; Lake and Abasht, 2020). This could explain the oxidative stress, metabolic alterations and myodegeneration characterizing woody fillets (Papah et al., 2018; Petracci et al., 2019).

Both metabolomic and gene expression studies have identified oxidative stress as one of the major pathways involved in WB occurrence and development (Mutryn et al., 2015; Abasht et al., 2016; Pampouille et al., 2019; Malila et al., 2020; Bordini et al., 2022). Electron microscopic examination has detected mitochondrial degeneration and dilation of the sarcoplasmic reticulum, which may underlie the oxidative damage observed in WB muscles (Papah et al., 2017). Bordini et al. (2021) have suggested that endoplasmic reticulum (ER) stress, probably resulting from the accumulation of aberrant proteins, may be one of the first factors triggering the cascade of molecular and biological events that lead to WB. It has been hypothesized that, at the endothelial and/or muscle fiber level, the abnormal accumulation of extracellular matrix components within the ER may play a role in impairing its functionality and inducing ER stress, thus leading to the inflammatory and apoptosis processes observed in WB (Lake and Abasht, 2020; Bordini et al., 2022). This hypothesis is supported by evidence that alterations in the ER compartment, along with subsequent activation of the unfolded protein response (UPR), were among the first ultrastructural and molecular changes associated with early onset of WB (Papah et al., 2017). Lake and Abasht (2020) have suggested that abnormal lipid accumulation at the perivascular level may induce phlebitis, resulting in ER stress and UPR activation, which could lead to the myodegeneration characterizing WB fillets.

Other theories deal with the compromised proliferation and differentiation of satellite cells (SCs), as well as a reduction in their number, probably due to exacerbation of SC activity (Daughtry et al., 2017; Phillips et al., 2020). Some authors have suggested that SCs attempt to repair and mitigate the muscle fiber degeneration process, resulting in the extensive fibrosis that characterizes WB fillets (Velleman, 2015; Zhang et al., 2015; Daughtry et al., 2017; Emami et al., 2021).

Despite the increase in knowledge of the complex etiology related to the occurrence of WB, the precise initial mechanism triggering the cascade of events leading to the onset of WB remains to be clarified. There is still a need to further explore pathophysiological mechanisms underlying WB occurrence and progression.

3 Main features characterizing wooden breast

Muscles affected by WB are characterized by focused or diffused firm/hardened consistency, which may be located in the cranial and/or caudal area of the muscle along with pale and swollen areas (Sihvo et al., 2017) (Fig. 1), resulting in inflexible muscle (Fig. 2). Severe WB fillets are often covered by viscous exudates and display petechial hemorrhages on their surface (Sihvo et al., 2014; Soglia et al., 2016; Papah et al., 2017).

Figure 1 Phenotype of a severe wooden breast (WB) muscle displaying an abnormally hardened consistency along with a prominent ridge in its cranial and caudal areas, respectively.

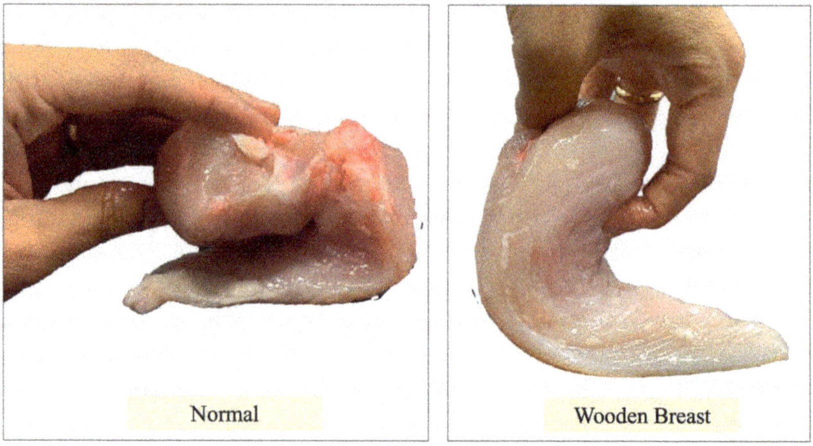

Figure 2 Flexibility of normal and WB *pectoralis major* muscle evidencing the distinctive stiffness associated with the occurrence of this abnormality.

At the early stage of WB development, isolated macroscopic lesions usually appear. As birds develop toward slaughter age, these diffuse throughout the whole fillet (Papah et al., 2017; Sihvo et al., 2017). Muscle firmness seems to originate at the cranial part of the breast and then extend to its full length (Kuttappan et al., 2016), thus determining the peculiar 'ridge-like bulge' at the caudal region of the PM muscle (Mudalal et al., 2015; Griffin et al., 2018). These features have been used to identify and evaluate WB severity (Papah et al., 2017; Petracci et al., 2019; Sihvo et al., 2017).

3.1 Criteria and methods for wooden breast evaluation

The most common methods to identify and classify WB breasts are subjective visual examination and/or palpation of muscles to determine the degree of hardness/stiffness and evaluate the presence of pale color, exudate and/or petechial hemorrhages (Petracci et al., 2019). This has allowed development of a 4-point scoring scale, in which breasts can be classified as (Petracci et al., 2019; Caldas-Cueva and Owens, 2020):

- normal (score 0 - no hardening areas and hemorrhages);
- mild (score 1 - isolated lightly firm areas diffused across the breast);
- moderate (score 2 - hardening and pale areas without hemorrhaging); and
- severe (score 3 - extremely hard and pale areas diffused throughout the muscle with hemorrhages).

Although these classification criteria are widely used worldwide, there is a need for more objective and raid techniques to identify and classify WB (Kuttappan et al., 2016; Caldas-Cueva et al., 2021a). Some authors have proposed near-infrared spectroscopy and computer vision systems as objective, rapid and non-destructive methods for WB detection and sorting (Geronimo et al., 2019). Caldas-Cueva et al. (2021b) and Sun et al. (2022a) have proposed the use of image analysis (e.g. image measurements regarding conformational changes in the breast region) to predict WB in broiler carcasses at the slaughterhouse. Another rapid, non-destructive technique during inline processing is measuring deformation of raw breast fillet surfaces from directing pressurized air on breasts (Sun et al., 2021).

Early in vivo detection of WB would greatly benefit the poultry industry since it could be used as a tool for preventative breeding selection programs. Kawasaki et al. (2016) have suggested assessing the ability of the birds to achieve back-to-back wing contact as a way to assess WB (Kawasaki et al., 2016; Hosotani et al., 2020). Simões et al. (2020) and Vieira et al. (2021) have proposed ultrasound as a diagnostic method (already used in human and veterinary medicine) to detect muscular injuries associated with WB in live birds. Castilho et al. (2021) have suggested the combination of ultrasound and muscular biopsy to accurately monitor WB occurrence and progression. Since the hardened muscle consistency of WB fillets is the clearest indicator of the defect, most methods used to assess its presence and severity are still based on features related to the hardness of WB fillets.

It has been shown that hardness associated with WB could be explained by the prominent fibrosis (i.e. proliferation and thickening of connective tissue; Sihvo et al., 2014) often observed in the chronic phase of WB (Petracci et al., 2019). Histological studies have helped to define the main microscopic features characterizing abnormal fillets.

3.2 Histological characteristics of abnormal wooden breasts

At the microscopic level, WB muscles exhibit variation in the size and shape of myofibers, showing evident polyphasic myodegeneration processes with inflammatory cell infiltration (Sihvo et al., 2017, 2014). Histologic changes characterizing WB fillets include extensive degeneration (up to lysis) of fibers, infiltration of lymphocytes and macrophages both at the perivascular (i.e. phlebitis) and myofiber (i.e. myositis) level, along with lipid deposition and fibrosis (Sihvo et al., 2017; Soglia et al., 2016, 2019; Chen et al., 2019). Myodegeneration is also accompanied by the occurrence of occasional polyphasic regeneration processes shown by nuclear rowing and myotube arrangement (Kuttappan et al., 2013; Papah et al., 2017; Soglia et al., 2020).

Early microscopic lesions (e.g. perivascular lipid accumulation and phlebitis) are detectable as early as from the first week after birth, suggesting that vascular inflammation may be the first alteration linked to WB (Abasht et al., 2021). Connective tissue accumulation and myofiber regeneration seem to be secondary to phlebitis (Sihvo et al., 2017). Histological lesions (e.g. myodegeneration, lipid deposition and fibrosis) tend to increase with severity of macroscopic changes as well as with age (Radaelli et al., 2017; Griffin et al., 2018). The skin-facing surface of the muscle seems to be more severely affected than the deeper layers (Soglia et al., 2016). There is an appreciable overlap between the microscopic features of the PM muscle of broilers affected by WB and their unaffected counterparts (Papah et al., 2017). Myodegeneration, inflammatory cell infiltration and a loss of normal muscle architecture have been frequently found as common features of skeletal muscles belonging to fast-growing broilers (Sihvo et al., 2017), including those having a macroscopically normal appearance (Papah et al., 2017; Radaelli et al., 2017; Sihvo et al., 2017).

The microscopic and biochemical changes characterizing WB fillets cause profound alterations in muscle composition, with detrimental effects on key quality traits of chicken meat (Soglia et al., 2019).

4 Effects of wooden breast on poultry meat quality

A great deal of recent research has investigated how WB affects poultry meat quality traits. Depending on severity, muscle abnormality results in an alteration of the chemical composition of chicken breast, negatively affecting the nutritional value and technological properties of both raw and processed meat (Mazzoni et al., 2014; Tasoniero et al., 2016; Bordignon et al., 2022) (see Fig. 3). Bordignon et al. (2022) have concluded that, among other growth-related

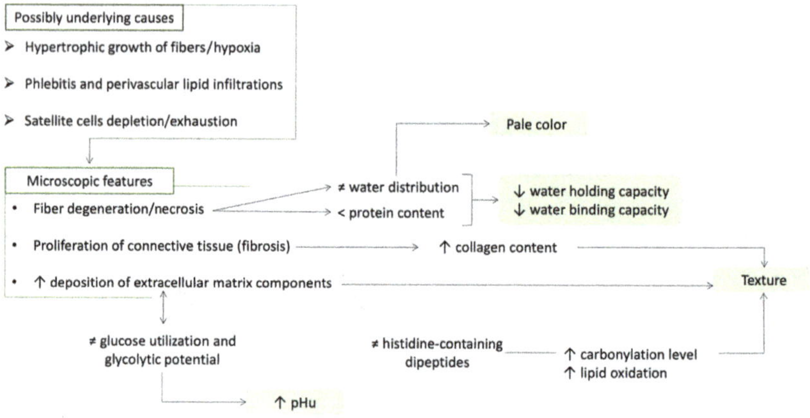

Figure 3 Schematic representation of the potential underlying causes and microscopic features leading to the impairment of the technological and quality traits characterizing WB.

abnormalities (i.e. white striping and spaghetti meat), WB most affects meat quality, with major changes to technological and sensory properties such as pH, cooking losses, texture, general appearance and nutritional value.

4.1 Implications on proximal composition and nutritional value of poultry meat

Chicken breast muscles affected by WB are characterized by a higher content of collagen, moisture and fat coupled with a lower protein content (Tasoniero et al., 2016; Tijare et al., 2016; Baldi et al., 2018; Dalle Zotte et al., 2020; Oliveira et al., 2021). This reduces the nutritional value of WB fillets (Kuttappan et al., 2021).

A number of studies highlight the significant reduction in high-quality protein content (Soglia et al., 2016; Wold et al., 2017; Cai et al., 2018), which seems to be mainly determined by the major alteration of muscle structure characterizing WB fillets. The significant myodegeneration affecting PM muscles, causing considerable degeneration of myofibrillar proteins, may be one of the main causes of decreased protein content in WB fillets (Petracci et al., 2014). Dalle Zotte et al. (2020) found that WB also influences the amino acid profile of affected PM muscles, decreasing the content of several amino acids (e.g. arginine, histidine, leucine, threonine, valine and others), resulting in a lower nutritional value of WB meat compared to normal and slow-growing chickens.

The elevated collagen-to-total protein ratio also contributes to significantly reducing the nutritional value of severely abnormal PM muscles (Petracci et al., 2014). Many studies have found a higher amount of collagen content in WB compared to normal breast meat (Clark and Velleman, 2017; Maharjan et al., 2020b; Oliveira et al., 2021; Geronimo et al., 2022), likely due to the profound fibrotic condition characterizing severely affected breasts. The low digestibility of collagen proteins together with the significantly reduced protein content are two major features responsible for the impaired nutritional value of affected breasts (Soglia et al., 2016).

It has also been shown that breast meat affected by WB exhibits an increased lipid content (Petracci et al., 2013a; Baldi et al., 2019; Thanatsang et al., 2020; Li et al., 2022). Increased fat content may be linked to the process of adipose tissue deposition (i.e. lipidosis) replacing degenerated WB muscle fibers (Velleman and Clark, 2015; Soglia et al., 2016). However, Oliveira et al. (2021) reported contradictory results by showing lower fat and cholesterol content in severe WB cases compared to normal samples.

Another aspect of the impaired nutritional value of WB meat is the oxidative damage affecting proteins and lipids (Estévez, 2015; Estévez and Luna, 2017). Protein oxidation contributes to a reduction of the functional properties and nutritional quality of proteins in chicken meat (Estévez, 2015; Thanatsang et al., 2020). The higher carbonylation level observed in WB-affected meat reflects more intense oxidative reactions affecting the protein fraction in these meats (Soglia et al., 2016). Besides, PM muscles affected by WB are more susceptible to lipid oxidation than normal meat (Soglia et al., 2016), as suggested by higher thiobarbituric acid reactive substances (TBARs) content (Li et al., 2022). Since lipid and protein contents seem to interact during oxidative processes taking place in the muscle (Estévez, 2015), it has been hypothesized that lipid oxidation may accelerate protein oxidation (Thanatsang et al., 2020), resulting in a general deterioration of the nutritional properties of WB muscles.

Overall, the onset of WB results in a significant reduction in the nutritional value of chicken breasts, caused both by the increased lipid and moisture content of the meat and by the impairment of protein quality and content (Soglia et al., 2019).

4.2 Impacts on technological properties of poultry meat

Severe WB reduces meat protein functionality, which affects subsequent processing (Oliveira et al., 2021). It is widely known that technological properties of poultry (such as water holding capacity – WHC) are significantly influenced by protein functionality. WHC influences quality characteristics such as final tenderness of the meat and/or meat products and the ability to form stable gels and emulsify lipids, which are essential for further processing (Pearce et al., 2011). Several studies show poor WHC in PM muscles affected by WB (Mudalal et al., 2015; Soglia et al., 2016; Tijare et al., 2016; Wold et al., 2017; Oliveira et al., 2021). This may be explained by the severe degeneration of muscle tissue affected by WB, particularly myofibrillar loss (Soglia et al., 2016). The degeneration of myofibrillar proteins appears to lead to a higher proportion of extra-myofibrillar water, which is less tightly bound compared to that observed in normal meat, contributing to the increased drip and cooking losses that characterize abnormal breast meat (Pang et al., 2020). Oxidative processes characterizing WB fillets may also impair the ability of meat proteins to hold water, contributing to poorer WHC (Estévez, 2015).

Breast muscles affected by growth-related abnormalities such as WB exhibit significantly higher ultimate pH (pH_u) (Baldi et al., 2020; Bordignon et al., 2022). This corroborates the role of protein degeneration/degradation in determining much poorer WHC observed in WB breasts. According to the so-called 'net charge effect', the higher pHu observed in WB should be associated with a negative charge with a repulsive effect, ultimately expanding the protein structure that water can infiltrate (Aberle et al., 2012).

Zambonelli et al. (2016) have shown that genes involved in glycolytic metabolisms are overexpressed in WB samples, even with increased pH_u.

These findings, together with changes in glycogen as well as concentrations of glycolysis intermediate products observed in WB (Abasht et al., 2016), were hypothesized to arise from alternative glucose utilization (e.g. pentose phosphate and glucuronic acid pathways) rather than from its availability. Other authors have suggested that reduced carbohydrate metabolism may lead to lower glycolytic potential, which, in turn, is responsible for higher pH_u found in WB samples (Kuttappan et al., 2017).

Mudalal et al. (2015) and Bowker et al. (2018) have found that WB negatively affects marination commonly used to improve the sensory and texture properties of meat (Tijare et al., 2016). Such evidence highlights the relevant detrimental effects of WB on the main quality and technological properties of chicken meat. This points out the need to develop strategies to minimize the negative effects of WB occurrence.

5 Potential strategies to mitigate wooden breast

WB results in major changes in the texture, appearance and functionality of affected meat. There are two main approaches to dealing with these quality issues:

- nutritional strategies to reduce WB occurrence and severity; and
- improving the sensory and technological properties of WB meat.

These are discussed in the following sections.

5.1 Dietary strategies

Numerous nutritional interventions have been considered to reduce the incidence and severity of WB in the poultry meat industry. Dietary strategies are mostly focused on (Kuttappan et al., 2021):

- slowing down growth rates; and
- using antioxidant-based diets to reduce negative effects of WB.

Some studies have proposed dietary restriction, reduced nutrient content and lysine depletion as potential strategies to reduce the occurrence of WB, with variable degrees of success (Cruz et al., 2017; Livingston et al., 2019; Meloche et al., 2018; Soglia et al., 2021b). These studies suggest that the reduction of WB development is possible but generally accompanied by decreased final slaughter weight and/or breast yield (Petracci et al., 2019; Kuttappan et al., 2021).

Looking at antioxidant-based approaches, there have been mixed results from supplementing diets with arginine, vitamin C, selenium (Se) and trace minerals (Sirri et al., 2016; Bodle et al., 2018). Vieira et al. (2022) showed that dietary supplementation of hydroxy-selenomethionine as an organic source of Se does not affect the occurrence and severity of WB. In contrast, it has been demonstrated that increasing diets with digestible arginine and vitamin C during the grower phase could mitigate WB (Bodle et al., 2018). However, the reduction in WB was also accompanied by a reduction in chicken body weight, feed conversion ratio and carcass yield. In contrast, Zampiga et al. (2018) showed that dietary supplementation with *L*-arginine seems to have positive effects in reducing incidence of WB, without significantly affecting production traits. Lackner et al. (2022) have shown that increasing the histidine-to-lysine ratio in broiler diets can decrease the incidence of WB, though high histidine supplementation slightly reduced broiler performance. Kuttappan et al. (2021) found a reduction of oxidative stress and a lower WB in fillet PM muscle by combining dietary antioxidants (i.e. ethoxyquin), chelate trace minerals (i.e. Zn, Cu and Mn) and organic Se, which may induce activation of endogenous antioxidant enzymes. However, these strategies do not significantly mitigate the incidence and negative effects of WB.

Given the importance of gut health, some authors have investigated the effect of dietary antibiotics and probiotics (e.g. *Bacillus subtilis*) on WB incidence. Interestingly, Jia et al. (2022) have shown that this type of dietary supplementation may increase WB in chicken broilers, probably due to increased growth performance.

Overall, to date, there is a lack of practical nutritional solutions allowing to achieve a relevant reduction in WB abnormality incidence without negatively affecting the live and slaughter performances of modern broiler chickens.

5.2 Processing strategies

Lima et al. (2022) found that papain, a heat-stable cysteine protease (Barekat and Soltanizadeh, 2019), can significantly improve the texture of WB meat, predominantly by degrading chemical bonds between myofibrillar and collagen proteins, thus reducing the hardness of affected breasts. However, low-frequency ultrasound (135 W at 40 kHz) (i.e. acoustic energy produced by waves with a vibration that can affect muscle integrity and modify collagen structure) was not found to improve the texture of WB fillets (Lima et al., 2022). In contrast, Li et al. (2021) reported that a 600 W with 20 kHz ultrasound treatment can effectively improve the texture of WB fillets.

Further processing solutions to alleviate the negative effects of WB include vacuum-tumbling marination to improve the texture of affected muscles (Bowker et al., 2018; Maxwell et al., 2018). In general, marination (i.e. the process of soaking meat with a solution that includes salt, phosphates and other additives) significantly reduces the toughness of meat and is a common method to add value to poultry meat (Bowker et al., 2018; Lima et al., 2022). However, it is well known that WB reduces the ability of meat to absorb a marinade solution (Tijare et al., 2016; U-chapaj et al., 2017; Maxwell et al., 2018; Bowker et al., 2018; Kuttappan et al., 2021). The impaired marinade uptake and reduced ability to hold water of WB fillets reduce the scope for processing strategies that, in normal conditions, could improve sensory and technological properties of poultry meat (e.g. texture).

Other technological strategies take into consideration the possible incorporation of WB meat into further processed chicken products, such as nuggets, sausages, patties and meatballs (Caldas-Cueva and Owens, 2020; Sun et al., 2022b). However, Chen et al. (2018) highlighted the difficulties in using WB meat in gel-type meat products, due to the severely impaired functional properties of WB meat (e.g. decreased ability in binding water), which lead to lower WHC and textural characteristics compared to processed products made with normal breasts. On the other hand, Sun et al. (2022b) recently demonstrated that using up to 25% WB meat could still produce chicken meatballs suitable for the industry and acceptable for consumers (Sun et al., 2022b).

In light of the above, and considering the noticeably adverse effects of WB on the principal technological properties of chicken meat, there are still challenges in finding processing strategies able to mitigate its impact.

6 Conclusion

WB mainly affects the PM muscles of meat-type chickens leading to significant economic losses for the poultry industry. The onset and progression of this defect results in profoundly altered meat quality traits and technological properties, which ultimately impair its suitability to be sold either as fresh meat or as raw material for further processing. Research in the past decade to elucidate the mechanisms leading to the development of WB has evidenced the potential implication of chickens' selection for fast-growing and high breast yield hybrids. However, its primary causes are still not fully understood and require further investigation. Although some dietary measures to reduce the incidence of WB and processing strategies to mitigate negative effects on meat quality have been tested, all can be considered only partially effective. One of the main challenges for the future is to develop markers for early identification of this defect.

7 Where to look for further information

The following articles provide a good overview of the subject:

- Velleman, S. G. 2015. Relationship of skeletal muscle development and growth to breast muscle myopathies: a review. *Avian Dis.* 59(4), 525–531. doi:10.1637/11223-063015-Review.1.

- Petracci, M., Soglia, F. and Berri, C. 2017. Muscle metabolism and meat quality abnormalities, in: Petracci, M. and Berri, C. (Eds.), *Poultry Quality Evaluation: Quality Attributes and Consumer Values*. 1st ed. Woodhead Publishing, pp. 51–75.
- Petracci, M., Soglia, F., Madruga, M., Carvalho, L., Ida, E. and Estévez, M. 2019. Wooden-breast, white striping, and spaghetti meat: causes, consequences and consumer perception of emerging broiler meat abnormalities. *Compr. Rev. Food Sci. Food Saf.* 18, pp. 565–583. doi: 10.1111/1541-4337.12431.
- Caldas-Cueva, J. P. and Owens, C. M. 2020. A review on the woody breast condition, detection methods, and product utilization in the contemporary poultry industry. *J. Animal Sci*. 98(7), 1–10. doi: 10.1093/jas/skaa207.
- Soglia, F., Petracci, M., Davoli, R., Zappaterra, M. 2021. A critical review of the mechanisms involved in the occurrence of growth-related abnormalities affecting broiler chicken breast muscles. *Poult. Sci.* 100, 101180. doi:10.1016/J.PSJ.2021.101180.

Key research in this area can be found at the following organizations:

- United States Department of Agriculture (USDA) (www.usda.gov).
- World's Poultry Science Association (WPSA) (www.wpsa.com).
- American Meat Science Association (www.meatscience.org).

8 References

Abasht, B., Mutryn, M. F., Michalek, R. D. and Lee, W. R. 2016. Oxidative stress and metabolic perturbations in wooden breast disorder in chickens. *PLoS ONE* 11(4), e0153750. https://doi.org/10.1371/JOURNAL.PONE.0153750.

Abasht, B., Papah, M. B. and Qiu, J. 2021. Evidence of vascular endothelial dysfunction in wooden breast disorder in chickens: insights through gene expression analysis, ultra-structural evaluation and supervised machine learning methods. *PLoS ONE* 16(1), e0243983. https://doi.org/10.1371/JOURNAL.PONE.0243983.

Aberle, E., Forrest, J., Gerrard, D., Merkel, R. A., Mills, E. and Judge, M. 2012. *Principles of Meat Science*. 5th edn, Kendall Hunt Publishing.

Alnahhas, N., Berri, C., Chabault, M., Chartrin, P., Boulay, M., Bourin, M. C. and Bihan-Duval, E. Le 2016. Genetic parameters of white striping in relation to body weight, carcass composition, and meat quality traits in two broiler lines divergently selected for the ultimate pH of the pectoralis major muscle. *BMC Genet.* 17, 61. https://doi.org/10.1186/s12863-016-0369-2.

Bailey, R. A., Watson, K. A., Bilgili, S. F. and Avendano, S. 2015. The genetic basis of pectoralis major myopathies in modern broiler chicken lines. *Poult. Sci.* 94(12), 2870–2879. https://doi.org/10.3382/ps/pev304.

Baldi, G., Soglia, F., Laghi, L., Tappi, S., Rocculi, P., Tavaniello, S., Prioriello, D., Mucci, R., Maiorano, G. and Petracci, M. 2019. Comparison of quality traits among breast meat affected by current muscle abnormalities. *Food Res. Int.* 115, 369–376. https://doi.org/10.1016/j.foodres.2018.11.020.

Baldi, G., Soglia, F., Mazzoni, M., Sirri, F., Canonico, L., Babini, E., Laghi, L., Cavani, C. and Petracci, M. 2018. Implications of white striping and spaghetti meat abnormalities on meat quality and histological features in broilers. *Animal* 12(1), 164–173. https://doi.org/10.1017/S1751731117001069.

Baldi, G., Yen, C. N., Daughtry, M. R., Bodmer, J., Bowker, B. C., Zhuang, H., Petracci, M. and Gerrard, D. E. 2020. Exploring the factors contributing to the high ultimate pH of broiler pectoralis major muscles affected by wooden breast condition. *Front. Physiol.* 11, 343. https://doi.org/10.3389/fphys.2020.00343.

Barbut, S. 2019. Recent myopathies in broiler's breast meat fillets. *World's Poult. Sci. J.* 75(4), 559–582. https://doi.org/10.1017/S0043933919000436.

Barekat, S. and Soltanizadeh, N. 2019. Application of high-intensity ultrasonic radiation coupled with papain treatment to modify functional properties of beef longissimus lumborum. *J. Food Sci. Technol.* 56(1), 224–232. https://doi.org/10.1007/s13197-018-3479-1.

Bodle, B. C., Alvarado, C., Shirley, R. B., Mercier, Y. and Lee, J. T. 2018. Evaluation of different dietary alterations in their ability to mitigate the incidence and severity of woody breast and white striping in commercial male broilers. *Poult. Sci.* 97(9), 3298–3310. https://doi.org/10.3382/PS/PEY166.

Bordignon, F., Xiccato, G., Boskovic Cabrol, M., Birolo, M. and Trocino, A. 2022. Factors affecting breast myopathies in broiler chickens and quality of defective meat: a meta-analysis. *Front. Physiol.* 0, 1322. https://doi.org/10.3389/FPHYS.2022.933235.

Bordini, M., Soglia, F., Davoli, R., Zappaterra, M., Petracci, M. and Meluzzi, A. 2022. Molecular pathways and key genes associated with breast width and protein content in white striping and wooden breast chicken pectoral muscle. *Front. Physiol.* 0, 1377. https://doi.org/10.3389/FPHYS.2022.936768.

Bordini, M., Zappaterra, M., Soglia, F., Petracci, M. and Davoli, R. 2021. Weighted gene co-expression network analysis identifies molecular pathways and hub genes involved in broiler white striping and wooden breast myopathies. *Sci. Rep.* 11(1), 1776. https://doi.org/10.1038/s41598-021-81303-7.

Bowker, B. C., Maxwell, A. D., Zhuang, H. and Adhikari, K. 2018. Marination and cooking performance of portioned broiler breast fillets with the wooden breast condition. *Poult. Sci.* 97(8), 2966–2970. https://doi.org/10.3382/PS/PEY144.

Brothers, B., Zhuo, Z., Papah, M. B. and Abasht, B. 2019. RNA-seq analysis reveals spatial and sex differences in pectoralis major muscle of broiler chickens contributing to difference in susceptibility to wooden breast disease. *Front. Physiol.* 10, 764. https://doi.org/10.3389/fphys.2019.00764.

Cai, K., Shao, W., Chen, X., Campbell, Y. L., Nair, M. N., Suman, S. P., Beach, C. M., Guyton, M. C. and Schilling, M. W. 2018. Meat quality traits and proteome profile of woody broiler breast (pectoralis major) meat. *Poult. Sci.* 97(1), 337–346. https://doi.org/10.3382/PS/PEX284.

Caldas-Cueva, J. P., Mauromoustakos, A., Sun, X. and Owens, C. M. 2021a. Detection of woody breast condition in commercial broiler carcasses using image analysis. *Poult. Sci.* 100(4), 100977. https://doi.org/10.1016/J.PSJ.2020.12.074.

Caldas-Cueva, J. P., Mauromoustakos, A., Sun, X. and Owens, C. M. 2021b. Use of image analysis to identify woody breast characteristics in 8-week-old broiler carcasses. *Poult. Sci.* 100(4), 100890. https://doi.org/10.1016/J.PSJ.2020.12.003.

Caldas-Cueva, J. P. and Owens, C. M. 2020. A review on the woody breast condition, detection methods, and product utilization in the contemporary poultry industry. *J. Anim. Sci.* 98(8) https://doi.org/10.1093/jas/skaa207.

Castilho, V. A. R. de, Komiyama, C. M., Burbarelli, M. F. C., Fernandes, A. R. M., Garcia, R. G., Barbosa, D. K., Przybulinski, B. B., Serpa, F. C., Valentim, J. K. and Leonardo, R. L. 2021. Techniques for in vivo assessment of myopathies in broiler chicken breasts using a biopsy as a support tool. *Avian Pathol.* 50(6), 477–489. https://doi.org/10.1080/03079457.2021.1970107.

Che, S., Wang, C., Varga, C., Barbut, S. and Susta, L. 2022. Prevalence of breast muscle myopathies (spaghetti meat, woody breast, white striping) and associated risk factors in broiler chickens from Ontario Canada. *PLoS ONE* 17(4), e0267019. https://doi.org/10.1371/journal.pone.0267019.

Chen, H., Wang, H. H., Qi, J., Wang, M., Xu, X. and Zhou, G. 2018. Chicken breast quality - normal, pale, soft and exudative (PSE) and woody - influences the functional properties of meat batters. *Int. J. Food Sci. Technol.* 53(3), 654–664. https://doi.org/10.1111/IJFS.13640.

Chen, L. R., Suyemoto, M. M., Sarsour, A. H., Cordova, H. A., Oviedo-Rondón, E. O., Wineland, M., Barnes, H. J. and Borst, L. B. 2019. Temporal characterization of wooden breast myopathy ("woody breast") severity and correlation with growth rate and lymphocytic phlebitis in three commercial broiler strains and a random-bred broiler strain. *Avian Pathol.* 48(4), 319–328. https://doi.org/10.1080/03079457.2019.1598541.

Clark, D. L. and Velleman, S. G. 2017. Spatial influence on breast muscle morphological structure, myofiber size, and gene expression associated with the wooden breast myopathy in broilers. *Poult. Sci.* 95(12), 2930–2945. https://doi.org/10.3382/ps/pew243.

Cruz, R. F. A., Vieira, S. L., Kindlein, L., Kipper, M., Cemin, H. S. and Rauber, S. M. 2017. Occurrence of white striping and wooden breast in broilers fed grower and finisher diets with increasing lysine levels. *Poult. Sci.* 96(2), 501–510. https://doi.org/10.3382/PS/PEW310.

da Rocha, T. C., Olegario, L. S., de Carvalho, L. M., Pereira, D. A., González-Mohino, A., Ventanas, S., Estévez, M. and Madruga, M. S. 2022. Consumer behaviour towards chicken breasts affected with myopathy (wooden breast): face-to-face vs. online tests. *Int. J. Food Sci. Technol.* 57(8), 5514–5522. https://doi.org/10.1111/ijfs.15892.

Dalle Zotte, A., Ricci, R., Cullere, M., Serva, L., Tenti, S. and Marchesini, G. 2020. Research Note. Research Note: Effect of chicken genotype and white striping-wooden breast condition on breast meat proximate composition and amino acid profile. *Poult. Sci.* 99(3), 1797–1803. https://doi.org/10.1016/J.PSJ.2019.10.066.

Dalle Zotte, A., Tasoniero, G., Puolanne, E., Remignon, H., Cecchinato, M., Catelli, E. and Cullere, M. 2017. Effect of "Wooden breast" appearance on poultry meat quality, histological traits, and lesions characterization. *Czech J. Anim. Sci.* 62(2), 51–57. https://doi.org/10.17221/54/2016-CJAS.

Daughtry, M. R., Berio, E., Shen, Z., Suess, E. J. R., Shah, N., Geiger, A. E., Berguson, E. R., Dalloul, R. A., Persia, M. E., Shi, H. and Gerrard, D. E. 2017. Satellite cell-mediated breast muscle regeneration decreases with broiler size. *Poult. Sci.* 96(9), 3457–3464. https://doi.org/10.3382/ps/pex068.

Emami, N. K., Cauble, R. N., Dhamad, A. E., Greene, E. S., Coy, C. S., Velleman, S. G., Orlowski, S., Anthony, N., Bedford, M. and Dridi, S. 2021. Hypoxia further exacerbates woody breast myopathy in broilers via alteration of satellite cell fate. *Poult. Sci.* 100(7), 101167. https://doi.org/10.1016/J.PSJ.2021.101167.

Estévez, M. 2015. Oxidative damage to poultry: from farm to fork. *Poult. Sci.* 94(6), 1368–1378. https://doi.org/10.3382/PS/PEV094.

Estévez, M. and Luna, C. 2017. Dietary protein oxidation: a silent threat to human health? *Crit. Rev. Food Sci. Nutr.* 57(17), 3781–3793. https://doi.org/10.1080/10408398.2016.1165182.

Geronimo, B. C., Mastelini, S. M., Carvalho, R. H., Barbon Júnior, S., Barbin, D. F., Shimokomaki, M. and Ida, E. I. 2019. Computer vision system and near-infrared spectroscopy for identification and classification of chicken with wooden breast, and physicochemical and technological characterization. *Infrared Phys. Technol.* 96, 303–310. https://doi.org/10.1016/J.INFRARED.2018.11.036.

Geronimo, B. C., Prudencio, S. H. and Soares, A. L. 2022. Biochemical and technological characteristics of wooden breast chicken fillets and their consumer acceptance. *J. Food Sci. Technol.* 59(3), 1185–1192. https://doi.org/10.1007/s13197-021-05123-3.

Griffin, J. R., Moraes, L., Wick, M. and Lilburn, M. S. 2018. Onset of white striping and progression into wooden breast as defined by myopathic changes underlying pectoralis major growth. Estimation of growth parameters as predictors for stage of myopathy progression. *Avian Pathol.* 47(1), 2–13. https://doi.org/10.1080/03079457.2017.1356908.

Havenstein, G. B., Ferket, P. R. and Qureshi, M. A. 2003. Growth, livability, and feed conversion of 1957 versus 2001 broilers when fed representative 1957 and 2001 broiler diets. *Poult. Sci.* 82(10), 1500–1508. https://doi.org/10.1093/PS/82.10.1500.

Hosotani, M., Kawasaki, T., Hasegawa, Y., Wakasa, Y., Hoshino, M., Takahashi, N., Ueda, H., Takaya, T., Iwasaki, T. and Watanabe, T. 2020. Physiological and pathological mitochondrial clearance is related to pectoralis major muscle pathogenesis in broilers with wooden breast syndrome. *Front. Physiol.* 11, 579. https://doi.org/10.3389/fphys.2020.00579.

Hoving-Bolink, A. H., Kranen, R. W., Klont, R. E., Gerritsen, C. L. M. and De Greef, K. H. 2000. Fibre area and capillary supply in broiler breast muscle in relation to productivity and ascites. *Meat Sci.* 56(4), 397–402. https://doi.org/10.1016/S0309-1740(00)00071-1.

Hubert, S. M. and Athrey, G. 2020. Energy metabolism and sources of oxidative stress in wooden breast-a review. *F1000Res* 9, 319. https://doi.org/10.12688/f1000research.23296.1.

Hubert, S. M., Williams, T. J. and Athrey, G. 2018. Insights into the molecular basis of wooden breast based on comparative analysis of fast- and slow-growth broilers. *bioRxiv*, 356683. https://doi.org/10.1101/356683.

Iwasaki, T., Watanabe, T., Hasegawa, Y., Hosotani, M. and Kawasaki, T. 2021. Nutrition during the early rearing period affects the incidence of wooden breasts in broilers. *J. Poult. Sci.* 58(3), 177–185. https://doi.org/10.2141/jpsa.0200034.

Jia, L., Zhang, X., Li, X., Schilling, W., David Peebles, E., Kiess, A. S., Zhai, W. and Zhang, L. 2022. Bacitracin, Bacillus subtilis, and Eimeria spp. challenge exacerbates woody breast incidence and severity in broilers. *Poult. Sci.* 101(1), 101512. https://doi.org/10.1016/j.psj.2021.101512.

Kang, K., Zhou, N., Peng, W., Peng, F., Ma, M., Li, L., Fu, F., Xiang, S., Zhang, H., He, X. and Song, Z. 2022. Multi-omics analysis of the microbiome and metabolome reveals the relationship Between the gut microbiota and wooden breast myopathy in broilers. *Front. Vet. Sci.* 9, 922516. https://doi.org/10.3389/fvets.2022.922516.

Kawasaki, T., Yoshida, T. and Watanabe, T. 2016. Simple method for screening the affected birds with remarkably hardened pectoralis major muscles among broiler chickens. *J. Poult. Sci.* 53(4), 291–297. https://doi.org/10.2141/jpsa.0160036.

Kuttappan, V. A., Lee, Y. S., Erf, G. F., Meullenet, J. F., McKee, S. R. and Owens, C. M. 2012. Consumer acceptance of visual appearance of broiler breast meat with varying degrees of white striping. *Poult. Sci.* 91(5), 1240–1247. https://doi.org/10.3382/ps.2011-01947. PMID: 22499884.

Kuttappan, V. A., Bottje, W., Ramnathan, R., Hartson, S. D., Coon, C. N., Kong, B. W., Owens, C. M., Vazquez-Añon, M. and Hargis, B. M. 2017. Proteomic analysis reveals changes in carbohydrate and protein metabolism associated with broiler breast myopathy. *Poult. Sci.* 96(8), 2992–2999. https://doi.org/10.3382/ps/pex069.

Kuttappan, V. A., Hargis, B. M. and Owens, C. M. 2016. White striping and woody breast myopathies in the modern poultry industry: a review. *Poult. Sci.* 95(11), 2724–2733. https://doi.org/10.3382/ps/pew216.

Kuttappan, V. A., Manangi, M., Bekker, M., Chen, J. and Vazquez-Anon, M. 2021. Nutritional intervention strategies using dietary antioxidants and organic trace minerals to reduce the incidence of wooden breast and other carcass quality defects in broiler birds. *Front. Physiol.* 12, 428. https://doi.org/10.3389/FPHYS.2021.663409/BIBTEX.

Kuttappan, V. A., Shivaprasad, H. L., Shaw, D. P., Valentine, B. A., Hargis, B. M., Clark, F. D., McKee, S. R. and Owens, C. M. 2013. Pathological changes associated with white striping in broiler breast muscles. *Poult. Sci.* 92(2), 331–338. https://doi.org/10.3382/ps.2012-02646.

Lackner, J., Hess, V., Stef, L. and Sauerwein, H. 2022. Effects of feeding different histidine to lysine ratios on performance, meat quality, and the occurrence of breast myopathies in broiler chickens. *Poult. Sci.* 101(2), 101568. https://doi.org/10.1016/j.psj.2021.101568.

Lahiri, S., Kim, H., Garcia-Perez, I., Reza, M. M., Martin, K. A., Kundu, P., Cox, L. M., Selkrig, J., Posma, J. M., Zhang, H., Padmanabhan, P., Moret, C., Gulyás, B., Blaser, M. J., Auwerx, J., Holmes, E., Nicholson, J., Wahli, W. and Pettersson, S. 2019. The gut microbiota influences skeletal muscle mass and function in mice. *Sci. Transl. Med.* 11(502). https://doi.org/10.1126/SCITRANSLMED.AAN5662.

Lake, J. A. and Abasht, B. 2020. Glucolipotoxicity: a proposed etiology for wooden breast and related myopathies in commercial broiler chickens. *Front. Physiol.* 11, 169. https://doi.org/10.3389/fphys.2020.00169.

Li, B., Dong, X., Puolanne, E. and Ertbjerg, P. 2022. Effect of wooden breast degree on lipid and protein oxidation and citrate synthase activity of chicken pectoralis major muscle. *LWT* 154, 112884. https://doi.org/10.1016/J.LWT.2021.112884.

Li, Z., Yang, Z., Zhang, Y., Lu, T., Zhang, X., Qi, Y., Wang, P. and Xu, X. 2021. Innovative characterization based on stress relaxation and creep to reveal the tenderizing effect of ultrasound on wooden breast. *Foods* 10(1), 195. https://doi.org/10.3390/FOODS10010195.

Lilburn, M. S., Griffin, J. R. and Wick, M. 2019. From muscle to food: oxidative challenges and developmental anomalies in poultry breast muscle. *Poult. Sci.* 98(10), 4255–4260. https://doi.org/10.3382/PS/PEY409.

Lima, J. L., Bezerra, T. K. A., Carvalho, L. M., Galvão, M. S., Lucena, L., Rocha, T. C., Estevez, M. and Madruga, M. S. 2022. Improving the poor texture and technological properties of chicken wooden breast by enzymatic hydrolysis and low-frequency ultrasound. *J. Food Sci.* 87(6), 2364–2376. https://doi.org/10.1111/1750-3841.16149.

Livingston, M. L., Landon, C., Barnes, H. J. and Brake, J. 2019. White striping and wooden breast myopathies of broiler breast muscle is affected by time-limited feeding, genetic background, and egg storage. *Poult. Sci.* 98(1), 217–226. https://doi.org/10.3382/ps/pey333.

Maharjan, P., Hilton, K., Weil, J., Suesuttajit, N., Beitia, A., Owens, C. M. and Coon, C. 2020a. Characterizing woody breast myopathy in a meat broiler line by heat production, microbiota, and plasma metabolites. *Front. Vet. Sci.* 6, 497. https://doi.org/10.3389/FVETS.2019.00497/BIBTEX.

Maharjan, P., Owens, C. M. and Coon, C. 2020b. In-vivo intramuscular collagen synthesis, muscle fiber growth and histomorphology of pectoralis major of a fast-growing broiler strain gallus gallus domesticus. *Front. Vet. Sci.* 6, 1–11. https://doi.org/10.3389/fvets.2019.00470.

Malila, Y., Uengwetwanit, T., Arayamethakorn, S., Srimarut, Y., Thanatsang, K. V., Soglia, F., Strasburg, G. M., Rungrassamee, W. and Visessanguan, W. 2020. Transcriptional profiles of skeletal muscle associated With increasing severity of white striping in commercial broilers. *Front. Physiol.* 11, 580. https://doi.org/10.3389/fphys.2020.00580.

Maxwell, A. D., Bowker, B. C., Zhuang, H., Chatterjee, D. and Adhikari, K. 2018. Descriptive sensory analysis of marinated and non-marinated wooden breast fillet portions. *Poult. Sci.* 97(8), 2971–2978. https://doi.org/10.3382/PS/PEY145.

Mazzoni, M., Petracci, M., Meluzzi, A., Cavani, C., Clavenzani, P. and Sirri, F. 2014. *Relationship between Pectoralis Major Muscle Histology and Quality Traits of Chicken Meat Bird Processing and Sample Collection Classification of Sample by Histology and Immunohistochemistry Evaluations The Specimens Were Oriented for the Transverse Fiber* 123–130.

Meloche, K. J., Fancher, B. I., Emmerson, D. A., Bilgili, S. F. and Dozier, W. A. 2018. Effects of reduced dietary energy and amino acid density on pectoralis major myopathies in broiler chickens at 36 and 49 days of age1. *Poult. Sci.* 97(5), 1794–1807. https://doi.org/10.3382/ps/pex454.

Mudalal, S., Lorenzi, M., Soglia, F., Cavani, C. and Petracci, M. 2015. Implications of white striping and wooden breast abnormalities on quality traits of raw and marinated chicken meat. *Animal* 9(4), 728–734. https://doi.org/10.1017/S175173111400295X.

Mutryn, M. F., Brannick, E. M., Fu, W., Lee, W. R. and Abasht, B. 2015. Characterization of a novel chicken muscle disorder through differential gene expression and pathway analysis using RNA-sequencing. *BMC Genomics* 16, 1–19. https://doi.org/10.1186/S12864-015-1623-0/TABLES/3.

OECD. *FAO Agricultural Outlook 2021–2030*. https://doi.org/10.1787/agr-outl-data.

Oliveira, R. F., de Mello, J. L. M., Ferrari, F. B., Cavalcanti, E. N. F., de Souza, R. A., Pereira, M. R., Giampietro-ganeco, A., Villegas-cayllahua, E. A., Fidelis, H. de A., Fávero, M. S., Amoroso, L., de Souza, P. A. and Borba, H. 2021. Physical, chemical and histological characterization of pectoralis major muscle of broilers affected by wooden breast myopathy. *Anim* 11, 596. https://doi.org/10.3390/ANI11030596.

Pampouille, E., Berri, C., Boitard, S., Hennequet-Antier, C., Beauclercq, S. A., Godet, E., Praud, C., Jégo, Y. and Le Bihan-Duval, E. 2018. Mapping QTL for white striping in relation to breast muscle yield and meat quality traits in broiler chickens. *BMC Genomics* 19(1), 202. https://doi.org/10.1186/s12864-018-4598-9.

Pampouille, E., Hennequet-Antier, C., Praud, C., Juanchich, A., Brionne, A., Godet, E., Bordeau, T., Fagnoul, F., Le Bihan-Duval, E. and Berri, C. 2019. Differential expression and co-expression gene network analyses reveal molecular mechanisms and candidate biomarkers involved in breast muscle myopathies in chicken. *Sci. Rep.* 9(1), 14905. https://doi.org/10.1038/s41598-019-51521-1.

Pang, B., Bowker, B., Zhang, J., Yang, Y. and Zhuang, H. 2020. Prediction of water holding capacity in intact broiler breast fillets affected by the woody breast condition using time-domain NMR. *Food Control* 118, 107391. https://doi.org/10.1016/J.FOODCONT.2020.107391.

Papah, M. B., Brannick, E. M., Schmidt, C. J. and Abasht, B. 2018. Gene expression profiling of the early pathogenesis of wooden breast disease in commercial broiler chickens using RNA-sequencing. *PLoS ONE* 13(12), e0207346. https://doi.org/10.1371/journal.pone.0207346.

Papah, M. B., Brannick, E. M., Schmidt, C. J. and Abasht, B. 2017. Evidence and role of phlebitis and lipid infiltration in the onset and pathogenesis of wooden breast disease in modern broiler chickens. *Avian Pathol.* 46(6), 623-643. https://doi.org/10.1080/03079457.2017.1339346.

Pearce, K. L., Rosenvold, K., Andersen, H. J. and Hopkins, D. L. 2011. Water distribution and mobility in meat during the conversion of muscle to meat and ageing and the impacts on fresh meat quality attributes - a review. *Meat Sci.* 89(2), 111-124. https://doi.org/10.1016/j.meatsci.2011.04.007.

Petracci, M., Mudalal, S., Babini, E. and Cavani, C. 2014. Effect of white striping on chemical composition and nutritional value of chicken breast meat. *Ital. J. Anim. Sci.* 13(1), 179-183. https://doi.org/10.4081/ijas.2014.3138.

Petracci, M., Mudalal, S., Bonfiglio, A. and Cavani, C. 2013a. Occurrence of white striping under commercial conditions and its impact on breast meat quality in broiler chickens. *Poult. Sci.* 92(6), 1670-1675. https://doi.org/10.3382/ps.2012-03001.

Petracci, M., Sirri, F., Mazzoni, M. and Meluzzi, A. 2013b. Comparison of breast muscle traits and meat quality characteristics in 2 commercial chicken hybrids. *Poult. Sci.* 92(9), 2438-2447. https://doi.org/10.3382/ps.2013-03087.

Petracci, M., Mudalal, S., Soglia, F. and Cavani, C. 2015. Meat quality in fast-growing broiler chickens. *World's Poult. Sci. J.* 71(2), 363-374. https://doi.org/10.1017/S0043933915000367.

Petracci, M., Soglia, F., Madruga, M., Carvalho, L., Ida, E. and Estévez, M. 2019. Wooden-breast, white striping, and spaghetti meat: causes, consequences and consumer perception of emerging broiler meat abnormalities. *Compr. Rev. Food Sci. Food Saf.* 18(2), 565-583. https://doi.org/10.1111/1541-4337.12431.

Phillips, C. A., Reading, B. J., Livingston, M., Livingston, K. and Ashwell, C. M. 2020. Evaluation via supervised machine learning of the broiler pectoralis major and liver transcriptome in association With the muscle myopathy wooden breast. *Front. Physiol.* 11, 101. https://doi.org/10.3389/fphys.2020.00101.

Praud, C., Jimenez, J., Pampouille, E., Couroussé, N., Godet, E., Le Bihan-Duval, E. and Berri, C. 2020. Molecular phenotyping of white striping and wooden breast myopathies in chicken. *Front. Physiol.* 11, 633. https://doi.org/10.3389/fphys.2020.00633.

Radaelli, G., Piccirillo, A., Birolo, M., Bertotto, D., Gratta, F., Ballarin, C., Vascellari, M., Xiccato, G. and Trocino, A. 2017. Effect of age on the occurrence of muscle fiber degeneration associated with myopathies in broiler chickens submitted to feed restriction. *Poult. Sci.* 96(2), 309-319. https://doi.org/10.3382/PS/PEW270.

Santos, M. N., Rothschild, D., Widowski, T. M., Barbut, S., Kiarie, E. G., Mandell, I., Guerin, M. T., Edwards, A. M. and Torrey, S. 2021. In pursuit of a better broiler: carcass traits and muscle myopathies in conventional and slower-growing strains of broiler chickens. *Poult. Sci.* 100(9), 101309. https://doi.org/10.1016/j.psj.2021.101309.

Sihvo, H. K., Immonen, K. and Puolanne, E. 2014. Myodegeneration With fibrosis and regeneration in the pectoralis major muscle of broilers. *Vet. Pathol.* 51(3), 619–623. https://doi.org/10.1177/0300985813497488.

Sihvo, H. K., Lindén, J., Airas, N., Immonen, K., Valaja, J. and Puolanne, E. 2017. Wooden breast myodegeneration of pectoralis major muscle over the growth period in broilers. *Vet. Pathol.* 54(1), 119–128. https://doi.org/10.1177/0300985816658099.

Simões, C. T., Vieira, S. L., Stefanello, C., Kindlein, L., Ferreira, T., Favero, A. and Xavier, B. 2020. An in vivo evaluation of the effects of feed restriction regimens on wooden breast using ultrasound images as a predictive tool. *Br. Poult. Sci.* 61(5), 583–589. https://doi.org/10.1080/00071668.2020.1764909.

Sirri, F., Maiorano, G., Tavaniello, S., Chen, J., Petracci, M. and Meluzzi, A. 2016. Effect of different levels of dietary zinc, manganese, and copper from organic or inorganic sources on performance, bacterial chondronecrosis, intramuscular collagen characteristics, and occurrence of meat quality defects of broiler chickens. *Poult. Sci.* 95(8), 1813–1824. https://doi.org/10.3382/ps/pew064.

Soglia, F., Baldi, G., Laghi, L., Mudalal, S., Cavani, C. and Petracci, M. 2018. Effect of white striping on turkey breast meat quality. *Animal* 12(10), 2198–2204. https://doi.org/10.1017/S1751731117003469.

Soglia, F., Mazzoni, M. and Petracci, M. 2019. Spotlight on avian pathology: current growth-related breast meat abnormalities in broilers. *Avian Pathol.* 48(1), 1–3. https://doi.org/10.1080/03079457.2018.1508821.

Soglia, F., Mazzoni, M., Zappaterra, M., Di Nunzio, M., Babini, E., Bordini, M., Sirri, F., Clavenzani, P., Davoli, R. and Petracci, M. 2020. Distribution and expression of vimentin and desmin in broiler pectoralis major affected by the growth-related muscular abnormalities. *Front. Physiol.* 10, 1581. https://doi.org/10.3389/fphys.2019.01581.

Soglia, F., Mudalal, S., Babini, E., Di Nunzio, M., Mazzoni, M., Sirri, F., Cavani, C. and Petracci, M. 2016. Histology, composition, and quality traits of chicken pectoralis major muscle affected by wooden breast abnormality. *Poult. Sci.* 95(3), 651–659. https://doi.org/10.3382/ps/pev353.

Soglia, F., Petracci, M., Davoli, R. and Zappaterra, M. 2021a. A critical review of the mechanisms involved in the occurrence of growth-related abnormalities affecting broiler chicken breast muscles. *Poult. Sci.* 100(6), 101180. https://doi.org/10.1016/J.PSJ.2021.101180.

Soglia, F., Zampiga, M., Baldi, G., Malila, Y., Thanatsang, K. V., Srimarut, Y., Tatiyaborworntham, N., Unger, O., Klamchuen, A., Laghi, L., Petracci, M. and Sirri, F. 2021b. Lysine depletion during different feeding phases: effects on growth performances and meat quality of broiler chickens. *Animals (Basel)* 11(6), 1499. https://doi.org/10.3390/ANI11061499.

Sun, X., Maynard, C. J., Caldas-Cueva, J. P., Bai, Y., You, J. and Dong, Y. 2022a. Use of image analysis to evaluate morphometric measurements of broilor breast fillets affected by the woody breast condition. *J. Food Sci. Technol.* 59(11), 4457–4465 https://doi.org/10.1007/s13197-022-05525-x.

Sun, X., You, J., Xu, L., Zhou, D., Cai, H., Maynard, C. J. and Caldas-Cueva, J. P. 2022b. Quality properties of chicken meatballs prepared with varying proportions of woody breast meat. *J. Food Sci. Technol.* 59(2), 666–676. https://doi.org/10.1007/s13197-021-05060-1.

Sun, X., Maynard, C. J., Caldas-Cueva, J. P., Coon, C. N. and Owens, C. M. 2021. Using air deformation of raw fillet surfaces to identify severity of woody breast myopathy in broiler fillets. *LWT* 141, 110904. https://doi.org/10.1016/j.lwt.2021.110904.

Tasoniero, G., Cullere, M., Cecchinato, M., Puolanne, E. and Dalle Zotte, A. 2016. Technological quality, mineral profile, and sensory attributes of broiler chicken breasts affected by white striping and wooden breast myopathies. *Poult. Sci.* 95(11), 2707–2714. https://doi.org/10.3382/PS/PEW215.

Thanatsang, K. V., Malila, Y., Arayamethakorn, S., Srimarut, Y., Tatiyaborworntham, N., Uengwetwanit, T., Panya, A., Rungrassamee, W. and Visessanguan, W. 2020. Nutritional properties and oxidative indices of broiler breast meat affected by wooden breast abnormality. *Animals (Basel)* 10(12), 1–16. https://doi.org/10.3390/ani10122272.

Tijare, V. V., Yang, F. L., Kuttappan, V. A., Alvarado, C. Z., Coon, C. N. and Owens, C. M. 2016. Meat quality of broiler breast fillets with white striping and woody breast muscle myopathies. *Poult. Sci.* 95(9), 2167–2173. https://doi.org/10.3382/ps/pew129.

Trocino, A., Piccirillo, A., Birolo, M., Radaelli, G., Bertotto, D., Filiou, E., Petracci, M. and Xiccato, G. 2015. Effect of genotype, gender and feed restriction on growth, meat quality and the occurrence of white striping and wooden breast in broiler chickens. *Poult. Sci.* 94(12), 2996–3004. https://doi.org/10.3382/ps/pev296.

U-Chupaj, J., Malila, Y., Petracci, M., Benjakul, S. and Visessanguan, W. 2017. Effect of tumbling marination on marinade uptake of chicken carcass and parts quality. *Rev. Bras. Cienc Avic* 19(1), 61–68. https://doi.org/10.1590/1806-9061-2016-0380.

Velleman, S. G. 2015. Relationship of skeletal muscle development and growth to breast muscle myopathies: a review. *Avian Dis.* 59(4), 525–531. https://doi.org/10.1637/11223-063015-Review.1.

Velleman, S. G. and Clark, D. L. 2015. Histopathologic and myogenic gene expression changes associated with wooden breast in broiler breast muscles. *Avian Dis.* 59(3), 410–418. https://doi.org/10.1637/11097-042015-Reg.1.

Vieira, S. L., Simões, C. T., Kindlein, L., Ferreira, T. Z., Soster, P. and Stefanello, C. 2021. Progressive in vivo detection of wooden breast in broilers as affected by dietary energy and protein. *Poult. Sci.* 100(6), 101120. https://doi.org/10.1016/j.psj.2021.101120.

Vieira, S. L., Teixeira, V. Q., Simões, C. T., Soster, P., Kindlein, L. and Stefanello, C. 2022. Broiler meat production as affected by dietary supplemental hydroxy-selenomethionine. *Livest. Sci.* 259. https://doi.org/10.1016/j.livsci.2022.104912.

Whitton, C., Bogueva, D., Marinova, D. and Phillips, C. J. C. 2021. Are we approaching peak meat consumption? Analysis of meat consumption from 2000 to 2019 in 35 countries and its rela-tionship to gross domestic product. *Animals* 11(12). https://doi.org/10.3390/ani11123466.

Wold, J. P., Veiseth-Kent, E., Høst, V. and Løvland, A. 2017. Rapid on-line detection and grading of wooden breast myopathy in chicken fillets by near-infrared spectroscopy. *PLoS ONE* 12(3). https://doi.org/10.1371/JOURNAL.PONE.0173384.

Xing, T., Zhao, X., Zhang, L., Li, J. L., Zhou, G. H., Xu, X. L. and Gao, F. 2020. Characteristics and incidence of broiler chicken wooden breast meat under commercial conditions in China. *Poult. Sci.* 99(1), 620–628. https://doi.org/10.3382/PS/PEZ560.

Zambonelli, P., Zappaterra, M., Soglia, F., Petracci, M., Sirri, F., Cavani, C. and Davoli, R. 2016. Detection of differentially expressed genes in broiler pectoralis major muscle affected by white striping - wooden breast myopathies. *Poult. Sci.* 95(12), 2771-2785. https://doi.org/10.3382/ps/pew268.

Zampiga, M., Laghi, L., Petracci, M., Zhu, C., Meluzzi, A., Dridi, S. and Sirri, F. 2018. Effect of dietary arginine to lysine ratios on productive performance, meat quality, plasma and muscle metabolomics profile in fast-growing broiler chickens. *J. Anim. Sci. Biotechnol.* 9, 1-14. https://doi.org/10.1186/S40104-018-0294-5/FIGURES/4.

Zampiga, M., Soglia, F., Baldi, G., Petracci, M., Strasburg, G. M. and Sirri, F. 2020. Muscle abnormalities and meat quality consequences in modern turkey hybrids. *Front. Physiol.* 11, 554. https://doi.org/10.3389/fphys.2020.00554.

Zanetti, M. A., Tedesco, D. C., Schneider, T., Teixeira, S. T. F., Daroit, L., Pilotto, F., Dickel, E. L., Santos, S. P. and Dos Santos, L. Rd 2018. Economic losses associated with wooden breast and white striping in broilers. *Semin. Cienc. Agrar.* 39(2), 887-892. https://doi.org/10.5433/1679-0359.2018v39n2p887.

Zhang, P., Liang, X., Shan, T., Jiang, Q., Deng, C., Zheng, R. and Kuang, S. 2015. mTOR is necessary for proper satellite cell activity and skeletal muscle regeneration. *Biochem. Biophys. Res. Commun.* 463(1-2), 102-108. https://doi.org/10.1016/J.BBRC.2015.05.032.

Zhang, X., Zhang, L., Li, X., White, S., Campbell, Y. L., Ruiz, A. A., To, K. V., Zhai, W. and Schilling, M. W. 2021. Cecal microbiota contribute to the development of woody breast myopathy. *Poult. Sci.* 100(6), 101124. https://doi.org/10.1016/J.PSJ.2021.101124.

Chapter 11

Quality defects associated with poultry muscle development: spaghetti meat

Giulia Tasoniero and Brian Bowker, USDA-ARS, USA

1 Introduction

Over the past 50 years, the poultry industry has achieved tremendous advancements in broiler growth rate through intense genetic selection and improvements in bird nutrition and management. Increasing consumer demands for poultry products and the corresponding shifts in how broiler meat is marketed have put a premium on breast muscle development and yield in modern broilers. Unfortunately, along with the gains in live production traits and breast meat yield have come increased problems with breast meat quality. The increasing occurrence of growth-related myopathies in the pectoralis major of modern broilers has led to significant problems with breast meat abnormalities and defects that cost the industry millions of dollars annually.

The effects of the white striping (WS) and woody breast (WB) myopathies on breast meat composition and quality have been widely studied and reported over the last decade. In recent years, an emerging breast myopathy known as spaghetti meat (SM) has become an increasing concern to the industry, as it can result in substantial downgrading of high-value breast meat. In the first known report on this condition, the SM myopathy was referred to as 'poor cohesion' (Sirri et al., 2016) and in later reports as 'stringy-spongy' (Anton et al., 2019). Relative to the WB and WS myopathies, less is known about the SM myopathy

http://dx.doi.org/10.19103/AS.2022.0112.12
Published by Burleigh Dodds Science Publishing Limited, 2023.

with regards to its underlying causes, impact on the muscle tissue, or its effects on meat quality and functionality characteristics. However, in the last few years, the SM myopathy has garnered increased research focus, and there is a growing body of scientific literature on this meat quality defect. The initial wave of SM research has focused primarily on characterizing the macroscopic and microscopic effects of the myopathy on the pectoralis major and measuring the impact on meat composition and quality (Table 1). Researchers are increasingly investigating factors associated with SM development in fast-growing broilers

Table 1 Studies reporting the characteristics of broiler pectoralis major muscles exhibiting the spaghetti meat myopathy

Reported traits	References
Histological features	Baldi et al. (2018); Mazzoni et al. (2020); Soglia et al. (2020); Sanden et al. (2021); Che et al. (2022a); Tasoniero et al. (2022)
Morphometric properties	
Breast weight, length, width, thickness	Baldi et al. (2018); Che et al. (2022a)
Chemical properties	
Proximate composition (moisture, protein, lipid, ash)	Baldi et al. (2018, 2019); Tasoniero et al. (2020); Pascual et al. (2021)
Collagen characteristics	Baldi et al. (2018, 2019); Mazzoni et al. (2020); Tasoniero et al. (2020); Sanden et al. (2021)
Lipid and protein oxidation (TBAR, protein carbonyl)	Baldi et al. (2018); Soglia et al. (2019)
Free amino acids and metabolite profiles	Soglia et al. (2019)
Fatty acids profiles	Baldi et al. (2019)
Protein profiles (SDS-PAGE)	Baldi et al. (2018); Soglia et al. (2020); Tasoniero et al. (2020, 2022)
Technological properties	
Meat pH	Baldi et al. (2018, 2019); Soglia et al. (2019); Campo et al. (2020); Tasoniero et al. (2020); Pascual et al. (2021)
Color (L^*, a^*, b^* color values)	Baldi et al. (2018, 2019); Soglia et al. (2019); Campo et al. (2020); Tasoniero et al. (2020); Yu et al. (2020); Pascual et al. (2021)
Texture measurements (compression, shear force)	Baldi et al. (2019); Campo et al. (2020); Yu et al. (2020); Pascual et al. (2021); Che et al. (2022a)
Water-holding capacity measurements	Soglia et al. (2019); Tasoniero et al. (2020); Yu et al. (2020); Pascual et al. (2021)
Myowater characteristics	Baldi et al. (2018, 2019); Soglia et al. (2019); Tasoniero et al. (2022)
Protein functionality	Baldi et al. (2018, 2019); Tasoniero et al. (2020)

Published by Burleigh Dodds Science Publishing Limited, 2023.

and the underlying mechanisms that cause the onset and development of this myopathy.

2 Characteristics of spaghetti meat myopathy

2.1 Macroscopic traits

The SM myopathy can have a profound impact on the visual and physical characteristics of the breast muscle in broilers. Similar to the WS and WB conditions, the SM myopathy most severely affects the ventral-cranial portion of the pectoralis major muscle. To date, there have been no published reports of the SM myopathy developing in other muscles in broilers. Macroscopically, the ventral-cranial portion of SM-affected pectoralis major muscles possesses areas exhibiting impaired muscle integrity and a stringy, soft consistency (Fig. 1). The tendency toward separation of the muscle fiber bundles on the breast muscle surface results in the affected areas having the appearance of the long, thin, and cylindrical pasta, and the distinct phenotype for which the myopathy is named (Baldi et al., 2018). In more mild cases of SM, the ventral surface on the cranial end of the breast fillet may remain mostly intact and the stringy appearance may not be fully expressed. However, upon palpation, the surface may feel loose and somewhat detached from the tissue below the surface. Multiple myopathies often occur in the same breast fillet (Bowker et al., 2019; Che et al., 2022a,b). Breast fillets with the SM myopathy often exhibit the white striations associated with WS on the ventral surface. Breast fillets simultaneously exhibiting the distinguishing macroscopic traits of both the SM and WB myopathy seem to be less common. However, the odds of observing both SM and WB phenotypes in the same breast fillet increase with fillet thickness and weight (Che et al., 2022a).

2.1.1 Spaghetti meat assessment

Currently, the incidence of the SM myopathy phenotype is determined solely by the visual presence and degree of severity of the distinguishing SM macroscopic characteristics exhibited in the pectoralis major. Detection and classification of the SM condition in breast meat are subjective as there are only limited objective criteria and no definitive markers for the myopathy. The typical method for identifying SM currently being used in both controlled research studies and for product sorting in commercial settings consists of visually and tactilely assessing any mushy or stringy consistency and the overall impaired integrity of the pectoralis major muscle, especially within the ventral-cranial part of the fillets. Much like other breast myopathies, the SM condition can vary widely in its degree of severity. Most of the studies published on SM have either used a three-level scoring system or just simply classified fillets as normal (show

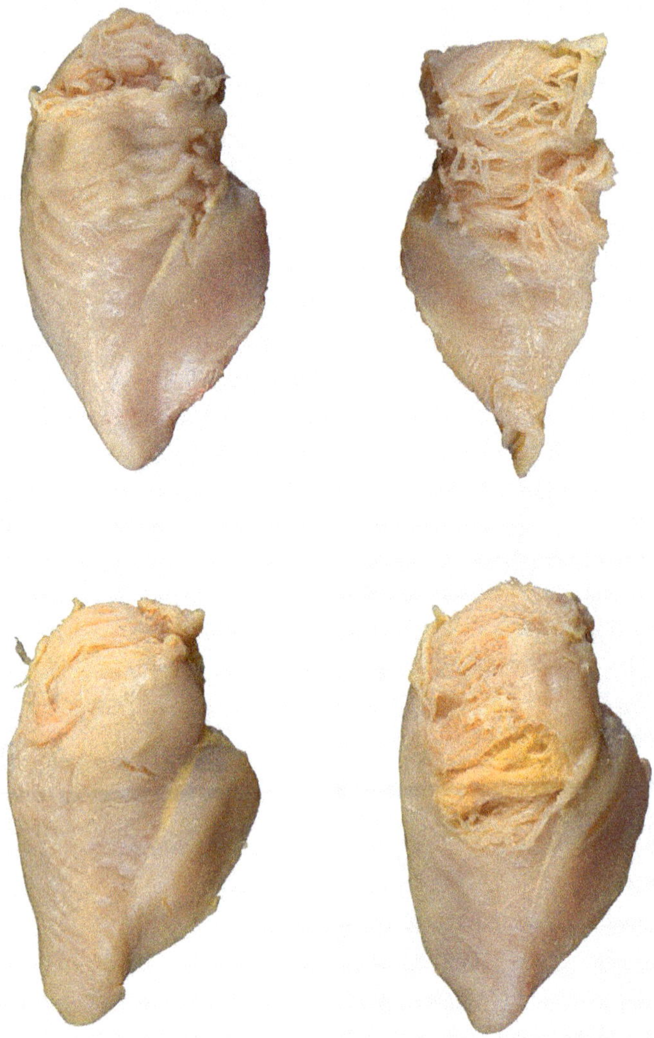

Figure 1 Macroscopic appearance of broiler pectoralis major muscles exhibiting severe spaghetti meat characteristics (overall impaired muscle integrity with soft spongy areas and muscle fiber bundle separation toward the cranial portion of the muscle).

no distinct SM characteristics) or SM (exhibit SM characteristics). With regard to the SM condition (referred to as poor cohesion), Sirri et al. (2016) distinguished moderate and severe degrees of the myopathy using a 0–2 scale: 0 (normal fillets with no SM characteristics), 1 (moderately loose structure perceived by pinching the muscle cranial surface), and 2 (extensive superficial lacerations). Wold and Løvland (2020) attributed scores of 0 to normal fillets, 1 to mild SM (limited areas of the fillet with loose/stringy consistency), and 2 to severe SM (large parts of the entire fillet with stringy/spongy consistency).

2.1.2 Morphometric traits

Although comprehensive studies covering a wide range of broiler lines, sizes, and slaughter ages have not been reported, currently available data suggest that breast fillets affected by the SM condition are typically associated with altered morphometric traits. When compared to normal breast fillets, those with SM or SM+WS defects are on average heavier and significantly thicker at the cranial, middle, and caudal regions of the muscle (Baldi et al., 2018; Tasoniero et al., 2020; Pascual et al., 2021). As for average breast fillet length, no differences have been reported between normal and SM fillets; however, SM+WS breasts were observed to be longer than both normal and SM fillets (Baldi et al., 2018). Similar breast fillet widths between SM and normal breast fillets have been reported (Baldi et al., 2018).

2.2 Microscopic traits and connective tissue characteristics

The histopathological lesions of SM are primarily observed on the superficial layer of the fillet and gradually decrease moving toward the deep portion of affected muscles. The histological changes observed in SM are similar to those of breast muscles exhibiting the WS and WB myopathies, as described by Kuttappan et al. (2013) and Sihvo et al. (2014). Muscle tissue from breast fillets with the SM myopathy exhibit active degeneration coupled with regenerative processes, loss of normal tissue architecture and poor fiber uniformity (presence of hypercontracted fibers and small-caliber regenerative fibers intermingled with abnormal fibers exhibiting variable dimensions, rounded profiles, and nuclear internalization), fiber necrosis and lysis, accumulation of fibrous tissue and fat, and abundant inflammatory cell infiltration. Consistent with macroscopic traits, the most severe histological lesions observed in SM are generally found in the superficial portion on the cranial end of the pectoralis major (Baldi et al., 2018). In general, microscopic lesion characteristics cannot be used to fully discriminate between the SM, WS, and WB myopathies. However, a histological trait of SM not typically observed in either WS or WB myopathies is a progressive rarefaction of the endomysial and perimysial connective tissue coupled with a loose, immature collagen deposition (Baldi et al., 2018; Mazzoni et al., 2020). Histologically, this results in increased spacing between the fibers and a reduced number of fibers per field of view. Macroscopically, these altered connective tissue characteristics lead to the detachment of the fibers and fiber bundles from each other, thus conferring the typical phenotype to the lesioned areas.

Several studies have reported more in-depth findings on the unique connective tissue characteristics of SM. Immunohistochemical analyses to assess the presence and distribution of procollagen and collagen type III showed a profound modification of the connective tissue architecture as well as an alteration of the collagen composing the endomysium and perimysium in SM fillets (Mazzoni et al., 2020). As type III collagen plays a key role in fibrillogenesis (Liu et al., 1997), these data suggest altered collagen turnover and synthesis in SM. Sanden et al. (2021) characterized the collagen structure of SM using histology, Fourier transform infrared (FTIR) microspectroscopy, and polarized FTIR imaging. Histological analyses revealed that while the perimysium appears dense in normal samples when stained, the perimysium of SM appears transparent and inhomogeneous. Compared to normal samples, most of the collagen fibers composing the perimysium are thinner, fewer, shorter, and poorly packed in SM. Consistent with microscopic images, FTIR spectra reveal differences in collagen structure and abundance between SM and normal samples (Sanden et al., 2021). While the perimysium of unaffected muscles consists of mature, thick, and tightly packed collagen fiber bundles, the collagen of lesioned breast muscles consists of more loosely bound α-helixes and the connective tissue possesses a higher amount of amorphous ground substances. In agreement with Baldi et al. (2019), who found a lower amount of hydroxylysylpyridinoline in the intramuscular connective tissue of SM's superficial portion, Sanden et al. (2021) found evidence of immature collagen consisting of reducible cross-links, as FTIR spectra indicated a lower presence of non-reducible cross-links in SM samples. Polarized FTIR-spectroscopy indicated that SM exhibited more randomly organized collagen fibers compared to the well-ordered structure observed in normal samples.

2.3 Chemical composition

Similar to other growth-related breast myopathies, the muscle degenerative/regenerative processes that cause the histopathological lesions observed with the SM abnormality directly affect the proximate composition (protein, moisture, fat, and ash contents) of the breast meat. Just as the myopathic lesions associated with SM are not uniformly distributed throughout the pectoralis major muscle, alterations in the chemical composition of SM breast fillets are not evenly distributed throughout the muscle. Differences in the proximate meat composition in breast meat with the SM condition are most apparent on the superficial portions of the fillets and decrease moving toward the deeper portions of the fillet (Baldi et al., 2018, 2019; Tasoniero et al., 2020). The changes in the proximate composition of the breast meat due to the SM myopathy have an overall negative impact on the nutritional quality of the meat.

The most notable change in the proximate composition of breast meat with the SM abnormality is a significant reduction in protein content compared to unaffected meat (Baldi et al., 2018, 2019; Tasoniero et al., 2020; Pascual et al., 2021). Although total protein content decreases with the SM abnormality, several studies have used sodium dodecyl sulfate-polyacrylamide gel electrophoresis (SDS-PAGE) approaches to demonstrate that the overall protein profiles of the myofibrillar and sarcoplasmic protein fractions in the affected muscle are very similar to that of the normal tissue, with only minor differences observed in the relative abundance of a few specific protein bands of <100 kDa (Baldi et al., 2018; Tasoniero et al., 2020, 2022). Increased numbers of high-molecular-weight protein bands (ranging in size from 100 kDa to 250 kDa) have been observed in both the myofibrillar and sarcoplasmic protein fractions from the superficial portion of breast meat exhibiting the SM, WS, or SM+WS myopathies (Baldi et al., 2018). The exact mechanisms causing alterations in protein content and composition in SM are yet to be determined. However, protein changes in SM may be related to myofibrillar protein degradation occurring during the structural breakdown of muscle fibers as part of the degenerative processes in the muscle (Harris et al., 2003) and/or due to increased enzyme efflux as a result of a compromised sarcolemma integrity (Mitchell, 1999; Sandercock and Mitchell, 2003).

Breast meat with the SM myopathy exhibits greater moisture content than normal breast meat (Baldi et al., 2018, 2019; Tasoniero et al., 2020; Pascual et al., 2021). The greater moisture content detected in SM-affected breasts likely reflects the presence of edema due to increased inflammatory processes in the tissue (Petracci et al., 2019). A reduction of the myowater fraction retained within the myofibrils as a consequence of changes in muscle microstructure may also play a role in increased moisture content measurements in SM fillets (Tasoniero et al., 2020).

Consistent with the increased lipidosis observed in breast muscle with growth-related myopathies (Sihvo et al., 2014; Soglia et al., 2016), breast meat with the SM abnormality can have greater fat content than normal meat. However, the fat content of SM seems to be intermediate between normal breast fillets and meat exhibiting the WS or WB conditions (Baldi et al., 2018, 2019; Pascual et al., 2021). Overall, the SM myopathy does not significantly influence the fatty acid profiles of the breast meat. The relative proportions of saturated fatty acids, monounsaturated fatty acids, polyunsaturated fatty acids, *n*-3 fatty acids, *n*-6 fatty acids, and the *n*-6/*n*-3 and polyunsaturated/saturated fat ratios are not altered in breast meat with the SM abnormality regardless of the sampling position (Baldi et al., 2019).

Published by Burleigh Dodds Science Publishing Limited, 2023.

Breast meat with the SM myopathy exhibits a slightly lower total ash content (Baldi et al., 2018, 2019; Tasoniero et al., 2020) and an altered mineral profile (Tasoniero et al., 2020) when compared to normal breast meat. The differences in the ash content and mineral profile observed in SM are more pronounced in the superficial portion of the muscle. Greater levels of calcium and sodium have been observed in SM fillets (Tasoniero et al., 2020). This same report, however, showed no differences between normal and SM fillets in magnesium and potassium levels.

Collagen content and characteristics have also been assessed in breast meat exhibiting the SM myopathy. Despite the endomysial and perimysial connective tissue rarefaction histologically observed in SM fillets, none of the available studies report differences in total intramuscular collagen content due to the SM myopathy (Baldi et al., 2018, 2019; Tasoniero et al., 2020). Location within the pectoralis major muscle has a more drastic effect on total collagen content than the SM condition. In both SM and normal fillets, greater total collagen content is present in the superficial portion of the muscle than in the deep portion of the muscle (Tasoniero et al., 2020). The relative proportions of soluble and insoluble collagen content in meat are often used as indirect indicators of the potential degree of collagen cross-linking. Similar to total collagen content, soluble collagen content is not different between normal and SM fillets; however, soluble collagen content is greater in the superficial portion of the muscle compared to the deep portion (Tasoniero et al., 2020). The data available on soluble collagen content relative to total collagen content indirectly suggest that the SM myopathy does not have a significant effect on collagen cross-linking. However, Baldi et al. (2019) observed a lower content of hydroxylysylpyridinoline, a non-reducible cross-link of muscular collagen, in the superficial portion of SM breasts. Thus, more research is needed to determine if the type of collagen or the degree of cross-linking is related to the unique connective tissue characteristics observed histologically in SM.

3 Impact of spaghetti meat myopathy on meat quality and functionality

Due to its impact on the chemical and physical properties of the muscle tissue, the SM myopathy has an overall negative influence on meat quality traits. As a result, the SM myopathy limits the options available for breast meat utilization. Breast meat with the fully expressed SM abnormality is typically not used for intact, whole muscle-type breast meat products due to the impaired structural and visual characteristics of the muscle. The softness and overall lack of muscle integrity in SM breasts also present considerable challenges in thin slicing and portioning the breast meat. As a result, breast meat with the SM defect is usually downgraded and included into further-processed meat products.

Published by Burleigh Dodds Science Publishing Limited, 2023.

Several studies have evaluated the influence of SM on traditional meat quality traits such as visual appearance and meat texture. Aside from the distinctive phenotype of muscle bundle separation, the SM myopathy does not seem to have a drastic effect on other visual characteristics of the breast meat such as color. Regardless of measurement location (superficial vs. deep portion, dorsal vs. ventral surface), the SM myopathy does not influence meat lightness (L^* values) or redness (a^* values) but is associated with greater meat yellowness (b^* values) in breast meat (Baldi et al., 2018; Soglia et al., 2019; Tasoniero et al., 2020; Campo et al., 2020). The increased yellowness of SM fillets may be related to the increased adipose accumulation that occurs among the muscle fibers, similar to what is postulated in WB-afflicted meat (Dalle Zotte et al., 2017). Differences in yellowness between normal and SM fillets, however, are not observed after freezing and thawing (Soglia et al., 2019).

Meat texture characteristics of both raw and cooked SM breast fillets have been assessed using various testing methods: compression test, texture profile analysis (TPA), Meullenet–Owens razor shear (MORS), Allo-Kramer (AK) shear force, and Warner–Bratzler shear force (WBSF). In the raw state, SM and normal fillets meat typically exhibit similar compression force measurements (Baldi et al., 2019; Pascual et al., 2021). However, one study using a modified compression test that hinders the muscle fibers from expanding transversally observed lower raw compression force in SM fillets when 80% compression heights were used (Campo et al., 2020). Considering that the resistance of the sample at higher compression rates is mainly attributable to the connective tissue (Sañudo et al., 2003), the texture analysis at those conditions likely reflected the progressive rarefaction of the endomysial and perimysial connective tissue observed in SM. In the cooked state, the SM myopathy does not affect compression force when measured on intact breast fillets (Pascual et al., 2021). Differences between normal and SM fillets have not been observed in either raw or cooked samples using the double-compression TPA test to assess variables such as hardness, adhesiveness, cohesiveness, springiness, gumminess, chewiness, and resilience (Campo et al., 2020). Texture characteristics are not uniform throughout the pectoralis major as SM only exhibits a lower cooked compression force in the superficial portion of the muscle than non-affected breasts (Baldi et al., 2019). Discrepancies in texture characteristics between layers within SM breasts may be ascribable to a lower amount of hydroxylysylpyridinoline, responsible for collagen cross-linking, found in the superficial layer of affected muscles (Baldi et al., 2019).

Findings from shear-based texture techniques, such as AK shear force and MORS, suggest a softer cooked texture in SM compared to normal breast meat (Pascual et al., 2021). However, these differences do not seem to be detectable using the WBSF technique (Campo et al., 2020). The discrepancies observed in the texture characteristics of cooked fillets using shear methods may be due to the different forces being applied to the sample surface by the various shear devices. Consequently, SM breast fillet samples probably experience diverse deformation ease and behavior under different test methods, which detect slightly different physical characteristics within the meat.

Similar to WB and WS, breast meat affected by the SM myopathy exhibits a higher ultimate pH than unaffected meat (Baldi et al., 2018, 2019; Soglia et al., 2019; Tasoniero et al., 2020) due to having an abbreviated postmortem muscle acidification, resulting in lower lactate concentration in the meat (Soglia et al., 2019).

With regard to water-holding capacity (WHC), available data generally suggest that the SM myopathy has a negative influence on the breast meat's ability to bind and hold intrinsic or added water. Intact breast fillets with the SM myopathy display greater drip loss (Tasoniero et al., 2020; Pascual et al., 2021) and cook loss (Pascual et al., 2021) than normal fillets. However, other studies did not observe statistically significant differences between SM and normal samples for expressible fluid measured via the filter press method (Yu et al., 2020) or various salt-induced water uptake assays (Soglia et al., 2019; Tasoniero et al., 2020). The discernable impact of the SM myopathy on breast meat WHC is likely dependent upon measurement methodology, sampling location, and degree of SM severity. In general, breast meat WHC is more negatively influenced by the WB myopathy than by the SM abnormality.

Time domain-nuclear magnetic resonance (TD-NMR) analysis has been utilized to investigate the myowater properties in breast meat with the SM myopathy. Myowater properties measured by NMR are known to be indicative of meat WHC traits (Bertram et al., 2002, 2006). NMR transverse relaxation data have indicated the presence of three water populations within the muscles: bond water (tightly bound to macromolecules), intra-myofibrillar water (located within the myofibrillar matrix), and extra-myofibrillar water (free water held only by capillary forces in the inter-myofibrillar spaces and the spaces between muscle fibers and fiber bundles). Available data suggest that compared to normal samples, SM breasts exhibit a lower proportion of bound and intra-myofibrillar water and a greater proportion of extra-myofibrillar water (Baldi et al., 2018, 2019; Soglia et al., 2019; Tasoniero et al., 2022). Longer relaxation times suggest greater water mobility in SM samples. Similar to other traits, position within the pectoralis major plays a significant role as myowater properties of the superficial portion of the muscle are more severely influenced by the occurrence of the SM abnormality than the deep portion, which exhibits characteristics similar to normal samples (Baldi et al., 2018). Studies using NMR analysis are generally in agreement that breast myopathies significantly alter myowater properties and increase the amount of loosely bound water in the muscle. The increase in the proportion of loosely bound water in SM samples is typically not as drastic as observed in WB (Baldi et al., 2019; Soglia et al.,

2019; Wold and Løvland, 2020). The higher mobility of the intra-myofibrillar water and greater amount of free water in SM fillets suggest impaired WHC in affected fillets.

Altered myowater distribution and mobility and reduced WHC in SM samples are primarily ascribable to the structural changes and histological lesions observed in the affected fillets: loss of normal tissue architecture, necrosis and lysis of fibers, edema, and immature collagen deposition coupled with a progressive rarefaction of endomysial and perimysial connective tissue that leads to muscle fiber detachment from each other and compromised fiber bundles cohesion (Baldi et al., 2018; Petracci et al., 2019). As suggested by the findings of Wold and Løvland (2020), altered myowater properties in SM may be due to the reduced protein content exhibited by affected fillets. Because most of the water in muscle tissue is typically held within the myofibrillar structure, with fewer myofibrils (as suggested by a lower muscle protein content) as a result of the myodegeneration and structural changes there is simply a lower potential for the meat to hold intrinsic or added water.

Impaired water retention ability in SM could also be explained at least in part by diminished muscle protein functionality. Protein solubility is defined as the amount of protein that dissolves in an aqueous solution under specific extraction conditions (Xiong 2004). As solubility reflects the equilibrium between protein and water, it is a useful indicator of meat protein functionality. The SM myopathy, either alone or in combination with WS, reduces total and myofibrillar protein solubility, especially in the superficial portion of the fillets (Baldi et al., 2018; Tasoniero et al., 2020). Decreased protein solubility is often attributed to protein denaturation and associated with decreased meat WHC and meat functionality. Differential scanning calorimetry data has shown similar protein denaturation patterns between normal and SM fillets (Baldi et al., 2019), suggesting that reduced protein solubility in SM may be predominantly a function of reduced muscle protein content in the tissue. However, adjusting protein solubility values for initial muscle protein content was found to not fully account for the reduced protein solubility in SM fillets (Tasoniero et al., 2020). Emulsions formed from myofibrillar and sarcoplasmic proteins extracted from SM fillets were shown to have diminished stability (Tasoniero et al., 2020), suggesting that the SM myopathy may negatively influence muscle protein functionality. In frozen/thawed breast meat, greater protein carbonylation has been observed in SM compared to normal samples (Soglia et al., 2019). However, similar protein carbonyl and TBARS levels have been reported for fresh SM and normal samples (Baldi et al., 2018; Soglia et al., 2019), suggesting that differentiating changes in oxidative processes and their influence on muscle proteins likely do not play a large part in the effects of SM on WHC. Discrepancies in oxidation data between frozen/thawed and fresh samples are likely due to the ice crystals in frozen samples exacerbating the profound histological lesions and disrupting cells and cellular organelles causing increased exposure of myofibrillar proteins to oxidative enzymes and pro-oxidant molecules (Xia et al., 2009). Data showing altered electrophoretic protein profiles, greater myofibrillar fragmentation index (Baldi et al., 2018), and a higher concentration of free amino acids (Soglia et al., 2019) in SM suggest that greater muscle protein degradation in the tissue likely plays a significant role in the reduced protein functionality and WHC in SM.

When compared to other myopathies, available data suggest that the occurrence of the SM defect has a more pronounced impact on meat quality traits (in terms of proximate composition and NMR traits) than WS. Except for a greater lipid percentage in the superficial portion of SM/WS breasts (Baldi et al., 2018), the concurrent presence of WS does not seem to further alter the proximate composition, collagen amount, or electrophoretic protein profiles compared to the occurrence of SM alone. Among the different emerging myopathies, the occurrence of WB exerts the most profound impact on meat quality (in terms of pH, color, chemical composition and NMR traits), while SM and WS generally result in more intermediate effects (Baldi et al., 2019; Soglia et al., 2019; Wold and Løvland, 2020).

4 Pathological mechanisms and predisposing factors for spaghetti meat

Currently, the etiology of the SM myopathy and its mechanistic relationship to the other breast myopathies in broilers is not fully understood. Similar histopathological traits and concomitant occurrence within the same breast fillets could suggest that these abnormalities might have a common causative basis or even represent a progression of the same condition. To this respect, Griffin et al. (2018) observed that indicators of SM onset can occur at 18 days post-hatch, coinciding with the onset of epimysium hemorrhages, shortly after the initial occurrence of WS at 16 days of age and prior to WB onset. The SM myopathy varies in its degree of severity and follows a similar pattern of manifestation within the pectoralis major as other breast abnormalities with regard to intra-fillet variations (Bailey et al., 2020). Although the appearance of these growth-related breast myopathies often coincides, the cause-and-effect relationships underlying their associations are not yet clear.

4.1 Pathological mechanisms

As with other breast myopathies, for SM there is a growing body of research on the underlying muscle tissue changes leading to the distinct phenotypic traits. However, the precise mechanisms or events triggering the onset of these changes have not been fully elucidated. For most of the tissue and molecular alterations that are currently described in the literature for SM, it is not clear yet if they represent the cause or the effect of the myopathy.

Research on the effects of the SM abnormality on the concentration of free amino acids, histidine-containing dipeptides, and main metabolites confirms the occurrence of muscle degenerative and regenerative processes, extracellular matrix (ECM) remodeling, altered redox homeostasis, and energy metabolism perturbation in affected breasts (Soglia et al., 2019). A remarkable increase in the concentration of free amino acids has been reported in SM (Soglia et al., 2019), suggesting that increased protein catabolism is likely taking place during the tissue degenerative processes that result in the observed histological lesions. Based on previous observations in dystrophic muscle (Griffin et al., 2001; Martins-Bach et al., 2012), it may also be hypothesized that higher concentrations of free amino acids such as glutamine and glutamate in SM represent an attempt at muscle tissue regeneration. The higher contents of valine, leucine, isoleucine, and proline reported in SM fillets (Soglia et al., 2019) may be considered indicators of ECM remodeling as they are in WB (Abasht et al., 2016). The depletion of histidine-containing dipeptides having antioxidant capacity such as anserine and carnosine in SM (Soglia et al, 2019) suggests that affected tissue might have required extra antioxidant protection that resulted in an altered redox homeostasis similar to WB (Abasht et al., 2016). Data on the concentration of metabolites (e.g. creatine, inosine-5-monophosphate, and lactate) in SM provide evidence of a perturbation of the energy metabolism in the tissue (Soglia et al., 2019).

Available data on the distribution of cytoskeletal proteins vimentin (VIM) and desmin (DES) and the expression of their corresponding genes has provided further insight into the development of growth-related broiler breast muscle abnormalities (Soglia et al., 2020). As these cytoskeletal proteins are considered markers for muscle regeneration within dystrophic muscle tissue (Gallanti et al., 1992; Fröhlich et al., 2016), it has been hypothesized that these two intermediate filaments might be involved in the manifestation of emerging muscular abnormalities in fast-growing broilers. During early myogenesis, DES and VIM coexist; however, VIM expression is gradually reduced as the development of myotubes proceeds (Tokuyasu et al., 1985). Immunohistochemistry has revealed the presence of many fibers that are diffusely immunoreactive to both VIM and DES at the sarcolemmal and sub-sarcolemmal levels in WS and WB samples compared to normal samples, suggesting early-stage muscle regenerative processes (Soglia et al., 2020). Intriguingly, SDS-PAGE and Western blot analyses demonstrated a reduction of both VIM and DES protein content in SM samples compared to normal samples, despite increased VIM gene expression in SM (Soglia et al., 2020). These data suggest that post-transcriptional mechanisms affecting VIM protein levels could play a role in SM development (Soglia et al., 2020). VIM is known to regulate the proliferation and differentiation of fibroblasts during regeneration processes following injury (Cheng et al., 2016). Thus, researchers hypothesized that the lack of translation of VIM mRNA into protein might lead to altered ECM synthesis, reduced fibroblasts within the perimysium, and the progressive perimysial connective tissue rarefaction observed in SM (Soglia et al., 2020). Further data are needed to confirm this mechanism.

It has also been hypothesized that chronic accumulation of lactic acid in the muscle tissue may contribute to the manifestation of the SM myopathy (Anton et al., 2019). Increased levels of lactic acid in the muscle are speculated to cause connective tissue degradation, leading to fiber and fiber bundle separation. Furthermore, increased lactic acid in the tissue can potentially inhibit protein synthesis, which may impact the maturity and integrity of the connective tissue. Further research is needed to confirm these hypotheses.

Changes in the mineral profile of affected meat (increased calcium and sodium) may also provide insight into the development of the SM myopathy. Prior to the recognition of growth-related breast myopathies such as SM, researchers hypothesized that a greater Na^+ level in the pectoralis major of modern broilers could be a triggering factor for imbalanced Ca^{2+} intake into the muscle fibers, which in turn could be responsible for altered cell membrane integrity through the activation of Ca^{2+}-dependent proteases and lipases (Sandercock and Mitchell, 2003, 2004; Sandercock et al., 2009). A connection between cation homeostasis disturbances and muscle cell injury development has been suggested for other emerging myopathies (Mutryn et al., 2015; Petracci et al., 2015; Soglia et al., 2016; Tasoniero et al., 2017) and may play a role in SM. However, additional data is needed to establish the role of cation homeostasis in SM development.

4.2 Predisposing factors

Many of the aforementioned changes provide a detailed picture of what is happening in the muscle during the development of the SM myopathy; however, it is still unclear if these alterations in the tissue represent the cause or the effect of the SM abnormality. One working hypothesis that ties together much data in growth-related breast myopathies such as SM is the predisposing factor of insufficient vascularization in the pectoralis major. Due to intense selection for rapid bird growth and high breast meat yield, the pectoralis major in modern broiler lines undergoes significant muscle hypertrophy. Furthermore, the breast muscle of modern broilers is comprised of nearly 100% fast-twitch glycolytic type muscle fibers, which by nature are characterized as having larger fiber diameters, lower capillary density, fewer mitochondria, and more glycolytic metabolism than other muscle fiber types. Insufficient vascularization in the muscle could reduce oxygenation of the muscle and impair metabolic waste product removal, leading to cellular stress and hypoxic conditions which may then trigger a cascade of downstream events in the tissue that result in the development of the muscle abnormality. This etiological chain of events has been widely suggested with the WB and WS conditions (Abasht et al., 2016; Malila et al., 2019; Greene et al., 2020) but less directly studied for the SM myopathy to date. Differences in vascularization may also explain intra-fillet variations in the various phenotypic traits between the superficial and deep portions of affected fillets.

4.2.1 Intrinsic factors

As with other breast myopathies, factors such as bird genetics, gender, and age as they relate to body weight and growth rate are speculated to play a role in SM occurrence. One study evaluated the incidence of breast meat abnormalities in 35-day-old female broilers belonging to two distinct light-size hybrids (Zampiga et al., 2019a) and found significant differences between genotypes in terms of total SM incidence (63% vs. 39%) as well as the incidence of severe SM cases (31% vs. 10%). Even though one of the genotypes exhibited higher weights during the first rearing phase (0–9 days), both hybrids achieved similar productive performances over the entire rearing cycle. These data indicate that genotype can play a role in the occurrence of SM and also show that the myopathy can occur in lighter birds slaughtered at a relatively early age. With regard to bird gender, Pascual et al. (2020) reported higher SM rates in females than in males (25% vs. 3.1%). This is different than for WB, which typically occurs at higher rates in males. It is speculated that the higher SM incidence in females is likely due to hormonal differences between the two genders, leading to variations in the regulation of biological processes such as metabolism, protein synthesis, and intramuscular fat deposition during the development of the myopathy condition (Baldi et al., 2021).

Although it is widely recognized that the rates of WB and WS generally increase in correspondence with increases in body weight, growth rate, breast yield, and age at slaughter, this trend has not been fully confirmed for SM. Bailey et al. (2020) reported a low heritability for SM (0.04) and only low to moderate genetic correlations between productive performances (body weight and breast yield) and the SM myopathy. Therefore, Bailey et al. (2020) concluded that only a small proportion of the genetic component influencing body weight and breast yield is shared with the genetic component of breast meat myopathies. These data suggest that the selection for higher body weight and breast yield are likely not the sole cause of the myopathy and that there might be a strong influence of nongenetic factors in the manifestation of SM.

4.2.2 Extrinsic factors

A growing body of scientific literature is increasingly describing the effects that broiler nutrition and dietary supplementation have on the severity and occurrence of the SM condition. Several studies have investigated the effects of manipulating dietary amino acid levels on breast myopathy development. Zampiga et al. (2019b) evaluated the effect of different digestible arginine-to-lysine ratios in broiler diets on the occurrence of breast myopathies. This study reported that dietary supplementation of L-arginine to increase the digestible arginine-to-lysine ratio by about 30% (digestible Arg:Lys ratio 1.35–1.37) compared to current practices (digestible Arg:Lys ratio 1.05–1.07) during the four different feeding phases (starter, grower I, grower II, and finisher) has a positive effect on SM and WS rates without negatively affecting productive performances and meat quality. This was hypothesized to occur because of the increase in the production of nitric oxide deriving from L-arginine that enhances vasodilation and promotes a better blood flow and thus oxygen supply to the muscles (Khajali and Wideman, 2010). In an earlier study, Zampiga et al. (2018) reported that L-arginine supplementation at a lower level (digestible Arg:Lys ratio 1.15–1.17) improved feed conversion ratio and increased body weight but had no impact on SM incidence or severity. Dietary lysine restriction during the grower feeding phases (10–20 days and 21–34 days) of broiler production was shown not to impact SM incidence or severity in 49-day-old broilers (Soglia et al., 2021b). Among the possible predisposing factors, Anton et al. (2019) hypothesized that inadequate dietary levels of amino acids critical for proper connective tissue development could play a role in myopathy development, especially when birds are fed with plant protein-based diets that are lower in proline. However, data is currently lacking to fully explain the effects of dietary amino acid levels on SM development.

Córdova-Noboa et al. (2018) tested the effects of guanidinoacetic acid (GAA) as a dietary supplement on breast myopathy development in broiler chickens fed either corn or sorghum-based diets. GAA is a metabolic intermediate synthesized from the amino acids glycine and arginine that can act as a precursor to creatine (DeGroot, 2014) and whose supply is low in diets with only vegetable protein sources (Vieira and Lima, 2005). Supplementation with GAA was reported to reduce WB occurrence, likely by affecting the concentration of metabolites involved in muscle energy metabolism, but GAA did not influence the incidence of SM or WS abnormalities at either 51 or 55 days of broiler age (Córdova-Noboa et al., 2018).

Using sodium butyrate, a short-chain fatty acid, as a poultry feed additive was previously found to have beneficial effects on gut health (Bortoluzzi et al., 2017; Guilloteau et al., 2010; Peng et al., 2007; Wu et al., 2020; Yang et al., 2018), to promote growth rate, to increase carcass and breast weight (Leeson et al., 2005; Namkung et al., 2011), to regulate cell growth and differentiation (Liang et al., 2010; Mali et al., 2010; Berni Canani et al., 2012), and to have a possible impact on muscle fibers (Wu et al., 2020). Pascual et al. (2020) investigated the effects of dietary supplementation with microencapsulated sodium butyrate (500 mg/kg; 30% Na-butyrate) on the occurrence of breast muscle abnormalities in male and female broiler chickens. With sodium butyrate supplementation, Pascual et al. (2020) did not observe changes in carcass traits and muscle growth but reported a substantial reduction in SM occurrence rate in females (control diet 33.3%, sodium butyrate diet 16.7%).

Trace minerals are important for animal growth and development as they serve as enzyme catalysts and are involved in many metabolic processes. However, the limited data currently available suggest that they likely do not play a key role in SM development. Sirri et al. (2016) reported that the incidence of the poor cohesion defect and collagen characteristics (intramuscular collagen and hydroxylysylpyridinoline amounts) in broiler pectoral muscles were not affected by different sources and doses of trace minerals (zinc, manganese, copper) administered in two different forms (organic vs. inorganic) at two different doses (high vs. low).

As with other growth-related breast myopathies, available evidence indicates that the SM abnormality manifests in the breast meat due to changes occurring in the muscle tissue throughout the life of the broiler. However, it is possible that events immediately pre- and post-slaughter may exacerbate muscle fiber bundle separation and the visual severity of the SM defect in the breast meat. It is currently unknown if antemortem factors such as bird handling, transportation stress, and live shackling play a role. Likewise, there is little to no data describing the potential role that postmortem processing steps such as scalding, defeathering, chilling, or deboning may play in the appearance or severity of the distinct characteristics of SM.

5 Conclusion

The occurrence of the SM myopathy in broiler breast meat is a growing concern within the poultry industry. Compared to other growth-related breast myopathies, such as WB and WS, there is relatively less known about the SM myopathy as it was more recently recognized. However, there is a growing body of scientific literature describing the characteristics and development of SM. The impacts of the SM myopathy on the physical and chemical properties of the pectoralis major have been well-characterized as have the impacts of the myopathy on basic meat quality and functionality traits. Although there are many overlaps in the histological lesion characteristics of SM with WB and WS, available data suggests that the SM myopathy is not as closely associated to broiler size and growth as the other breast myopathies. The underlying etiological events that lead to the unique phenotypic features of SM are not fully understood and merit further research. The severely impaired structural integrity of breast fillets exhibiting the SM myopathy significantly limits their usage in different types of breast meat products causing logistical and economic burdens for the poultry industry. Although much has been learned about SM in a relatively short period of time, additional research is needed to understand the underlying factors that trigger the myopathy and to develop effective strategies to prevent and better manage the problems created by this breast meat abnormality.

6 Where to look for further information

The body of scientific literature available on the SM myopathy has grown rapidly in recent years. Several excellent review articles have been published on growth-related myopathies in broilers that cover the topic of SM (see Petracci et al., 2019; Baldi et al., 2021; and Soglia et al., 2021a).

Although research on SM is being conducted across many different institutes around the world some of the key institutes focusing on SM include the following:

- Department of Agricultural and Food Sciences, University of Bologna, Italy.
- Department of Food Science, University of Guelph, Canada.
- Department of Poultry Science, University of Arkansas, USA.
- U.S. National Poultry Research Center, USDA-Agricultural Research Service, USA.

The World's Poultry Science Association (WPSA) is an organization that seeks to advance all aspects of poultry science. Many researchers around the world working on SM and other myopathies are part of WPSA and present findings on SM at numerous conferences and meetings organized by WPSA national branches. Information on and links to many of these meetings can be found on the WPSA website (www.wpsa.com).

7 References

Abasht, B., Mutryn, M. F., Michalek, R. D. and Lee, W. R. (2016). Oxidative stress and metabolic perturbations in wooden breast disorder in chickens, *PLoS ONE* 11(4): e0153750.

Anton, A., Avendano, S., Bailey, R., Bilgili, S., Canela, L., Corzo, A., Fancher, B., French, N., Nicholson, D., Pearson, D., Rossi, L., and Thomson, S. (2019). Breast muscle myopathies, Available at: http://eu.aviagen.com/tech-center/dowload/1282/Breast-Muscle-Myopathies-2019-EN.pdf (accessed 11 May 2021).

Bailey, R. A., Souza, E. and Avendano, S. (2020). Characterising the influence of genetics on breast muscle myopathies in broiler chickens, *Front. Physiol.* 11: 1041.

Baldi, G., Soglia, F., Laghi, L., Tappi, S., Rocculi, P., Tavaniello, S., Prioriello, D., Mucci, R., Maiorano, G. and Petracci, M. (2019). Comparison of quality traits among breast meat affected by current muscle abnormalities, *Food Res. Int.* 115: 369–376.

Baldi, G., Soglia, F., Mazzoni, M., Sirri, F., Canonico, L., Babini, E., Laghi, L., Cavani, C. and Petracci, M. (2018). Implications of white striping and spaghetti meat abnormalities on meat quality and histological features in broilers, *Animal* 12(1): 164–173.

Baldi, G., Soglia, F. and Petracci, M. (2021). Spaghetti meat abnormality in broilers: current understanding and future research directions, *Front. Physiol.* 12: 684497.

Berni Canani, R., Di Costanzo, M. and Leone, L. (2012). The epigenetic effects of butyrate: potential therapeutic implications for clinical practice, *Clin. Epigenet.* 4(1): 4.

Bertram, H. C., Purslow, P. P. and Andersen, H. J. (2002). Relationship between meat structure, water mobility, and distribution: a low-field nuclear magnetic resonance study, *J. Agric. Food Chem.* 50(4): 824–829.

Bertram, H. C., Wu, Z., van den Berg, F. and Andersen, H. J. (2006). NMR relaxometry and differential scanning calorimetry during meat cooking, *Meat Sci.* 74(4): 684–689.

Bortoluzzi, C., Pedroso, A. A., Mallo, J. J., Puyalto, M., Kim, W. K. and Applegate, T. J. (2017). Sodium butyrate improved performance while modulating the cecal microbiota and regulating the expression of intestinal immune-related genes of broiler chickens, *Poult. Sci.* 96(11): 3981–3993.

Bowker, B. C., Zhuang, H., Yoon, S. C., Tasoniero, G. and Lawrence, K. C. (2019). Relationship between attributes of woody breast and white striping myopathies in commercially processed broiler breast meat, *J. Appl. Poult. Res.* 28(2): 490–496.

Campo, M. D. M., Mur, L., Guerrero, A., Barahona, M., Resconi, V. C., Magalhaes, D. R., Linsbinski, E., Boito, B., Moraes de Oliveira, I. M. and Olleta, J. L. (2020). Differentiating breast myopathies through color and texture analyses in broiler, *Foods* 9(6): 824.

Che, S., Wang, C., Iverson, M., Varga, C., Barbut, S., Bienzle, D. and Susta, L. (2022a). Characteristics of broiler chicken breast myopathies (spaghetti meat, woody breast, white striping) in Ontario, Canada, *Poult. Sci.* 101(4): 101747.

Che, S., Wang, C., Varga, C., Barbut, S. and Susta, L. (2022b). Prevalence of breast muscle myopathies (spaghetti meat, woody breast, white striping) and associated risk factors in broiler chickens from Ontario Canada, *PLoS ONE* 17(4): e0267019. doi: 10.1371/journal.pone.0267019.

Cheng, F., Shen, Y., Mohanasundaram, P., Lindström, M., Ivaska, J., Ny, T. and Eriksson, J. E. (2016). Vimentin coordinates fibroblast proliferation and keratinocyte differentiation in wound healing via TGF-b–Slug signaling, *Proc. Natl. Acad. Sci. U. S. A.* 113(30): E4320–E4327.

Córdova-Noboa, H. A., Oviedo-Rondón, E. O., Sarsour, A. H., Barnes, J., Ferzola, P., Rademacher-Heilshorn, M. and Braun, U. (2018). Performance, meat quality and pectoral myopathies of broilers fed either corn or sorghum-based diets supplemented with guanidinoacetic acid, *Poult. Sci.* 97(7): 2479–2493.

Dalle Zotte, A., Tasoniero, G., Puolanne, E., Remignon, H., Cecchinato, M., Catelli, E. and Cullere, M. (2017). Effect of "Wooden Breast" appearance on poultry meat quality, histological traits, and lesions characterization, *Czech J. Anim. Sci.* 62(2): 51–57.

DeGroot, A. A. (2014). Efficacy of dietary guanidinoacetic acid in broilers chicks, Master Thesis. University of Illinois, Urbana-Champaign, IL.

Fröhlich, T., Kemter, E., Flenkenthaler, F., Klymiuk, N., Otte, K. A., Blutke, A., Krause, S., Walter, M. C., Wanke, R., Wolf, E. and Arnold, G. J. (2016). Progressive muscle proteome changes in a clinically relevant pig model of Duchenne muscular dystrophy, *Sci. Rep.* 6: 33362.

Gallanti, A., Prelle, A., Moggio, M., Ciscato, P., Checcarelli, N., Sciacco, M., Comini, A. and Scarlato, G. (1992). Desmin and vimentina s markers of regeneration on muscle diseases, *Acta Neuropathol.* 85(1): 88–92.

Greene, E., Cauble, R., Dhamad, E. E., Kidd, M. T., Kong, B., Howard, S. M., Castro, H. F., Campagna, S. R., Bedford, M. and Dridi, S. (2020). Muscle metabolome profiles in woody breast-(un)affected broilers: effects of quantum blue phytase-enriched diet, *Front. Vet. Sci.* 7:458.

Griffin, J. R., Moraes, L., Wick, M. and Lilburn, M. S. (2018). Onset of white striping and progression into wooden breast as defined by myopathic changes underlying pectoralis major growth. Estimation of growth parameters as predictors for stage of myopathy progression, *Avian Pathol.* 47(1): 2–13.

Griffin, J. L., Williams, H. J., Sang, E., Clarke, K., Rae, C. and Nicholson, J. K. (2001). Metabolic profiling of genetic disorders: a multitissue ^{1}H nuclear magnetic resonance spectroscopic and pattern recognition study into dystrophic tissue, *Anal. Biochem.* 293(1): 16–21.

Guilloteau, P., Martin, L., Eeckhaut, V., Ducatelle, R., Zabielski, R. and Van Immerseel, F. (2010). From the gut to the peripheral tissues: the multiple effects of butyrate, *Nutr. Res. Rev.* 23(2): 366–384.

Harris, J. B., Vater, R., Wilson, M. and Cullen, M. J. (2003). Muscle fibre breakdown in venom-induced muscle degeneration, *J. Anat.* 202(4): 363–372.

Khajali, F. and Wideman, R. F. (2010). Dietary arginine: metabolic, environmental, immunological and physiological interrelationships, *Worlds Poult. Sci. J.* 66(4): 751–766.

Kuttappan, V. A., Shivaprasad, H. L., Shaw, D. P., Valentine, B. A., Hargis, B. M., Clark, F. D., McKee, S. R. and Owens, C. M. (2013). Pathological changes associated with white striping in broiler breast muscles, *Poult. Sci.* 92(2): 331–338.

Leeson, S., Namkung, H., Antongiovanni, M. and Lee, E. H. (2005). Effect of butyric acid on the performance and carcass yield of broiler chickens, *Poult. Sci.* 84(9): 1418–1422.

Liang, G., Taranova, O., Xia, K. and Zhang, Y. (2010). Butyrate promotes induced pluripotent stem cell generation, *J. Biol. Chem.* 285(33): 25516–25521.

Liu, X., Wu, H., Byrne, M., Krane, S. and Jaenish, R. (1997). Type III collagen is crucial for collagen I fibrillogenesis and for normal cardiovascular development, *Proc. Natl. Acad. Sci. U. S. A.* 94(5): 1852–1856.

Mali, P., Chou, B. K., Yen, J., Ye, Z., Zou, J., Dowey, S., Brodsky, R. A., Ohm, J. E., Yu, W., Baylin, S. B., Yusa, K., Bradley, A., Meyers, D. J., Mukherjee, C., Cole, P. A. and Cheng, L. (2010). Butyrate greatly enhances derivation of human induced pluripotent stem cells by promoting epigenetic remodeling and the expression of pluripotency-associated genes, *Stem Cells* 28(4): 713–720.

Malila, Y., Thanatsang, K., Arayamethakorn, S., Uengwetwanit, T., Srimarut, Y., Petracci, M., Strasburg, G., Rungrassamee, W. and Visessanguan, W. (2019). Absolute expressions of hypoxia-inducible factor-1 alpha (HIF1A) transcript and the associated genes in chicken skeletal muscle with white striping and wooden breast myopathies, *PLoS ONE* 14(8):e0220904.

Martins-Bach, A. B., Bloisea, A. C., Vainzofb, M. and Rabbania, S. R. (2012). Metabolic profile of dystrophic mdx mouse muscles analyzed with in vitro magnetic resonance spectroscopy (MRS), *Magn. Reson. Imaging* 30(8): 1167–1176.

Mazzoni, M., Soglia, F., Petracci, M., Sirri, F., Lattanzio, G. and Clavenzani, P. (2020). Fiber metabolism, procollagen and collagen type III immunoreactivity in broiler pectoralis major affected by muscle abnormalities, *Animals (Basel)* 10(6): 1081.

Mitchell, M. A. (1999). Muscle abnormalities–pathophysiological mechanisms. In: Richardson, R. I. and Mead, G. C. (Eds), *Poultry Meat Science*. CABI Publishing, Oxon, pp. 65-98.

Mutryn, M. F., Brannick, E. M., Fu, W., Lee, W. R. and Abasht, B. (2015). Characterization of a novel chicken muscle disorder through differential gene expression and pathway analysis using RNA sequencing, *BMC Genomics* 16: 399.

Namkung, H., Yu, H., Gong, J. and Leeson, S. (2011). Antimicrobial activity of butyrate glycerides toward *Salmonella typhimurium* and *Clostridium perfringens*, *Poult. Sci.* 90(10): 2217-2222.

Pascual, A., Trocino, A., Birolo, M., Cardazzo, B., Bordignon, F., Ballarin, C., Carraro, L. and Xiccato, G. (2020). Dietary supplementation with sodium butyrate: growth, gut response at different ages, and meat quality of female and male broiler chickens, *Ital. J. Anim. Sci.* 19(1): 1134-1145.

Pascual, A., Trocino, A., Susta, L. and Barbut, S. (2021). Comparing three textural measurements of chicken breast fillets affected by severe wooden breast and Spaghetti Meat, *Ital. J. Anim. Sci.* 20(1): 465-471.

Peng, L., He, Z., Chen, W., Holzman, I. R. and Lin, J. (2007). Effects of butyrate on intestinal barrier function in a Caco-2 cell monolayer model of intestinal barrier, *Pediatr. Res.* 61(1): 37-41.

Petracci, M., Mudalal, S., Soglia, F. and Cavani, C. (2015). Meat quality in fast-growing broiler chickens, *World Poult. Sci. J.* 71(2): 363-374.

Petracci, M., Soglia, F., Madruga, M., Carvalho, L., Ida, E. and Estévez, M. (2019). Wooden-breast, white striping, and spaghetti meat: causes, consequences and consumer perception of emerging broiler meat abnormalities, *Compr. Rev. Food Sci. F* 2: 565-583.

Sanden, K. W., Böcker, U., Ofstad, R., Pedersen, M. E., Høst, V., Afseth, N. K., Rønning, S. B. and Plesko, N. (2021). Characterization of collagen structure in normal, Wooden Breast and Spaghetti Meat chicken fillets by FTIR microspectroscopy and histology, *Foods* 10(3): 548.

Sandercock, D. A., Barker, Z. E., Mitchell, M. A. and Hocking, P. M. (2009). Changes in muscle cell cation regulation and meat quality traits are associated with genetic selection for high body weight and meat yield in broiler chickens, *Genet. Sel. Evol.* 41: 8.

Sandercock, D. A. and Mitchell, M. A. (2003). Myopathy in broiler chickens: a role for Ca21-activated phospholipase A2?, *Poult. Sci.* 82(8): 1307-1312.

Sandercock, D. A. and Mitchell, M. A. (2004). The role of sodium ions in the pathogenesis of skeletal muscle damage in broiler chickens, *Poult. Sci.* 83(4): 701-706.

Sañudo, C., Alfonso, M., Sanchez, A., Berge, P., Dransfield, E., Zygoyiannis, D., Stamataris, C., Thorkelsson, G., Valdimarsdottir, T., Piasentier, E., Mills, C., Nute, G. R. and Fischer, A. V. (2003). Meat texture of lambs from different European production systems, *Aust. J. Agr. Res.* 54(6): 551-560.

Sihvo, H. K., Immonen, K. and Puolanne, E. (2014). Myodegeneration with fibrosis and regeneration in the pectoralis major muscle of broilers, *Vet. Pathol.* 51(3): 619-623.

Sirri, F., Maiorano, G., Tavaniello, S., Chen, J., Petracci, M. and Meluzzi, A. (2016). Effect of different levels of dietary zinc, manganese, and copper from organic or inorganic sources on performance, bacterial chondronecrosis, intramuscular collagen characteristics, and occurrence of meat quality defects of broiler chickens, *Poult. Sci.* 95(8): 1813–1824.

Soglia, F., Mazzoni, M., Zappaterra, M., Di Nunzio, M., Babini, E., Bordini, M., Sirri, F., Clavenzani, P., Davoli, R. and Petracci, M. (2020). Distribution and expression of vimentin and desmin in broiler pectoralis major affected by the growth-related muscular abnormalities, *Front. Physiol.* 10: 1581.

Soglia, F., Mudalal, S., Babini, E., Di Nunzio, M., Mazzoni, M., Sirri, F., Cavani, C. and Petracci, M. (2016). Histology, composition, and quality traits of chicken pectoralis major muscle affected by wooden breast abnormality, *Poult. Sci.* 95(3): 651–659.

Soglia, F., Petracci, M., Davoli, R. and Zappaterra, M. (2021a). A critical review of the mechanisms involved in the occurrence of growth-related abnormalities affecting broiler chicken breast muscles, *Poult. Sci.* 100:101180.

Soglia, F., Zampiga, M., Baldi, G., Malila, Y., Thanatsang, K. V., Srimarut, Y., Tatiyaborworntham, N., Unger, O., Klamchuen, A., Laghi, L., Petracci, M. and Sirri, F. (2021b). Lysine depletion during different feeding phases: effects on growth performances and meat quality of broiler chickens, *Animals (Basel)* 11(6): 1499. doi: 10.3390/ani11061499.

Soglia, F., Silva, A. K., Lião, L. M., Laghi, L. and Petracci, M. (2019). Effect of broiler breast abnormality and freezing on meat quality and metabolites assessed by H-NMR spectroscopy, *Poult. Sci.* 98(12): 7139–7150.

Tasoniero, G., Bertram, H. C., Young, J. F., Dalle Zotte, A. and Puolanne, E. (2017). Relationship between hardness and myowater properties in Wooden Breast affected chicken meat: a nuclear magnetic resonance study, *LWT* 86: 20–24.

Tasoniero, G., Zhuang, H. and Bowker, B. (2022). Biochemical and physicochemical changes in spaghetti meat during refrigerated storage of chicken breast, *Front. Physiol.* 13: 894544.

Tasoniero, G., Zhuang, H., Gamble, G. R. and Bowker, B. C. (2020). Effect of spaghetti meat abnormality on broiler chicken breast meat composition and technological quality, *Poult. Sci.* 99(3): 1724–1733.

Tokuyasu, K. T., Maher, P. A. and Singer, S. J. (1985). Distribution of vimentin and desmin in developing chick myotubes in vivo. II. Immunoelectron microscopic study, *J. Cell Biol.* 100(4): 1157–1166.

Vieira, S. L. and Lima, I. L. (2005). Live performance, water intake and excreta characteristics of broilers fed all vegetable diets based on corn and soybean meal, *Int. J. Poult. Sci.* 4(6): 365–368.

Wold, J. P. and Løvland, A. (2020). NIR spectroscopic techniques for quality and process control in the meat industry, *Meat Muscle Biol.* 4(2): 8: 1-8.

Wu, Y., Wang, Y., Wu, W., Yin, D., Sun, X., Guo, X., Chen, J., Mahmood, T., Yan, L. and Yuan, J. (2020). Effects of nicotinamide and sodium butyrate on meat quality and muscle ubiquitination degradation genes in broilers reared at a high stocking density, *Poult. Sci.* 99(3): 1462–1470.

Xia, X., Kong, B., Liu, Q. and Liu, J. (2009). Physicochemical change and protein oxidation in porcine longissimus dorsi as influenced by different freeze-thaw cycles, *Meat Sci.* 83(2): 239–245.

Xiong, Y. (2004). Muscle protein. In: Yada, R. Y. (Ed.), *Proteins in Food Processing*. Woodhead Publishing Limited, Oxford, pp. 100–122.

Yang, X., Yin, F., Yang, Y., Lepp, D., Yu, H., Ruan, Z., Yang, C., Yin, Y., Hou, Y., Leeson, S. and Gong, J. (2018). Dietary butyrate glycerides modulate intestinal microbiota composition and serum metabolites in broilers, *Sci. Rep.* 8(1): 4940.

Yu, X., Feng, Y., Bowker, B. and Zhuang, H. (2020). Expressible fluid measurements of broiler breast meat affected by emerging muscle abnormalities, *LWT* 133: 110110.

Zampiga, M., Laghi, L., Petracci, M., Zhu, C., Meluzzi, A., Dridi, S. and Sirri, F. (2018). Effect of dietary arginine to lysine ratios on productive performance, meat quality, plasma and muscle metabolomics profile in fast-growing broiler chickens, *J. Anim. Sci. Biotechnol.* 9: 79. doi: 10.1186/s40104-018-0294-5.

Zampiga, M., Meluzzi, A., Pignata, S. and Sirri, F. (2019a). Occurrence of breast meat abnormalities and foot pad dermatitis in light-size broiler chicken hybrids, *Animals (Basel)* 9(10): 706. doi: 10.3390/ani9100706.

Zampiga, M., Soglia, F., Petracci, M., Meluzzi, A. and Sirri, F. (2019b). Effect of different arginine-to-lysine ratios in broiler chicken diets on the occurrence of breast myopathies and meat quality attributes, *Poult. Sci.* 98(6): 2691–2697.

Chapter 12

Factors affecting shelf life of poultry meat

Alberto González-Mohíno and Mario Estévez, Universidad de Extremadura, Spain

1 Introduction

According to the last report from the Organisation for Economic Co-operation and Development (OECD) and the Food and Agricultural Organization (FAO) from the United Nations in 2021, poultry meat consumption is expected to grow steadily over the following decades since consumers perceive poultry meat as healthier and more environmentally friendly than beef and pork (OECD/FAO, 2021). Poultry meat remains the most consumed meat in the world, with around 15.1 kg being consumed per capita, followed by pork (11.8 kg/capita) and beef/veal (6.3 kg/capita). Providing poultry meat to a growing population that is safe, nutritious and of good sensory quality is challenging given the number of chemical and biological hazards threatening poultry meat quality and safety.

Once harvested, processed and packed, the commercial life of poultry products is limited by the onset of microbial spoilage and chemical reactions such as oxidation. Commercial 'shelf life' has been defined as the period of time after processing and packaging during which the food product preserves a minimum level of acceptable quality and, hence, may be consumed (Nicoli,

http://dx.doi.org/10.19103/AS.2022.0112.13

2012). While microbiological stability is a primary concern in raw (uncooked) poultry meat and products, the onset of oxidation and rancidity is a major cause of deterioration in processed poultry products. Extending the shelf life of these products requires details of these spoilage mechanisms and the ways they work.

Microbial spoilage is a major concern in raw (unprocessed) poultry meat and poultry meat products. The shelf life of all types of fresh food products is typically set by criteria relating to microbiological stability (O'Sullivan, 2016). While both spoilage bacteria and pathogens are potential threats to poultry meat quality and safety, there has been less study of the former compared to the latter (Rouger et al., 2017). Microbial spoilage has a crucial impact on poultry meat shelf life with negative effects on colour, texture, odour and flavour properties as well as on safety (Vihavainen and Björkroth, 2010).

Oxidative reactions are inherent to biological systems and take place at all stages in the poultry supply chain between the farm (living animal) and the fork (consumption by the consumer), including intermediate stages such as processing, storage and culinary preparation (Estevez, 2015). Modern broiler and turkey hybrids are particularly sensitive to oxidative stress owing to genetic selection towards fast growth rates and large breast muscles (Petracci et al., 2019; Zampiga et al., 2020). Such stress is known to impair physiological processes in poultry birds, contributing to poorer health, production rates and meat quality (Carvalho et al., 2017a). Recent evidence indicates that persistent oxidative stress may be an important factor in the onset of quality defects in chicken breast meat such as wooden breast (WB) and white striping (WS) (Petracci et al., 2019) (covered in other chapters in this book).

Muscle tissue is deprived of physiological antioxidant defences after slaughter and, hence, cell components are exposed to the pro-oxidant action of reactive oxygen species (ROS) (Estévez et al., 2020). Meat lipids and proteins, including heme pigments such as myoglobin, undergo oxidative damage as poultry meat is handled, processed and stored. These oxidative modifications are responsible for the loss of nutritional value (i.e. depletion of essential fatty acids and amino acids), deterioration of sensory properties (i.e. onset of rancidity and increased toughness) and accretion of potentially toxic compounds such as malondialdehyde (MDA) and assorted oxidized amino acids (Estevez, 2015; Estévez and Xiong, 2019).

Apart from appropriate hygienic practices to prevent microbial contamination during handling, processing and storage of poultry meat, there remains a need for effective strategies to control the growth of spoilage microorganisms in poultry meat products. There is also a need for effective strategies to control oxidation processes (Estévez, 2021). This has resulted in a range of innovative solutions in both areas. As an example, the use of plant-based antimicrobials is of increasing interest owing to growing consumer concern about traditional antimicrobial additives (Beya et al., 2021; Estévez et al., 2020). Other solutions include modifying animal feed, strategies involving muscle food formulation and innovative packaging solutions (Estévez, 2021). This chapter analyses the most recent advances in the field of microbial spoilage and oxidative reactions as major threats to the shelf life of poultry and poultry products. Innovative means of protection against microbial growth and poultry oxidation are also addressed.

2 Understanding microbial spoilage of poultry meat

Raw poultry meat is highly susceptible to microbial spoilage and this is the primary factor limiting shelf life (Fig. 1). Poultry meat has a high amount of water content and contains variable nutrients, including carbohydrates (glycogen, glucose, ribose, etc.), nucleotides, amino acids, minerals and vitamins to support the growth of diverse microbial populations (Vihavainen and Björkroth, 2010).

More than 21 phyla and 280 genera of microorganisms have been found to colonize broiler meat, potentially compromising its safety, quality and/or consumer appeal (Kim et al., 2017). The predominant spoilage microorganisms are Gram-positive and Gram-negative bacteria, which can occur in relatively large numbers (Mead, 2004). Among other bacteria genera, *Pseudomonas spp.* and microorganisms from the Enterobacteriaceae family are frequently identified as spoilage colonizers of poultry meat (Nychas and Drosinos, 2014). Fungi generally play a minor role in poultry spoilage. However, when bacterial growth is inhibited by chemical and/or physical means, fungi can become a major cause of spoilage, with *Aspergillus* and *Penicillium* being some of the dominant genera (Cerveny et al., 2009; Nasser, 2015).

It is important also to highlight the important role of pathogenic bacteria, such as *Campylobacter* or *Salmonella* bacteria, in the quality and safety of the final product (Marmion et al., 2021). Among these bacteria, it is worth highlighting *Salmonella* genera, as chicken meat is a good reservoir for this microorganism (Wessels et al., 2021), *Campylobacter* genera (particularly *Campylobacter jejuni*) owing to its common incidence in the poultry production chain (Thornval and Hoorfar, 2021) and *Escherichia coli* which is also a frequent colonizer of poultry carcasses (Belluco et al., 2016).

In the post-mortem period, the initial bacterial load can result from sources such as contaminated work surfaces, equipment, water or avian gut microbiota during the slaughter process (Odeyemi et al., 2020). The extent to which microbial growth affects poultry meat depends on the type and counts of the bacterial population, which determine, in turn, the nature and intensity of the bacterial metabolic processes which eventually spoil poultry meat.

The level of susceptibility of chicken meat to spoilage is also influenced by other various intrinsic and extrinsic factors, such as the occurrence of enzymatic

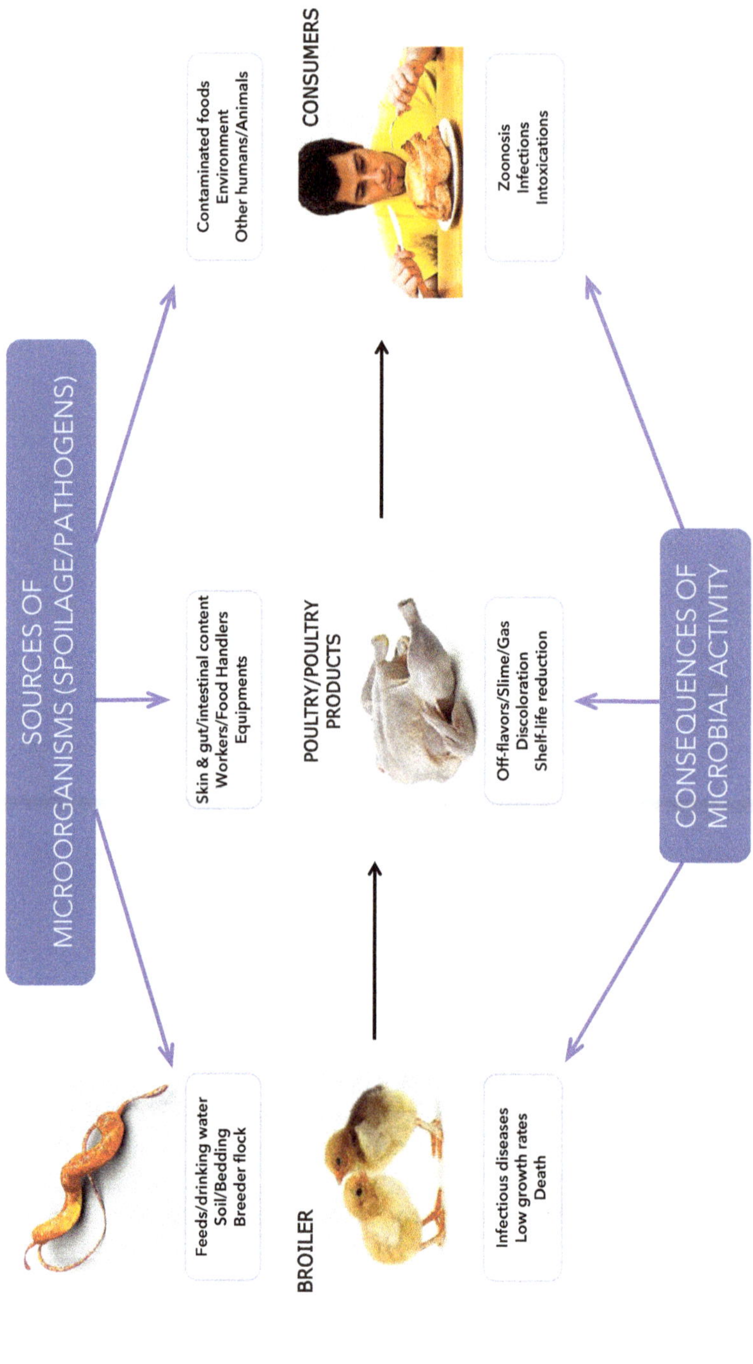

Figure 1 Main sources of microorganisms and consequences of their growth and activity at each of the stages of the food chain.

and chemical reactions, pH, water activity (aW), temperature and oxygenation. The pH of meat is variable, depending on the occurrence of ante-mortem stress and the type of poultry meat cut, with normal pH in chicken meat ranging from 5.7–5.9 in breast muscle, to 6.2 in leg muscles (Vihavainen and Björkroth, 2010). Spoilage bacterial growth is promoted when pH is above 6.2. Factors such as feed regimes and stress in the period prior to and during slaughter can affect pH and other meat characteristics (Petracci et al., 2010; Addis, 2015). Water activity also influences microbial growth, e.g. lowering aW levels can slow the proliferation of *Pseudomonas* and *Enterobacteriaceae* genera, two of the main spoilage bacteria in poultry meat (Odeyemi et al., 2020). Temperature and degree of oxygenation also play a key role in the rate and degree of spoilage with higher CO_2 levels, e.g. exercising a bacteriostatic effect on aerobic bacteria such as *Pseudomonas*.

Two bacterial groups are responsible for meat spoilage. The first group, known as specific spoilage organisms (SSOs), initiates the deterioration process. The growth of bacteria from the second group, known as spoilage-associated organisms (SAOs), is facilitated by SSOs (Boziaris and Parlapani, 2017; Saenz-García et al., 2020). As described by Saenz-García et al. (2020), residual glucose, lactic acid and water-soluble proteins are initially consumed by fast-growing bacteria of the SAO group. Subsequently, SSOs start a slower proliferation in this nutrient-depleted environment by obtaining nutrients from muscle tissue, initiating the spoilage process.

Spoilage microbiota become established in poultry meat within 4 days after slaughter. They cause noticeable visual damage and odour changes when bacteria levels reach around 10^7 cells/cm^2 and 10^8 cells/cm^2, respectively. Some species, such as *Brocothrix thermosphacta*, cause noticeable spoilage effects (e.g. undesirable odour) at levels as low as 10^5 cells/cm^2 (Lauritsen et al., 2019). Microbial spoilage is mainly perceived as the development of undesirable odours and flavours, changes in colour (discolouration, presence of blue-green colours, etc.), structural changes and formation of slime on meat surfaces. At a molecular level, these changes are caused by the degradation of muscle components by bacteria and the subsequent production of volatiles and exopolysaccharides (Rouger et al., 2017). As putrefaction progresses, many other flavour and odour-active compounds may be formed, such as aldehydes, sulphur-containing compounds and organic acids (Marmion et al., 2021).

Cold storage and packaging gas composition are two key techniques for slowing bacterial spoilage (Marmion et al., 2021). Duqué et al. (2021), e.g. found low levels of *Campylobacter spp.* in chicken legs chilled for 6 days at 6°C, although the most efficient means to control *Campylobacter spp.* in poultry meat is by hurdle strategies combining chilling with other techniques (Stella et al., 2021). However, the growth of microbiota in fresh poultry meat is a continuous process even under cold storage, eventually leading to the onset of irreversible and undesirable changes. As cold storage progresses, microbiota diversity decreases, but the overall bacterial load continues to grow (Chaillou et al., 2015; Höll et al., 2016). Samapundo et al. (2019) reported complete meat putrefaction within 10 days of chilled storage at 7°C. The occurrence of off-odours is the most noticeable effect of microbial growth during refrigerated storage of chicken meat, resulting from spoilage compounds such as nitrogenous and sulphur-containing compounds, along with acid lactic and ethanol (Katiyo et al., 2020).

Some microorganisms are adapted to cold storage conditions. These include *Aeromonas*, *Buttiauxella*, *Enterobacter*, *Pseudomonas*, *Psychrobacter*, *Brochothrix*, *Hafnia*, *Yersinia*, *Lactococcus*, *Serratia*, *Carnobacterium* and *Shewanella* genera (Rouger et al., 2017; Li et al., 2020; Yu et al., 2021). *Shewanella putrefaciens* is frequently found on chilled poultry carcasses (Hinton et al., 2004). *Pseudomonas* is the largest genera of raw chicken microflora, increasing in numbers during chilled storage (Cerveny et al., 2009). *Pseudomonas fluorescens*, *Serratia liquefaciens* and *Aeromonas salmonicida* have been found in poultry meat stored at 8°C over 4 days.

Freezing is effective in significantly slowing both spoilage and pathogen growth (Yu et al., 2019). Yu et al. (2021) found *Pseudomonas spp*, *Staphylococcus spp* and *Psychrobacter spp* to be the predominant bacteria in frozen broiler carcasses. Research suggests that bacteria of the genus of *Brochothrix* spp., *Lactococcus spp*, *Serratia spp* and *Carnobacterium spp* are effectively controlled by industrial freezing (Kluth et al., 2021).

3 Understanding oxidative spoilage of poultry meat

The high levels of polyunsaturated fatty acids in muscle lipids make poultry meat highly susceptible to oxidative reactions (Estevez, 2015; Domínguez et al., 2019). Such reactions are typically manifested as a chain reaction in which lipids, and other muscle components are degraded by ROS-mediated mechanisms. The processes of lipid and protein oxidation in post-mortem muscles have been described for different species such as chicken broilers (Salles et al., 2019), spent laying hens (Liu et al., 2020a), ducks (He et al., 2019), geese (Nemati et al., 2021), turkeys (Wu et al., 2021), partridges (Jingting et al., 2017), pigeon (Dal Bosco et al., 2005) and ostriches (Leygonie et al., 2011).

While microbial spoilage is a major threat to the shelf life of fresh (unprocessed) poultry products, oxidative reactions and their undesirable effects are particularly significant in processed meat products (Estevez, 2015). Meat manipulation and industrial processing practices such as mincing, slicing, salting, curing, drying and high temperatures, among many others, promote the oxidative degradation of lipid and proteins in poultry meat and meat products (Soladoye et al., 2015).

3.1 Pre-slaughter oxidative stress

The onset of oxidative stress starts on the farm before slaughter (Fig. 2). Growing broilers are especially sensitive to oxidative stress owing to high metabolic activity in their muscles (Estevez, 2015; Kuttappan et al., 2017). Oxidative stress occurs whenever biologically relevant molecules are affected by an uncontrolled pro-oxidative environment. Such oxidative damage leads to altered physiological processes which are reflected in increased vulnerability to disease, impaired animal growth and performance (Fellenberg and Speisky, 2006). Pre-slaughter oxidative stress may also reduce the shelf life and quality characteristics of chicken meat (Carvalho et al., 2017a). Oxidative stress has been identified as a crucial biochemical mechanism in the onset of chicken breast myopathies such as WB and WS which significantly reduce meat quality (Petracci et al., 2019; Carvalho et al., 2021).

Environmental conditions at the farm and during transportation, as well as diet, are the main sources of oxidative stress affecting broilers (Brannan et al., 2021; Kpomasse et al., 2021). The combination of a hot climate and the intake of oxidized feeds has been found to promote oxidative stress in growing chickens (Bayraktar et al., 2011; Ismail et al., 2013). Tarkhan et al. (2020) found that a hot environment affected the expression of genes decoding for antioxidant enzymes such as catalase (CAT), NADPH oxidase 4 and superoxide dismutase (SOD) 2. Delles et al (2014) showed that providing broilers with oxidized fats favoured the onset of oxidative stress via various mechanisms, including a decreased activity of antioxidant enzymes such as CAT, glutathione-peroxidase (GSH-Px) and SOD. Bayraktar et al. (2011) found more intense GSH-Px activity in broilers fed oxidised oil, possibly due to muscle tissue attempting to counteract the oxidative stress induced by the intake of oxidized feeds. The fact that tocopherol supplementation led to a decreased GSH-Px activity in broilers fed oxidised oil supports this hypothesis.

The occurrence of severe oxidative stress in domestic birds prior to slaughter modifies crucial traits such as pH and water-holding capacity (WHC) which significantly affect both microbial and chemical (oxidative) spoilage (as reviewed by Estevez, 2015; and Qaisrani et al., 2021). Avoiding oxidative stress is thus important for animal welfare, productivity and meat quality (Wasti et al., 2020). Oxidative stress can be controlled, e.g. through diet (He et al., 2018; Bacou et al., 2021). Such strategies include supplementing feeds with antioxidant additives, micronutrients and probiotics.

3.2 Post-slaughter oxidative stress

Physicochemical modifications occurring in post-mortem muscle, such as the collapse of endogenous antioxidant defences and the accretion of lactate,

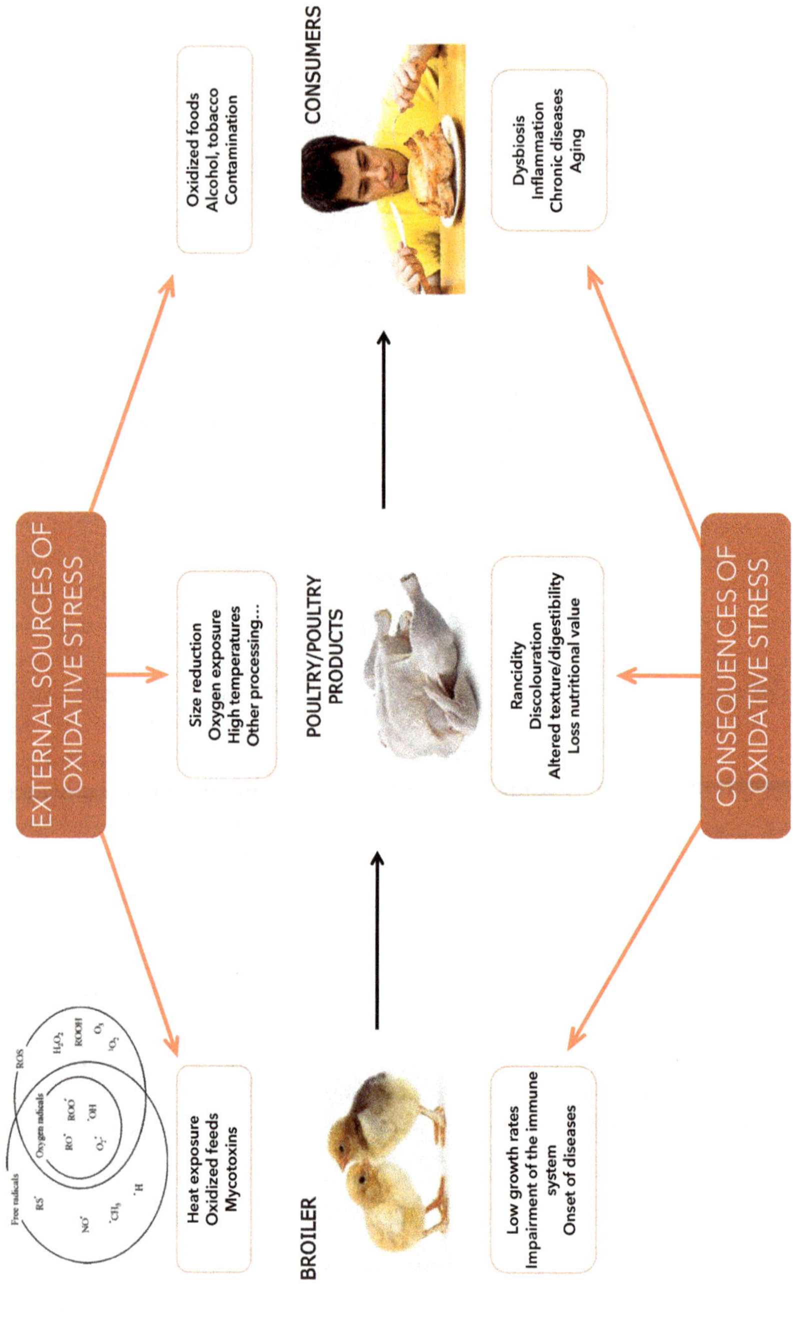

Figure 2 Main sources of oxidative stress and its consequences at each of the stages of the food chain.

facilitate the onset of oxidative reactions in poultry meat immediately after slaughter. Lactate accumulation and pH decline in post-rigour meat facilitate the pro-oxidation action of myoglobin and the subsequent generation of ROS (Estévez, 2011, 2015). Metmyoglobin acts as a pro-oxidant at acidic pH and in the presence of hydroperoxides, while free-catalytic iron is also involved in the generation of free radicals and other pro-oxidative species in post-mortem muscle (Faustman et al., 2010; Domínguez et al., 2019). Zhang et al. (2021) have investigated the role of myoglobin, hemin and non-heme iron in the onset of lipid oxidation and other quality traits in chicken meat. Meat from other domestic birds, such as duck, ostrich, pigeon or ostrich, are known to have similar heme and non-heme iron concentrations as red meat and thus play a role as promoters of oxidative reactions (Min et al., 2008).

Post-mortem ROS-mediated changes affect both lipid oxidation and meat proteins (Xiong and Guo, 2020). The depletion of sulphur amino acids and tryptophan and the accretion of protein carbonyls cause oxidative damage to meat proteins (Estévez et al., 2020; Xiong and Guo, 2020). pH decline and subsequent protein denaturation facilitate the oxidation of meat proteins, which is also affected by the pro-oxidant action of lipid oxidation products (Estévez, 2021). MDA is known to covalently bind to protein residues, leading to the formation of secondary protein carbonyls (Estévez, 2011; Soglia et al., 2020). The Maillard reaction has also been described as a source of oxidative damage to proteins (Zhu et al., 2020). Reducing sugars and their degradation products (e.g. dicarbonyls such as glyoxal) react with alkaline amino acids such as lysine to induce the formation of primary carbonyls in meat proteins (Luna and Estévez, 2019; Zhu et al., 2020; Estévez et al., 2022).

In contrast to microbial spoilage, freezing has less effect on lipid and particularly protein oxidation (Leygonie et al., 2011; Utrera and Estévez, 2013). Chicken and turkey meat have been found to suffer more severe oxidative damage than pork and beef during frozen storage (reviewed by Leygonie et al., 2012; Utrera and Estévez, 2013). A detailed description of the occurrence and consequences of the oxidation of proteins in chicken meat subjected to long-term frozen storage was recently carried out by Soncu (2020). In addition to the accretion of protein carbonyls and loss of thiols, a severe denaturation and aggregation of oxidized proteins promoted the loss of solubility and WHC, suggesting frozen chicken meat should not be stored for longer than 3 months. The application of a consecutive freeze-thaw cycle to chicken meat significantly increases the extent of lipid and protein oxidation with severe consequences in terms of protein functionality and meat quality (Ali et al., 2015, 2016).

It is, however, during the processing of poultry meat that lipid and protein oxidation become the primary threat to meat quality (Soladoye et al., 2015; Domínguez et al., 2019). Whilst they combat microbial hazards, techniques such as salting, curing, drying and cooking promote oxidative degradation of poultry meat lipids and proteins. Silva et al. (2017) reported that the application of salt and subsequent drying of preserved layer hen meat promoted lipid oxidation and rancidity. Qu et al. (2020) have studied the impact of NaCl on lipid and protein oxidation in dry-cured chicken. In a subsequent study of the sensory properties of jerky chicken, certain muscle types (thigh vs breast) and muscles from older animals (spent hen vs broiler) were found particularly susceptible to oxidation (Silva et al., 2018). Choi and Chin (2021) have investigated the structural changes caused by salt and phosphates on chicken meat.

Cooking methods accelerate protein oxidation in jerky chicken due to the depletion of essential amino acids and accretion of protein oxidation products such as carbonyls and cross-links (Silva et al., 2016). Ferreira et al. (2016) subjected ready-to-eat chicken patties to initial cooking, subsequent chilled storage and microwave reheating. Lipid and protein oxidation are accelerated during chilled storage. Pre-cooking methods such as boiling (lower temperatures/longer time) caused more severe oxidation than grilling and roasting (higher temperatures/shorter time). Xiong et al. (2020) recently studied the underlying mechanisms of pro-oxidative effect of heating on chicken meat and concluded that the release of heme iron and depletion of antioxidant enzymes increase oxidative degradation of lipids at temperatures above 70°C. A recent study by Sobral et al. (2020) found that cooking affects both the formation of toxic lipid oxidation products in chicken burger patties and digestibility.

While oxidative damage is affected by the severity of processing, the properties of the raw material have also been found to play a role. Chicken breasts affected by myopathies such as the WB or the WS are more susceptible to oxidation than normal chicken breasts (Petracci et al., 2019). Carvalho et al. (2021) have highlighted the role played by oxidative stress in the onset of WS. Thanatsang et al. (2020) also found increased oxidation in WB muscles. This is significant because oxidative instability complicates the processing of chicken WB and WS meat in products such as sausages (Cavalcante da Rocha et al., 2020), burger patties (de Oliveira et al., 2020), marinated poultry products (U-Chupaj et al., 2021) and chicken nuggets (Bordignon et al., 2020).

4 Impact of oxidative stress on poultry meat quality and shelf life

The negative consequences of oxidative stress on poultry meat quality are noticeable even if oxidation occurred *in vivo*, prior to animal slaughter (Estevez, 2015). *In vivo* oxidative stress increases the occurrence of chicken breast abnormalities such as WS (Carvalho et al., 2021), WB (Sihvo et al., 2014), spaghetti meat (Petracci et al., 2019) and pink colour defect (Holownia et al., 2003). *In vivo* protein oxidation of muscle tissue has been reported to be responsible for impaired WHC in chicken meat (Zhang et al., 2011). Recent findings from Estévez and Petracci (2019) in broilers subjected to oxidative stress corroborate the connection between *in vivo* protein oxidation and meat protein functionality.

The onset of rancidity and off-odours as a result of severe lipid oxidation during the processing of poultry meat products remains a major concern (Min and Ahn, 2005; Domínguez et al., 2019). Oxidation before cooking leads to the generation of oxidation precursors that contribute to the onset of off-flavours during heating. The most intense rancidity has been detected in ultra-processed poultry meat products, which are typically subjected to processing steps such as cold storage of raw material, precooking, cold storage of precooked meat and reheating. Lipid oxidation has been found to occur during initial storage while precooking creates a pro-oxidative environment which facilitates the onset of intense flavour deterioration during subsequent chilled storage (known as warmed-over-flavour (WOF)) (Pegg and Shahidi, 2012). WOF in processed poultry meat products causes significant economic losses (Pegg and Shahidi, 2012).

A high degree of fat unsaturation and susceptibility to oxidative reactions makes poultry meat particularly prone to WOF (Pegg and Shahidi, 2012). Pro-oxidant enzymes such as lipoxygenases and heme and non-heme iron could also contribute to the oxidative degradation of poultry meat lipids (Min et al., 2008; Utrera and Estévez, 2013; Domínguez et al., 2019). Min et al. (2008) and Utrera et al. (2014) found that the susceptibility of poultry meat products to lipid oxidation was higher when subjected to oxidation induced during ultra-processing. WOF is linked to an imbalance between the formation of certain lipid-derived volatiles (pentanal, hexanal and alkadienals) which impart unpleasant flavour notes and those from Maillard reaction (pyrazines, pyrroles and furans) which contribute positive flavour notes (Jayasena et al., 2013). WOF has been identified as a factor leading to the rejection of products such as jerky chicken (Silva et al., 2017) and ready-to-eat chicken patties (Ferreira et al., 2016).

Lipid oxidation is traditionally linked to myoglobin oxidation and colour deterioration, which limits shelf life in red meats (Liu et al., 2020b; Ramanathan et al., 2020). Discolouration of red poultry meat such as duck or ostrich also limits shelf life during refrigerated storage (Zheng et al., 2020; Seydim et al., 2006). Lipid oxidation is also a source of toxic compounds such as MDA and cholesterol oxidation products (Estevez, 2015; Estévez et al., 2017). There has been significant research into understanding and preventing these harmful compounds from developing in poultry products which are particularly susceptible to them (Choe et al., 2018; Hashari et al., 2020; Hsu and Chen, 2020).

Though much less studied than lipid oxidation, protein oxidation also impacts poultry meat. There is a well-established link between protein oxidation and altered protein functionality (including emulsifying, gelling and water-holding properties) (Bao et al., 2018; Xia et al., 2018; Luna and Estévez, 2019). Higher water loss during processing or upon thawing of frozen chicken meat also affects other quality attributes such as colour and texture (Utrera et al., 2014).

Bao and Ertbjerg (2019) have explored the connection between protein oxidation, protein functionality and rheological properties of meat products. Protein oxidation is identified as responsible for increased meat toughness via two mechanisms:

- impaired tenderization during ageing due to calpain inactivation (Huff-Lonergan et al., 2010) and
- formation of oxidative protein cross-links which strengthen meat structure (Lund et al., 2011).

Protein oxidation promotes toughening of refrigerated turkey meat (Smaoui et al., 2017), jerky chicken (Silva et al., 2018), emulsified (Santos et al., 2020) and ready-to-eat chicken patties (Ferreira et al., 2018), dry-cured chicken (Qu et al., 2020) and chicken breast muscle subjected to high-oxygen atmosphere packaging (Jongberg et al., 2014).

Besides impacting sensory quality, protein oxidation also reduces nutritional value due to oxidative degradation of essential amino acids and impaired digestibility of oxidized proteins (Soladoye et al., 2015). These effects have been reported in chicken products subjected to severe processing conditions (Ferreira et al., 2018). Intake of oxidized amino acids and proteins has also been linked to physiological impairments, including alteration of microbiota, dysfunction of the pancreas, thyroid and liver and loss of spatial memory in experimental animals (reviewed by Estévez and Xiong, 2019). *In vitro* and *in vivo* studies indicate that red meats (i.e. beef) may be more harmful than white meats (chicken, turkey) since the former are more prone to protein oxidation during digestion (Goethals et al., 2020; Van Hecke et al., 2021).

4.1 Oxidation and the appearance of poultry meat

The appearance of poultry meat and meat products plays a crucial role in consumer preferences and purchase intentions (Carvalho et al., 2017a). Oxidation of meat pigments such as myoglobin causes undesirable modifications of meat colour (Ramanathan et al., 2020). The colour of poultry meat is affected by a number of factors including the concentration and chemical state of myoglobin, concentration of moisture and intramuscular lipids, age, gender, genetic and feeding backgrounds, pre-slaughter conditions and slaughter variables such as stunning method (Totosaus et al., 2007; Liu et al.,

2020b; Ramanathan et al., 2020). Colour defects commonly found in chicken meat and meat products include bruising due to abnormal post-mortem glycogen metabolism (Carvalho et al., 2017a).

The colour displayed by normal fresh chicken meat can fluctuate from grey-white to slight pinkish and yellowish tones. These variations are caused by factors affecting myoglobin concentration (muscle type, age, exercise) or the accumulation of dietary carotenoids (Brown, 2018). Deviating from this normal range of colours is commonly caused by altered post-mortem glycogen metabolism (e.g. in pale, soft and exudative [PSE] meat; or dark, firm and dry [DFD] meat) or by the onset of myopathies of multifactorial aetiology (e.g. deep pectoral myopathy, WB, WS and Spaghetti meat [SM]).

Most of these abnormalities are related to selecting genetic strains with fast growth and large meat yields (Carvalho et al., 2017b; Petracci et al., 2019). These selected strains are prone to suffer severe pre-slaughter stress, which leads to abnormal glycogen metabolism and altered pH decline in post-mortem muscles. Abnormalities caused by these conditions include PSE-like and DFD-like meats with lighter and darker colours, respectively (Petracci et al., 2009; Dadgar et al., 2012; Estevez, 2015). The incidence of PSE and PSE-like chicken meat is, at a molecular level, associated with lipid degradation mechanisms mediated by enzymes such as phospholipase A2 as well as protein denaturation, degradation and oxidation mediated by the proteasome and ROS (Soares et al., 2003; Desai et al., 2016; Carvalho et al., 2017b).

The appearance of chicken breasts affected by the WB and WS myopathies is noticeably different from that displayed by normal chicken breasts. As reported by Petracci et al. (2019), these visual alterations are caused by histological impairments suffered by living breast muscles during the onset of myopathy. WB is characterized by the occurrence of raised areas in which pale and dark hues are contrasted owing to congestion, inflammation and visualization of petechiae. WS is distinguished by white striations of variable thickness parallel to the muscle fibres. These pathological features are manifested in higher lightness (L*) (Dalle Zotte et al., 2017) and yellowness (b*) (Kuttappan et al., 2017) compared to normal chicken breasts. Baldi et al. (2018) observed that chicken breast muscles affected by the WS myopathy displayed more intense yellowness and less intense redness than their normal counterparts. de Carvalho et al. (2020) have shown that consumers are able to identify WS muscles by their altered appearance, affecting their intention to purchase. Further processing of WB breasts causes undesirable colour changes as cooked WB fillets were found to be darker, redder and more yellow than normal chicken breasts (Zhuang and Bowker, 2018). It is worth noticing that colour variations in chicken breasts affected by WB and WS conditions are reported to occur within the first 0.3 cm of meat surface (Baldi et al., 2018).

Other causes of chicken meat discolouration are linked to processing, most notably the pink colour defect. This condition appears right after cooking or during storage of cooked chicken meat (King and Whyte, 2006) and features an unexpected pink colour which may cause consumer rejection because it gives the appearance of non-fully cooked meat (Holownia et al., 2003). The occurrence of this chicken colour defect has been associated with a variety of factors including pre-slaughter (feeding background, stunning method) and processing conditions (nitrite contamination, cooking method and/or irradiation). Recent studies have underlined the role played by common additives such as NaCl and phosphates. According to Bae et al. (2020a), pre-salting (2% NaCl) chicken breasts for more than 5 days prior to cooking showed a high incidence of pink colour, while the application of 0.5% sodium tripolyphosphate prevented the defect. Reducing the time between the addition of the additives and cooking decreased the incidence of the colour defect (Bae et al., 2020b).

5 Strategies for extending poultry meat shelf life: dietary strategies

The serious consequences of the onset of oxidative reactions and bacterial growth on the quality and shelf life of poultry meat require the design and application of targeted and efficient antioxidant and antimicrobial strategies. Figure 3 summarizes the sources of preservatives used for controlling both chemical and biological threats and the most common means of application. Tables 1 and 2 summarize recent antimicrobial and antioxidant applications in poultry. This section focuses on dietary strategies

Feed formulation has been demonstrated as a feasible way to improve the redox status of domestic birds and the oxidative stability of their products. Certain feed supplements can even be effective at mitigating the effects of oxidative stress on broilers caused by stresses such as heat (He et al., 2018; Wasti et al., 2020). The antioxidant protection of broiler meat through dietary means can be achieved using a limited number of supplements:

- α-tocopherol;
- ascorbate;
- methionine and analogues;
- selenium;
- magnesium; and
- phytogenics.

These are discussed in more detail later.

Dietary administration of α-tocopherol (~200 mg tocopherol/kg) and ascorbate (up to 500 mg ascorbate/kg) is known to alleviate *in vivo* oxidative

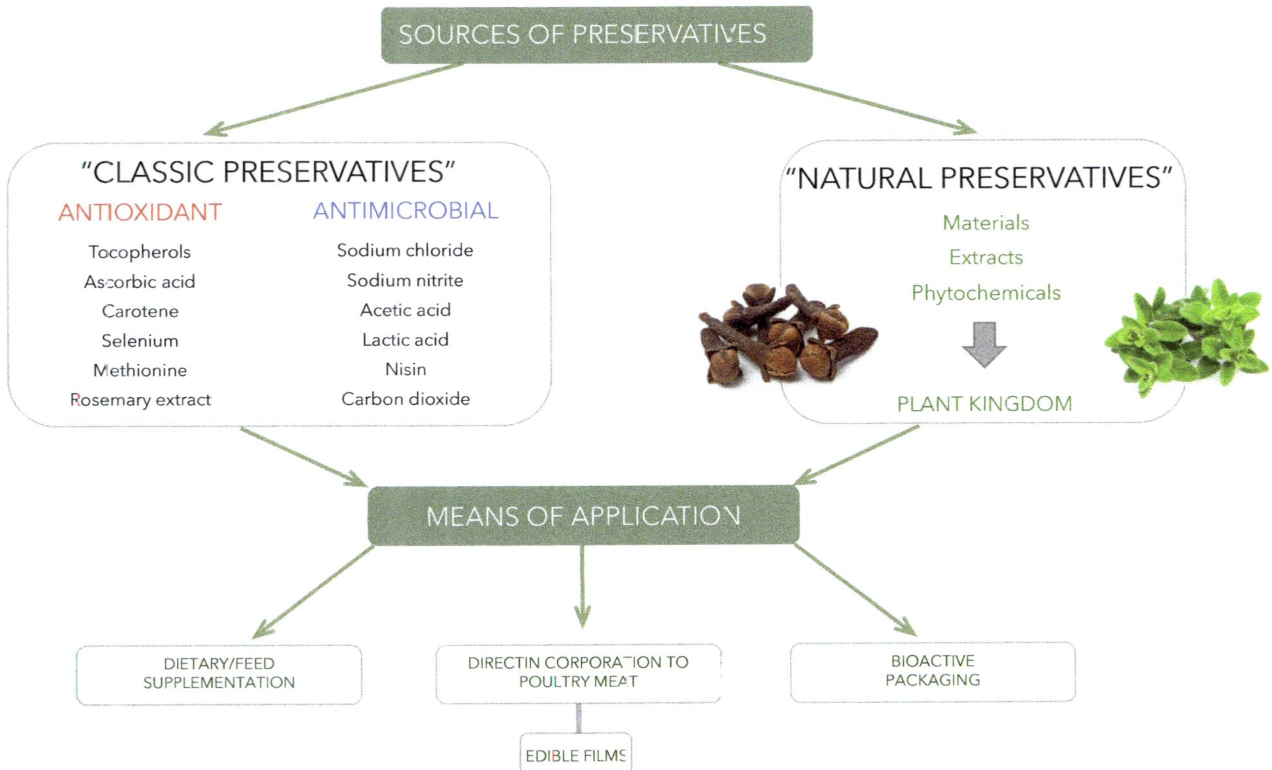

Figure 3 Antioxidant and antimicrobial protection of poultry and poultry meat: main sources of preservatives and means of application.

Table 1 Summary of antimicrobial strategies applied to poultry and poultry meat

Antimicrobial approach	Poultry system	Main antimicrobial effect	Reference
Dietary supplementation			
Cinnamon bark oil	Broiler chickens	Antioxidant status improved, microbiota reduction	Chowdhury et al. (2018)
Lavender oil	Broiler chickens	Antioxidant status improved, microbiota reduction	Adaszyńska and Szczerbińska (2019)
Curcumin + yucca extract	Broiler chickens	Anticoccidial and antibacterial effects	Galli et al. (2020)
Oregano essential oil	Broiler chickens	Reduction of total viable counts and *Salmonella* spp. in carcass	Aksit et al. (2006)
Carvacrol oil	Broiler chickens	Microbiota reduction (*Salmonella* spp. *Escherichia coli*, *Campylobacter jejuni*)	Liu et al. (2018), Wagle et al. (2020)
Gudichi extract	Assorted poultry	Antioxidant, and antimicrobial (*Escherichia coli*, *Salmonella enteritidis*) activities	Saeed et al. (2020)
Tannins	Assorted poultry	Antimicrobial activity (*Escherichia* spp. *Pseudomonas* spp.)	Choi and Kim (2020)
Product formulation			
Drumstick tree extract	Chicken sausages	Antioxidant protection and antimicrobial activity	Jayawardana et al. (2015)
Clove, rosemary spice extracts	Raw chicken meat	Antioxidant and antimicrobial protection (*Pseudomonas* spp.)	Zhang et al. (2016)
Rosemary, oregano, ginger essential oils	Assorted poultry	Growth inhibition of *Listeria monocytogenes*	Yousefi et al. (2020)
Olive leaf extract	Chicken breast	Broad antimicrobial effect	Saleh et al. (2020)
Monocaprin + carvacrol	Chicken breast	Antimicrobial protection (*Escherichia coli* O157:H7 and *Salmonella enterica* serovar Typhimurium)	Ma et al. (2021)
Rosemary in edible film	Chicken breast fillets	Antimicrobial protection (coliforms)	de Melo et al. (2012)
Acerola in edible film	Chicken thigh	Antioxidant protection, microbial spoilage preservation	Portugal Zegarra et al. (2018)

Antimicrobial approach	Poultry system	Main antimicrobial effect	Reference
Lauric arginate in edible film	Raw chicken, turkey	Reduction of pathogenic microbiota	Hassan and Cutter (2020)
Pediocin in edible film	Chicken mortadella	Growth inhibition of *Listeria monocytogenes*	Espitia et al. (2013)
Clove oil in edible film	Ground chicken meat	Extension of shelf life during chilled storage	Muppalla et al. (2014)
Grape juice in edible film	Chicken breast fillets	Antimicrobial protection (aerobic mesophilic, psychrophilic and *Enterobacteriaceae*)	Yıldırım-Yalçın et al. (2021)
Camellia oil in edible film	Chicken breast	Extension of shelf life during chilled storage	Zhou et al. (2021)
Nanoencapsulated sour tea	Chicken nuggets	Antioxidant protection Antimicrobial activity	Bahrami Feridoni and Khademi Shurmasti (2020)
Packaging atmosphere			
Oxygen exclusion in MAP	Assorted poultry	Antioxidant, and antimicrobial (e.g. *Pseudomonas* spp) activities	Karumathil et al. (2016)
CO_2 enriched MAP	Pre-cooked chicken breast	Extension of shelf life during chilled storage	Patsias et al. (2006)
30% N_2, 70% CO_2 MAP	Chicken breast meat	Antioxidant and antimicrobial protection during chilled storage	Jaspal et al. (2021)
Low O_2 MAP	Assorted poultry	Antimicrobial activity	Marmion et al. (2021)
Bioactive packaging			
Film + oregano oil	Chicken fillets	Delay *Escherichia coli* growth	Cestari et al. (2015)
Film + Ag-Cu	Chicken meat	Antimicrobial activity (*Salmonella enterica* serovar Typhimurium, *Campylobacter jejuni*)	Ahmed et al. (2018)
Bio-based film + ginger	Assorted poultry	Extension of shelf life during chilled storage	Souza et al. (2018)
Film + rosemary and ginger	Assorted poultry	Antioxidant, and antimicrobial activities	Pires et al. (2018)
Film + thyme	Chicken meat	Antimicrobial activity (*Campylobacter jejuni*)	Lin et al. (2018)
Film + clove essential oil	Chicken meat	Antimicrobial activity (*Salmonella enterica* serovar Typhimurium *and Listeria monocytogenes*)	Mulla et al. (2017)

Table 2 Summary of antioxidant strategies applied to poultry and poultry meat

Antioxidant approach	Poultry system	Main antioxidant effect	Reference
Dietary supplementation			
α-Tocopherol + ascorbate	Broiler chickens	Counteract oxidative stress caused by oxidized feeds	Leskovec et al. (2018)
α-Tocopherol	Broiler chickens	Counteract lipid and protein oxidation caused by oxidized feeds	Sałek et al. (2021)
Methionine	Broiler chickens	Counteract oxidative stress caused by heat stress	De Freitas Dionizio et al. (2021)
Methionine	Broiler chickens	Decreased cholesterol oxidation in cooked thigh muscle	Zeitz et al. (2020)
Selenoproteins	Broiler chickens	Increase muscle Se concentration and enhanced stability	Mohamed et al. (2020)
Organic selenium	Geese	Prolonged shelf life of liver and meat	Nemati et al. (2021)
Magnesium	Broiler chickens	Decreased incidence of myopathy and improved meat quality	Estévez and Petracci (2019)
Apple peel	Broiler chickens	Counteract oxidative stress caused by heat stress	Heidarisafar et al. (2016)
Tomato extract	Broiler chickens	Increased redox status	Vossen et al. (2011)
Clove essential oil	Broiler chickens	Counteract oxidative stress caused by heat stress	Mohammadi (2021)
Willow bark	Broiler chickens	Counteract oxidative stress caused by heat stress	Saracila et al. (2021)
Rosemary + blackcurrant extract	Broiler chickens	Protection of frozen thigh against lipid oxidation	Sierżant et al. (2021)
Grape skin + seed meals	Broiler chickens	Protection of chilled breast against lipid oxidation	Romero et al. (2021)
Probiotics (*Lactobacillus spp.*)	Broiler chickens	Counteract oxidative stress caused by mycotoxins	Souza et al. (2020)
Resveratrol	Broiler chickens	Counteract oxidative stress caused by heat stress	Wang et al. (2021a)
Product formulation			
Rosemary	Chicken nuggets	Antioxidant protection during frozen storage	Rocío Teruel et al. (2015)

Antioxidant approach	Poultry system	Main antioxidant effect	Reference
Lactic acid bacteria isolated	Experimental sausage	Antioxidant protection during chilled storage	Chen et al. (2015)
Green cabbage	Chicken meatballs	Antioxidant protection during chilled storage	Verma et al. (2016)
Acorn extract	Ready-to-eat chicken patties	Antioxidant protection during ultraprocessing	Ferreira et al. (2017)
Caja bagasse fruit	Ready-to-eat chicken patties	Antioxidant protection during ultraprocessing	Santana Neto et al. (2021)
Grape seed extract	Roast chicken meat	Protection against lipid oxidation during chilled storage	Guo et al. (2020)
Garlic in edible film	Ready-to-eat chicken meat	Extension of shelf life during chilled storage	Diao et al. (2020)
Lemon verbena essential oil in edible film	Chicken breast	Extension of shelf life during chilled storage	Hosseini et al. (2021)
Coffee silver skin	Chicken burger patty	Protection against lipid oxidation during chilled storage	Martusrcelli et al. (2021)
Stingless bee honey	Ready-to-eat chicken meat	Protection against lipid and protein oxidation during chilled storage	Nepomuceno-Irene et al. (2020)
Packaging atmosphere			
High O_2 MAP	Chicken thigh	Pro-oxidant	Jongberg et al. (2014)
High O_2 MAP	Duck, goose meat	Preservation of oxymyoglobin during chilled storage	Orkusz et al. (2013)
Vacuum packaging	Chicken thigh	Antioxidant protection during chilled storage	Xiao et al. (2011)
Vacuum packaging	Ratites meat	Antioxidant protection during chilled storage	Horbańczuk, and Wierzbicka (2017)
Vacuum packaging	Turkey meat	Antioxidant protection during chilled storage	Michalczyk et al. (2021)
Vacuum packaging	Chicken breast meat	Antioxidant protection during frozen storage	Wang et al. (2021b)
100% N_2 MAP	Turkey meat	Antioxidant protection during chilled storage	Fraqueza and Barreto (2011)
30% N_2, 70% CO_2 MAP	Chicken breast meat	Antioxidant protection during chilled storage	Jaspal et al. (2021)

stress (Shakeri et al., 2020) and protect against lipid and protein oxidation in meat (Leskovec et al., 2018). The benefits of dietary α-tocopherol as an inhibitor of meat oxidation and discolouration have been known for decades (Buckley et al., 1995). A recent study showed that increasing dietary tocopherol levels increased feed lipid unsaturation in broilers' diet without affecting the sensory properties of chicken meat (Sałek et al., 2021).

As a lipid-soluble molecule, α-tocopherol is naturally located in biological membranes and its antioxidant activity is essential to maintain the integrity and functionality of the phospholipid bilayer. As a water-soluble molecule, ascorbate is typically located in the cytosol where it acts as a scavenger of aqueous radicals. Both compounds act synergistically as ascorbate reduces tocopheroxyl radicals, returning tocopherol to its active form (Doba et al., 1985). Both biologically active compounds are involved in various physiological processes including the strengthening of the immune system (Shakeri et al., 2020).

Under conditions of severe oxidative stress, such as those created in hot environments, broilers may be supplemented with additional antioxidant supplements such as methionine and selenium. Both micronutrients have been found to protect chicken meat against lipid oxidation, discolouration and the loss of WHC linked to impaired protein functionality (de Freitas Dionizio et al., 2021; Mohamed et al., 2020; Shakeri et al., 2020; Zeitz et al., 2020). Methionine and its analogues are precursors of potent endogenous antioxidants such as glutathione and taurine (Estévez et al., 2020). It has also been able to modulate the expression of genes encoding for antioxidant enzymes (Gasparino et al., 2018). Selenium is a crucial component of various selenoproteins with many of them having antioxidant properties such as GSH-Px (Shakeri et al., 2020). Selenium supplementation has been observed to improve the quality and shelf life of goose meat and liver (Nemati et al., 2021).

Supplementation with other micronutrients such as magnesium has also been reported to strengthen endogenous antioxidant defences, reduce lipid and protein oxidation, improve WHC and reduce the incidence of myopathies in broiler chicken breasts (Estévez and Petracci, 2019). Amongst other dietary supplements to improve the antioxidant status of broiler muscles and meat quality, phytogenics are of increasing interest owing to their natural occurrence in the plant kingdom, safety, availability, low price as well as their bioactive properties (Surai, 2014; Mohammadi Gheisar and Kim, 2018). Numerous phenolic-rich plant materials have been tested as dietary supplements to control *in vivo* oxidative stress and the oxidative stability and quality of chicken meat. These include:

- apple peel (Heidarisafar et al., 2016);
- tomato extract (Vossen et al., 2011);
- clove (*Syzygium aromaticum* L.);

- essential oil (Mohammadi, 2021);
- pea seeds (Laudadio and Tufarelli, 2011);
- willow bark (Saracila et al., 2021);
- rosemary (*Rosmarinus officinalis*) (Sierżant et al., 2021);
- blackcurrant (*Ribes nigrum*) (Sierżant et al., 2021); and
- grape skin and seed (Romero et al., 2021).

The antioxidant actions of these plant materials may be due to the dietary delivery of tocopherols and other bioactive compounds (Romero et al., 2021). In addition to meat quality and stability, some research points to improved growth performance, health status and intestinal microflora in broiler chickens supplemented, e.g. with orange and grapefruit peel (Vlaicu et al., 2020). Other authors emphasise the dual antioxidant and antimicrobial effects of dietary phytogenics. Galli et al. (2020), e.g. has shown supplementation with curcumin and yucca extract had antioxidant, anticoccidial and antibacterial effects, promoting health, growth and meat quality in broilers. On this basis, dietary interventions based on phenolic-rich plant materials or pure phytochemicals have been suggested as alternatives to classic antibiotics and growth promoters in broilers (Mohammadi Gheisar and Kim, 2018).

Some research suggests dietary phytogenics could improve the microbial composition of broiler carcases. Aksit et al. (2006) reported that dietary oregano (*Origanum onites*) essential oil (15 g/kg of feed) reduced pathogens such as *Salmonella spp.* on broiler carcases. Phenolic compounds have shown inhibitory activities against Gram-negative and Gram-positive bacteria (Mahfuz et al., 2021) due to the presence and the location of hydroxyl groups in their structure. Tannins, a polyphenolic compound, have potential antimicrobial effects against *Salmonella*, *Escherichia* and *Pseudomonas* genera. However, as shown by Choi and Kim (2020), the dose of dietary tannins has to be carefully set to avoid their well-known antinutritional effects when fed to poultry in high quantities. Oregano oils have been found to inhibit *Pseudomonas spp.* in turkey meat (Govaris et al., 2005). Other essential oils, such as lavender (Adaszyńska-Skwirzyńska and Szczerbińska, 2019) and cinnamon bark oil (Chowdhury et al., 2018), have shown the ability to reduce *Escherichia* genera and coliforms in poultry meat. Carvacrol oil reduces levels of *Salmonella* spp, *Escherichia coli* (Liu et al., 2018) and *Campylobacter jejuni* (Wagle et al., 2020) in broilers. An antioxidant extract from gudichi (*Tinospora cordifolia*) exerts significant antibacterial activity against *Escherichia coli* and *Salmonella enteritidis* in poultry broilers (Saeed et al., 2020).

Other innovative dietary strategies to enhance broiler health and meat quality include feed from alternative protein sources such as insects or microalgae (Altmann et al., 2020), supplementation with probiotics (Souza et al., 2020) or the use of pure phytochemicals such as resveratrol (Wang et al., 2021a).

6 Strategies for extending poultry meat shelf life: antioxidants and antimicrobials in meat products

The meat industry has been using additives with antimicrobial and antioxidant activities for several decades to assure the quality and safety of meat and meat products. Increasing consumer concern about the use of traditional additives has led to the search for alternative sources of preservatives from the plant kingdom (Estévez, 2021). These are perceived by consumers as 'natural' and therefore safer (Beya et al., 2021). There is now a significant body of research on the use of plant extracts, dried herbs, essential oils, powders or pure phytochemicals as preservatives in poultry meat products. These natural preservatives may be added as such ingredients in minced poultry products or (more recently) incorporated in edible films (Umaraw et al., 2020).

Plant materials and extracts with antioxidant and antimicrobial activity include clove (*Syzygium aromaticum)*, cinnamon (*Cinnamomum cassia)*, marjoram (*Origanum vulgare)* and mustard (*Brassica nigra)*. Radha Krishnan et al. (2014) have studied the efficiency of spices as natural antimicrobials against *Enterobacteriaceae* and *Pseudomonas* genera, in addition to raw chicken meat retarding microbial growth and lipid oxidation, thus increasing shelf life. Zhang et al. (2016) have shown the effectiveness of clove and rosemary in inhibiting the growth of *Enterobacteriaceae* and *Pseudomonas* genera and slowing the onset of lipid oxidation. Olive leaf extract has been shown to have antimicrobial properties in chicken meat stored at refrigerated temperatures, reducing aerobic bacterial, psychrotrophic, Enterobacteriaceae, staphylococcal, mould and yeast counts (Saleh et al., 2020). The addition of essential oils such as those from rosemary, ginger or oregano has the ability to inhibit the growth of pathogens such as *Listeria monocytogenes* in poultry meat (Yousefi et al., 2020). Drumstick tree (*Moringa oleifera*) leaves in chicken sausages (Jayawardana et al., 2015) demonstrated antioxidant and antimicrobial activities against *E. coli*, *S. aureus*, *Pseudomonas aeruginosa* and *Enterobacter aerogenes*.

The number of articles specifically devoted to study the effects of plant-based antioxidants in poultry products has increased significantly in the last three decades (Ribeiro et al., 2021; Estévez, 2021). These include studies testing the antioxidant properties of:

- acorn (*Quercus ilex* L.) extracts (Ferreira et al., 2017);
- caja bagasse fruit (*Spondias mombin L.*) (Santana Neto et al., 2021);
- grape seed extract (Guo et al., 2020);
- garlic extract (Diao et al., 2020);
- lemon verbena (*Aloysia citriodora*) essential oil (Hosseini et al., 2021);
- coffee silver skin (Martusrcelli et al., 2021);
- *Cudrania tricuspidata* fruit powder (Kim and Chin, 2021); and

- stingless bee (*Scaptotrigona mexicana*) honey (Nepomuceno-Irene et al., 2020).

The antioxidant properties of these natural antioxidants are typically compared with those of traditional antioxidant additives and/or synthetic antioxidants. However, further research is needed to translate this largely experimental data, e.g. by fully understanding modes of action, establishing appropriate doses and developing commercial applications with appropriate regulatory approval (Estévez, 2021). Natural antioxidants have also been used to counteract the pro-oxidant effects of technologies such as irradiation (Nisar et al., 2019), ultra-processing (cooking-chilling-reheating) (Ferreira et al., 2017; Santana Neto et al., 2021) or high-hydrostatic pressure (Alves et al., 2012). Other innovative approaches to protect poultry meat products against oxidative damage include terpenes (Ma et al., 2021), advanced products from the Maillard reaction (Miranda et al., 2012), infusion and osmotic treatment with water-soluble antioxidants (Giannakourou et al., 2021) and inoculation with lactic acid bacteria (Chen et al., 2015).

The incorporation of natural preservatives in edible coatings is an approach of growing interest (Ribeiro et al., 2021). There is now significant research on antimicrobial effects. Chicken breast in contact with cellulose acetate-based film with rosemary essential oil was found to show effective control of coliforms (de Melo et al., 2012). Chitosan film with 1% antioxidant extract of acerola agro-industrial by-product prolonged the shelf life of chicken thigh for more than 20 days (Portugal Zegarra et al., 2018). The use of an antimicrobial film with lauric arginate significantly reduced the microbiota population in foods such as raw chicken and turkey breast over 28 days of cold storage (Hassan and Cutter, 2020). Cellulosic antimicrobial films incorporated with 50% of the peptide pediocin, a bacteriocin produced by *Pediococcus acidilactici*, showed a significant reduction of *Listeria monocytogenes* counts in chicken mortadella (Espitia et al., 2013). Sacacin A-containing pullulan film also reduced *Listeria monocytogenes* on the surface of turkey breast (Trinetta et al., 2010). Muppalla et al. (2014) packed ground chicken meat with edible films made of carboxymethyl cellulose and polyvinyl alcohol and impregnated with clove oil in different ratios and found that treated samples had a shelf life of 12 days at 4°C while the control counterparts spoiled within 4 days under the same storage conditions.

The positive effects of edible films impregnated with natural antioxidants against oxidation of poultry meat are also well documented. Yıldırım-Yalçın et al. (2021) have assessed the antioxidant effects of an edible film based on grape juice and cross-linked maize starch on chicken breast fillets during refrigerated storage. The authors reported a significant reduction in lipid oxidation as well as finding aerobic mesophilic, psychrophilic and

Enterobacteriaceae counts diminished in treated samples by 1.58, 0.94 and 0.85 log, respectively. Hosseini et al. (2021) preserved chicken breast during refrigerated storage using sodium alginate coating incorporated with lemon verbena and clove essential oils in combination with modified atmosphere packaging (MAP) (65% CO_2, 30% N_2 and 5% O_2). This strategy showed lipid antioxidant effects and antibacterial effects against both Gram-positive and Gram-negative bacteria, increasing the shelf life of chicken breast to 15 days at 4°C. Zhou et al. (2021) evaluated the preservative effect of a konjac glucomannan/carrageenan edible coating impregnated with camellia oil on chicken breast stored at 4°C and were able to prolong shelf life for 10 days compared to control treatments. Bahrami Feridoni and Khademi Shurmasti (2020) were able to delay oxidative and microbial spoilage in raw chicken nuggets by applying a carboxymethylcellulose coating treated with nanoencapsulated sour tea (*Hibiscus sabdariffa* L.).

7 Strategies for extending poultry meat shelf life: packaging

The shelf life of poultry meat can efficiently be prolonged by using packaging to protect food from both microbial growth and oxidative reactions. Packaging strategies are based on either modifying the composition of the packaging atmosphere or incorporating antioxidant/antimicrobial components in the packaging material.

7.1 Packaging atmosphere

Vacuum packaging has been used to inhibit protein and lipid oxidation in chilled meat from broilers (Xiao et al., 2011), ratites (Horbańczuk, and Wierzbicka, 2017) and turkeys (Michalczyk et al., 2021). Vacuum packaging has also been used to preserve poultry meat and meat products against oxidative deterioration during frozen storage (Wang et al., 2021b). Extensive improvements have also been made recently in the application of MAP strategies to reduce microbial spoilage and oxidative reactions in poultry meat quality. Carbon dioxide (CO_2), oxygen (O_2) and nitrogen (N_2) are the most common gases used in MAP of poultry meat.

Carbon dioxide has long been recognized for its antimicrobial properties (Daniels et al., 1985), and it is particularly effective against Gram-negative aerobes such as *Pseudomonas* spp. (Kropf and Mancini, 2014), one of the main contributors to poultry meat spoilage. Karumathil et al. (2016) have shown the use of CO_2 in MAP leads to an extension of the lag phase of growth and a reduction in the growth rate during the logarithmic growth phase, thus extending shelf life. Effectiveness depends on the concentration of the gas, the overall composition of the MAP, initial bacterial count and temperature of storage (Kropf and Mancini, 2014). Although high concentrations of CO_2 can lead to the formation of metmyoglobin and can result in the darkening of meat, this limitation is not applicable to white poultry meat due to a low myoglobin concentration (Narasimha Rao and Sachindra, 2002).

Oxygen is typically used at high concentrations (70–80%) in MAP of red meat to guarantee the formation and conservation of oxymyoglobin, responsible for the desirable bright red colour required for retail display (McMillin, 2008). High-oxygen MAP (80% O_2, 20% CO_2) has been used for red meat from birds such as turkeys, geese or ducks, preserving red colour for 11 days of chilled storage (Orkusz et al., 2013). However, since this benefit is not applicable to white poultry meat, and since O_2 facilitates the onset of oxidative reactions and the growth of aerobic microorganisms (Arvanitoyannis and Stratakos, 2012), its use is limited to inhibiting the growth of strictly anaerobic bacteria. The use of high-oxygen MAP in chicken thighs increased protein oxidation and meat toughness (Jongberg et al., 2014).

Nitrogen also contributes to controlling lipid and protein oxidation as well as inhibiting the growth of aerobic microorganisms. Fraqueza and Barreto (2011) reported that 100% N_2 MAP was the most effective strategy for preventing lipid oxidation in turkey meat. However, anoxic atmospheres created by 100% N_2 or vacuum packaging also favour the growth of anaerobic, aerotolerant lactobacilli (McMillin, 2008).

The combination of CO_2 and N_2 is most widely used in MAP of white poultry meat in prolonging shelf life (Rouger et al., 2017). Jaspal et al. (2021) recently combined MAP (30% N_2, 70% CO_2) with 1.25% lactic acid treatment to effectively preserve meat colour and keep low oxidation rates and microbial load throughout 12 days of refrigerated storage at 4°C. For processed poultry meat products such as ready-to-eat chicken patties, oxygen exclusion is the best option, possibly combined with other interventions such as the addition of antioxidant additives (Ferreira et al., 2017; Guo et al., 2020; de Santana Neto et al., 2021).

7.2 Bioactive packaging materials

In addition to the physical barrier effect of packaging material and the antimicrobial and antioxidant effects of MAP, diverse antioxidant additives can be incorporated into wrapping materials to provide supplementary protection to poultry meat products (Domínguez et al., 2018). This approach requires a detailed understanding of the migration kinetics of the bioactive compounds from the impregnated material to the poultry meat product. Effective antioxidant protection involves controlled migration and penetration of the bioactive components into poultry meat (Ahmed et al., 2017; Domínguez et al., 2018).

The ability of immobilized bioactive compounds to scavenge ROS or interact with spoilage or pathogenic bacteria to effectively extend poultry meat shelf life is a matter of debate (Huang et al., 2019). Several types of natural and synthetic compounds with antimicrobial properties have been investigated such as antimicrobial peptides, bacteriocins, organic acids (and salts), fatty acid esters, plant essential oils and metals (Motelica et al., 2020). Bioactive linear polyethylene films mixed with bimetallic nanoparticles (Ag-Cu) were found to be effective in the inhibition of *Salmonella enterica* serovar Typhimurium and *Campylobacter jejuni* in chicken meat stored at refrigerated temperature for 21 days (Ahmed et al., 2018). Essential oils are also effective when incorporated into packaging materials to protect poultry meat against microbial spoilage and oxidative reactions. The incorporation of ginger essential oil in bio-based films of chitosan/montmorillonite demonstrated antimicrobial and antioxidant properties, extending the shelf life of fresh poultry meat for 15 days (Souza et al., 2018). The use of oregano oil films for packaging chicken fillets also delayed *Escherichia coli* growth during frozen storage (Cestari et al., 2015). The addition of rosemary and ginger essential oil into packaging materials significantly extended shelf life of fresh poultry meat (Pires et al., 2018). Encapsulation of thyme in packaging material showed antimicrobial activity against *Campylobacter jejuni* on chicken samples as a promising strategy in poultry meat preservation (Lin et al., 2018). Mulla et al. (2017) investigated the efficiency of packaging material produced with linear low-density polyethylene modified by chromic acid and coated with clove essential oil in fresh chicken meat. The packaging material exhibited antimicrobial effects against *Salmonella enterica* serovar Typhimurium and *Listeria monocytogenes* for 21 days at 4°C.

8 Conclusion

Poultry meat and meat products are highly susceptible to spoilage by both chemical and biological agents. Natural preservatives may be a feasible means to protect meat against oxidative and microbial deterioration and fulfil consumer expectations for 'free from' or clean-label poultry products. While enormous advances have been made in post-mortem strategies (formulation and packaging), more attention should be given to the application of dietary strategies based on using plant materials rich in bioactive compounds (e.g. biowaste materials) to feed domestic birds. The use of phytogenics in poultry nutrition has the potential to emerge as a safe, affordable and efficient strategy to protect animals against stress and disease as well as preserve poultry meat against spoilage.

9 Where to look for further information

- **Poultry meat oxidation:** Updated references from key scientists Skibsted, Leif H. (Scopus ID: 35496601800); Ahn, D. U. (Scopus ID: 55701286500) and De Smet, Stefaan (Scopus ID: 7003308815).
- **Poultry microbial spoilage:** Updated references from key scientists Kontominas, Michael George (Scopus ID: 7004189818); Kreyenschmidt, Judith (Scopus ID: 7801480276) and Savvaidis, Ioannis N. (Scopus ID: 6603787850).
- **Poultry nutrition and meat quality:** Updated references from key scientists Suraї, P.F. (Scopus ID: 7005199156) and Kidd, Michael T. (Scopus ID: 7102389338).
- **Chicken myopathies and meat quality:** Updated references from key scientists Petracci, Massimiliano (Scopus ID: 12808314200) and Owens, Casey M. (Scopus ID: 7102511635).
- **Antioxidants and meat quality:** Updated references from key scientists Wang, Tian (Scopus ID: 7405564454) and Lorenzo, José M. (Scopus ID: 57195195602).

10 References

Adaszyńska-Skwirzyńska, M. and Szczerbińska, D. (2019). The effect of lavender (*Lavandula angustifolia*) essential oil as a drinking water supplement on the production performance, blood biochemical parameters, and ileal microflora in broiler chickens. *Poultry Science* 98(1), 358–365.

Addis, M. (2015). Major causes of meat spoilage and preservation techniques: a review. *Food Science and Quality Management* 41, 101–114.

Ahmed, I., Lin, H., Zou, L., Brody, A. L., Li, Z., Qazi, I. M., Pavase, T. R. and Lv, L. (2017). A comprehensive review on the application of active packaging technologies to muscle foods. *Food Control* 82, 163–178.

Ahmed, J., Mulla, M., Arfat, Y. A., Bher, A., Jacob, H. and Auras, R. (2018). Compression molded LLDPE films loaded with bimetallic (Ag-Cu) nanoparticles and cinnamon essential oil for chicken meat packaging applications. *LWT* 93, 329–338.

Aksit, M., Goksoy, E., Kok, F., Ozdemir, D. and Ozdogan, M. (2006). The impacts of organic acid and essential oil supplementations to diets on the microbiological quality of chicken carcasses. *Archiv Fur Geflugelkunde* 70(4), 168–173.

Ali, S., Rajput, N., Li, C., Zhang, W. and Zhou, G. (2016). Effect of freeze-thaw cycles on lipid oxidation and myowater in broiler chickens. *Revista Brasileira de Ciência Avícola* 18(1), 35–40.

Ali, S., Zhang, W., Rajput, N., Khan, M. A., Li, C. B. and Zhou, G. H. (2015). Effect of multiple freeze-thaw cycles on the quality of chicken breast meat. *Food Chemistry* 173, 808–814.

Altmann, B. A., Wigger, R., Ciulu, M. and Mörlein, D. (2020). The effect of insect or microalga alternative protein feeds on broiler meat quality. *Journal of the Science of Food and Agriculture* 100(11), 4292–4302.

Alves, A. B., Bragagnolo, N., da Silva, M. G., Skibsted, L. H. and Orlien, V. (2012). Antioxidant protection of high-pressure processed minced chicken meat by industrial tomato products. *Food and Bioproducts Processing* 90(3), 499–505.

Arvanitoyannis, I. S. and Stratakos, A. C. (2012). Application of modified atmosphere packaging and active/smart technologies to red meat and poultry: a review. *Food and Bioprocess Technology* 5(5), 1423–1446.

Bacou, E., Walk, C., Rider, S., Litta, G. and Perez-Calvo, E. (2021). Dietary oxidative distress: a review of nutritional challenges as models for poultry, swine and fish. *Antioxidants* 10(4), 525.

Bae, S. M., Cho, M. G. and Jeong, J. Y. (2020a). Presalting condition effects on the development of pink color in cooked ground chicken breasts. *Food Science of Animal Resources* 40(2), 197–208.

Bae, S. M., Cho, M. G. and Jeong, J. Y. (2020b). The effects of addition timing of NaCl and sodium tripolyphosphate and cooking rate on pink color in cooked ground chicken breasts. *Food Science of Animal Resources* 40(2), 231–241.

Bahrami Feridoni, S. and Khademi Shurmasti, D. (2020). Effect of the nanoencapsulated sour tea (*Hibiscus sabdariffa* L.) extract with carboxymethylcellulose on quality and shelf life of chicken nugget. *Food Science and Nutrition* 8(7), 3704–3715.

Baldi, G., Soglia, F., Mazzoni, M., Sirri, F., Canonico, L., Babini, E., Laghi, L., Cavani, C. and Petracci, M. (2018). Implications of white striping and spaghetti meat abnormalities on meat quality and histological features in broilers. *Animal* 12(1), 164–173.

Bao, Y., Boeren, S. and Ertbjerg, P. (2018). Myofibrillar protein oxidation affects filament charges, aggregation and water-holding. *Meat Science* 135, 102–108.

Bao, Y. and Ertbjerg, P. (2019). Effects of protein oxidation on the texture and water-holding of meat: a review. *Critical Reviews in Food Science and Nutrition* 59(22), 3564–3578.

Bayraktar, H., Altan, Ö., Açıkgöz, Z., Baysal, Ş. H. and Şeremet, Ç. (2011). Effects of oxidised oil and vitamin E on performance and some blood traits of heat-stressed male broilers. *South African Journal of Animal Science* 41(3), 288–296.

Belluco, S., Barco, L., Roccato, A. and Ricci, A. (2016). Escherichia coli and Enterobacteriaceae counts on poultry carcasses along the slaughterline: a systematic review and meta-analysis. *Food Control* 60, 269–280.

Beya, M. M., Netzel, M. E., Sultanbawa, Y., Smyth, H. and Hoffman, L. C. (2021). Plant-based phenolic molecules as natural preservatives in comminuted meats: a review. *Antioxidants* 10(2), 263.

Bordignon, S., Stefani, L. M. and Boiago, M. M. (2020). The use of white striped chicken breasts on the quality of nuggets and hamburgers. *Food Science and Technology* 41(3), 570–575.

Boziaris, I. S. and Parlapani, F. F. (2017). Specific spoilage organisms (SSOs) in fish. In: *The Microbiological Quality of Food*. Woodhead Publishing, Elsevier, Amsterdam, Netherlands, pp. 61–98.

Brannan, K. E., Livingston, M. L., Wall, B., Livingston, K. A. and Jansen van Rensburg, C. (2021). Embryonic thermal manipulation and dietary fat source during acute heat stress: 2. Effect on broiler carcass characteristics and breast muscle myopathies. *Journal of Applied Poultry Research* 30(3), 100174.

Brown, A. C. (2018). *Understanding Food: Principles and Preparation*. Cengage Learning, Boston.

Buckley, D. J., Morrissey, P. A. and Gray, J. I. (1995). Influence of dietary vitamin E on the oxidative stability and quality of pig meat. *Journal of Animal Science* 73(10), 3122–3130.

Carvalho, L. M., Delgado, J., Madruga, M. S. and Estévez, M. (2021). Pinpointing oxidative stress behind the white striping myopathy: depletion of antioxidant defenses, accretion of oxidized proteins and impaired proteostasis. *Journal of the Science of Food and Agriculture* 101(4), 1364–1371.

Carvalho, L. M., Ventanas, S., Olegario, L. S., Madruga, M. S. and Estévez, M. (2020). Consumers awareness of white-striping as a chicken breast myopathy affects their purchasing decision and emotional responses. *LWT* 131, 109809.

Carvalho, R., Shimokomaki, M. and Estévez, M. (2017a). Chapter 6: poultry meat color and oxidation. In: *Poultry Quality Evaluation*. Woodhead Publishing, Elsevier, Amsterdam, Netherlands, pp. 133–157.

Carvalho, R. H., Ida, E. I., Madruga, M. S., Martínez, S. L., Shimokomaki, M. and Estévez, M. (2017b). Underlying connections between the redox system imbalance, protein oxidation and impaired quality traits in pale, soft and exudative (PSE) poultry meat. *Food Chemistry* 215, 129–137.

Cavalcante da Rocha, T., de Carvalho, L. M., Soares, A. J., Coutinho, D. G., Olegario, L. S., de Sousa Galvão, M., Estévez, M. and Madruga, M. S. (2020). Impact of chicken wooden breast on quality and oxidative stability of raw and cooked sausages subjected to frozen storage. *Journal of the Science of Food and Agriculture* 100(6), 2630–2637.

Cerveny, J., Meyer, J. D. and Hall, P. A. (2009). Microbiological spoilage of meat and poultry products. In: *Compendium of the Microbiological Spoilage of Foods and Beverages*. Springer-Verlag, New York, pp. 69–86.

Cestari, L. A., Gaiotto, R. C., Antigo, J. L., Scapim, M. R. S., Madrona, G. S., Yamashita, F., Pozza, M. S. S. and Prado, I. N. (2015). Effect of active packaging on low-sodium restructured chicken steaks. *Journal of Food Science and Technology* 52(6), 3376–3382.

Chaillou, S., Chaulot-Talmon, A., Caekebeke, H., Cardinal, M., Christieans, S., Denis, C., Desmonts, M. H., Dousset, X., Feurer, C., Hamon, E., Joffraud, J. J., La Carbona, S., Leroi, F., Leroy, S., Lorre, S., Macé, S., Pilet, M. F., Prévost, H., Rivollier, M., Roux, D., Talon, R., Zagorec, M. and Champomier-Vergès, M. C. (2015). Origin and ecological selection of core and food-specific bacterial communities associated with meat and seafood spoilage. *The ISME Journal* 9(5), 1105–1118.

Chen, Q., Kong, B., Sun, Q., Dong, F. and Liu, Q. (2015). Antioxidant potential of a unique LAB culture isolated from Harbin dry sausage: in vitro and in a sausage model. *Meat Science* 110, 180–188.

Choe, J., Min, J. S., Lee, S. O., Khan, M. I., Yim, D. G., Lee, M. and Jo, C. (2018). Influence of cooking, storage period, and re-heating on production of cholesterol oxides in chicken meat. *Korean Journal for Food Science of Animal Resources* 38(3), 433–441.

Choi, J. and Kim, W. K. (2020). Dietary application of tannins as a potential mitigation strategy for current challenges in poultry production: a review. *Animals : an Open Access Journal From MDPI* 10(12), 2389.

Choi, J. S. and Chin, K. B. (2021). Structural changes of meat protein of chicken sausages with various levels of salt and phosphate and their effects on in vitro digestion. *International Journal of Food Science and Technology* 56(10), 5250–5258.

Chowdhury, S., Mandal, G. P., Patra, A. K., Kumar, P., Samanta, I., Pradhan, S. and Samanta, A. K. (2018). Different essential oils in diets of broiler chickens: 2. Gut microbes and morphology, immune response, and some blood profile and antioxidant enzymes. *Animal Feed Science and Technology* 236, 39-47.

Dadgar, S., Crowe, T. G., Classen, H. L., Watts, J. M. and Shand, P. J. (2012). Broiler chicken thigh and breast muscle responses to cold stress during simulated transport before slaughter. *Poultry Science* 91(6), 1454-1464.

Dal Bosco, A., Castellini, C. and Cardinali, R. (2005). Effect of dietary administration of rosemary extract on the oxidative stability of pigeon meat. *Italian Journal of Food Science* 17(4), 419-428.

Dalle Zotte, A., Tasoniero, G., Puolanne, E., Remignon, H., Cecchinato, M., Catelli, E. and Cullere, M. (2017). Effect of "wooden breast" appearance on poultry meat quality, histological traits, and lesions characterization. *Czech Journal of Animal Science* 62(2), 51-57.

Daniels, J. A., Krishnamurthi, R. and Rizvi, S. S. H. (1985). A review of effects of carbon dioxide on microbial growth and food quality. *Journal of Food Protection* 48(6), 532-537.

de Oliveira, R. F., Fávero, M. S., Mello, J. L. M., Ferrari, F. B., Cavalcanti, E. N. F., Souza, R. A., Pereira, M. R., Giampietro-Ganeco, A., Villegas-Cayllahua, E. A., Fidelis, H. A., Souza, P. A. and Borba, H. (2020). Physicochemical properties and consumer acceptance of hamburgers processed with chicken meat affected by wooden breast myopathy. *Animals: An Open Access Journal From MDPI* 10(12), 2330.

de Melo, A. A., Geraldine, R. M., Silveira, M. F., Torres, M. C., E Rezende, C. S., Fernandes, T. H. and de Oliveira, A. N. (2012). Microbiological quality and other characteristics of refrigerated chicken meat in contact with cellulose acetate-based film incorporated with rosemary essential oil. *Brazilian Journal of Microbiology* 43(4), 1419-1427

Delles, R. M., Xiong, Y. L., True, A. D., Ao, T. and Dawson, K. A. (2014). Dietary antioxidant supplementation enhances lipid and protein oxidative stability of chicken broiler meat through promotion of antioxidant enzyme activity. *Poultry Science* 93(6), 1561-1570.

Desai, M. A., Jackson, V., Zhai, W., Suman, S. P., Nair, M. N., Beach, C. M. and Schilling, M. W. (2016). Proteome basis of pale, soft, and exudative-like (PSE-like) broiler breast (pectoralis major) meat. *Poultry Science* 95(11), 2696-2706.

Diao, X., Huan, Y. and Chitrakar, B. (2020). Extending the shelf life of ready-to-eat spiced chicken meat: garlic aqueous extracts-carboxymethyl chitosan ultrasonicated coating solution. *Food and Bioprocess Technology* 13(5), 786-796.

Doba, T., Burton, G. W. and Ingold, K. U. (1985). Antioxidant and co-antioxidant activity of vitamin C. the effect of vitamin C, either alone or in the presence of vitamin E or a water-soluble vitamin E analogue, upon the peroxidation of aqueous multilamellar phospholipid liposomes. *Biochimica et Biophysica Acta* 835(2), 298-303.

Domínguez, R., Barba, F. J., Gómez, B., Putnik, P., Kovačević, D. B., Pateiro, M., Santos, E. M. and Lorenzo, J. M. (2018). Active packaging films with natural antioxidants to be used in meat industry: a review. *Food Research International* 113, 93-101.

Domínguez, R., Pateiro, M., Gagaoua, M., Barba, F. J., Zhang, W. and Lorenzo, J. M. (2019). A comprehensive review on lipid oxidation in meat and meat products. *Antioxidants* 8(10), 429.

Duqué, B., Canon, J., Haddad, N., Guillou, S. and Membré, J. M. (2021). Quantitative approach to assess the compliance to a performance objective (PO) of

Campylobacter jejuni in poultry meat in France. *International Journal of Food Microbiology* 336, 108916.

Espitia, P. J. P., Pacheco, J. J. R., Melo, N. R., Soares, N. F. and Durango, A. M. (2013). Packaging properties and control of Listeria monocytogenes in bologna by cellulosic films incorporated with pediocin. *Brazilian Journal of Food Technology* 16(3), 226–235.

Estévez, M. (2011). Protein carbonyls in meat systems: a review. *Meat Science* 89(3), 259–279.

Estevez, M. (2015). Oxidative damage to poultry: from farm to fork. *Poultry Science* 94(6), 1368–1378.

Estévez, M. (2021). Critical overview of the use of plant antioxidants in the meat industry: opportunities, innovative applications and future perspectives. *Meat Science* 181, 108610.

Estévez, M., Díaz-Velasco, S. and Martínez, R. (2022). Protein carbonylation in food and nutrition: a concise update. *Amino Acids* 54(4), 559–573.

Estévez, M., Geraert, P. A., Liu, R., Delgado, J., Mercier, Y. and Zhang, W. (2020). Sulphur amino acids, muscle redox status and meat quality: more than building blocks-invited review. *Meat Science* 163, 108087.

Estévez, M., Li, Z., Soladoye, O. P. and Van-Hecke, T. (2017). Health risks of food oxidation. In: *Advances in Food and Nutrition Research*. Academic Press, Elsevier, Amsterdam, Netherlands, Vol. 82, pp. 45–81.

Estévez, M., Padilla, P., Carvalho, L., Martín, L., Carrapiso, A. and Delgado, J. (2019). Malondialdehyde interferes with the formation and detection of primary carbonyls in oxidized proteins. *Redox Biology*, 26, 101277.

Estevez, M. and Petracci, M. (2019). Benefits of magnesium supplementation to broiler subjected to dietary and heat stress: improved redox status, breast quality and decreased myopathy incidence. *Antioxidants* 8(10), 456.

Estévez, M. and Xiong, Y. (2019). Intake of oxidized proteins and amino acids and causative oxidative stress and disease: recent scientific evidences and hypotheses. *Journal of Food Science* 84(3), 387–396.

Faustman, C., Sun, Q., Mancini, R. and Suman, S. P. (2010). Myoglobin and lipid oxidation interactions: mechanistic bases and control. *Meat Science* 86(1), 86–94.

Fellenberg, M. A. and Speisky, H. (2006). Antioxidants: their effects on broiler oxidative stress and its meat oxidative stability. *World's Poultry Science Journal* 62(1), 53–70.

Ferreira, V. C. S., Morcuende, D., Hérnandez-López, S. H., Madruga, M. S., Silva, F. A. P. and Estévez, M. (2017). Antioxidant extracts from acorns (*Quercus ilex* L.) effectively protect ready-to-eat (RTE) chicken patties irrespective of packaging atmosphere. *Journal of Food Science* 82(3), 622–631.

Ferreira, V. C., Morcuende, D., Madruga, M. S., Hernández-López, S. H., Silva, F. A., Ventanas, S. and Estévez, M. (2016). Effect of pre-cooking methods on the chemical and sensory deterioration of ready-to-eat chicken patties during chilled storage and microwave reheating. *Journal of Food Science and Technology* 53(6), 2760–2769.

Ferreira, V. C. S., Morcuende, D., Madruga, M. S., Silva, F. A. P. and Estévez, M. (2018). Role of protein oxidation in the nutritional loss and texture changes in ready-to-eat chicken patties. *International Journal of Food Science and Technology* 53(6), 1518–1526.

Fraqueza, M. J. and Barreto, A. S. (2011). Gas mixtures approach to improve turkey meat shelf life under modified atmosphere packaging: the effect of carbon monoxide. *Poultry Science* 90(9), 2076–2084.

Freitas Dionizio, A., de Souza Khatlab, A., Alcalde, C. R., Gasparino, E. and Feihrmann, A. C. (2021). Supplementation with free methionine or methionine dipeptide improves meat quality in broilers exposed to heat stress. *Journal of Food Science and Technology* 58(1), 205–215.

Galli, G. M., Griss, L. G., Boiago, M. M., Petrolli, T. G., Glombowsky, P., Bissacotti, B. F., Copetti, P. M., da Silva, A. D., Schetinger, M. R., Sareta, L., Mendes, R. E., Mesadri, J., Wagner, R., Gundel, S., Ourique, A. F. and Da Silva, A. S. (2020). Effects of curcumin and yucca extract addition in feed of broilers on microorganism control (anticoccidial and antibacterial), health, performance and meat quality. *Research in Veterinary Science* 132, 156–166.

Gasparino, E., Del Vesco, A. P., Khatlab, A. S., Zancanela, V., Grieser, D. O. and Silva, S. C. C. (2018). Effects of methionine hydroxy analogue supplementation on the expression of antioxidant-related genes of acute heat stress-exposed broilers. *Animal* 12(5), 931–939.

Giannakourou, M. C., Poulis, S., Konteles, S. J., Dipla, A., Lougovois, V. P., Kyrana, V., Proestos, C. and Sinanoglou, V. J. (2021). Combined effect of impregnation with an *Origanum vulgare* infusion and osmotic treatment on the shelf life and quality of chilled chicken fillets. *Molecules* 26(9), 2727.

Goethals, S., Van Hecke, T., Vossen, E., Vanhaecke, L., Van Camp, J. and De Smet, S. (2020). Commercial luncheon meat products and their in vitro gastrointestinal digests contain more protein carbonyl compounds but less lipid oxidation products compared to fresh pork. *Food Research International* 136, 109585.

Govaris, A., Botsoglou, E., Florou-Paneri, P., Moulas, A. and Papageorgiou, G. (2005). Dietary supplementation of oregano essential oil and α-tocopheryl acetate on microbial growth and lipid oxidation of turkey breast fillets during storage. *International Journal of Poultry Science* 4(12), 969–975.

Guo, Y., Huang, J., Chen, Y., Hou, Q. and Huang, M. (2020). Effect of grape seed extract combined with modified atmosphere packaging on the quality of roast chicken. *Poultry Science* 99(3), 1598–1605.

Hashari, S. Z., Rahim, A. A., Meng, G. Y. and Ramiah, S. K. (2020). Quantification of cooking method effect on COP content in meat types using triple quadrupole GC-MS/MS. *Molecules* 25(21), 4978.

Hassan, A. H. A. and Cutter, C. N. (2020). Development and evaluation of pullulan-based composite antimicrobial films (CAF) incorporated with nisin, thymol and lauric arginate to reduce foodborne pathogens associated with muscle foods. *International Journal of Food Microbiology* 320, 108519.

He, J., Xia, C., He, Y., Pan, D., Cao, J., Sun, Y. and Zeng, X. (2019). Proteomic responses to oxidative damage in meat from ducks exposed to heat stress. *Food Chemistry* 295, 129–137.

He, S. P., Arowolo, M. A., Medrano, R. F., Li, S., Yu, Q. F., Chen, J. Y. and He, J. H. (2018). Impact of heat stress and nutritional interventions on poultry production. *World's Poultry Science Journal* 74(4), 647–664.

Heidarisafar, Z., Sadeghi, G., Karimi, A. and Azizi, O. (2016). Apple peel waste as a natural antioxidant for heat-stressed broiler chickens. *Tropical Animal Health and Production* 48(4), 831–835.

Hinton Jr., A., Cason, J. A. and Ingram, K. D. (2004). Tracking spoilage bacteria in commercial poultry processing and refrigerated storage of poultry carcasses. *International Journal of Food Microbiology* 91(2), 155–165.

Höll, L., Behr, J. and Vogel, R. F. (2016). Identification and growth dynamics of meat spoilage microorganisms in modified atmosphere packaged poultry meat by MALDI-TOF MS. *Food Microbiology* 60, 84–91.

Holownia, K., Chinnan, M. S. and Reynolds, A. E. (2003). Pink color defect in poultry white meat as affected by endogenous conditions. *Journal of Food Science* 68(3), 742–747.

Horbańczuk, O. K. and Wierzbicka, A. (2017). Effects of packaging methods on shelf life of ratite meats. *Journal of Veterinary Research* 61(3), 279–285.

Hosseini, M., Jamshidi, A., Raeisi, M. and Azizzadeh, M. (2021). Effect of sodium alginate coating containing clove (*Syzygium aromaticum*) and lemon verbena (*Aloysia citriodora*) essential oils and different packaging treatments on shelf life extension of refrigerated chicken breast. *Journal of Food Processing and Preservation* 45(3), e14946.

Hsu, K. Y. and Chen, B. H. (2020). Analysis and reduction of heterocyclic amines and cholesterol oxidation products in chicken by controlling flavorings and roasting condition. *Food Research International* 131, 109004.

Huang, T., Qian, Y., Wei, J. and Zhou, C. (2019). Polymeric antimicrobial food packaging and its applications. *Polymers* 11(3), 560.

Huff-Lonergan, E., Zhang, W. and Lonergan, S. M. (2010). Biochemistry of postmortem muscle–lessons on mechanisms of meat tenderization. *Meat Science* 86(1), 184–195.

Ismail, I. B., Al Busadah, K. A. and El-Bahr, S. M. (2013). Oxidative stress biomarkers and biochemical profile in broilers chicken fed zinc bacitracin and ascorbic acid under hot climate. *American Journal of Biochemistry and Molecular Biology* 3(2), 202–214.

Jaspal, M. H., Ijaz, M., Haq, H. Au, Yar, M. K., Asghar, B., Manzoor, A., Badar, I. H., Ullah, S., Islam, M. S. and Hussain, J. (2021). Effect of oregano essential oil or lactic acid treatments combined with air and modified atmosphere packaging on the quality and storage properties of chicken breast meat. *LWT* 146, 111459.

Jayasena, D. D., Ahn, D. U., Nam, K. C. and Jo, C. (2013). Factors affecting cooked chicken meat flavour: a review. *World's Poultry Science Journal* 69(3), 515–526.

Jayawardana, B. C., Liyanage, R., Lalantha, N., Iddamalgoda, S. and Weththasinghe, P. (2015). Antioxidant and antimicrobial activity of drumstick (*Moringa oleifera*) leaves in herbal chicken sausages. *LWT - Food Science and Technology* 64(2), 1204–1208.

Jingting, S., Qin, X., Yanju, S., Ming, Z., Yunjie, T., Gaige, J., Zhongwei, S. and Jianmin, Z. (2017). Oxidative and glycolytic skeletal muscles show marked differences in gene expression profile in Chinese Qingyuan partridge chickens. *PLoS ONE* 12(8), e0183118.

Jongberg, S., Wen, J., Tørngren, M. A. and Lund, M. N. (2014). Effect of high-oxygen atmosphere packaging on oxidative stability and sensory quality of two chicken muscles during chill storage. *Food Packaging and Shelf Life* 1(1), 38–48.

Karumathil, D. P., Upadhyay, A. and Venkitanarayanan, K. (2016). *Antimicrobial Packaging for Poultry*. Antimicrobial Food Packaging, Elsevier, Amsterdam, Netherlands.

Katiyo, W., de Kock, H. L., Coorey, R. and Buys, E. M. (2020). Sensory implications of chicken meat spoilage in relation to microbial and physicochemical characteristics during refrigerated storage. *LWT* 128, 109468.

Kim, H. E. and Chin, K. B. (2021). Effects of different drying methods on antioxidant activities of *Cudrania tricuspidata* fruit powder and its effects on the product quality of marinated chicken breast. *Journal of Food Processing and Preservation* 45(10), e15830.

Kim, S. A., Park, S. H., Lee, S. I., Owens, C. M. and Ricke, S. C. (2017). Assessment of chicken carcass microbiome responses during processing in the presence of commercial antimicrobials using a next generation sequencing approach. *Scientific Reports* 7(1), 1.

King, N. J. and Whyte, R. (2006). Does it look cooked? A review of factors that influence cooked meat color. *Journal of Food Science* 71(4), 31-40.

Kluth, I. K., Teuteberg, V., Ploetz, M. and Krischek, C. (2021). Effects of freezing temperatures and storage times on the quality and safety of raw turkey meat and sausage products. *Poultry Science* 100(9), 101305.

Kpomasse, C. C., Oke, O. E., Houndonougbo, F. M. and Tona, K. (2021). Broiler production challenges in the tropics: a review. *Veterinary Medicine and Science* 7(3), 831-842.

Kropf, D. H. and Mancini, R. A. (2014). Packaging|modified and controlled atmosphere. In: *Encyclopedia of Meat Sciences*. Academic Press, Elsevier, New York, pp. 9-12.

Kuttappan, V. A., Owens, C. M., Coon, C., Hargis, B. M. and Vazquez-Anon, M. (2017). Incidence of broiler breast myopathies at 2 different ages and its impact on selected raw meat quality parameters. *Poultry Science* 96(8), 3005-3009.

Laudadio, V. and Tufarelli, V. (2011). Pea (*Pisum sativum* L.) seeds as an alternative dietary protein source for broilers: influence on fatty acid composition, lipid and protein oxidation of dark and white meats. *Journal of the American Oil Chemists' Society* 88(7), 967-973.

Lauritsen, C. V., Kjeldgaard, J., Ingmer, H., Bisgaard, M. and Christensen, H. (2019). Microbiota encompassing putative spoilage bacteria in retail packaged broiler meat and commercial broiler abattoir. *International Journal of Food Microbiology* 300, 14-21.

Leskovec, J., Levart, A., Nemec Svete, A., Peric, L., Đukic Stojcic, M., Žikic, D., Salobir, J. and Rezar, V. (2018). Effects of supplementation with α-tocopherol, ascorbic acid, selenium, or their combination in linseed oil-enriched diets on the oxidative status in broilers. *Poultry Science* 97(5), 1641-1650.

Leygonie, C., Britz, T. J. and Hoffman, L. C. (2011). Oxidative stability of previously frozen ostrich *Muscularis iliofibularis* packaged under different modified atmospheric conditions. *International Journal of Food Science and Technology* 46(6), 1171-1178.

Leygonie, C., Britz, T. J. and Hoffman, L. C. (2012). Impact of freezing and thawing on the quality of meat: review. *Meat Science* 91(2), 93-98.

Li, S., Mann, D. A., Zhang, S., Qi, Y., Meinersmann, R. J. and Deng, X. (2020). Microbiome-informed food safety and quality: longitudinal consistency and cross-sectional distinctiveness of retail chicken breast microbiomes. *mSystems* 5(5), e00589-20.

Lin, L., Zhu, Y. and Cui, H. (2018). Electrospun thyme essential oil/gelatin nanofibers for active packaging against Campylobacter jejuni in chicken. *LWT* 97, 711-718.

Liu, B., Jiang, J., Yu, D., Lin, G. and Xiong, Y. L. (2020a). Effects of supplementation of microalgae (*Aurantiochytrium* sp.) to laying hen diets on fatty acid content, health lipid indices, oxidative stability, and quality attributes of meat. *Foods* 9(9), 1271.

Liu, W., Luo, X., Yang, X., Zhang, Y., Zhu, L., Mao, Y., Liang, R., Han, M. and Cheng, H. (2020b). Recent progress in research on the effect of lipid oxidation on meat color. *Food Science* 41(21), 238-247.

Liu, S., Song, M., Yun, W., Lee, J., Lee, C., Kwak, W., Han, N., Kim, H. and Cho, J. (2018). Effects of oral administration of different dosages of carvacrol essential oils on intestinal barrier function in broilers. *Journal of Animal Physiology and Animal Nutrition* 102(5), 1257–1265.

Luna, C. and Estévez, M. (2019). Formation of allysine in β-lactoglobulin and myofibrillar proteins by glyoxal and methylglyoxal: impact on water-holding capacity and in vitro digestibility. *Food Chemistry* 271, 87–93.

Lund, M. N., Heinonen, M., Baron, C. P. and Estévez, M. (2011). Protein oxidation in muscle foods: a review. *Molecular Nutrition and Food Research* 55(1), 83–95.

Ma, M., Zhao, J., Yan, X., Zeng, Z., Wan, D., Yu, P., Xia, J., Zhang, G. and Gong, D. (2021). Synergistic effects of monocaprin and carvacrol against *Escherichia coli* O157. *Food Control* 132, 108480.

Mahfuz, S., Shang, Q. and Piao, X. (2021). Phenolic compounds as natural feed additives in poultry and swine diets: a review. *Journal of Animal Science and Biotechnology* 12(1), 48.

Marmion, M., Ferone, M. T., Whyte, P. and Scannell, A. G. M. (2021). The changing microbiome of poultry meat; from farm to fridge. *Food Microbiology* 99, 103823.

Martuscelli, M., Esposito, L. and Mastrocola, D. (2021). The role of coffee silver skin against oxidative phenomena in newly formulated chicken meat burgers after cooking. *Foods* 10(8), 1833.

McMillin, K. W. (2008). Where is MAP going? A review and future potential of modified atmosphere packaging for meat. *Meat Science* 80(1), 43–65.

Mead, G. C. (2004). Microbiological quality of poultry meat: a review. *Revista Brasileira de Ciência Avícola* 6(3), 135–142.

Michalczyk, M., Satora, P., Banaś, J. and Fiutak, G. (2021). Effect of adding essential oils of caraway and rosemary on volatile aroma compounds derived from stored vacuum packaged minced Turkey meat. *Annals of Animal Science* 21(3), 1133–1147.

Min, B. and Ahn, D. U. (2005). Mechanism of lipid peroxidation in meat and meat products - a review. *Food Science and Biotechnology* 14, 152–163.

Min, B., Nam, K. C., Cordray, J. and Ahn, D. U. (2008). Endogenous factors affecting oxidative stability of beef loin, pork loin, and chicken breast and thigh meats. *Journal of Food Science* 73(6), C439–C446.

Miranda, L. T., Rakovski, C. and Were, L. M. (2012). Effect of Maillard reaction products on oxidation products in ground chicken breast. *Meat Science* 90(2), 352–360.

Mohamed, D. A., Sazili, A. Q., Teck Chwen, L. and Samsudin, A. A. (2020). Effect of microbiota-selenoprotein on meat selenium content and meat quality of broiler chickens. *Animals: An Open Access Journal from MDPI* 10(6), 981.

Mohammadi, F. (2021). Effect of different levels of clove (*Syzygium aromaticum* L.) essential oil on growth performance and oxidative/nitrosative stress biomarkers in broilers under heat stress. *Tropical Animal Health and Production* 53(1), 84.

Mohammadi Gheisar, M. and Kim, I. H. (2018). Phytobiotics in poultry and swine nutrition - a review. *Italian Journal of Animal Science* 17(1), 92–99.

Motelica, L., Ficai, D., Ficai, A., Oprea, O. C., Kaya, D. A. and Andronescu, E. (2020). Biodegradable antimicrobial food packaging: trends and perspectives. *Foods* 9(10), 1438.

Mulla, M., Ahmed, J., Al-Attar, H., Castro-Aguirre, E., Arfat, Y. A. and Auras, R. (2017). Antimicrobial efficacy of clove essential oil infused into chemically modified LLDPE film for chicken meat packaging. *Food Control* 73, 663–671.

Muppalla, S. R., Kanatt, S. R., Chawla, S. P. and Sharma, A. (2014). Carboxymethyl cellulose-polyvinyl alcohol films with clove oil for active packaging of ground chicken meat. *Food Packaging and Shelf Life* 2(2), 51-58.

Narasimha Rao, D. and Sachindra, N. M. (2002). Modified atmosphere and vacuum packaging of meat and poultry products. *Food Reviews International* 18(4), 263-293.

Nasser, L. A. (2015). Molecular identification of isolated fungi, microbial and heavy metal contamination of canned meat products sold in Riyadh, Saudi Arabia. *Saudi Journal of Biological Sciences* 22(5), 513-520.

Nemati, Z., Alirezalu, K., Besharati, M., Holman, B. W. B., Hajipour, M. and Bohrer, B. M. (2021). The effect of dietary supplementation with inorganic or organic selenium on the nutritional quality and shelf life of goose meat and liver. *Animals: An Open Access Journal From MDPI* 11(2), 261.

Nepomuceno-Irene, R., Hernández Cázares, A. S., Vivar Vera, MdlÁ, Hidalgo Contreras, J. V., Contreras Oliva, A., Real Luna, N., Pérez Sato, J. A. and Velasco Velasco, J. (2020). Efecto de la adición de miel de abeja sin aguijón (scaptotrigona mexicana) en la conservación de carne de pollo. *Agrociencia* 54(7), 885-896.

Neto, D. C., Cordeiro, Â. M. T. M., Meireles, B. R. L. A., Araújo, ÍB. S., Estévez, M., Ferreira, V. C. S. and Silva, F. A. P. (2021). Inhibition of protein and lipid oxidation in ready-to-eat chicken patties by a *Spondias mombin* L. bagasse phenolic-rich extract. *Foods* 10(6), 1338.

Nicoli, M. C. (2012). *Shelf Life Assessment of Food*. CRC Press, Taylor & Francis Group, New York.

Nisar, M. F., Arshad, M. S., Yasin, M., Arshad, M. U. and Nadeem, M. T. (2019). Influence of irradiation and Moringa leaf powder on the amino acid and fatty acid profiles of chicken meat stored under various packaging materials. *Journal of Food Processing and Preservation* 43(9), e14166.

Nychas, G. J. E. and Drosinos, E. H. (2014). Meat and poultry: spoilage of meat. In: *Encyclopedia of Food Microbiology* (2nd edn., vol. 2). Academic Press, Elsevier, Amsterdam, Netherlands, pp. 514-519.

Odeyemi, O. A., Alegbeleye, O. O., Strateva, M. and Stratev, D. (2020). Understanding spoilage microbial community and spoilage mechanisms in foods of animal origin. *Comprehensive Reviews in Food Science and Food Safety* 19(2), 311-331.

OECD (2021). *FAO Agricultural Outlook*. OECD, pp. 2021-2030.

Orkusz, A., Wołoszyn, J., Haraf, G. and Okruszek, A. (2013). Effect of protective atmosphere on color of goose meat. *Poultry Science* 92(8), 2188-2194.

O'Sullivan, M. G. (2016). The stability and shelf life of meat and poultry. In: *The Stability and Shelf Life of Food*. Woodhead Publishing, Elsevier, Amsterdam, Netherlands, pp. 521-543.

Patsias, A., Chouliara, I., Badeka, A., Savvaidis, I. N. and Kontominas, M. G. (2006). Shelf-life of a chilled precooked chicken product stored in air and under modified atmospheres: microbiological, chemical, sensory attributes. *Food Microbiology* 23(5), 423-429.

Pegg, R. B. and Shahidi, F. (2012). Off flavors and rancidity in foods. In: *Handbook of Meat, Poultry and Seafood Quality*. John Wiley & Sons, Inc., Hoboken, NJ, USA, pp. 127-139.

Petracci, M., Bianchi, M. and Cavani, C. (2009). The European perspective on pale, soft, exudative conditions in poultry. *Poultry Science* 88(7), 1518-1523.

Petracci, M., Bianchi, M. and Cavani, C. (2010). Pre-slaughter handling and slaughtering factors influencing poultry product quality. *World's Poultry Science Journal* 66(1), 17-26.

Petracci, M., Soglia, F., Madruga, M., Carvalho, L., Ida, E. and Estévez, M. (2019). Wooden-breast, white striping, and spaghetti meat: causes, consequences and consumer perception of emerging broiler meat abnormalities. *Comprehensive Reviews in Food Science and Food Safety* 18(2), 565-583.

Pires, J. R. A., de Souza, V. G. L. and Fernando, A. L. (2018). Chitosan/montmorillonite bionanocomposites incorporated with rosemary and ginger essential oil as packaging for fresh poultry meat. *Food Packaging and Shelf Life* 17, 142-149.

Portugal Zegarra, M. C. C., Santos, A. M. P., Silva, A. M. A. D. and Melo, E. A. (2018). Chitosan films incorporated with antioxidant extract of acerola agroindustrial residue applied in chicken thigh. *Journal of Food Processing and Preservation* 42(4), e13578.

Qaisrani, S. N., Rizwan, M., Yaseen, G., Bibi, F., Sarfraz, M. A., Khan, N. A., Naveed, S. and Pasha, T. N. (2021). Effects of dietary oxidized oil on growth performance, meat quality and biochemical indices in poultry - a review. *Annals of Animal Science* 21(1), 29-46.

Qu, C., He, Z., Wang, Z., Li, S., Wang, Z. and Li, H. (2020). Effects of NaCl content and drying temperature on lipid oxidation, protein oxidation, and physical properties of dry-cured chicken. *Journal of Food Science* 85(6), 1651-1660.

Radha krishnan, K., Babuskin, S., Azhagu Saravana Babu, P., Sasikala, M., Sabina, K., Archana, G., Sivarajan, M. and Sukumar, M. (2014). Antimicrobial and antioxidant effects of spice extracts on the shelf life extension of raw chicken meat. *International Journal of Food Microbiology* 171, 32-40.

Ramanathan, R., Suman, S. P. and Faustman, C. (2020). Biomolecular interactions in postmortem skeletal muscles governing fresh meat color: a review. *Journal of Agricultural and Food Chemistry* 68(46), 12779-12787.

Ribeiro, A. M., Estevinho, B. N. and Rocha, F. (2021). Preparation and incorporation of functional ingredients in edible films and coatings. *Food and Bioprocess Technology* 14(2), 209-231.

Romero, C., Nardoia, M., Arija, I., Viveros, A., Rey, A. I., Prodanov, M. and Chamorro, S. (2021). Feeding broiler chickens with grape seed and skin meals to enhance α-and γ-tocopherol content and meat oxidative stability. *Antioxidants* 10(5), 699.

Rouger, A., Tresse, O. and Zagorec, M. (2017). Bacterial contaminants of poultry meat: sources, species, and dynamics. *Microorganisms* 5(3), 50.

Saeed, M., Naveed, M., Leskovec, J., Ali Kamboh, A., Kakar, I., Ullah, K., Ahmad, F., Sharif, M., Javaid, A., Rauf, M., Abd El-Hack, M. E., Abdel-Latif, M. A. and Chao, S. (2020). Using Guduchi (*Tinospora cordifolia*) as an eco-friendly feed supplement in human and poultry nutrition. *Poultry Science* 99(2), 801-811.

Saenz-García, C. E., Castañeda-Serrano, P., Mercado Silva, E. M., Alvarado, C. Z. and Nava, G. M. (2020). Insights into the identification of the specific spoilage organisms in chicken meat. *Foods* 9(2), 225.

Saleh, E., Morshdy, A. E., El-Manakhly, E., Al-Rashed, S., Hetta, H. F., Jeandet, P., Yahia, R., El-Saber Batiha, G. and Ali, E. (2020). Effects of olive leaf extracts as natural preservative on retailed poultry meat quality. *Foods* 9(8), 1017.

Sałek, P., Konieczka, P., Przybylski, W., Jaworska, D., Rosikiewicz, A., Olszewska, K., Montowska, M., Szkopek, D., Kinsner, M. and Cieślak, A. (2021). Gradually increasing vitamin E dose allow to increase dietary polyunsaturation level while maintaining the oxidation status of lipids and proteins in chicken breast muscle. *Annals of Animal Science* 21, 1523–1541.

Salles, G. B. C., Boiago, M. M., Silva, A. D., Morsch, V. M., Gris, A., Mendes, R. E., Baldissera, M. D. and da Silva, A. S. (2019). Lipid peroxidation and protein oxidation in broiler breast fillets with white striping myopathy. *Journal of Food Biochemistry* 43(4), e12792.

Samapundo, S., de Baenst, I., Aerts, M., Cnockaert, M., Devlieghere, F. and Van Damme, P. (2019). Tracking the sources of psychrotrophic bacteria contaminating chicken cuts during processing. *Food Microbiology* 81, 40–50.

Santos, M. M. F., Lima, D. A. S., Madruga, M. S. and Silva, F. A. P. (2020). Lipid and protein oxidation of emulsified chicken patties prepared using abdominal fat and skin. *Poultry Science* 99(3), 1777–1787.

Saracila, M., Panaite, T. D., Papuc, C. P. and Criste, R. D. (2021). Heat stress in broiler chickens and the effect of dietary polyphenols, with special reference to willow (*Salix* spp.) bark supplements—a review. *Antioxidants* 10(5), 686.

Seydim, A. C., Acton, J. C., Hall, M. A. and Dawson, P. L. (2006). Effects of packaging atmospheres on shelf-life quality of ground ostrich meat. *Meat Science* 73(3), 503–510.

Shakeri, M., Oskoueian, E., Le, H. H. and Shakeri, M. (2020). Strategies to combat heat stress in broiler chickens: unveiling the roles of selenium, vitamin E and vitamin C. *Veterinary Sciences* 7(2), 71.

Sierżant, K., Korzeniowska, M., Orda, J., Wojdyło, A., Gondret, F. and Półbrat, T. (2021). The effect of rosemary (*Rosmarinus officinalis*) and blackcurrant extracts (*Ribes nigrum*) supplementation on performance indices and oxidative stability of chicken broiler meat. *Animals: An Open Access Journal From MDPI* 11(4), 1155.

Sihvo, H. K., Immonen, K. and Puolanne, E. (2014). Myodegeneration with fibrosis and regeneration in the pectoralis major muscle of broilers. *Veterinary Pathology* 51(3), 619–623.

Silva, F. A. P., Estévez, M., Ferreira, V. C. S., Silva, S. A., Lemos, L. T. M., Ida, E. I., Shimokomaki, M. and Madruga, M. S. (2017). Characterization of preserved meat from spent hen and broiler by salting and forced-air drying. *Journal of Food Processing and Preservation* 41(4), e13048.

Silva, F. A. P., Estévez, M., Ferreira, V. C. S., Silva, S. A., Lemos, L. T. M., Ida, E. I., Shimokomaki, M. and Madruga, M. S. (2018). Protein and lipid oxidations in jerky chicken and consequences on sensory quality. *LWT* 97, 341–348.

Silva, F. A. P., Ferreira, V. C. S., Madruga, M. S. and Estévez, M. (2016). Effect of the cooking method (grilling, roasting, frying and sous-vide) on the oxidation of thiols, tryptophan, alkaline amino acids and protein cross-linking in jerky chicken. *Journal of Food Science and Technology* 53(8), 3137–3146.

Smaoui, S., Ennouri, K., Chakchouk-Mtibaa, A., Karray-Rebai, I., Hmidi, M., Bouchaala, K. and Mellouli, L. (2017). Relationships between textural modifications, lipid and protein oxidation and sensory attributes of refrigerated turkey meat sausage treated with bacteriocin BacTN635. *Food and Bioprocess Technology* 10(9), 1655–1667.

Soares, A. L., Ida, E. I., Miyamoto, S., Hernández-Blazquez, F. J., Olivo, R., Pinheiro, J. W. and Shimokomaki, M. (2003). Phospholipase A2 activity in poultry PSE, pale, soft, exudative, meat. *Journal of Food Biochemistry* 27(4), 309–320.

Sobral, M. M. C., Casal, S., Faria, M. A., Cunha, S. C. and Ferreira, I. M. P. L. V. O. (2020). Influence of culinary practices on protein and lipid oxidation of chicken meat burgers during cooking and in vitro gastrointestinal digestion. *Food and Chemical Toxicology* 141, 111401.

Soglia, F., Baldi, G. and Petracci, M. (2020). Effect of the exposure to oxidation and malondialdehyde on turkey and rabbit meat protein oxidative stability. *Journal of Food Science* 85(10), 3229–3236.

Soladoye, O. P., Juárez, M. L., Aalhus, J. L., Shand, P. and Estévez, M. (2015). Protein oxidation in processed meat: mechanisms and potential implications on human health. *Comprehensive Reviews in Food Science and Food Safety* 14(2), 106–122.

Soncu, E. D. (2020). Protein oxidation and subsequent changes in chicken breast and thigh meats during long-term frozen storage. *Agricultural and Food Science* 29(5), 505–514.

Souza, M., Baptista, A. A. S., Valdiviezo, M. J. J., Justino, L., Menck-Costa, M. F., Ferraz, C. R., da Gloria, E. M., Verri, W. A., Jr. and Bracarense, A. P. F. R. L. (2020). *Lactobacillus* spp. reduces morphological changes and oxidative stress induced by deoxynivalenol on the intestine and liver of broilers. *Toxicon* 185, 203–212.

Souza, V. G. L., Pires, J. R. A., Vieira, É. T., Coelhoso, I. M., Duarte, M. P. and Fernando, A. L. (2018). Shelf life assessment of fresh poultry meat packaged in novel bionanocomposite of chitosan/montmorillonite incorporated with ginger essential oil. *Coatings* 8(5), 177.

Stella, S., Tirloni, E., Bernardi, C. and Grilli, G. (2021). Evaluation of effect of chilling steps during slaughtering on the *Campylobacter* sp. counts on broiler carcasses. *Poultry Science* 100(3), 100866.

Surai, P. F. (2014). Polyphenol compounds in the chicken/animal diet: from the past to the future. *Journal of Animal Physiology and Animal Nutrition* 98(1), 19–31.

Tarkhan, A. H., Saleh, K. M. M. and Al-Zghoul, M. B. (2020). Heat exposure affects the mRNA levels of antioxidant enzymes in embryonic and adult broiler chickens. *Jordan Journal of Biological Sciences* 13(4), 431–435.

Teruel, M. R., Garrido, M. D., Espinosa, M. C. and Linares, M. B. (2015). Effect of different format-solvent rosemary extracts (Rosmarinus officinalis) on frozen chicken nuggets quality. *Food Chemistry*, 172, 40–46.

Thanatsang, K. V., Malila, Y., Arayamethakorn, S., Srimarut, Y., Tatiyaborworntham, N., Uengwetwanit, T., Panya, A., Rungrassamee, W. and Visessanguan, W. (2020). Nutritional properties and oxidative indices of broiler breast meat affected by wooden breast abnormality. *Animals: An Open Access Journal From MDPI* 10(12), 2272.

Thornval, N. R. and Hoorfar, J. (2021). Progress in detection of Campylobacter in the food production chain. *Current Opinion in Food Science* 39, 16–21.

Totosaus, A., Pérez-Chabela, M. L. and Guerrero, I. (2007). Color of fresh and frozen poultry. In: *Handbook of Meat, Poultry and Seafood Quality*. John Wiley & Sons, Inc., Hoboken, NJ, USA, pp. 455–465.

Trinetta, V., Floros, J. D. and Cutter, C. N. (2010). Sakacin a-containing pullulan film: an active packaging system to control epidemic clones of Listeria monocytogenes in ready-to-eat foods. *Journal of Food Safety* 30(2), 366–381.

U-Chupaj, J., Malila, Y., Gozzi, G., Vannini, L., Dellarosa, N., Laghi, L., Petracci, M., Benjakul, S. and Visessanguan, W. (2021). Influence of non-phosphate and low-sodium salt marination in combination with tumbling process on properties of chicken breast meat affected by white striping abnormality. *Journal of Food Science* 86(2), 319–326.

Umaraw, P., Munekata, P. E. S., Verma, A. K., Barba, F. J., Singh, V. P., Kumar, P. and Lorenzo, J. M. (2020). Edible films/coating with tailored properties for active packaging of meat, fish and derived products. *Trends in Food Science and Technology* 98, 10–24.

Utrera, M. and Estévez, M. (2013). Oxidative damage to poultry, pork, and beef during frozen storage through the analysis of novel protein oxidation markers. *Journal of Agricultural and Food Chemistry* 61(33), 7987–7993.

Utrera, M., Morcuende, D. and Estévez, M. (2014). Fat content has a significant impact on protein oxidation occurred during frozen storage of beef patties. *LWT - Food Science and Technology* 56(1), 62–68.

Van Hecke, T., Vossen, E., Goethals, S., Boon, N., De Vrieze, J. and De Smet, S. (2021). In vitro and in vivo digestion of red cured cooked meat: oxidation, intestinal microbiota and fecal metabolites. *Food Research International* 142, 110203.

Verma, A. K., Pathak, V., Singh, V. P. and Umaraw, P. (2016). Storage study of chicken meatballs incorporated with green cabbage (Brassica olerecea) at refrigeration temperature (4±1 C) under aerobic packaging. *Journal of applied animal research*, 44(1), 409–414.

Vihavainen, E. J. and Björkroth, J. (2010). Microbial ecology and spoilage of poultry meat and poultry meat products. In: *Handbook of Poultry Science and Technology* 2(2). John Wiley & Sons, Inc., Hoboken, NJ, USA, pp. 485–493.

Vlaicu, P. A., Untea, A. E., Panaite, T. D. and Turcu, R. P. (2020). Effect of dietary orange and grapefruit peel on growth performance, health status, meat quality and intestinal microflora of broiler chickens. *Italian Journal of Animal Science* 19(1), 1394–1405.

Vossen, E., Ntawubizi, M., Raes, K., Smet, K., Huyghebaert, G., Arnouts, S. and De Smet, S. (2011). Effect of dietary antioxidant supplementation on the oxidative status of plasma in broilers. *Journal of Animal Physiology and Animal Nutrition* 95(2), 198–205.

Wagle, B. R., Donoghue, A. M., Shrestha, S., Upadhyaya, I., Arsi, K., Gupta, A., Liyanage, R., Rath, N. C., Donoghue, D. J. and Upadhyay, A. (2020). Carvacrol attenuates *Campylobacter jejuni* colonization factors and proteome critical for persistence in the chicken gut. *Poultry Science* 99(9), 4566–4577.

Wang, C., Zhao, F., Li, Z., Jin, X., Chen, X., Geng, Z., Hu, H. and Zhang, C. (2021a). Effects of resveratrol on growth performance, intestinal development, and antioxidant status of broilers under heat stress. *Animals: An Open Access Journal from MDPI* 11(5), 1427.

Wang, K., Li, Y., Bai, F., Gao, R., Wang, J., Xu, P., Zhao, Y. and Wang, R. (2021b). Effects of α-tocopherol and vacuum packaging on the quality of sturgeon surimi incorporated with chicken meat during frozen storage. *Food Science* 42(11), 195–204.

Wasti, S., Sah, N. and Mishra, B. (2020). Impact of heat stress on poultry health and performances, and potential mitigation strategies. *Animals: An Open Access Journal from MDPI* 10(8), 1266.

Wessels, K., Rip, D. and Gouws, P. (2021). Salmonella in chicken meat: consumption, outbreaks, characteristics, current control methods and the potential of bacteriophage use. *Foods* 10(8), 1742.

Wu, H., Xiao, S., Yin, J., Zhang, J. and Richards, M. P. (2021). Mechanisms involved in the inhibitory effects of free fatty acids on lipid peroxidation in turkey muscle. *Food Chemistry* 342, 128333.

Xia, T., Cao, Y., Chen, X., Zhang, Y., Xue, X., Han, M., Li, L., Zhou, G. and Xu, X. (2018). Effects of chicken myofibrillar protein concentration on protein oxidation and water holding capacity of its heat-induced gels. *Journal of Food Measurement and Characterization* 12(4), 2302–2312.

Xiao, S., Zhang, W. G., Lee, E. J., Ma, C. W. and Ahn, D. U. (2011). Effects of diet, packaging, and irradiation on protein oxidation, lipid oxidation, and color of raw broiler thigh meat during refrigerated storage. *Poultry Science* 90(6), 1348–1357.

Xiong, Q., Zhang, M., Wang, T., Wang, D., Sun, C., Bian, H., Li, P., Zou, Y. and Xu, W. (2020). Lipid oxidation induced by heating in chicken meat and the relationship with oxidants and antioxidant enzymes activities. *Poultry Science* 99(3), 1761–1767.

Xiong, Y. L. and Guo, A. (2020). Animal and plant protein oxidation: chemical and functional property significance. *Foods* 10(1), 40.

Yıldırım-Yalçın, M., Sadıkoğlu, H. and Şeker, M. (2021). Characterization of edible film based on grape juice and cross-linked maize starch and its effects on the storage quality of chicken breast fillets. *LWT* 142, 111012.

Yousefi, M., Khorshidian, N. and Hosseini, H. (2020). Potential application of essential oils for mitigation of *Listeria monocytogenes* in meat and poultry products. *Frontiers in Nutrition* 7, 577287.

Yu, Z., Joossens, M., Kerkhof, P. J. and Houf, K. (2021). Bacterial shifts on broiler carcasses at retail upon frozen storage. *International Journal of Food Microbiology* 340, 109051.

Yu, Z., Peruzy, M. F., Dumolin, C., Joossens, M. and Houf, K. (2019). Assessment of food microbiological indicators applied on poultry carcasses by culture combined MALDI-TOF MS identification and 16S rRNA amplicon sequencing. *Food Microbiology* 82, 53–61.

Zampiga, M., Soglia, F., Baldi, G., Petracci, M., Strasburg, G. M. and Sirri, F. (2020). Muscle abnormalities and meat quality consequences in modern Turkey hybrids. *Frontiers in Physiology* 11, 554.

Zeitz, J. O., Ehbrecht, T., Fleischmann, A., Most, E., Gessner, D. K., Friedrichs, S., Sparenberg, M., Failing, K., Whelan, R., Lütjohann, D. and Eder, K. (2020). Effect of DL-methionine supplementation on tissue and plasma antioxidant status and concentrations of oxidation products of cholesterol and phytosterols in heat-processed thigh muscle of broilers. *Animals: An Open Access Journal From MDPI* 10(11), 2050.

Zhang, H., Wu, J. and Guo, X. (2016). Effects of antimicrobial and antioxidant activities of spice extracts on raw chicken meat quality. *Food Science and Human Wellness* 5(1), 39–48.

Zhang, M., Yan, W., Wang, D. and Xu, W. (2021). Effect of myoglobin, hemin, and ferric iron on quality of chicken breast meat. *Animal Bioscience* 34(8), 1382–1391.

Zhang, W., Xiao, S., Lee, E. J. and Ahn, D. U. (2011). Consumption of oxidized oil increases oxidative stress in broilers and affects the quality of breast meat. *Journal of Agricultural and Food Chemistry* 59(3), 969–974.

Zheng, N., Zhang, L., Xin, Q., Miao, Z., Zhu, Z., Li, L. and Huang, Y. (2020). Changes in antioxidant enzymes in postmortem muscle and effects on meat quality from three duck breeds during cold storage. *Canadian Journal of Animal Science* 100(2), 234–243.

Zhou, X., Zong, X., Zhang, M., Ge, Q., Qi, J., Liang, J., Xu, X. and Xiong, G. (2021). Effect of konjac glucomannan/carrageenan-based edible emulsion coatings with camellia oil on quality and shelf-life of chicken meat. *International Journal of Biological Macromolecules* 183, 331–339.

Zhu, Z., Fang, R., Ali, I. and Huang, M. (2020). Impact of methylglyoxal modification of chicken sarcoplasmic protein emulsions on emulsifying properties, rheological behavior and advanced glycation end products. *Journal of the Science of Food and Agriculture* 100(11), 4208–4216.

Zhuang, H. and Bowker, B. (2018). The wooden breast condition results in surface discoloration of cooked broiler pectoralis major. *Poultry Science* 97(12), 4458–4461.

Index

CPSIA information can be obtained
at www.ICGtesting.com
Printed in the USA
BVHW051409010223
657370BV00010BA/28

9 781801 461030